Ma...rs
and Applied Scientists

C

Mathematics for Engineers and Applied Scientists

STANLEY C. LENNOX

B.Sc. (Dunelm), M.Sc. (Dunelm), Ph.D. (Edin.), F.I.M.A.
Senior Lecturer in Engineering Mathematics,
University of Newcastle upon Tyne

MARY CHADWICK

B.Sc. (Liverpool)
Formerly Lecturer in Mathematics, King's College,
University of Durham and
Lecturer in Mathematics at Norwich City College

HEINEMANN

LONDON

0259216
Heinemann Educational Books Ltd
22 Bedford Square, London WC1B 3HH

LONDON   EDINBURGH   MELBOURNE   AUCKLAND
HONG KONG   SINGAPORE   KUALA LUMPUR   NEW DELHI
IBADAN   NAIROBI   JOHANNESBURG
EXETER(NH)   KINGSTON   PORT OF SPAIN

ISBN 0 435 71282 9

Set in 10/11 pt and 9/10 pt Times New Roman, printed and bound in Great
Britain at The Pitman Press, Bath

Preface to the Second Edition

We have been pleased with the success of our book and its adoption as a working text by first year students in many engineering courses at Universities and higher technological Institutes. Following our own ideas, but also in response to requests, we have, in this edition, included new material. In Chapter 9 we have added two new sections, one on eigenvalues and eigenvectors and the other an introduction to the idea of the gradient of a scalar field. Both these topics are often met by engineering students during the first or second year courses in their own engineering disciplines. In Chapter 11 a new section on isoclines has been added and the ideas of the numerical integration of differential equations form a new section in Chapter 13. Two other new sections of Chapter 13 give an elementary introduction to error analysis. Now that all students have electronic pocket calculators it seems to us important that the concept of accuracy, propagation of error, and error bounds should be dealt with early in the curriculum. Additional examples have been added to the relevant Exercises.

This edition also contains a completely new Chapter 17 on Applications. The teacher of mathematics is often asked by engineering students to explain why it is that they have to be taught topics such as eigenvalues, vectors, complex numbers, partial differentiation. Chapter 17 goes some little way towards showing students that these topics (and others) have application in the physical and industrial world. We would have liked this chapter to be much larger and to have covered more topics but are necessarily restricted by limitation on the size (and price) of the book. We will be satisfied if the student finds something of interest which will induce further study of mathematics and if the teachers of mathematics to engineers can be encouraged to investigate further applications themselves. One of us (SCL) is indebted to his colleagues at Newcastle upon Tyne for much discussion on the applications but we both accept responsibility for any mistakes that have appeared. We would like to thank all those readers who have written to us with suggestions and also those who have indicated errors in the first edition. We hope we have incorporated all corrections and ideas in this edition. We would be pleased to have any further comments.

Finally our thanks are due to our publishers who have allowed us to amend the text as we wished and have made an excellent job of putting together the new edition and have tried to keep the costing to a minimum.

Stanley C. Lennox   *Mary Chadwick*
1977

Preface to the First Edition

This book is the outcome of many years of teaching mathematics to engineering students particularly in the University of Newcastle upon Tyne. It is intended to cover the mathematical content of the first year to eighteen months of a three year course in a University or Institute of higher technological education. For certain disciplines the contents will meet the complete mathematical requirements of the entire three year course while for others the contents cover the basic requirements before proceeding to more advanced mathematical topics. It covers the mathematical syllabus for the Part I examination of the Council of Engineering Institutions (CEI). We have assumed a knowledge roughly equivalent to a pass in A-level single subject Pure and Applied Mathematics, a Higher National Certificate, or an equivalent qualification.

Although primarily written for engineering students the text should also be of value to science students whose requirements demand a manipulative skill in mathematical processes. Where necessary a motivation, usually within an engineering context, is used to introduce the reader to a new concept before a proof is introduced. Where such a proof is omitted it is simply because we believe that its introduction, at that point, would unduly delay the development of the subject and may well confuse the student. We do not subscribe to the view that the engineer or non-mathematical student should never be exposed to mathematical rigour but think that at this particular level it is more important that he should be able to use and appreciate the place of mathematics in his own subject, and to this end it is necessary that the student possesses manipulative skill as well as a broad understanding. To give an analogy it is not necessary to have a complete and detailed knowledge of the workings of an automobile before attempting to drive. Indeed an attempt to teach such detail to some would-be drivers may well lead them to abandon the whole process in despair.

In writing the book we have not made any distinction between different kinds of engineering student but consider, no matter what discipline, that all should have a common mathematical core syllabus. It is now accepted that all engineers should be exposed to the mathematical ideas of probability, statistics, and numerical analysis and an important feature of this book is the introduction of these subjects alongside the more common curricula subjects of analysis and algebra. Chapters 1 to 7 cover analysis, including functions, differential and integral calculus, convergence, and Taylor series. Chapters 8 to 11 deal with the algebra of complex numbers, linear algebra, elementary vector analysis, and ordinary differential equations. Chapters 12 to 14 are devoted to the analytical and numerical solutions of algebraic and transcendental equations, the numerical solution of a set of linear equations, an introduction to finite differences and polynomial interpolation, and numerical differentiation and integration. Chapters 15 and 16 provide a basic statistics course covering distributions, probability, sampling, and simple tests of significance.

The book is, however, presented as a whole and it is not intended that branches of mathematics be put into different compartments. For convenience it is divided into chapters but that does not mean that any series of lectures taken from the book need follow the chapter order. Certain chapters obviously demand knowledge of the work covered previously, for instance it would be difficult to understand Taylor's theorem in Chapter 5 without having read the previous four chapters. On the other hand the two chapters on statistics demand a minimum of prior knowledge while the linear algebra of Chapter 8 and the vector analysis of Chapter 9 may be started after reading Chapter 1. The numerical work of Chapters 12, 13, and 14 can be read in isolation providing that differentiation and integration is covered before reaching the end of Chapter 14. The sketch gives a visual indication of the chapters covered by four possible series of lectures and shows the linkage points between the series.

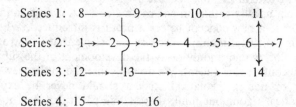

This book can be used as working text and to encourage both understanding and skill a large number of examples is given and solved within the text. In addition an exercise is presented at the end of each chapter. Every example set has been provided with an answer and hints are given when it is considered necessary. Most of the examples are original but some have been taken from those collected and amended over many years of teaching and examining for the Universities of Newcastle upon Tyne, Durham, Edinburgh, Glasgow, and London together with the Heriot-Watt and Strathclyde Universities. The answers to the exercises are collected together at the end of the book and while they have all been checked we will be amazed if some are not in error. We would be grateful to any reader who is prepared to indicate the point of error for future amendment.

It is inevitable that certain items have had to be omitted and others curtailed but a compromise has to be made between various factors such as the size of the book, the time devoted to mathematical studies, and the extent of the syllabus. Throughout, however, it has been our aim to achieve a satisfactory balance between these factors.

Finally, we are indebted first to Professor D. V. Lindley and Dr. J. C. P. Miller for permission to reproduce extracts from their book of Elementary Statistical Tables, and second to Professor L. Maunder both for encouraging us to write the book and for providing us with many helpful and constructive criticisms.

Stanley C. Lennox   *Mary Chadwick*

Contents

Mathematics for Engineers and Applied Scientists

1
Introduction

1.1 FUNCTIONS OF ONE VARIABLE

When two quantities are so related that one of them is determined uniquely when the other is known, then the first quantity is a *function* of the second quantity. That one quantity is a simple function of another does not necessarily mean that there is a formula connecting them. A function can be expressed as a table of values, in the form of a graph, as a formula, and in other ways.

1.1.1 Notation and definitions

When two variables x and y are related so that y is determined uniquely when x is known, the functional relationship is written $y = f(x)$ where $f(x)$ is the rule prescribing the value of y for any given value of x. Other notations such as $y = g(x)$, $y = \phi(x)$, $y = F(x)$, $\phi(x, y) = 0$ will be used, and $y = y(x)$, $x = x(t)$ are used to mean that y is a function of x and x is a function of t.

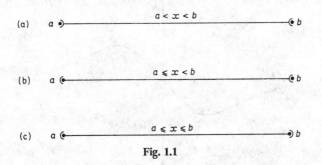

Fig. 1.1

Suppose $y = f(x)$, then the variable x is called the *argument* or *independent variable* and $f(x)$ or y the *dependent variable* or value of f for the argument x. Values of $y = f(x)$ are found for any given value of x by replacing x in $f(x)$ by the given value. Thus the value of $f(x)$ when $x = 2$ is denoted by $f(2)$.

The set of numbers over which x may vary is called the *range* of x and is usually the set of points in an *interval* on the x-axis. This interval may be the whole line from $-\infty$ to $+\infty$. The set of all real numbers x between two fixed numbers a and b, written $a < x < b$ means that x may take any value which is simultaneously greater than a and less than b. This set of numbers is called an *open* interval. See Fig. 1.1(a).

The correct description of a mathematical *real number* requires more precise detail than is given here. It is sufficient to recognize that a real number is either a *rational* number of the form p/q, where p and q are positive or negative integers and $q \neq 0$, e.g. 0, $\frac{1}{2}$, -1, $+2$, $-\frac{3}{7}$; or an *irrational* number that cannot be expressed in the form p/q, e.g. $\sqrt{2}$, π, $\sqrt[3]{(-\frac{1}{3})}$. To each and every point of the x-axis there corresponds a rational or irrational number x, called simply a real number.

The set $a \leqslant x < b$ of real numbers x consists of all numbers in the open interval plus the left end point $x = a$. See Fig. 1.1(b). The set $a \leqslant x \leqslant b$ is the set of all real numbers between a and b including both end points. This is called a closed interval. See Fig. 1.1(c). If x may take all values in a range it is called a *continuous* variable.

The functional relationship $y = f(x)$ now determines the set of values y may take. In general the function may be represented graphically by those points whose coordinates (x, y) satisfy $y = f(x)$ for all those values which x may take.

Example (i) $y = x/(x^2 + 1)$.

Here x may be any real number and the graph of the function is sketched in Fig. 1.2.

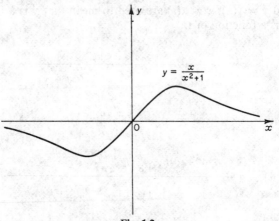

$$y = \frac{x}{x^2+1}$$

Fig. 1.2

Example (ii) $y = \sqrt{(x - 1)}$.

The range of values of x consists in those values for which the function has a meaning, so that in this case, $x \geqslant 1$. See Fig. 1.3.

Example (iii) Let $f(x)$ be the sum of x terms of the geometric series

$$1 + \tfrac{1}{2} + \tfrac{1}{4} + \tfrac{1}{8} + \cdots$$

Graphically $y = f(x)$ is represented by a set of points which may not be joined by a continuous line. Values of x are positive integers only, and the value of the function for non-integral values of x is meaningless. The successive sums $f(1), f(2), f(3), \ldots$ are illustrated in Fig. 1.4.

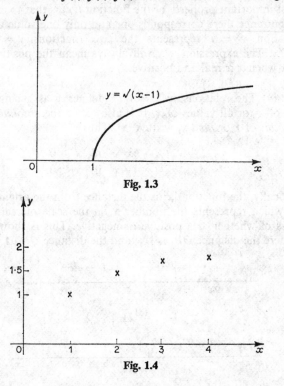

$$y = \sqrt{(x-1)}$$

Fig. 1.3

Fig. 1.4

Example (iv) It is possible to define a function by different formulae for different values of x, as

$$f(x) = \begin{cases} \sqrt{(x-1)} & \text{if} \quad 1 \leqslant x < 5 \\ 4 + 1/(x-4) & \text{if} \quad x > 5. \end{cases}$$

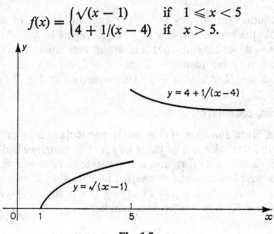

$$y = 4 + 1/(x-4)$$

$$y = \sqrt{(x-1)}$$

Fig. 1.5

This function is defined for all values of $x \geqslant 1$ except for $x = 5$. Its graph is sketched in Fig. 1.5.

A most important property of the function $f(x)$ is that to each value of x in the range there corresponds one and only one value of f. Thus the equation $y^2 = x$ represents the two functions $y = \sqrt{x}$ and $y = -\sqrt{x}$. The expression $\sqrt{x}$ will always mean the positive square root of x when x is real and positive.

Example (v) The relationship $y = \sqrt{x^2}$ defines y as a single-valued function of x for all values of x. It is also called the *absolute* value of x or *modulus* of x or *mod x*, written $|x|$. Thus

$$\sqrt{(x)^2} = |x| = \begin{cases} x & \text{if } x \geqslant 0 \\ -x & \text{if } x < 0. \end{cases}$$

Geometrically, the function $|x|$ is the distance from the origin O to the point P which represents the number x on the scale of real numbers regardless of whether x is positive or negative. This is shown in Fig. 1.6(a) where the distance OP_1 is $|x_1|$ and the distance OP_2 is $|x_2|$.

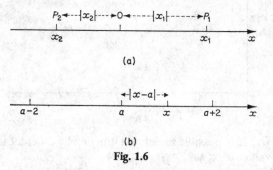

(a)

(b)

Fig. 1.6

The modulus symbol may be used to express a range of values of x, for example the range $-1 < x < 1$ is equivalent to $|x| < 1$.

Also, $|x - a| < 2$ means that x is within two units of a. This means that x may vary between $a - 2$ and $a + 2$, so that $|x - a| < 2$ is the same as $a - 2 < x < a + 2$. This is shown in Fig. 1.6(b).

1.1.2 Limits, continuity

Consider a given function $f(x)$ at some particular value of $x = a$. As x approaches the value a, the value of $f(x)$ will usually change steadily and approach the value $f(a)$, and it is possible to choose a value of $f(x)$ 'as near as we please' to the value $f(a)$ by choosing x sufficiently near to the value a. A different way of saying this is that $f(x)$ tends to $f(a)$ as x tends to a. This is written in symbols,

$$f(x) \rightarrow f(a) \quad \text{as} \quad x \rightarrow a.$$

The value of $f(a)$ is the *limit* of $f(x)$ as $x \to a$, or

$$\operatorname*{Lim}_{x \to a} f(x) = f(a).$$

If x increases towards the value a, or $x \to a$ from the left, the notation

$$f(x) \to f(a - 0) \quad \text{as} \quad x \to a-$$

or

$$\operatorname*{Lim}_{x \to a-} f(x) = f(a - 0)$$

is used. Similarly if x decreases towards the value a, or x approaches a from the right,

$$f(x) \to f(a + 0) \quad \text{as} \quad x \to a+$$

or

$$\operatorname*{Lim}_{x \to a+} f(x) = f(a + 0).$$

Often it does not matter whether x approaches the fixed value a from the left or from the right since in both cases the limit will be $f(a)$, the value of the function at $x = a$. For example, consider the graph in Fig. 1.5. If the value of a is between 1 and 5 or greater than 5,

$$f(a - 0) = f(a) = f(a + 0). \tag{1}$$

When equations (1) are satisfied, the function is said to be *continuous* at the point $x = a$. For continuity at the point $x = a$ it is not sufficient that the limits of $f(x)$ as x tends to a from the left and right should both exist and be equal, for the common value may not be $f(a)$. For example, consider the function defined by

$$f(x) = 0 \text{ when } x \neq 0, \qquad f(x) = 1 \text{ when } x = 0.$$

Then

$$\operatorname*{Lim}_{x \to 0-} f(x) = \operatorname*{Lim}_{x \to 0+} f(x) = 0,$$

but this limit is not equal to $f(0)$ which is 1, and the function is not continuous at $x = 0$.

Consider the function sketched in Fig. 1.5. For this function,

$$\operatorname*{Lim}_{x \to 5-} f(x) = 2, \qquad \operatorname*{Lim}_{x \to 5+} f(x) = 5,$$

i.e. equations (1) are not satisfied at the point $x = 5$ and the function is not continuous at this point. There is a 'jump' in the value of the function as x increases through the value 5.

Consider also the way different functions behave when the argument x increases indefinitely. If the value of the function increases indefinitely the function is said to 'tend to infinity' as x tends to infinity, i.e. $f(x) \to \infty$ as $x \to \infty$.

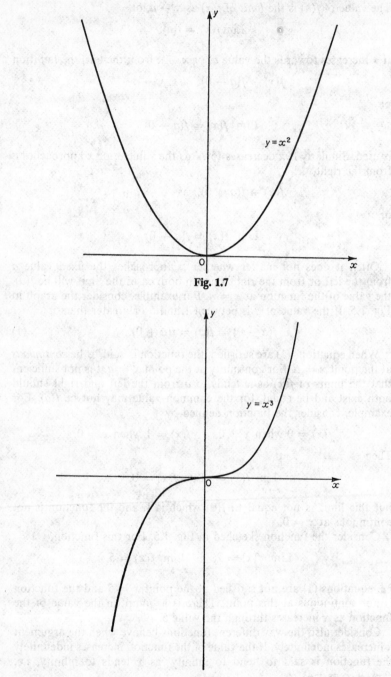

Fig. 1.7

Fig. 1.8

Example (i) Consider the function $f(x) = x^2$. For all values of x this function is continuous and as $x \rightarrow \pm\infty$, $f(x) \rightarrow \infty$. The graph of $y = x^2$ is sketched in Fig. 1.7.

The graphs of $y = x^{2n}$ where n is a positive integer are similar to that of $y = x^2$.

Example (ii) Consider the function $f(x) = x^3$. This function is continuous for all values of x and as $x \rightarrow +\infty$, $x^3 \rightarrow +\infty$, as $x \rightarrow -\infty$, $x^3 \rightarrow -\infty$. The graph of $y = x^3$ is sketched in Fig. 1.8.

The graphs of $y = x^{2n+1}$ for positive integers n are similar to that of $y = x^3$.

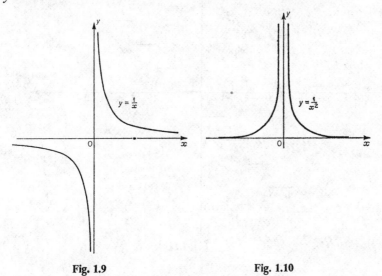

Fig. 1.9   Fig. 1.10

Example (iii) The function $f(x) = 1/x$ is defined and continuous for all values of x except $x = 0$. This example demonstrates two properties not mentioned so far;

(a) as $x \rightarrow 0+$, $f(x) \rightarrow +\infty$,

     as $x \rightarrow 0-$, $f(x) \rightarrow -\infty$,

i.e. the 'jump' in the value of $y = f(x)$ as x increases through $x = 0$ is infinite;

(b) as $x \rightarrow \infty$, $1/x \rightarrow 0$, yet zero is not a value of $f(x)$, i.e. the limit of a function is not necessarily a value of the function, $y = 1/x$ is never zero.

Similarly the function $f(x) = 1/x^2$ is defined and continuous for all values of x except $x = 0$. For this function, $f(x) \rightarrow +\infty$ as $x \rightarrow 0+$ and as $x \rightarrow 0-$. See Fig. 1.9 and 1.10.

Example (iv) Consider the function $f(x) = [x]$ where $[x]$ is defined as the integral part of x. Thus $[3 \cdot 2] = 3$, $[-4 \cdot 2] = -5$ and when x is an integer, $[x] = x$.

This is an example of a *step function* and at each integer value of x there is a point where the graph is not continuous. This function is illustrated in Fig. 1.11.

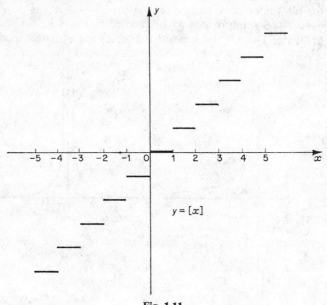

$y = [x]$

Fig. 1.11

Example (v) Consider the function $f(x)$ defined by

$$f(x) = \begin{cases} x & \text{when} \quad 0 \leqslant x \leqslant 1 \\ 2 - x & \text{when} \quad 1 < x \leqslant 2. \end{cases}$$

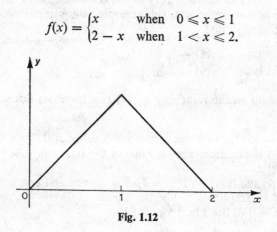

Fig. 1.12

Here $y = f(x)$ is defined only over the interval $0 \leqslant x \leqslant 2$ and

$$f(1 - 0) = f(1) = f(1 + 0) = 1.$$

Hence equations (1) are satisfied and the function is continuous at the point $x = 1$. See Fig. 1.12.

A function which is continuous at all points in a particular range is a continuous function in that range. Examples (i) and (ii) are examples of functions which are continuous for all values of x. The functions $1/x$, $1/x^2$ of Example (iii) are continuous for all values of x except $x = 0$; the function defined in Example (v) is continuous for all values of x for which the function is defined, i.e. in $0 \leqslant x \leqslant 2$; the function $[x]$ of Example (iv) is not continuous at any integer value of x.

1.1.3 Odd and even functions

The function $f(x)$ is an even function if $f(-x) = f(x)$ for all values of x. The graph of an even function is symmetrical about the y-axis, as x^2, $1/x^2$, $\cos x$.

The function $f(x)$ is an odd function if $f(-x) = -f(x)$ for all values of x. The graph of an odd function is symmetrical about the origin, as x^3, $1/x$, $\sin x$.

Any function can be written as the sum of an odd and an even function, for

$$f(x) = \tfrac{1}{2}\{f(x) + f(-x)\} + \tfrac{1}{2}\{f(x) - f(-x)\}.$$

1.1.4 Monotonic functions

The function $f(x)$ is a *monotonically increasing* function, or $f(x)$ increases *monotonically* in an interval of values of x if $f(x)$ increases steadily as x increases through the interval, i.e. if $f(x_2) \geqslant f(x_1)$ whenever $x_2 > x_1$.

Similarly $f(x)$ is a *monotonically decreasing* function if $f(x)$ decreases steadily as x increases through an interval, i.e. if $f(x_2) \leqslant f(x_1)$ whenever $x_2 > x_1$.

1.2 INVERSE FUNCTIONS

Suppose $y = f(x)$ is a continuous function of x, then it is sometimes possible to express x as a function of y. For example, the formula $y = mx + c$ determines y uniquely when x is given, so that y is a function of x from the definition of function given in section 1.1. Also $x = (y - c)/m$ is a function of y.

Consider the function given by $y = x^2$; y is determined uniquely when x is given, for all values of x. If x is expressed in terms of y there are two possibilities, $x = \pm\sqrt{y}$ and x is not uniquely determined when y is given. Hence from the definition of a function, x cannot be expressed as a function of y but is expressed as two functions of y for all positive values of y.

If a given function $y = f(x)$ can be replaced by a function $x = \phi(y)$, then $x = \phi(y)$ is called the *inverse function* of $y = f(x)$. For this to be possible, to each value of x there is one and only one corresponding value of y and to each value of y there is one and only one corresponding value of x.

Example The function $y = x^2$, $-2 \leqslant x \leqslant 2$, has no unique inverse function since $x = \pm\sqrt{y}$ gives two functions in the range.

The function $y = x^2$, $0 < x < 2$, has the unique inverse function $x = \sqrt{y}$. These functions are sketched in Fig. 1.13 and 1.14.

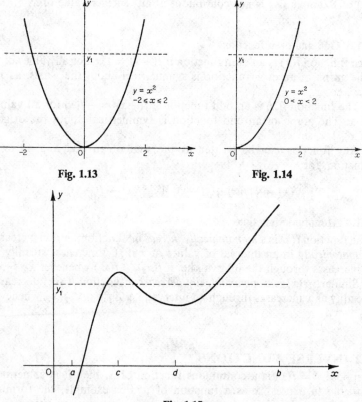

Fig. 1.13   Fig. 1.14

Fig. 1.15

Suppose $y = f(x)$ is defined in the range $a \leqslant x \leqslant b$ and has a graph as in Fig. 1.15. For each value of x in the range y is uniquely determined but to each value of y there may be more than one corresponding value of x. Hence this function cannot be inverted in the whole range $a \leqslant x \leqslant b$.

In the range $a < x < c$, x is uniquely determined when y is given, $f(x)$ is monotonically increasing and hence x can be expressed as a function of y in $a < x < c$. Similarly x can be expressed as a (different) function of y in the range $c < x < d$ where the function is monotonically decreasing and in the range $d < x < b$ where the function is monotonically increasing.

More generally if $y = f(x)$ is continuous and monotonic in an interval then it is possible to define the inverse function $x = \phi(y)$. Consider the graphs of $y = f(x)$ and $x = \phi(y)$ where $y = f(x)$ is continuous and monotonically increasing in $a \leqslant x \leqslant b$, and $\alpha \leqslant y \leqslant \beta$. These graphs are sketched in Figs. 1.16 (a) and (b).

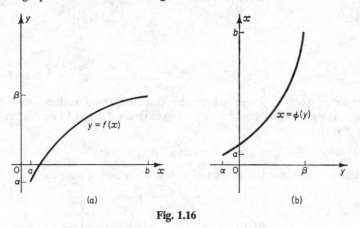

(a)   (b)

Fig. 1.16

Geometrically the graph of the inverse function is drawn by reflecting the graph of $y = f(x)$ in the line $y = x$, or by 'rotating' the axes of the graph of $y = f(x)$ through $90°$ and reflecting in the new x-axis.

1.2.1 The inverse circular functions

The function $y = \sin^{-1} x$ or $y = \arc \sin x$ is defined as the value of y such that

$$x = \sin y \quad \text{and} \quad -\tfrac{1}{2}\pi \leqslant y \leqslant \tfrac{1}{2}\pi. \tag{2}$$

That is the function $x = \sin y$ is inverted only in that range where there is a one to one correspondence between values of x and values of y. There are many ranges for y where this is possible. The range chosen in equation (2) is that one which includes $y = 0$.

Note that if $x = \sin y$ there are many values of y which satisfy the equation for a given value of x. For example if $\sin y = \tfrac{1}{2}$, then $y = n\pi + (-1)^n \tfrac{1}{6}\pi$, $n = 0, \pm 1, \pm 2, \ldots$. But $\sin^{-1}(\tfrac{1}{2}) = \tfrac{1}{6}\pi$. Figure 1.17 shows sketch graphs of $x = \sin y$ and $y = \sin^{-1} x$.

Similarly the expression $x = \cos y$ is many-valued and to obtain the full range of values of $\cos y$ once and once only, the range is

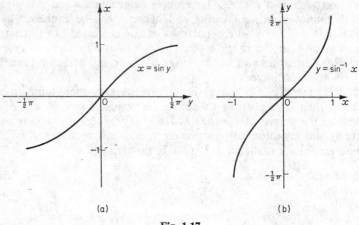

Fig. 1.17

restricted to $0 \leqslant y \leqslant \pi$ where the function $\cos y$ is monotonically decreasing, i.e. $y = \cos^{-1} x$ or $y = \arccos x$ is defined to be that value of y such that

$$x = \cos y \quad \text{and} \quad 0 \leqslant y \leqslant \pi. \tag{3}$$

Figure 1.18 shows the graphs of $y = \cos^{-1} x$ and $x = \cos y$.

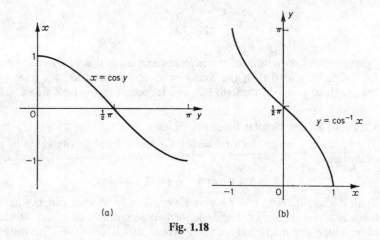

Fig. 1.18

The expression $x = \tan y$ is many-valued but takes all values once only in the range $-\tfrac{1}{2}\pi \leqslant y \leqslant \tfrac{1}{2}\pi$ where it is monotonically increasing. Hence the inverse function $y = \tan^{-1} x$ or $y = \arctan x$ means

$$x = \tan y \quad \text{and} \quad -\tfrac{1}{2}\pi \leqslant y \leqslant \tfrac{1}{2}\pi. \tag{4}$$

Figure 1.19 shows the graph of $y = \tan^{-1} x$.

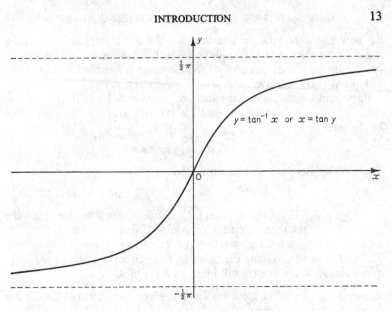

Fig. 1.19

The difference between $\sin^{-1} x$ and $(\sin x)^{-1} = \operatorname{cosec} x$ should be noted.

Example Prove that $\tan^{-1} 2 + \tan^{-1} 3 = \frac{3}{4}\pi$. Is it true that $\tan^{-1} 2 + \tan^{-1} 3 = \tan^{-1}(-1)$?

Solution Let $\tan^{-1} 2 = \alpha$, $\tan^{-1} 3 = \beta$, then

$$\tan(\alpha + \beta) = \frac{\tan\alpha + \tan\beta}{1 - \tan\alpha\tan\beta} = \frac{2 + 3}{1 - 6} = -1. \qquad \text{(i)}$$

Now by definition of the inverse tangent, $-\frac{1}{2}\pi \leqslant \alpha \leqslant \frac{1}{2}\pi$ and $-\frac{1}{2}\pi \leqslant \beta \leqslant \frac{1}{2}\pi$. Moreover since $\tan\alpha > 1$ and $\tan\beta > 1$ both α and β are greater than $\frac{1}{4}\pi$, i.e. $\frac{1}{4}\pi < \alpha < \frac{1}{2}\pi$ and $\frac{1}{4}\pi < \beta < \frac{1}{2}\pi$ hence $\frac{1}{2}\pi < \alpha + \beta < \pi$, and this with equation (i) means $\alpha + \beta = \frac{3}{4}\pi$.

By definition, $\tan^{-1}(-1) = -\frac{1}{4}\pi$ so that $\tan^{-1} 2 + \tan^{-1} 3 \neq \tan^{-1}(-1)$.

1.3 COORDINATE SYSTEMS

1.3.1 Coordinate systems in the plane

The position of a point P in the plane is fixed when its *rectangular Cartesian coordinates* (x, y) are known. These are coordinates referred to two perpendicular axes Ox, Oy through an arbitrarily chosen point O called the *origin*. The x-coordinate is called the *abscissa* and the y-coordinate the *ordinate* of the point.

The position of the point P in the plane is also fixed if its distance r from a fixed point O in the plane and the angle θ the line OP makes with a fixed straight line in the plane are known. The point O is called the *pole*, the fixed straight line is called the *initial line*. The distance r is always positive and the angle is measured positively, or anti-clockwise, from the initial line so that the direction is specified by the angle. The point P is then given the polar coordinates (r, θ).

Polar and Cartesian coordinates are transformed one to the other if the origin is used as the pole and the positive x-axis is the initial line by the equations

$$x = r \cos \theta, \qquad y = r \sin \theta.$$

These equations may also be expressed

$$r = \surd(x^2 + y^2), \qquad \cos \theta = x/r, \qquad \sin \theta = y/r.$$

See Fig. 1.20. It is not sufficient to consider $\tan \theta = y/x$ since the value of θ is not then uniquely determined. For example, if both the x- and y-coordinates are negative the point is in the third quadrant, yet $\tan \theta$ is positive giving the point (r, θ) in either the first or the third quadrant. A rough sketch will often prove useful.

Example (i) The point P whose Cartesian coordinates are $(1, \surd 3)$ has polar coordinates $(2, \frac{1}{3}\pi)$.

Example (ii) The point Q whose Cartesian coordinates are $(-1, \surd 3)$ has polar coordinates $(2, \frac{2}{3}\pi)$. See Fig. 1.21.

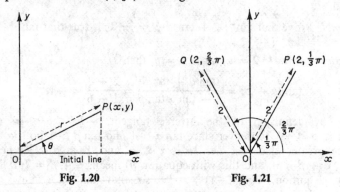

Fig. 1.20   Fig. 1.21

1.3.2 Coordinates in three dimensions

(i) *Cartesian coordinates* A point P in three-dimensional space is specified by the coordinates (x, y, z) referred to three mutually perpendicular axes through an origin O. The set of numbers specifying the coordinates of P may also be written (x_1, x_2, x_3). See Fig. 1.22.

(ii) *Cylindrical polar coordinates* In three dimensions the point P is fixed when the cylindrical polar coordinates (ρ, ϕ, z) are known, where z is the perpendicular distance of P from the plane (xy) and (ρ, ϕ) are the polar coordinates of the projection N of P in the (xy) plane. See Fig. 1.23.

The relations connecting cylindrical polar coordinates with Cartesian coordinates are

$$x = \rho \cos \phi,$$
$$y = \rho \sin \phi,$$
$$z = z.$$

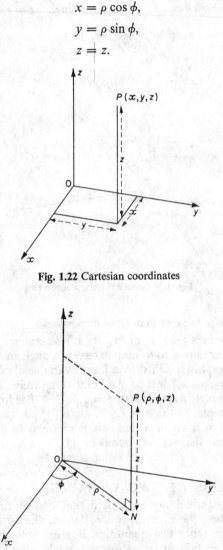

Fig. 1.22 Cartesian coordinates

Fig. 1.23 Cylindrical polar coordinates

(iii) *Spherical polar coordinates* The point P in three-dimensional space is fixed by the spherical polar coordinates (r, θ, ϕ), where r is the distance of P from the origin O, θ is the angle made by OP with the positive z-axis and ϕ is the angle made by the projection ON of

OP on to the (xy) plane with the positive x-axis. The transformation from (x, y, z) coordinates to (r, θ, ϕ) coordinates is given by

$$x = r \sin \theta \cos \phi, \quad y = r \sin \theta \sin \phi, \quad z = r \cos \theta.$$

Note that $ON = r \sin \theta = \rho$. See Fig. 1.24.

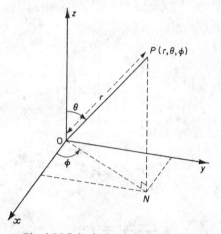

Fig. 1.24 Spherical polar coordinates

1.3.3 Coordinates in more than three dimensions

The set of numbers (x, y, z) or (x_1, x_2, x_3) determines the position of a point in space referred to rectangular axes through an arbitrary origin. Suppose the point is specified by a fourth variable also, then the set of numbers (x_1, x_2, x_3, x_4) will fix the point completely. This set may be written more shortly as (x_i), where $i = 1, 2, 3, 4$. The fourth coordinate could, for example, be the time t.

It may be that a set of n numbers is required to describe a point uniquely. Then the set of numbers $(x_1, x_2, x_3, \ldots, x_n)$ or (x_i), $i = 1, 2, 3, \ldots, n$ describes the point completely.

1.4 FUNCTIONS OF TWO VARIABLES

When three quantities are so related that one of them is uniquely determined when the other two are known, the first quantity is a function of the other two quantities. For example the volume of a cylinder is a function of its radius and height given by the formula $V = \pi r^2 h$. The variables r and h are independent; if one depends on the other, then V is a function of one variable only.

Consider the function $\omega = f(x, y)$, where x and y are independent variables, i.e. ω is uniquely determined when x and y are given, and is defined over some range R of values of (x, y). If a geometrical interpretation of $\omega = f(x, y)$ is required, then the number pair (x, y) may be represented by a point N of rectangular coordinates (x, y) in the region R of the xy-plane. The value of the function ω or $f(x, y)$ is then represented by a line NP drawn perpendicular to this plane at the point N of length equal to $f(x, y)$. Thus as N varies over the region R the corresponding point $\omega = f(x, y)$ varies over a surface consisting of all points P with Cartesian coordinates $(x, y, f(x, y))$.

1.5 RULES FOR INEQUALITIES

(i) If $a \leqslant b$ and $b < c$ then $a < c$.

(ii) If $a < b$ then $a + x < b + x$ for every x, i.e. the inequality relationship is unchanged when the same number (positive or negative) is added to both sides.

(iii) If $a < b$ then $pa < pb$ when p is positive and $qa > qb$ when q is negative. In particular if $a < b$ then $-a > -b$. This is easily seen if a and b are considered as points on the x-axis.

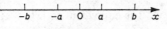

Fig. 1.25

In Fig. 1.25 the inequality $a < b$ means that a is to the left of b and $-a$ is to the right of $-b$ on the line in which numbers increase from left to right.

For example, $3 < 4$, $-3 > -4$;

                     $6 < 8$, $-3/2 > -4/2$.

Thus the inequality is unchanged if both sides are multiplied by a positive quantity; but the inequality is changed when both sides are multiplied by a negative quantity.

(iv) If $a < b$, and ab is positive, then $1/a > 1/b$.

For example, $3 < 4$ and $\tfrac{1}{3} > \tfrac{1}{4}$;

                  $-3 < -2$ and $-\tfrac{1}{3} > -\tfrac{1}{2}$.

EXERCISE 1

1 Find the values of $f(2), f(-1), f(1/x), f(x + h)$ when $f(x) = x^2 + 1/x$.

2 Find the values of $f(0), f(\tfrac{1}{2}\pi), f(3\pi/2)$ when $f(x) = \sin x + \cos x$.

3 Solve the following equations for x in terms of y and discuss the possible ranges of x and y.

(i) $y = \dfrac{x - 1}{x + 1}$;   (iii) $x^2 + xy + y^2 = 3$;

(ii) $y^2 = \dfrac{x}{x - 1}$;   (iv) $y = \dfrac{x^2}{x^2 - 1}$

4 Investigate the continuity of the following functions at $x = 0$, and draw a rough sketch of the graph of $y = f(x)$ in each case:

(i) $f(x) = |x|$;   (ii) $f(x) = 1/(x^2 + 2x)$;

(iii) $f(x) = \begin{cases} 0 & \text{when } x \leqslant 0 \\ 1 & \text{when } x > 0; \end{cases}$

(iv) $f(x) = \begin{cases} x & \text{when } x \leqslant 0 \\ 1 - x & \text{when } x > 0. \end{cases}$

5 Show that the function $x + [x]$ is not continuous at any integer value of x and draw a rough sketch of the graph of $y = x + [x]$.

6 (i) Sketch the graph of $y = 1/(x - 1)$.

(ii) Sketch the graph of $y = f(x)$ where $f(x)$ is defined by

$$f(x) = \begin{cases} x - 1 & \text{when } 0 \leqslant x \leqslant 2 \\ 3 - x & \text{when } 2 < x < 3. \end{cases}$$

7 Show that (i) $x^2 < 1$ is equivalent to $|x| < 1$;

(ii) if $x^2 > 1$ then $x > 1$ or $x < -1$.

8 Describe the range of the variable x without the symbol 'modulus' in each of the following:

(i) $|x| < 3$;   (iii) $|x - 2| \leqslant 4$;

(ii) $|x| \geqslant 3$;   (iv) $|x - 3| < 3$ and at the same time $|x - 4| \geqslant 2$.

9 Given that $\theta = \sin^{-1}(-\tfrac{1}{2})$, find the values of $\cos \theta$, $\tan \theta$, $\sec \theta$, $\operatorname{cosec} \theta$.

10 Given that $\theta = \cos^{-1}(-\tfrac{1}{2})$, find the values of $\sin \theta$, $\tan \theta$, $\sec \theta$, $\operatorname{cosec} \theta$.

11 Evaluate   (i) $\sin^{-1}(1) - \sin^{-1}(-1)$;

(ii) $\tan^{-1}(1) - \tan^{-1}(-1)$;

(iii) $\sec^{-1}(2) - \sec^{-1}(-2)$.

12 Describe and sketch the locus of points P whose rectangular Cartesian coordinates are (x, y, z) and which satisfy the following pairs of simultaneous equations:

(i) $x = \text{constant}, y = \text{constant}$;   (iii) $x^2 + y^2 = 9, z = -3$;

(ii) $y = x, z = 4$;   (iv) $y = 0, x^2/a^2 + z^2/b^2 = 1$.

13 Describe and sketch the locus of points P whose cylindrical polar coordinates (ρ, ϕ, z) satisfy the following pairs of simultaneous equations:

(i) $\rho = 3, z = 2$;   (ii) $\phi = \tfrac{1}{3}\pi, z = \rho$.

14 Describe and sketch the locus of points P whose spherical polar coordinates (r, θ, ϕ) satisfy the following pairs of simultaneous equations:

(i) $r = 5, \phi = \frac{1}{4}\pi$; (iii) $\phi = \frac{1}{4}\pi, \theta = \frac{1}{4}\pi$;

(ii) $r = 4, \theta = \frac{1}{4}\pi$; (iv) $\phi = \frac{1}{2}\pi, r = 4\cos\theta$.

15 Transform each of the following equations from the given co-ordinate system into forms which are appropriate to the other two systems:

(i) $x^2 + y^2 + z^2 = 4$; (iii) $z^2 = \rho^2$;

(ii) $x^2 + y^2 + z^2 = 9z$; (iv) $r = 4\cos\theta$.

2
Differentiation

2.1 DEFINITION OF THE DERIVATIVE

2.1.1 Formal definition

Let $y = f(x)$ be defined and continuous for all values of x in the interval $a \leqslant x \leqslant b$ and let x_0 be a value of x in this interval. Then the *derivative* or *differential coefficient* of the function at x_0 is defined as

$$\underset{h \to 0}{\text{Lim}} \frac{f(x_0 + h) - f(x_0)}{h} \tag{1}$$

when this limit exists. It is denoted by $f'(x_0)$. At each point x_0 where the limit is finite the function is said to be *differentiable* at the point. When the derivative exists over the whole range of values of x the function is differentiable in that range.

The value x_0 may be any allowable value of x, consequently it is usual to write x instead of x_0 in equation (1) and define the derivative of f as

$$f'(x) = \underset{h \to 0}{\text{Lim}} \frac{f(x + h) - f(x)}{h} \tag{2}$$

remembering that x is to be held constant, while h varies and approaches zero.

Example Find from first principles the derivative of

$$f(x) = x^3 - 3x + 3,$$

where $-\infty < x < \infty$.

Solution $\qquad\qquad f(x + h) = (x + h)^3 - 3(x + h) + 3$

so that $\qquad f(x + h) - f(x) = 3x^2h + 3xh + h^3 - 3h$

and $\qquad \dfrac{f(x + h) - f(x)}{h} = (3x^2 - 3) + h(3x + h).$

As $h \to 0$, $3x^2 - 3$ is held constant while $h(3x + h) \to 0$ and

$$f'(x) = 3x^2 - 3.$$

2.1.2 Geometrical interpretation

Let $P(x_0, f(x_0))$ be any point on the curve $y = f(x)$ and

$$Q(x_0 + h, f(x_0 + h))$$

be another point on the same curve. Then in Fig. 2.1 the slope of the line joining the points P and Q is QR/PR where

$$\frac{QR}{PR} = \frac{f(x_0 + h) - f(x_0)}{h}.$$

If the value of this quotient approaches a finite limit as $Q \to P$ along the curve, P being held fixed, then the slope of PQ tends to the slope of a line through P which is defined to be the tangent at P. Then

$$f'(x_0) = \operatorname*{Lim}_{h \to 0} \frac{f(x_0 + h) - f(x_0)}{h}$$

is the slope of the tangent to the curve at P, or more briefly, the slope of the curve at P.

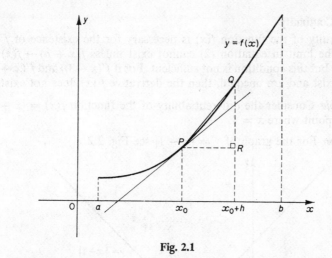

Fig. 2.1

2.1.3 Notation

It is sometimes convenient to denote an *increment* in the value of x by $\delta x \, (= h)$, an increment of the function f by δf or of y by δy, so that

$$\delta f(x_0) = f(x_0 + \delta x) - f(x_0), \tag{3}$$

and substituting in equation (1)

$$f'(x_0) = \operatorname*{Lim}_{\delta x \to 0} \frac{\delta f(x_0)}{\delta x} \tag{4}$$

$$= \operatorname*{Lim}_{\delta x \to 0} \left(\frac{\delta y}{\delta x} \right)_{x = x_0} \tag{5}$$

The derivative of $y = f(x)$ when it exists is a function of x and is written $f'(x)$ or y'. From the right hand side of equations (4) and (5) is derived the notation

$$\frac{d}{dx} f(x) \quad \text{or} \quad \frac{dy}{dx}$$

where x_0 is replaced by the more general x. The quotient form dy/dx derives from equation (5) and is a limit of a quotient. It is not $dy \div dx$. The symbol d/dx, regarded as an operator which when applied to $f(x)$ gives $f'(x)$, is sometimes written D or D_x and must operate on some function of x. Hence Dy means dy/dx. Thus $x\,Dy$ and $y\,Dx$ are entirely different, for $x\,Dy = x\,dy/dx$ and $y\,Dx = y\,d(x)/dx = y$.

The limit $f'(x - 0)$ if it exists is called the derivative on the left, and similarly $f'(x + 0)$ is called the derivative on the right.

2.1.4 Continuity

Continuity of the function $f(x)$ is necessary for the existence of $f'(x)$ since the limit in equation (2) cannot exist unless $f(x + h) \rightarrow f(x)$ as $h \rightarrow 0$, but this condition is not sufficient. For if $f'(x - 0)$ and $f'(x + 0)$ both exist and are unequal, then the derivative $f'(x)$ does not exist.

Example Consider the differentiability of the function $f(x) = |x - 1|$ at the point where $x = 1$.

Solution For the graph of $y = |x - 1|$ see Fig. 2.2.

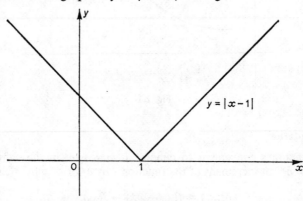

$y = |x - 1|$

Fig. 2.2

The function is continuous for all values of x including $x = 1$ but $f'(1 - 0) = -1$ and $f'(1 + 0) = 1$; i.e. the derivative at $x = 1$ does not exist and the function is continuous but not differentiable at $x = 1$.

2.1.5 Higher derivatives

If $f'(x)$ exists then it is a function of x which may or may not be differentiated. If $f'(x)$ has a derivative it is denoted by $f''(x)$ or d^2y/dx^2

or D^2y and it is the second order derivative of $y = f(x)$. Similarly third, fourth, up to nth order derivatives may be defined:

If $y = f(x)$, then

$$Dy \text{ or } \frac{dy}{dx} = f'(x),$$

$$D^2y \text{ or } \frac{d^2y}{dx^2} = f''(x) = \frac{d}{dx}\left(\frac{dy}{dx}\right),$$

$$\cdot \quad \cdot \quad \cdot$$
$$\cdot \quad \cdot \quad \cdot$$
$$\cdot \quad \cdot \quad \cdot$$

and $\quad D^ny \text{ or } \frac{d^ny}{dx^n} = f^{(n)}(x) = \frac{d}{dx}\left(\frac{d^{n-1}y}{dx^{n-1}}\right).$

2.2 STANDARD DERIVATIVES AND RULES

2.2.1 The derivative of a constant

Let $f(x) = c$ where c is a constant, then $f(x + h) = c$ and

$$f(x + h) - f(x) = 0.$$

Hence $\quad f'(x) = \lim_{h \to 0}\frac{f(x + h) - f(x)}{h} = 0,$

i.e. the derivative of a constant is zero.

2.2.2 The derivative of x^n

Let $f(x) = x^n$ where n is any positive integer, then

$$f(x + h) = (x + h)^n = x^n + nx^{n-1}h + h^2 \text{ (terms in } x \text{ and } h).$$

If $h \neq 0$, $\{f(x + h) - f(x)\}/h = nx^{n-1} + h$ (terms in x and h). Now let $h \to 0$, then $f'(x) = nx^{n-1}$.

2.2.3 The derivative of $\sin x$, $\cos x$

(i) $\dfrac{d}{dx}(\sin x) = \cos x.$

From the definition,

$$\frac{\sin (x + h) - \sin x}{h} = \frac{2 \cos (x + \frac{1}{2}h) \sin \frac{1}{2}h}{h}$$

$$= \cos (x + \tfrac{1}{2}h) \frac{\sin \frac{1}{2}h}{\frac{1}{2}h}. \tag{6}$$

Since $\cos (x + \frac{1}{2}h) \to \cos x$ as $h \to 0$ and since

$$\lim_{\theta \to 0}\frac{\sin \theta}{\theta} = 1, \quad \frac{\sin \frac{1}{2}h}{\frac{1}{2}h} \to 1 \quad \text{as} \quad h \to 0$$

then the limit as $h \to 0$ of equation (6), or the derivative of $\sin x$, is $\cos x$.

(ii) $\dfrac{d}{dx}(\cos x) = -\sin x$.

From the definition,

$$\frac{\cos(x+h) - \cos x}{h} = \frac{-2\sin(x + \tfrac{1}{2}h)\sin\tfrac{1}{2}h}{h}$$

$$= -\sin(x + \tfrac{1}{2}h)\frac{\sin\tfrac{1}{2}h}{\tfrac{1}{2}h}$$

$$\to -\sin x \,.\, 1 \quad \text{as} \quad h \to 0.$$

2.2.4 The derivative of $cf(x)$

If $f(x)$ is a differentiable function of x and c is a constant, then

$$\frac{d}{dx}\{cf(x)\} = cf'(x).$$

Let $\phi(x) = cf(x)$, then for $h \neq 0$

$$\frac{\phi(x+h) - \phi(x)}{h} = \frac{cf(x+h) - cf(x)}{h}$$

$$= c\left[\frac{f(x+h) - f(x)}{h}\right]$$

Let $h \to 0$, then $\qquad \phi'(x) = cf'(x).$

2.2.5 Differentiation of sums

The derivative of the sum of a finite number of differentiable functions of x is equal to the sum of their derivatives, and more generally if $f(x)$ and $g(x)$ are differentiable and a and b are constants,

$$\frac{d}{dx}\{af(x) + bg(x)\} = af'(x) + bg'(x).$$

This is easily proved using section 2.2.4 and the definition of a derivative, and can be extended to a linear function of any number of differentiable functions.

Example $f(x) = x^3 + 7x^2 - 5x + 4$

$$f'(x) = \frac{d}{dx}(x^3) + 7\frac{d}{dx}(x^2) - 5\frac{d}{dx}(x) + \frac{d}{dx}(4)$$

$$= 3x^2 + 14x - 5.$$

2.2.6 Differentiation of products

The derivative of the product of two differentiable functions $f(x)$ and $g(x)$ is $f'(x)g(x) + f(x)g'(x)$. For let $\phi(x) = f(x)g(x)$ and $h \neq 0$ then,

$$\frac{\phi(x + h) - \phi(x)}{h}$$

$$= \frac{f(x + h)\, g(x + h) - f(x)\, g(x)}{h}$$

$$= \frac{f(x + h)\, g(x + h) - f(x + h)\, g(x) + f(x + h)\, g(x) - f(x)\, g(x)}{h}$$

$$= f(x + h)\left[\frac{g(x + h) - g(x)}{h}\right] + g(x)\left[\frac{f(x + h) - f(x)}{h}\right].$$

Since $f(x)$ and $g(x)$ are differentiable and hence continuous, as $h \to 0$, it follows that

$$\phi'(x) = f(x)g'(x) + g(x)f'(x).$$

2.2.7 The derivative of the reciprocal $1/g(x)$

The derivative of the reciprocal of $g(x)$ is $-g'(x)/\{g(x)\}^2$ at a point where $g(x) \neq 0$, for, let $\phi(x) = 1/g(x)$, $h \neq 0$, then

$$\frac{\phi(x + h) - \phi(x)}{h} = \frac{1}{h}\left[\frac{1}{g(x + h)} - \frac{1}{g(x)}\right]$$

$$= \frac{-[g(x + h) - g(x)]}{hg(x + h)g(x)}$$

$$= \frac{-1}{g(x + h)g(x)}\left[\frac{g(x + h) - g(x)}{h}\right].$$

As $h \to 0$, $$\phi'(x) = \frac{-g'(x)}{[g(x)]^2}.$$

The proofs of the formulae in sections 2.2.6 and 2.2.7 depend on the assumption that the limit of a product (or quotient) is the product (or quotient) of the limits.

Example Show that the derivative of x^n is nx^{n-1} when n is a negative integer.

Solution Let $n = -m$ where m is a positive integer, then

$$y = x^n = x^{-m} = \frac{1}{x^m}.$$

Applying the above result and using section 2.2.2, gives

$$\frac{dy}{dx} = \frac{-mx^{m-1}}{x^{2m}}$$

$$= (-m)x^{-m-1}$$

$$= nx^{n-1}.$$

Hence

$$\frac{d}{dx} x^n = nx^{n-1}.$$

2.2.8 Differentiation of a quotient

The derivative of the quotient of two differentiable functions $f(x)$ and $g(x)$ follows by applying the formulae in sections 2.2.6 and 2.2.7 for, let

$$\phi(x) = \frac{f(x)}{g(x)} = f(x) \times \frac{1}{g(x)}$$

then at a point where $g(x) \neq 0$,

$$\phi'(x) = \frac{g(x) f'(x) - f(x) g'(x)}{[g(x)]^2}$$

Example (i) $\dfrac{d}{dx} (\tan x) = \sec^2 x$

since $\qquad \tan x = \dfrac{\sin x}{\cos x}$

$$(\tan x)' = \frac{\cos x \cos x + \sin x \sin x}{(\cos x)^2}$$

$$= \sec^2 x.$$

Similarly $\qquad (\cot x)' = - \operatorname{cosec}^2 x.$

Example (ii) $\dfrac{d}{dx} (\sec x) = \sec x \tan x$

since $\qquad \sec x = 1/\cos x$

$$(\sec x)' = \frac{\sin x}{(\cos x)^2} \quad \text{using section 2.2.7}$$

$$= \sec x \tan x.$$

Similarly $\qquad (\operatorname{cosec} x)' = - \operatorname{cosec} x \cot x.$

2.2.9 Differentiation of a function of a function

Suppose f is a function of g and g is a function of x then f is a function of a function of x and

$$\frac{d}{dx} [f\{g(x)\}] = f'\{g(x)\}g'(x) \tag{7}$$

at a point x where $g(x)$ is differentiable and f is differentiable at $g(x)$. The symbol $f'\{g(x)\}$ means the derivative of the function f with respect to the bracket $\{\ \}$ or $g(x)$, at $g(x)$.

Writing $g(x) = t$, then $y = f(t)$ and equation (7) may be written

$$\frac{dy}{dx} = \frac{dy}{dt}\frac{dt}{dx}. \tag{8}$$

For if $\delta x \neq 0$, $\qquad \delta t = g(x + \delta x) - g(x)$,

$$\delta y = f(t + \delta t) - f(t),$$

and $\qquad \dfrac{\delta y}{\delta x} = \dfrac{\delta y}{\delta t}\dfrac{\delta t}{\delta x} \quad$ if $\quad \delta t \neq 0$

or $\qquad \dfrac{\delta y}{\delta x} = 0 \quad$ if $\quad \delta t = 0$.

If $\delta t \neq 0$, as $\delta x \to 0$

$$\frac{dy}{dx} = \frac{dy}{dt}\frac{dt}{dx}.$$

If $\delta t = 0$ but $\delta x \neq 0$

$$\frac{\delta t}{\delta x} = 0 \qquad \text{and} \qquad \frac{dy}{dx} = 0.$$

A rigorous proof of this may be found in a more advanced text-book. It follows that the derivative of $f(ax + b)$, a and b are constants, is

$$af'(ax + b). \tag{9}$$

Example (i) If $\qquad y = \cos(3x + 2)$,

then $\qquad y' = -3\sin(3x + 2)$.

Example (ii) If $f(t)$ is an arbitrary differentiable function and $g(x)$ is such that

$$y = f[g(x)] = x$$

then (8) becomes

$$1 = (dx/dt)(dt/dx)$$

i.e. $\qquad dt/dx = 1/(dx/dt)$, provided $dx/dt \neq 0$

(see also section 2.6).

Example (iii) If $\qquad y = \tan^2 x = (\tan x)^2$,

then $\qquad y' = 2(\tan x)\sec^2 x$.

Example (iv) If $\qquad y = \cos^2(\sin 3x) = \{\cos(\sin 3x)\}^2$,

then $\qquad y' = 2\{\cos(\sin 3x)\}[-\sin(\sin 3x)] 3\cos 3x$

using equation (8) repeatedly.

2.2.10 Parametric differentiation

When $y = f(x)$ and x and y are both functions of a single variable, say t, called a *parameter*, i.e. $x = x(t)$, $y = y(t)$, and x and y are differentiable then,

$$\frac{dy}{dx} = \frac{dy/dt}{dx/dt} \quad \text{provided} \quad \frac{dx}{dt} \neq 0. \tag{10}$$

This follows from (8) and Example (ii) of section 2.2.9 and is useful when a curve is given *parametrically* in the form

$$x = x(t), \qquad y = y(t).$$

Example Find dy/dx and d^2y/dx^2 as functions of t when

$$x = a(t - \sin t), \quad y = a(1 - \cos t).$$

Solution

$$x = a(t - \sin t)$$

$$dx/dt = a(1 - \cos t)$$

and

$$y = a(1 - \cos t)$$

$$dy/dt = a \sin t,$$

and hence,

$$\frac{dy}{dx} = \frac{dy/dt}{dx/dt}$$

$$= \frac{a \sin t}{a(1 - \cos t)}$$

$$= \frac{\sin t}{1 - \cos t}$$

provided $1 - \cos t \neq 0$.

This may be simplified using the half-angle formulae of trigonometry to

$$\frac{dy}{dx} = \frac{2 \sin \tfrac{1}{2}t \cos \tfrac{1}{2}t}{2 \sin^2 \tfrac{1}{2}t} = \cot \tfrac{1}{2}t.$$

Differentiating again and using section 2.2.9,

$$\frac{d^2y}{dx^2} = \frac{d}{dx}\left(\frac{dy}{dx}\right) = \frac{d}{dt}(\cot \tfrac{1}{2}t)\frac{dt}{dx}$$

$$= -\tfrac{1}{2}\operatorname{cosec}^2 \tfrac{1}{2}t \frac{1}{a(1 - \cos t)}$$

$$= -\frac{1}{4a}\operatorname{cosec}^4 \tfrac{1}{2}t.$$

It is important to notice the procedure for finding $\dfrac{d^2y}{dx^2}$ as $\dfrac{d}{dt}\{y'(t)\}\dfrac{dt}{dx}$. The expression $\dfrac{d}{dx}$ (function of t) must be calculated as $\dfrac{d}{dt}$ (function of t) $\dfrac{dt}{dx}$.

2.2.11 Implicit functions

These are given in the form $f(x, y) = 0$ or $f(x, y, c) = 0$ where c is some number which may be a parameter, that is as equations which involve x and y in such a way that y is not given explicitly in terms of x, e.g. $x^2 + y^2 = 1$; $x^5 + 4xy^3 - 3y^5 = 2$.

It is possible to calculate dy/dx from such an equation by implicit differentiation using the previous rules.

Example (i) Find dy/dx at any point (x, y) on the curve given by $x^5 + 4xy^3 - 3y^5 = 2$.

Solution Differentiate both sides of this equation with respect to x, let D be the operator d/dx

$$D(x^5) + D(4xy^3) - D(3y^5) = D(2)$$
$$5x^4 + 4\{y^3 + xDy^3\} - 3D(y^5) = 0$$
$$5x^4 + 4y^3 + 4x3y^2Dy - 3 \times 5y^4Dy = 0.$$

Hence at points (x, y) where $15y^4 - 12xy^2 \neq 0$,

$$Dy \quad \text{or} \quad \frac{dy}{dx} = \frac{5x^4 + 4y^3}{15y^4 - 12xy^2}.$$

Example (ii) Find dy/dx if $y^3 = 2x - 1$.

Solution

(a) Differentiating both sides of the equation with respect to x, then

$$3y^2 \frac{dy}{dx} = 2$$

or

$$\frac{dy}{dx} = \frac{2}{3y^2}.$$

(b) Write $\quad x = \frac{1}{2}y^3 + \frac{1}{2}, \quad$ then

$$\frac{dx}{dy} = \frac{3y^2}{2}$$

$$\frac{dy}{dx} = \frac{2}{3y^2}.$$

Thus dy/dx can be calculated as the reciprocal of dx/dy provided y is a differentiable function of x and x is a differentiable function of y.

Example (iii) Show that the derivative of x^n is nx^{n-1} when n is a rational, non-zero fraction.

Solution If n is rational it is of the form p/q where p and q are integers and $q \neq 0$. If $p \neq 0$ and p, q have no common factor then n is of the required form.

Then

$$y = x^n = x^{p/q}$$

and

$$y^q = x^p$$

where p and q are non-zero integers.

Differentiating with respect to x,

$$qy^{q-1}\frac{dy}{dx} = px^{p-1}$$

or
$$\frac{dy}{dx} = \left(\frac{p}{q}\right) x^{p-1-(p/q)(q-1)}$$

$$= \left(\frac{p}{q}\right) x^{p/q-1}.$$

Hence
$$\frac{d}{dx} x^n = nx^{n-1}.$$

2.3 REPEATED DIFFERENTIATION, LEIBNITZ' THEOREM

2.3.1

Consider first some examples.

Example (i) Show that the nth derivative of $\sin ax$ is $a^n \sin (ax + n\pi/2)$.

Solution Let $f(x) = \sin ax$

$$f'(x) = a \cos ax$$
$$= a \sin (ax + \tfrac{1}{2}\pi)$$

i.e. $f^{(n)}(x) = a^n \sin (ax + n\pi/2)$ is true when $n = 1$. The method of induction is used for the proof. Suppose that for some fixed integer value of n,

$$f^{(n)}(x) = a^n \sin (ax + n\pi/2)$$

Differentiating both sides,

$$f^{(n+1)}(x) = a^{n+1} \cos (ax + n\pi/2)$$
$$= a^{n+1} \sin (ax + n\pi/2 + \pi/2)$$
$$= a^{n+1} \sin \{ax + (n + 1)\pi/2\}$$

i.e. the result is true when n is replaced by $n + 1$. Since it is true when $n = 1$ it is true for all integer values of $n \geqslant 1$.

Example (ii) Find the nth derivative of $\sin 2x$ when $x = 0$.

Solution $D^n (\sin 2x) = 2^n \sin (2x + n\pi/2)$ using example above. When $x = 0$, $D^n (\sin 2x) = 2^n \sin (n\pi/2)$ and this has different values according as n is even or odd.

When n is even, write $n = 2k$:

at $x = 0$, $D^{2k} (\sin 2x) = 2^{2k} \sin k\pi = 0.$

When n is odd, write $n = 2k + 1$:

at $x = 0$, $D^{2k+1} (\sin 2x) = 2^{2k+1} \sin (2k + 1)\pi/2$
$$= 2^{2k+1} (-1)^k.$$

2.3.2 Leibnitz' theorem

Let f and g be functions of x which may be differentiated n times. Then

$$D(fg) = fDg + gDf.$$

Repeatedly applying the rule for differentiating products and using $D^{(n)}$ for the operator d^n/dx^n,

$$D^2(fg) = fD^2g + 2DfDg + gD^2f$$
$$D^3(fg) = fD^3g + 3DfD^2g + 3D^2fDg + gD^3f$$
$$D^4(fg) = fD^4g + 4DfD^3g + 6D^2fD^2g + 4D^3fDg + gD^4f.$$

These coefficients are binomial coefficients and the result

$$D^{(n)}(fg) = fD^{(n)}g + \binom{n}{1} Df\, D^{n-1}g + \ldots + \binom{n}{r} D^{(r)}fD^{(n-r)}g$$
$$+ \ldots + gD^{(n)}f \qquad (11)$$

for any positive integer n is implied and may be proved by induction. It is called *Leibnitz' theorem*.

Example Find $D^{(n)}\{x^2 \sin x\}$.

Solution

$$D^{(n)}\{x^2 \sin x\} = x^2 D^{(n)} (\sin x) + nD(x^2)D^{n-1} (\sin x)$$
$$+ \tfrac{1}{2}n(n-1)D^2(x^2)D^{n-2}(\sin x)$$
$$= x^2 \sin (x + \tfrac{1}{2}n\pi) + n2x \sin \{x + \tfrac{1}{2}(n-1)\pi\}$$
$$+ \tfrac{1}{2}n(n-1)2 \sin \{x + \tfrac{1}{2}(n-2)\pi\},$$

since $D^{(n)}(x^2) = 0$ when $n > 2$ and $D^{(n)} (\sin x) = \sin (x + \tfrac{1}{2}n\pi)$ from example (i) of section 2.3.1.

2.4 THE INCREMENT OF A FUNCTION

Consider an estimate of the change δy produced in a function $y = f(x)$ when x changes by a small amount δx. Let P be the point (x, y) on the curve $y = f(x)$ where the function is differentiable, and let Q be the neighbouring point $(x + \delta x, y + \delta y)$ where $\delta x \neq 0$. Then the slope of the chord PQ is $\delta y/\delta x$ which approaches the limiting value dy/dx as $\delta x \to 0$. Hence the difference between $\delta y/\delta x$ and dy/dx is numerically small when $|\delta x|$ is small. Let this difference be denoted by ε, i.e.

$$\frac{\delta y}{\delta x} - \frac{dy}{dx} = \varepsilon. \qquad (12)$$

Then the statement

$$\lim_{\delta x \to 0} \frac{\delta y}{\delta x} = \frac{dy}{dx} = f'(x) \qquad (13)$$

is equivalent to the statement

$$\underset{\delta x \to 0}{\text{Lim}}\ \varepsilon = 0 \tag{14}$$

or $\qquad \dfrac{\delta y}{\delta x} = \dfrac{dy}{dx} + \varepsilon \quad$ where $\quad \varepsilon \to 0 \quad$ as $\quad \delta x \to 0,$ $\qquad$ (15)

and multiplying by δx

$$\delta y = \frac{dy}{dx}\,\delta x + \varepsilon \delta x \tag{16}$$

Although equation (16) was derived under the assumption that $\delta x \neq 0$, it is still true even when $\delta x = 0$, since it is only necessary to define $\varepsilon = 0$ when $\delta x = 0$.

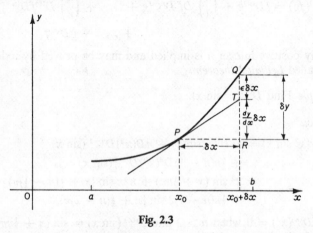

Fig. 2.3

Consider geometrically the graph of $y = f(x)$ as sketched in Fig. 2.3. The first term on the right hand side of equation (16) is represented by the length TR, since the slope of the tangent PT is dy/dx, and this is the change produced in y along the tangent at P. When δx is small, this term is usually large compared with the second term $\varepsilon \delta x$. The difference between $QR = \dfrac{dy}{dx}\,\delta x + \varepsilon \delta x$ and $TR = \dfrac{dy}{dx}\,\delta x$ is $\varepsilon \delta x$ which tends to zero more rapidly than δx does when δx approaches zero.

For example, if $y = f(x) = x^2$, $dy/dx = 2x = f'(x)$ and

$$\frac{dy}{dx}\,\delta x = 2x\,\delta x.$$

On the other hand the exact value of δy is given by

$$\delta y = 2x\,\delta x + (\delta x)^2 = f(x + \delta x) - f(x).$$

Comparing this with equation (16), for this particular example, $\varepsilon\delta x = (\delta x)^2$. When δx is small, $(\delta x)^2$ is smaller than δx, and δy is approximately equal to $2x\,\delta x$. This example may be illustrated geometrically as in Fig. 2.4.

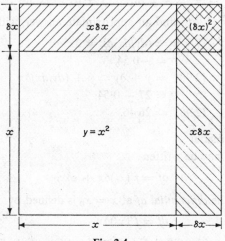

Fig. 2.4

The original square has side of length x and hence area $y = x^2$. When the sides are increased to $x + \delta x$ the area is increased by $\delta y = 2x\,\delta x + (\delta x)^2$. When δx is small compared with x most of the change in y is given by the two rectangular areas and only a very small part is given by the small square of area $(\delta x)^2$.

Hence if the original square has side 2 cm say and the larger square side 2·01 cm, then $x = 2$, $x + \delta x = 2·01$, $\delta x = 0·01$. The area of the original square is $y = x^2 = 4$ and the area of the larger square is

$$y + \delta y = (x + \delta x)^2$$
$$= x^2 + 2x\delta x + (\delta x)^2$$
$$= 4 + 0·04 + 0·0001.$$

The increment $\delta y = 0·04 + 0·0001$. Hence if $2x\delta x = 0·04$ is used as an approximation to δy there is an error of $0·0001$ square centimetres which is just under $\frac{1}{4}\%$ of δy.

Since the term $(dy/dx)\delta x$ usually gives a good approximation to δy when δx is small compared with dy/dx, it is customary to use the approximation

$$\delta y \simeq \frac{dy}{dx}\,\delta x \qquad (17)$$

or $$\delta f \simeq f'(x_0)\delta x \qquad (18)$$

in numerical calculations.

Example Using the approximation in equation (17) determine a reasonable approximation to $(2 \cdot 98)^3$.

Solution Let $y = x^3$ and start from the point $x = 3$, $y = 27$,

$$x + \delta x = 2 \cdot 98, \qquad y + \delta y = (2 \cdot 98)^3.$$

Using equation (17) to find an approximate answer, with $\delta x = -0 \cdot 02$

at $x = 3$,
$$\delta y \simeq (dy/dx)\delta x = 3x^2\delta x$$
$$= -0 \cdot 54$$

hence
$$(2 \cdot 98)^3 = y + \delta y \simeq y + (dy/dx)\delta x$$
$$= 27 - 0 \cdot 54$$

or
$$(2 \cdot 98)^3 \simeq 26 \cdot 46.$$

2.4.1 The differential

Equation (16) may be written

$$\delta f = f'(x_0)\delta x + \varepsilon \delta x. \tag{19}$$

If $f'(x_0) \neq 0$, the *differential df* at $x = x_0$ is defined by

$$df = f'(x_0)\delta x \tag{20}$$

and is represented by TR in Fig. 2.3. The differential of the function x is

$$dx = \delta x \tag{21}$$

so that equation (20) may be written

$$df = f'(x_0)dx$$
or
$$dy = f'(x_0)dx. \tag{22}$$

Thus the differential dy at $x = x_0$ is the *amount* of change in y along the tangent line at the point x_0 which would be produced by a change dx in x; dy/dx is the *rate* of change of y per unit change in x.

2.5 APPLICATIONS OF DIFFERENTIATION TO CURVE SKETCHING

2.5.1 The sign of $f'(x)$

Let $y = f(x)$ be continuous and differentiable in the range $a \leqslant x \leqslant b$.

If $f(x)$ increases monotonically in this range, for any value of x in $a < x < b$, $[f(x + h) - f(x)]/h$ has the sign of $f'(x)$ provided h is small enough, and hence $f'(x) \geqslant 0$ in the range.

Similarly if $f(x)$ decreases monotonically, then $f'(x) \leqslant 0$ in $a < x < b$. See section 1.1.4 and Figs. 2.5(a) and 2.5(b).

2.5.2 Concavity

Let $f'(x)$ and $f''(x)$ both exist in the range $a \leqslant x \leqslant b$. The graph of $y = f(x)$ is said to be *concave downwards* if $f'(x)$ decreases, i.e. $f''(x) < 0$ as x increases through the range. The graph of $y = f(x)$ is concave upwards if $f'(x)$ increases, i.e. $f''(x) > 0$ as x increases throughout the range. See Figs. 2.6 and 2.7.

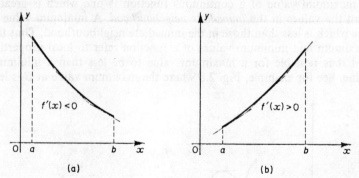

Fig. 2.5

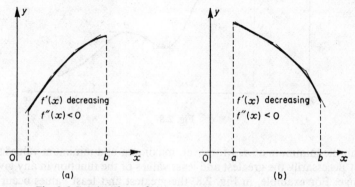

Fig. 2.6 Graphs of $y = f(x)$ concave downwards in $a \leqslant x \leqslant b$

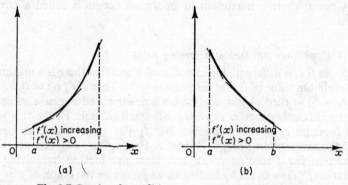

Fig. 2.7 Graphs of $y = f(x)$ concave upwards in $a \leqslant x \leqslant b$

When the graph is concave downwards the tangent at each point is 'above' the curve, and when the graph is concave upwards the tangent is 'below' the curve.

2.5.3 Maxima, minima

A maximum value of a continuous function is one which is greater than the values in the *immediate neighbourhood*. A minimum value is one which is less than those in the immediate neighbourhood. Thus the maximum and minimum values of a function refer to local properties and it is possible for a maximum value to be less than a minimum value. See for example, Fig. 2.8 where the maximum value at B is less

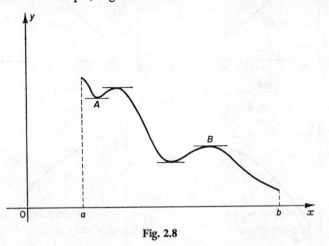

Fig. 2.8

than the minimum at A. Moreover, maximum and minimum values are not necessarily the greatest and least values of the function in any given range. For example, in Fig. 2.8, the greatest and least values occur at the beginning and end of the range.

A point where a maximum or minimum occurs is called a *turning point*.

2.5.4 Conditions satisfied at a turning point

Suppose $f(x)$ is differentiable in $a \leqslant x \leqslant b$ and that there is a maximum or minimum value of $f(x)$ at the point $x = x_0$. Then if $f'(x_0) \neq 0, h > 0$, $\{f(x_0 + h) - f(x_0)\}/h$ can assume both positive and negative values in the neighbourhood of x_0 provided h is small enough. Hence at $x = x_0$ the function is not differentiable. But $f(x)$ is differentiable at $x = x_0$ and the assumption $f'(x_0) \neq 0$ is false. Hence $f'(x_0) = 0$ is a necessary condition for a turning point, and these are located by solving the equation $f'(x) = 0$, i.e. by finding those points on the graph of $y = f(x)$ where the tangent is parallel to the x-axis.

This condition is not sufficient to discriminate between a maximum and a minimum value. If there exists x_0 such that $f'(x_0) = 0$ then usually $f'(x)$ changes sign as x increases through the value x_0. If $f'(x)$ changes sign from positive to negative then $f(x_0)$ is a maximum value, and if $f'(x)$ changes sign from negative to positive then $f(x_0)$ is a minimum value of the function. See Fig. 2.9.

The use of the sign of $f'(x)$ to the immediate left and immediate right of a root of $f'(x) = 0$ is an infallible method of distinguishing between maximum and minimum.

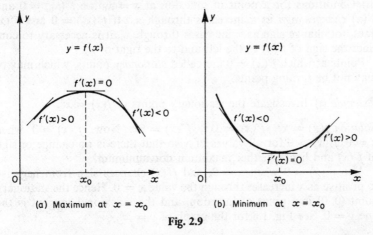

(a) Maximum at $x = x_0$ (b) Minimum at $x = x_0$

Fig. 2.9

Alternatively the sign of $f''(x_0)$ may be investigated in order to discriminate between maximum and minimum. At a maximum turning point the curve is concave downwards, i.e. $f'(x)$ decreases as x increases through the value x_0, or $f''(x_0) < 0$. Similarly at a minimum turning value the graph is concave upwards, $f'(x)$ increases as x increases through the value x_0, i.e. $f''(x_0) > 0$.

2.5.5 Conditions for a point of inflexion

Suppose $f''(x_0) = 0$, then usually $f''(x)$ changes sign as x increases through x_0. If $f''(x)$ changes sign from positive to negative, $x = x_0$ is a transition between $f''(x) > 0$ (i.e. $f'(x)$ increasing, $f(x)$ concave upwards, and tangent 'below' the curve) and $f''(x) < 0$ (i.e. $f'(x)$ decreasing, $f(x)$ concave downwards, and tangent 'above' the curve). Such a point of transition is called a *point of inflexion* and the tangent there which 'crosses' the curve is called the *inflexional* tangent.

Similarly if $f''(x)$ changes sign from negative to positive as x increases through x_0, then $(x_0, f(x_0))$ is a point where the graph of $y = f(x)$ changes from being concave downwards to concave upwards and there is a point of inflexion at $x = x_0$.

2.5.6 Summary

Let $y = f(x)$, then
(i) Conditions for a turning point at $x = x_0$ are $f'(x_0) = 0$ and $f'(x)$ *changes sign* as x increases through x_0,

(a) $f'(x)$ changes from positive to negative (or $f''(x_0) < 0$) gives a maximum at $x = x_0$;

(b) $f'(x)$ changes from negative to positive (or $f''(x_0) > 0$) gives a minimum at $x = x_0$.

(ii) Conditions for a point of inflexion at $x = x_0$ are $f''(x_0) = 0$ and $f''(x)$ *changes sign* as x increases through x_0. If $f''(x_0) = 0$ and $f''(x)$ does not change sign as x increases through x_0, it is necessary to consider the sign of $f'(x)$ to the left and to the right of x_0.

Points at which $f'(x) = 0$ are called *stationary* points, which may or may not be turning points.

Example (i) Investigate the stationary points of $f(x) = x^3$.

Solution $f(x) = x^3$, $f'(x) = 3x^2$, $f''(x) = 6x$. Now, $f'(x) = 0$ when $x = 0$, $f'(x) > 0$ for all values of x so that there is no change of sign of $f'(x)$ and hence neither maximum nor minimum.

Also $f''(x) = 0$ when $x = 0$, and $f''(x)$ changes sign from negative to positive as x increases through the value $x = 0$. Hence the stationary point $(0, 0)$ is a point of inflexion, and the inflexional tangent is the line $y = 0$. See Fig. 1.8 for the graph of $y = x^3$.

Example (ii) Investigate the stationary points of $f(x) = x^4$.

Solution $f(x) = x^4$, $f'(x) = 4x^3$, $f''(x) = 12x^2$. Now $f'(x) = 0$ when $x = 0$ and $f''(0) = 0$. But when $x = 0$ there is a minimum value of $f(x)$, since $f'(x)$ changes sign from negative to positive as x increases through the value zero.

Example (iii) Investigate the inflexion point on the curve $y = \sin x$, $-\frac{1}{4}\pi \leqslant x \leqslant \frac{1}{4}\pi$.

Solution $y = \sin x$, $y' = \cos x$, $y'' = -\sin x$. Now $y' = 0$ when $\cos x = 0$ which has no solution in the given range. Also $y'' = 0$ when $\sin x = 0$ and $x = 0$ is the only value in the range. Moreover y'' changes sign from positive to negative as x increases through the value $x = 0$. Hence $(0, 0)$ is a point of inflexion. The slope of the inflexional tangent is 1, since $y'(0) = 1$, and its equation is $y = x$. See Fig. 1.17(a).

2.6 DIFFERENTIATION OF INVERSE FUNCTIONS

Let $y = f(x)$ be differentiable and monotonically increasing in $a \leqslant x \leqslant b$. Then the inverse function $x = \phi(y)$ can be shown to be differentiable for all value of y between $f(a)$ and $f(b)$, provided that $f'(x) \neq 0$.

In Fig. 2.10, let P be the point $(x_1, f(x_1))$, then $x_1 = \phi(y_1)$ and P can be considered as the point $(\phi(y_1), y_1)$ on the graph of $x = \phi(y)$. In effect the same graph holds for both functions. If $f'(x) \neq 0$, the tangent at P to $y = f(x)$ is not parallel to the x-axis, which also means that the tangent at P to $x = \phi(y)$ is not perpendicular to the y-axis, and $\phi(y)$ is differentiable at P.

Let $\tan \psi$ be the gradient of the tangent at P so that $f'(x_1) = \tan \psi$. Let θ be the angle made by the tangent at P to the y-axis so that $\phi'(y_1) = \tan \theta$. Since $\theta = \frac{1}{2}\pi - \psi$, $\tan \theta = 1/\tan \psi$, or

$$\frac{dy}{dx} = \frac{1}{(dx/dy)}.$$

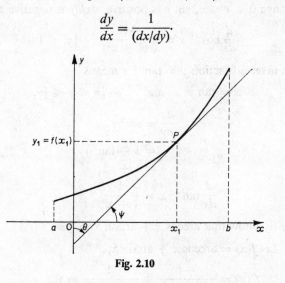

Fig. 2.10

2.6.1 Differentiation of the inverse circular functions

(i) The inverse function $y = \sin^{-1} x$ means

$$x = \sin y \quad \text{and} \quad -\tfrac{1}{2}\pi \leqslant y \leqslant \tfrac{1}{2}\pi$$

and hence,

$$\frac{dx}{dy} = \cos y$$

$$= \pm\sqrt{(1 - \sin^2 y)}$$

$$= \pm\sqrt{(1 - x^2)}.$$

In the range $-\frac{1}{2}\pi \leqslant y \leqslant \frac{1}{2}\pi$, $\cos y$ is positive so that

$$\frac{dy}{dx} = \frac{1}{\sqrt{(1 - x^2)}}$$

or

$$\frac{d}{dx} \sin^{-1} x = \frac{1}{\sqrt{(1 - x^2)}}, \; |x| < 1.$$

(ii) The inverse function $y = \cos^{-1} x$ means

$$x = \cos y \quad \text{and} \quad 0 \leqslant y \leqslant \pi,$$

and
$$\frac{dx}{dy} = -\sin y$$
$$= \mp\sqrt{(1 - \cos^2 y)}$$
$$= \mp\sqrt{(1 - x^2)}.$$

In the range $0 \leqslant y \leqslant \pi$, $\sin y$ is positive, dx/dy is negative and

$$\frac{d}{dx}\cos^{-1} x = \frac{-1}{\sqrt{(1 - x^2)}} \qquad |x| < 1$$

(iii) The inverse function $y = \tan^{-1} x$ means

$$x = \tan y \qquad \text{and} \qquad -\tfrac{1}{2}\pi \leqslant y \leqslant \tfrac{1}{2}\pi,$$

and
$$\frac{dx}{dy} = \sec^2 y$$
$$= 1 + \tan^2 y$$
$$= 1 + x^2$$

and
$$\frac{d}{dx}\tan^{-1} x = \frac{1}{1 + x^2}.$$

Example (i) Show that $\arccos x + \arcsin x = \tfrac{1}{2}\pi$.

Solution Let $f(x) = \arccos x + \arcsin x$,

then
$$f'(x) = \frac{-1}{\sqrt{(1 - x^2)}} + \frac{1}{\sqrt{(1 - x^2)}} = 0.$$

Hence $f(x) = c$ for all values of x, where c is some constant. Writing $x = 0$,

$$f(0) = \tfrac{1}{2}\pi + 0 = c,$$

i.e. $\arccos x + \arcsin x = \tfrac{1}{2}\pi$.

Example (ii) Differentiate (a) $\arcsin x^2$, (b) $\arcsin (\cos x)$.

Solution

(a) $\dfrac{d}{dx}\{\arcsin x^2\} = \dfrac{1}{\sqrt{\{1 - (x^2)^2\}}} \times 2x$

$$= \frac{2x}{\sqrt{(1 - x^4)}}$$

(b) $\dfrac{d}{dx}\{\arcsin (\cos x)\} = \dfrac{d}{dx}\{\arcsin t\} \qquad \text{where } t = \cos x,$

$$= \frac{d}{dt}\{\arcsin t\} \times \frac{dt}{dx}$$

$$= \frac{1}{\sqrt{(1 - t^2)}} \times \frac{dt}{dx}$$

$$= \frac{-\sin x}{\sqrt{(1 - \cos^2 x)}}$$

$$= \frac{-\sin x}{|\sin x|}$$

$$= \begin{cases} -1 & \text{if } \sin x > 0 \\ 1 & \text{if } \sin x < 0. \end{cases}$$

2.7 THE PARTIAL DERIVATIVE

The derivative of a function of two variables with respect to either is called the *partial* derivative. Suppose $\omega = f(x, y)$ is a function of the two independent variables x and y defined over the region R in the (xy) plane and consider the expression

$$\frac{f(x_0 + h, y_0) - f(x_0, y_0)}{h}.$$

If this tends to a finite limit as $h \to 0$, its value is denoted by $f_x(x_0, y_0)$. It may be called the rate of change of $f(x, y)$ with respect to x when y is held constant at the point (x_0, y_0) in R. The function $f_x(x, y)$ is called the *partial derivative* of $f(x, y)$ with respect to the variable x, i.e.

$$\lim_{h \to 0} \frac{f(x + h, y) - f(x, y)}{h} = f_x(x, y)$$

when this limit exists. Other notations are $\left(\frac{\partial f}{\partial x}\right)_{y \text{ constant}}$ or $\left(\frac{\partial f}{\partial x}\right)_y$ to emphasize y is fixed, $\frac{\partial f}{\partial x}$, and $f_1(x, y)$ this last meaning the partial derivative of f with respect to the first variable mentioned. If no ambiguity can arise, the notation $\partial \omega / \partial x$ where $\omega = f(x, y)$ is also used. Similarly

$$f_y(x, y) \quad \text{or} \quad \frac{\partial f}{\partial y} \quad \text{or} \quad \left(\frac{\partial f}{\partial y}\right)_x = \lim_{k \to 0} \frac{f(x, y + k) - f(x, y)}{k}$$

is the partial derivative of f with respect to y and may be regarded as the rate of change of f with respect to y when x is considered to be held fixed. Then the ordinary rules of differentiation apply.

Example (i) $\dfrac{\partial}{\partial x} \sin (2x^2 - y) = 4x \cos (2x^2 - y)$

$$\frac{\partial}{\partial y} \sin (2x^2 - y) = -\cos (2x^2 - y).$$

Example (ii) $x = r \cos \theta$, r and θ being independent variables

$$\frac{\partial x}{\partial r} = \cos \theta \ (\theta \text{ considered constant}),$$

$$\frac{\partial x}{\partial \theta} = -r \sin \theta \ (r \text{ considered constant}).$$

Suppose also that $y = r \sin \theta$, then

$$\frac{\partial y}{\partial r} = \sin \theta, \qquad \frac{\partial y}{\partial \theta} = r \cos \theta.$$

In this example, $r = \sqrt{(x^2 + y^2)}$ (x, y independent) so that

$$\frac{\partial r}{\partial x} = \frac{x}{\sqrt{(x^2 + y^2)}} \quad (y \text{ considered constant})$$

$$= \frac{x}{r}$$

$$= \cos \theta$$

i.e.
$$\frac{\partial x}{\partial r} \neq \frac{1}{(\partial r / \partial x)}$$

In calculating $\partial x / \partial r$, x is a function of the independent variables r and θ and θ is considered as held fixed; in calculating $\partial r / \partial x$, r is a function of the independent variables x and y and y is considered fixed. Thus $(\partial x / \partial r)_\theta$ and $(\partial r / \partial x)_y$ are not necessarily reciprocals.

Note however, that from $x = r \cos \theta$ if θ is held fixed,

$$\left(\frac{\partial x}{\partial r}\right)_\theta = \cos \theta \quad \text{and} \quad \left(\frac{\partial r}{\partial x}\right)_\theta = \frac{1}{\cos \theta}$$

so that

$$\left(\frac{\partial x}{\partial r}\right)_\theta = \frac{1}{(\partial r / \partial x)_\theta}.$$

2.7.1 Higher derivatives

Since f_x and f_y may themselves be functions of x and y they may under certain circumstances be differentiated partially with respect to x and y.

If $\dfrac{\partial}{\partial x}\left(\dfrac{\partial f}{\partial x}\right)$ exists as a limit it is denoted by

$$f_{xx}(x, y) \quad \text{or} \quad \frac{\partial^2 f}{\partial x^2} \quad \text{or} \quad f_{11}(x, y)$$

Similarly

$$\frac{\partial}{\partial y}\left(\frac{\partial f}{\partial y}\right) = \frac{\partial^2 f}{\partial y^2} \quad \text{or} \quad f_{yy}(x, y) \quad \text{or} \quad f_{22}(x, y)$$

Also

$$\frac{\partial}{\partial y}\left(\frac{\partial f}{\partial x}\right) = \frac{\partial^2 f}{\partial y\,\partial x} \quad \text{or} \quad f_{yx}(x, y) \quad \text{or} \quad f_{21}(x, y)$$

$$\frac{\partial}{\partial x}\left(\frac{\partial f}{\partial y}\right) = \frac{\partial^2 f}{\partial x\,\partial y} \quad \text{or} \quad f_{xy}(x, y) \quad \text{or} \quad f_{12}(x, y)$$

These last two are often called 'mixed' derivatives.

2.7.2 Extension to functions of three or more variables

The definition of the partial derivative of a function of two variables is extended to a function of three or more variables. For, let $V(x, y, z)$ be a function of the independent variables x, y, and z, then

$$\frac{\partial V}{\partial x} = \underset{h \to 0}{\text{Lim}} \frac{V(x + h, y, z) - V(x, y, z)}{h}$$

i.e. V is differentiated partially with respect to x; y, and z being considered constant. This may be written

$$\left(\frac{\partial V}{\partial x}\right)_{y,z \text{ constant}} \quad \text{or} \quad \left(\frac{\partial V}{\partial x}\right)_{y,z}$$

Similarly if $V(x_1, x_2, x_3, \ldots, x_n)$ is a function of the independent variables $x_1, x_2, x_3, \ldots, x_n$,

$$\frac{\partial V}{\partial x_1} = \underset{h \to 0}{\text{Lim}} \frac{V(x_1 + h, x_2, x_3, \ldots, x_n) - V(x_1, x_2, x_3, \ldots, x_n)}{h}$$

i.e. V is differentiated partially with respect to x_1, the other $n - 1$ variables being held constant.

Example (i) $f(x, y) = x^2 \sin (2x^2 - y)$

$$\frac{\partial f}{\partial x} = 2x \sin (2x^2 - y) + x^2 4x \cos (2x^2 - y)$$

$$\frac{\partial f}{\partial y} = -x^2 \cos (2x^2 - y)$$

$$\frac{\partial^2 f}{\partial x^2} = \frac{\partial}{\partial x}\left(\frac{\partial f}{\partial x}\right) = 2 \sin (2x^2 - y) + 2x4x \cos (2x^2 - y)$$
$$+ 12x^2 \cos (2x^2 - y) - 4x^3 4x \sin (2x^2 - y)$$
$$= (2 - 16x^4) \sin (2x^2 - y) + 20x^2 \cos (2x^2 - y)$$

$$\frac{\partial^2 f}{\partial y^2} = \frac{\partial}{\partial y}\left(\frac{\partial f}{\partial y}\right) = -x^2 \sin (2x^2 - y)$$

$$\frac{\partial^2 f}{\partial y\,\partial x} = \frac{\partial}{\partial y}\left(\frac{\partial f}{\partial x}\right) = -2x \cos (2x^2 - y) + 4x^3 \sin (2x^2 - y)$$

$$\frac{\partial^2 f}{\partial x\,\partial y} = \frac{\partial}{\partial x}\left(\frac{\partial f}{\partial y}\right) = -2x \cos (2x^2 - y) + 4x^3 \sin (2x^2 - y).$$

In this example the mixed derivatives f_{yx} and f_{xy} are equal, which suggests that the operations $\partial/\partial x$ and $\partial/\partial y$ are commutative. This is not always true, but it is possible to prove that when f_{yx} and f_{xy} are both continuous they are equal.

Example (ii) If z is a function of x and y determined by the equation

$$z^3 - 3yz - 3x = 0,$$

show that

(a) $z \dfrac{\partial z}{\partial x} = \dfrac{\partial z}{\partial y}$,   and   (b) $z \left[\dfrac{\partial^2 z}{\partial x \, \partial y} + \left(\dfrac{\partial z}{\partial x} \right)^2 \right] = \dfrac{\partial^2 z}{\partial y^2}$.

(Assume $z^2 - y \neq 0$.)

Solution $z^3 - 3yz - 3x = 0$ (i)

(a) Differentiate (i) partially with respect to x to give

$$3z^2 \frac{\partial z}{\partial x} - 3y \frac{\partial z}{\partial x} - 3 = 0$$

i.e.

$$\frac{\partial z}{\partial x} = \frac{1}{z^2 - y}$$

and with respect to y to give

$$3z^2 \frac{\partial z}{\partial y} - 3y \frac{\partial z}{\partial y} - 3z = 0$$

i.e.

$$\frac{\partial z}{\partial y} = \frac{z}{z^2 - y}$$

hence

$$z \frac{dz}{dx} = \frac{\partial z}{dy}.$$ (ii)

(b) Differentiate equation (ii) partially with respect to y

$$\frac{\partial z}{\partial y} \frac{\partial z}{\partial x} + z \frac{\partial^2 z}{\partial y \, \partial x} = \frac{\partial^2 z}{\partial y^2}$$

substitute for $\partial z/\partial y$ from equation (ii) and

$$z \left(\frac{\partial z}{\partial x} \right)^2 + z \frac{\partial^2 z}{\partial y \, \partial x} = \frac{\partial^2 z}{\partial y^2}$$

which gives the required result when $\partial^2 z/\partial y \, \partial x$ is replaced by $\partial^2 z/\partial x \, \partial y$.

Notice that ordinary rules of differentiation of constants, of products, and of function of a function are used.

EXERCISE 2

1 Find $f'(1)$, $f'(0)$ for the function $f(x) = x^2 + 1/(1 + x^2)$.

2 Find $f'(0)$, $f'(\tfrac{1}{2}\pi)$ for the function $\sin 2x + \cos^3 x$.

3 Differentiate

(i) $\dfrac{1 + \tan x}{1 - \tan x}$;   (ii) $\sqrt{(1 + \sin^2 x)}$;   (iii) $\sin^2 x \cot 2x$.

4 Show that the gradient of the curve $y = x^3 + 6x^2 + 15x + 28$ is positive for all values of x. Show that the curve has a point of inflexion when $x = -2$ and state the gradient of the curve at this point. Sketch the graph of $y = x^3 + 6x^2 + 15x + 28$.

5 Find dy/dx as a function of t if $x = t + \sin t$ and $y = 1 - \cos t$. Show that $d^2y/dx^2 = 1/(1 + \cos t)^2$.

6 Find dy/dx in terms of t when

$$x = \frac{1 - t}{1 + 2t}, \qquad y = \frac{1 - 2t}{1 + t}$$

and show that $\dfrac{d^2y}{dx^2} = -\dfrac{2}{3}\left(\dfrac{1 + 2t}{1 + t}\right)^3$.

7 Find dy/dx at the point $(2, 1)$ on the curve given by
$$x^3 - 3xy + y^3 = 3.$$

8 Find the equation of the tangent at the point $(1, 0)$ on the curve
$$3x^3 + 2xy^2 = 2y + 3.$$

9 Find the coordinates of the points on the curve $x^2 - xy + y^2 = 12$ at which the gradient is zero.

10 Show that the nth derivative of $\cos x$ is $\cos (x + \tfrac{1}{2}n\pi)$, and find the nth derivative of $(x^3 + 1) \cos 3x$ when $x = 0$.

11 Find the coordinates of the turning points on the curve $y = x + 1 - 2 \cos x$ in the range $0 \leqslant x \leqslant 2\pi$, distinguishing between the maximum and the minimum. Show that the tangent to the curve where $x = 0$ is also the tangent at the point where $x = 2\pi$. Draw a sketch of the curve in the range $0 \leqslant x \leqslant 2\pi$.

12 Find the values of f_x, f_y, f_{xx}, and f_{yy}, and verify that $f_{xy} = f_{yx}$ for the following functions.

(i) x/y;   (ii) $(x^2 + y^2)^{-\frac{1}{2}}$;   (iii) $x \cos y + y \cos x$.

In (i) and (ii) assume $x \neq 0$, $y \neq 0$.

13 Show that, if $u = 1/r^n$ and $r^2 = x^2 + y^2 + z^2$ then

$$\frac{\partial^2 u}{\partial x^2} + \frac{\partial^2 u}{\partial y^2} + \frac{\partial^2 u}{\partial z^2} = \frac{n(n - 1)}{r^{n+2}}, \qquad r \neq 0.$$

14 If u and v are functions of the independent variables x and y which satisfy the equations $u^2 - v^2 + 2x + 3y = 0$ and $uv + x - y = 0$ and $u \neq 0$, $v \neq 0$, find $\partial u/\partial x$, $\partial u/\partial y$, $\partial v/\partial x$ and $\partial v/\partial y$ in terms of u and v.

15 Show that if $V(x, y, z) = x/(x^2 + y^2 + z^2)^{3/2}$

then $x\dfrac{\partial V}{\partial x} + y\dfrac{\partial V}{\partial y} + z\dfrac{\partial V}{\partial z} = -2V, \qquad x^2 + y^2 + z^2 \neq 0.$

16 If $V = x^n f(y)$ where n is a constant and

$$\frac{\partial^2 V}{\partial x^2} + \frac{2}{x}\frac{\partial V}{\partial x} + \frac{1}{x^2}\frac{\partial^2 V}{\partial y^2} = 0$$

show that

$$\frac{d^2 f(y)}{dy^2} + n(n + 1)f(y) = 0, \qquad (x \neq 0).$$

3
Infinite Series and Convergence

3.1 LIMIT OF A SEQUENCE

Consider an infinite sequence of numbers denoted by $\{a_n\}$ or $a_1, a_2, a_3,$ $\ldots, a_n, \ldots$ where n is a positive integer. The value of any term depends on its position in the sequence, and there is no last term.

Example (i) $0, \frac{1}{2}, \frac{2}{3}, \frac{3}{4}, \ldots, 1 - \dfrac{1}{n}, \ldots, n \geqslant 1.$

Example (ii) $1, 1\frac{1}{2}, 1\frac{3}{4}, \ldots, 2 - 1/2^n, \ldots, n \geqslant 0.$

Example (iii) $1, \frac{1}{2}, \frac{1}{4}, \ldots, 1/2^n, \ldots, n \geqslant 0.$

Example (iv) $y, y^2, y^3, \ldots, y^n, \ldots, n \geqslant 1.$

Example (v) $1, -2, 3, -4, \ldots, (-1)^{n+1}n, \ldots, n \geqslant 1.$

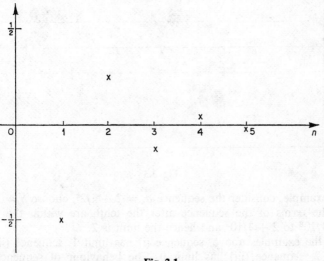

Fig. 3.1

Graphically each of these examples may be represented as points on a straight line, or as functions of n by a 'graph' which is a set of isolated points defined for integer values of n only. Figure 3.1 shows the 'graph' of $\{y^n\}$ when $y = -\frac{1}{2}$ and Fig. 3.2 shows the graph of the sequence $\{1 - 1/n\}$.

A sequence approaches a limit l if when any positive number, say h, however small, is chosen, then eventually all the terms of the sequence lie in the range $l - h$ to $l + h$, i.e. $a_n \rightarrow l$ as $n \rightarrow \infty$ if it is possible to choose h so that all the terms of the sequence after the Nth, lie in the range $l - h$ to $l + h$, where N depends on the value of h chosen. Such a sequence is illustrated in Fig. 3.3.

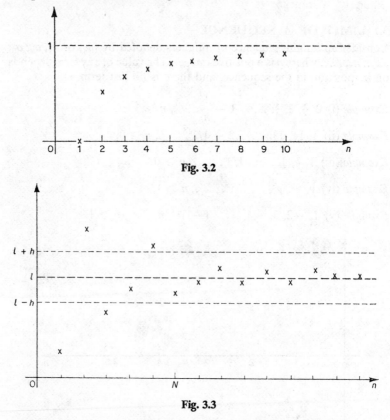

Fig. 3.2

Fig. 3.3

For example, consider the sequence $a_n = 2 - 1/2^n$, choose $h = 1/10^3$. All the terms of the sequence after the tenth are within the range $2 - 1/10^3$ to $2 + 1/10^3$ and hence the limit is 2.

In the examples above, sequence (i) has limit 1, sequence (ii) has limit 2, sequence (iii) has limit 0. The behaviour of sequence (iv) depends on the value of y; it will be shown in section 3.3.1 that the limit is 0 if $-1 < y < 1$, but that if y is outside this range there is no limit. Sequence (v) has no limit.

A sequence of positive terms $\{a_n\}$ is said to tend to infinity if for every choice of a number A, however large, a_n is greater than A for all large values of n. This is written $a_n \rightarrow \infty$ as $n \rightarrow \infty$. Similarly a sequence of negative terms $a_n \rightarrow -\infty$ as $n \rightarrow \infty$ if it is possible to choose a number, say $-B$ where B is any positive number, such that all the terms of the sequence beyond a certain point are less than $-B$. A sequence which does not tend to a limit or to $\pm\infty$ is an oscillating sequence. For example, if $a_n = y^n$ and $y = -1$, the sequence $-1, 1, -1, 1, -1, \ldots$ is an oscillating sequence.

A sequence is monotonic increasing if $a_{n+1} \geqslant a_n$ for all values of n. Similarly a sequence is monotonic decreasing if $a_{n+1} \leqslant a_n$ for all values of n.

A sequence is *bounded above* if there is some number C, such that $a_n \leqslant C$ for all values of n, and it is *bounded below* if there is some number E, such that $a_n \geqslant E$ for all n. Sequence (i),

$$0, \tfrac{1}{2}, \tfrac{2}{3}, \ldots, 1 - 1/n, \ldots$$

is monotonic increasing and is bounded below by 0 and above by 1. Sequence (v), $\{(-1)^{n+1}n\}$ is bounded neither above nor below.

The most fundamental theorem on monotonic sequences is the following, which is stated without proof. An increasing sequence which is bounded above tends to a limit; a decreasing sequence bounded below tends to a limit. An increasing sequence which is not bounded above tends to infinity as n tends to infinity, a decreasing sequence which is not bounded below tends to minus infinity as n tends to infinity.

Suppose l and l' are the limits of the sequences $\{a_n\}$ and $\{b_n\}$ respectively. Then $a_n = l + k_n$ and $b_n = l' + k'_n$ where n depends on the values of k_n, k_n' chosen. It is not difficult to show that $\lambda l + \mu l'$ is the limit of the sequence $\{\lambda a_n + \mu b_n\}$, where λ, μ are any constants. For it is possible to choose a positive number h, however small, so that for sufficiently large n, $|\lambda k_n| < \tfrac{1}{2}h$ and $|\mu k'_n| < \tfrac{1}{2}h$, and hence

$$|\lambda a_n + \mu b_n - (\lambda l + \mu l')| = |\lambda k_n + \mu k'_n| < h.$$

that is, $\qquad \lambda a_n + \mu b_n \to \lambda l + \mu l'$ as $n \to \infty$

In particular, if $\lambda = 1, \mu = -1$ then $a_n - b_n \to l - l'$ that is, the limit of the sequence formed from the difference of two sequences is the difference of the limits.

Example (i) Examine the sequence $\{a_n\} = \{1 - 1/n\}$.

Solution Since $\qquad n + 1 > n,$

$$\frac{1}{n+1} < \frac{1}{n},$$

and $\qquad 1 - \dfrac{1}{n+1} > 1 - \dfrac{1}{n},$

then $\qquad a_{n+1} > a_n$ all n.

Hence the sequence is monotonic increasing. Also $a_n < 1$ for all n and so the sequence is bounded above and tends to a limit as $n \to \infty$. Moreover $\lim_{n \to \infty} a_n = 1$.

Example (ii) Examine $a_n = 1/2^n$.

Solution Here, $a_{n+1} < a_n$, since $1/2^{n+1} < 1/2^n$, all n, and $a_n > 0$ for all n. That is, the sequence $\{a_n\}$ is decreasing, bounded below, and hence has a limit. The limit is zero.

Example (iii) Examine $a_n = n$.

Solution Here $a_{n+1} > a_n$, the sequence is monotonic increasing bounded below by zero and is not bounded above. Hence $a_n \to \infty$ as $n \to \infty$.

Example (iv) Examine the sequence $\{a_n\} = \{(1 + 1/n)^n\}$.

Solution The general term is

$$a_n = 1 + \frac{n}{n} + \frac{n(n-1)}{1 \cdot 2} \frac{1}{n^2} + \frac{n(n-1)(n-2)}{1 \cdot 2 \cdot 3} \frac{1}{n^3} + \ldots + \frac{1}{n^n}$$

$$= 1 + 1 + \frac{1}{1 \cdot 2} \left(1 - \frac{1}{n}\right) + \frac{1}{1 \cdot 2 \cdot 3} \left(1 - \frac{1}{n}\right) \left(1 - \frac{2}{n}\right) + \ldots$$

$$+ \frac{1}{1 \cdot 2 \cdot 3 \ldots n} \left(1 - \frac{1}{n}\right) \left(1 - \frac{2}{n}\right) \ldots \left(1 - \frac{n-1}{n}\right).$$

Consider any value $p > n$, then

$$a_p = 1 + 1 + \frac{1}{1 \cdot 2} \left(1 - \frac{1}{p}\right) + \frac{1}{1 \cdot 2 \cdot 3} \left(1 - \frac{1}{p}\right) \left(1 - \frac{2}{p}\right) + \ldots$$

$$\ldots + \frac{1}{1 \cdot 2 \cdot 3 \ldots p} \left(1 - \frac{1}{p}\right) \left(1 - \frac{2}{p}\right) \ldots \left(1 - \frac{p-1}{p}\right).$$

Now $a_p > a_n$, since a_p consists of more terms than a_n which are all positive, and in those which correspond, those in a_p are the greater, for example:

$$\frac{1}{1 \cdot 2} \left(1 - \frac{1}{p}\right) > \frac{1}{1 \cdot 2} \left(1 - \frac{1}{n}\right)$$

$$\left(\text{since } p > n, \frac{1}{p} < \frac{1}{n}, 1 - \frac{1}{p} > 1 - \frac{1}{n}\right).$$

Therefore $\{a_n\}$ is an increasing sequence. Moreover the sequence is bounded above since

$$a_n < 1 + 1 + \frac{1}{1 \cdot 2} + \frac{1}{1 \cdot 2 \cdot 3} + \ldots + \frac{1}{1 \cdot 2 \cdot 3 \ldots n}$$

$$< 1 + 1 + \frac{1}{2} + \frac{1}{2^2} + \ldots + \frac{1}{2^{n-1}}$$

$$= 1 + \frac{1 - (\frac{1}{2})^n}{1 - \frac{1}{2}} = 1 + 2 \left(1 - (\tfrac{1}{2})^n\right)$$

$$< 3 \text{ for all values of } n.$$

Hence a_n approaches a limit since it is an increasing sequence bounded above. This limit is denoted by the constant e.

3.2 CONVERGENCE OF SERIES

3.2.1 Sum of a series

A *series* is the sum of terms of a sequence. The series $\sum\limits_{1}^{n} a_n$ having a finite number of terms is called a *finite series* and has a finite sum. The series $\sum\limits_{1}^{\infty} a_n$ having an infinite number of terms is called an *infinite series* and may or may not have a finite 'sum'.

The sum of the first n terms of a series is denoted by s_n and is called the nth *partial sum*. Thus $s_1 = a_1$, $s_2 = a_1 + a_2$, $s_3 = a_1 + a_2 + a_3$, $\ldots$, $s_n = a_1 + a_2 + a_3 + \ldots + a_n$. For example, consider the infinite series

$$1 + \frac{1}{2} + \frac{1}{4} + \frac{1}{8} + \ldots + \frac{1}{2^{n-1}} + \ldots,$$

then, $\quad s_1 = 1, s_2 = 1\frac{1}{2}, s_3 = 1\frac{3}{4}, \ldots, s_n = 2 - (\frac{1}{2})^{n-1}.$

These partial sums themselves form a new sequence which has the limit $S = 2$ as $n \to \infty$, and this limit of partial sums is called the 'sum to infinity', or just the *sum* of the infinite series. This is written

$$1 + \frac{1}{2} + (\frac{1}{2})^2 + \ldots + (\frac{1}{2})^n + \ldots = 2.$$

The 'equals' sign is used here to mean the limit of the partial sums. It does not mean that a large number of terms of the series may be added to give the sum 2, for there is no value of n such that the sum of n terms is 2, nor is 2 a member of the series.

When the sequence of partial sums $\{s_n\}$ tends to a limit S as $n \to \infty$, the series is said to be *convergent*, or to *converge* and the limit S is called the *sum* of the series. When the sequence of partial sums does not converge, the series is *divergent*, that is if $s_n \to \infty$ or $s_n \to -\infty$ or if s_n oscillates.

3.2.2 Conditional and absolute convergence

When the series

$$a_1 + a_2 + a_3 + \ldots + a_n + \ldots \tag{1}$$

consists of some positive and some negative terms it may be *conditionally* convergent or *absolutely* convergent.

If the series

$$|a_1| + |a_2| + |a_3| + \ldots + |a_n| + \ldots \tag{2}$$

converges then, as shown below, so also does the series (1), and (1) is then said to be absolutely convergent.

If series (1) converges and series (2) does not converge, then (1) is said to be conditionally convergent. For example the series

$$1 - \tfrac{1}{2} + \tfrac{1}{3} - \tfrac{1}{4} + \ldots$$

is conditionally convergent, since it converges (see section 3.5.3), yet the series

$$1 + \tfrac{1}{2} + \tfrac{1}{3} + \tfrac{1}{4} + \ldots$$

may be shown to be divergent (see section 3.3.2). The series

$$1 - \frac{1}{2^2} + \frac{1}{3^2} - \frac{1}{4^2} + \ldots$$

is absolutely convergent since the series

$$1 + \frac{1}{2^2} + \frac{1}{3^2} + \frac{1}{4^2} + \ldots$$

may be shown to be convergent. See example (iii) of section 3.5.1.

To show that if (2) converges then so does (1), write $u_n = \tfrac{1}{2}(|a_n| + a_n)$ and $v_n = \tfrac{1}{2}(|a_n| - a_n)$. When $a_n > 0$, $u_n = a_n$ and $v_n = 0$; when $a_n < 0$, $u_n = 0$ and $v_n = -a_n$, so that whatever the sign of a_n, $u_n \geqslant 0$ and $v_n \geqslant 0$. Also $|a_n| = u_n + v_n$, $a_n = u_n - v_n$, $u_n \leqslant |a_n|$, and $\sum_1^\infty |a_n|$ is convergent.

Suppose $\sum_1^\infty |a_n| = A$, then $\sum_1^n u_r \leqslant A$ for all values of n. Hence the partial sums $\sum_1^n u_r$ tend to a limit, i.e. $\sum_1^\infty u_n$ is convergent. Similarly $\sum_1^\infty v_n$ is convergent. Using the result that the limit of a sequence $\{u_n - v_n\}$ is the difference of the limits, leads to the result that $\sum_1^\infty a_n = \sum_1^\infty (u_n - v_n)$ is convergent.

3.3 SOME STANDARD INFINITE SERIES

3.3.1 The geometric series

$$1 + y + y^2 + \ldots + y^n + \ldots = \sum_{n=0}^\infty y^n.$$

The nth partial sum, s_n is given by

$$s_n = 1 + y + y^2 + \ldots + y^{n-1}$$
$$= \frac{1 - y^n}{1 - y}$$
$$= \frac{1}{1 - y} - \frac{1}{1 - y} \cdot y^n.$$

When $-1 < y < 1$, then $y^n \to 0$ as $n \to \infty$ and $s_n \to 1/(1 - y)$, so that, by the definition of convergence of series in section 3.2.1 the geometric series $\sum_{n=0}^\infty y^n$ is convergent and has the sum $1/(1 - y)$.

If $y \leqslant -1$ or $y > 1$, y^n does not tend to a limit as $n \to \infty$ and hence neither does s_n.

If $y = 1$ the series is $1 + 1 + 1 + \ldots$, and $s_n = n \to \infty$ with n.

Hence, if $y \leqslant -1$ or $y \geqslant 1$ the geometric series $\sum\limits_{n=0}^{\infty} y^n$ is divergent, if $-1 < y < 1$ it is convergent.

3.3.2 The harmonic series
The series

$$1 + \tfrac{1}{2} + \tfrac{1}{3} + \tfrac{1}{4} + \ldots = \sum_{n=1}^{\infty} 1/n$$

is sometimes called the harmonic series. This is divergent, for although $a_n = 1/n \to 0$ as $n \to \infty$, it is possible to prove s_n increases beyond all limit. The sum of the first four terms

$$1 + \tfrac{1}{2} + \tfrac{1}{3} + \tfrac{1}{4} > 1 + \tfrac{1}{2} + \tfrac{1}{4} + \tfrac{1}{4} = 1 + \tfrac{1}{2} + \tfrac{1}{2}.$$

The sum of the next four terms

$$\tfrac{1}{5} + \tfrac{1}{6} + \tfrac{1}{7} + \tfrac{1}{8} > \tfrac{1}{8} + \tfrac{1}{8} + \tfrac{1}{8} + \tfrac{1}{8} = \tfrac{4}{8} = \tfrac{1}{2}.$$

The sum of the next eight terms is $> 8/16 = \tfrac{1}{2}$ and so on. Hence

$$s_n > 1 + \tfrac{1}{2} + \tfrac{1}{2} + \tfrac{1}{2} + \ldots$$

to k terms where $n = 2^{k-1}$ and increases beyond all limit as $n \to \infty$. Hence the series is divergent.

3.3.3 The series $\sum\limits_{n=1}^{\infty} 1/n(n+1)$

The series $\sum\limits_{n=1}^{\infty} 1/n(n+1)$ is convergent and its sum can be found. The nth partial sum is given by

$$s_n = \frac{1}{1 \cdot 2} + \frac{1}{2 \cdot 3} + \frac{1}{3 \cdot 4} + \ldots + \frac{1}{n(n+1)}$$

and since

$$\frac{1}{n(n+1)} = \frac{1}{n} - \frac{1}{n+1}$$

then

$$s_n = 1 - \frac{1}{2} + \frac{1}{2} - \frac{1}{3} + \frac{1}{3} - \frac{1}{4} + \ldots + \frac{1}{n} - \frac{1}{n+1}$$

$$= 1 - \frac{1}{n+1}$$

and

$$s_n \to 1 \quad \text{as} \quad n \to \infty.$$

Hence the series converges and its sum is 1.

3.4 SOME RULES

3.4.1 Rule 1

Let
$$a_1 + a_2 + a_3 + \ldots \tag{3}$$

be convergent and its sum be S. Then $\underset{n \to \infty}{\text{Lim}}\, a_n = 0$ since $a_n = s_n - s_{n-1}$ and s_n and s_{n-1} have the same limit S as $n \to \infty$,

$$\underset{n \to \infty}{\text{Lim}}\, a_n = S - S = 0.$$

This property must *not* be taken as a test of convergence. If a series converges it is necessary that the nth term tends to zero, and if the nth term does not tend to zero there is no need to test further for convergence. That is, the converse of this rule is a test of divergence. The series $\sum_{n=1}^{\infty} a_n$ is divergent if a_n does not tend to zero as n tends to infinity.

For example, the series $\sum_{n=1}^{\infty} 1/n$ was shown to be divergent in section 3.3.2, yet $a_n = 1/n \to 0$ as $n \to \infty$. The series

$$\sum_{n=1}^{\infty} \frac{n}{n+1} = \frac{1}{2} + \frac{2}{3} + \frac{3}{4} + \ldots$$

diverges since

$$a_n = \frac{n}{n+1} = \frac{1}{1 + 1/n} \to 1 \quad \text{as} \quad n \to \infty.$$

3.4.2 Rule 2

Consider the remainder R_m, after m terms of the convergent series (3). This is defined to be the remainder when the first m terms are removed and is

$$R_m = \sum_{n=m+1}^{\infty} a_n = a_{m+1} + a_{m+2} + a_{m+3} + \ldots \tag{4}$$

The series (4) converges for all integer values of m and its sum is $S - s_m$. This means that a finite number of terms at the beginning of a series may be removed without altering the convergence of the series. The proof depends on the fact that the nth partial sum of the series (4), plus the terms removed, is a partial sum of the original series (3), which has the limit S. Hence the limit of the nth partial sum of (4) is $S - s_m$, and hence (4) is convergent.

Similarly if a finite number of terms is added to the convergent series (3) the resulting series is convergent and its sum is S plus the sum of the terms added.

As a corollary, if $\sum_{}^{\infty} a_n$ is convergent,

$$\underset{m \to \infty}{\text{Lim}}\, R_m = \underset{m \to \infty}{\text{Lim}}\, (a_{m+1} + a_{m+2} + a_{m+3} + \ldots) = 0.$$

since $R_m = S - s_m$ which $\to 0$ as $m \to \infty$.

3.5 TESTS FOR CONVERGENCE

In order to prove that a series converges it is *sometimes* possible to find its sum. This is not *always* possible and tests have been devised to discover whether a given series is convergent or not.

3.5.1 The comparison test

Let $\sum\limits^{\infty} a_n$ and $\sum\limits^{\infty} b_n$ be two series of positive terms. If the second series is convergent and $a_n \leqslant cb_n$ for all values of n where c is any constant, then the first series is also convergent. For if

$$b_1 + b_2 + b_3 + \ldots = B,$$

then $\qquad b_1 + b_2 + b_3 + \ldots + b_n \leqslant B$ for all n,

and $\qquad a_1 + a_2 + a_3 + \ldots + a_n \leqslant cB.$

Hence the partial sums of the first series tend to a limit and $\sum\limits^{\infty} a_n$ is convergent.

If the series $\sum\limits^{\infty} b_n$ is divergent and $a_n \geqslant cb_n$ for all values of n and c is a constant, then $\sum\limits^{\infty} a_n$ also diverges.

Example (i) Prove $\sum\limits_{n=1}^{\infty} 1/\sqrt{n}$ is divergent.

Solution Write $a_n = 1/\sqrt{n}$, choose $b_n = 1/n$, then since $\sqrt{n} \leqslant n$ for all values of n, $\dfrac{1}{\sqrt{n}} \geqslant \dfrac{1}{n}$, or $a_n \geqslant b_n$ for all n. $\sum\limits^{\infty} b_n$ is known to be divergent and hence $\sum\limits^{\infty} a_n$ is also divergent.

Example (ii) Prove that $\sum\limits^{\infty} \dfrac{1}{n^p}$ is divergent when $p \leqslant 1$.

Solution Let $a_n = 1/n^p$, $p \leqslant 1$ and choose $b_n = 1/n$, then since $n^p \leqslant n$ for all values of n when $p \leqslant 1$, $a_n \geqslant b_n$ for all values of n. But $\sum\limits^{\infty} b_n$ is divergent and hence $\sum\limits^{\infty} a_n$ is divergent, i.e. $\sum\limits^{\infty} \dfrac{1}{n^p}$ is divergent when $p \leqslant 1$.

Example (iii) Prove that $\sum\limits_{n=1}^{\infty} 1/n^2$ is convergent.

Solution Write $a_n = \dfrac{1}{(n+1)^2}$ and choose $b_n = \dfrac{1}{n(n+1)}$. Since $(n+1)^2 \geqslant n(n+1)$ for all values of n, then $\dfrac{1}{(n+1)^2} \leqslant \dfrac{1}{n(n+1)}$ for all values of n. It is known that $\sum\limits_{n=1}^{\infty} \dfrac{1}{n(n+1)}$ is convergent (see section 3.3.3), hence $\sum\limits_{n=1}^{\infty} \dfrac{1}{(n+1)^2}$ is convergent. Now the given series can be written

$$\sum_{n=1}^{\infty} \frac{1}{n^2} = 1 + \frac{1}{2^2} + \frac{1}{3^2} + \ldots = 1 + \sum_{n=1}^{\infty} \frac{1}{(n+1)^2}.$$

Hence, using Rule 2 of section 3.4.2, $\sum\limits_{1}^{\infty} 1/n^2$ is also convergent.

Example (iv) Prove that the series $\sum\limits_{n=1}^{\infty} 1/n^p$ is convergent if $p \geqslant 2$.

Solution Let $a_n = 1/n^p$, $p \geqslant 2$, and choose $b_n = 1/n^2$. Since

$$n^p \geqslant n^2, p \geqslant 2,$$

then $\qquad\qquad \dfrac{1}{n^p} \leqslant \dfrac{1}{n^2}, p \geqslant 2$ and all n.

Hence, since $\sum\limits_{1}^{\infty} 1/n^2$ is convergent, so also is $\sum\limits_{1}^{\infty} 1/n^p$ when $p \geqslant 2$.

It is also possible to prove that $\sum\limits_{}^{\infty} 1/n^p$ is convergent when $1 < p < 2$, using a test for convergence which is not included in this book. The series $\sum\limits_{n=1}^{\infty} 1/n^p$, which is convergent for $p > 1$ and divergent for $p \leqslant 1$, is used as a standard series in the comparison test. The choice of standard series in this test is made from common sense and experience. For example, if $a_n = (2n-1)/(n^4+1)$, a_n is *of the order* of $1/n^3$, since when n is large, $a_n \simeq 2n/n^4 = 2/n^3$, and the series $\sum\limits_{}^{\infty} a_n$ is compared with the convergent series $\sum\limits_{}^{\infty} 1/n^3$, as follows:

$$a_n = \frac{2n-1}{n^4+1} = \frac{2/n^3 - 1/n^4}{1 + 1/n^4} < \frac{2}{n^3} = 2 \times \frac{1}{n^3}.$$

$\sum\limits_{}^{\infty} 1/n^3$ converges, and hence so also does $\sum\limits_{}^{\infty} a_n$.

3.5.2 The ratio test

If $\left| \dfrac{a_{n+1}}{a_n} \right| \to r < 1$ as $n \to \infty$ then $\sum\limits_{1}^{\infty} a_n$ is absolutely convergent and hence convergent. If $r > 1$ the series diverges; if $r = 1$ the test fails.

For suppose $r < 1$ then it is possible to choose h such that all the terms of the sequence $|a_{n+1}/a_n|$ after the Nth lie in the range $r - h$ to $r + h$. Choose $h = \frac{1}{2}(1 - r)$, then

$$r + h = \tfrac{1}{2}(1 + r) < 1$$

and $\qquad\qquad \left| \dfrac{a_{N+1}}{a_N} \right| < r + h, \quad \left| \dfrac{a_{N+2}}{a_{N+1}} \right| < r + h \quad$ etc.

or $|a_{N+1}| < (r + h)|a_N|, |a_{N+2}| < (r + h)|a_{N+1}| < (r + h)^2|a_N|$ etc.

and $\sum\limits_{n=N}^{\infty} |a_n| = |a_N| + |a_{N+1}| + |a_{N+2}| + \dots$
$$< |a_N|\{1 + (r + h) + (r + h)^2 + \dots\}.$$

Hence $\sum\limits_{n=N}^{\infty} |a_n|$ is convergent by comparison with the geometric series $(r + h < 1)$. Adding to this convergent series the finite number of terms $|a_1| + |a_2| + |a_3| + \dots + |a_{N-1}|$ and using Rule 2 (section 3.4.2) shows that $\sum\limits_{1}^{\infty} |a_n|$ is convergent. From section 3.2.2 it follows that $\sum\limits_{1}^{\infty} a_n$ is absolutely convergent and hence convergent.

If $r > 1$ then, for n sufficiently large, $|a_{n+1}| > |a_n|$ and the magnitude of the terms increase with n. Hence $\mathrm{Lim}\limits_{n \to \infty} a_n \neq 0$, and by the converse of Rule 1 (section 3.4.1) the series $\sum\limits_{1}^{\infty} a_n$ is divergent.

Example (i) Prove that the series $\sum\limits_{1}^{\infty} n^2/3^n$ is convergent.

Solution Write $a_n = n^2/3^n$, then

$$\left|\frac{a_{n+1}}{a_n}\right| = \frac{(n + 1)^2}{3^{n+1}} \cdot \frac{3^n}{n^2}$$
$$= \frac{1}{3}\left(1 + \frac{1}{n}\right)^2 \to \frac{1}{3} \quad \text{as} \quad n \to \infty.$$

$r = \frac{1}{3}$ and the series $\sum\limits_{1}^{\infty} a_n$ is convergent.

Example (ii) Prove that the series $\sum\limits_{n=0}^{\infty} 1/n!$ converges.

Solution Write $a_n = 1/n!$, then

$$\left|\frac{a_{n+1}}{a_n}\right| = \frac{n!}{(n + 1)!}$$
$$= \frac{1}{n + 1} \to 0 \quad \text{as} \quad n \to \infty.$$

Hence the series $\sum\limits_{n=0}^{\infty} \frac{1}{n!} = 1 + 1 + \frac{1}{2!} + \frac{1}{3!} + \dots$ is convergent.

Example (iii) Prove that the series $\sum\limits_{1}^{\infty} 2^n/n$ is divergent.

Solution Write $a_n = 2^n/n$, then

$$\left|\frac{a_{n+1}}{a_n}\right| = \frac{2^{n+1}}{n + 1} \cdot \frac{n}{2^n}$$
$$= \frac{2}{1 + 1/n} \to 2 \quad \text{as} \quad n \to \infty.$$

Hence $\sum\limits_{1}^{\infty} a_n$ diverges.

Example (iv) Consider the convergence or divergence of the series $\sum\limits^{\infty} (n + 2)/(n^2 - 1)$.

Solution Write $a_n = (n + 2)/(n^2 - 1)$, then

$$\left| \frac{a_{n+1}}{a_n} \right| = \frac{n + 3}{n^2 + 2n} \cdot \frac{n^2 - 1}{n + 2}$$

$$= \frac{(1 + 3/n)(1 - 1/n^2)}{(1 + 2/n)(1 + 2/n)} \to 1 \quad \text{as} \quad n \to \infty.$$

That is, $r = 1$ and the test fails. This series diverges since

$$a_n = \frac{n + 2}{n^2 - 1} > \frac{1}{n}, n > 1$$

$\sum\limits^{\infty} 1/n$ diverges, hence $\sum\limits^{\infty} a_n$ diverges.

Example (v) Consider the convergence or divergence of the series $\sum\limits^{\infty} (-1)^{n+1}/n$.

Solution Write $a_n = (-1)^{n+1}/n$

$$\left| \frac{a_{n+1}}{a_n} \right| = \frac{n}{n + 1}$$

$$= \frac{1}{1 + 1/n} \to 1 \quad \text{as} \quad n \to \infty.$$

That is, $r = 1$ and the test fails. This series converges as is proved in section 3.5.3 following.

The examples (iv) and (v) emphasize the fact that when $r = 1$ there is a failure of the ratio test and that resort must be made to some other test.

3.5.3 The alternating series

Consider a series with alternating sign, i.e. a series in which the signs are alternatively positive and negative. Let the series be $a_1 - a_2 + a_3 - a_4 + - \ldots$ where a_i is positive for all i. Then if

$$\lim_{n \to \infty} a_n = 0 \quad \text{and} \quad q_1 > a_2 > a_3 \ldots > a_n > 0,$$

the series converges.

Let s_{2n} be the sum of the first $2n$ terms.

$$s_{2n} = (a_1 - a_2) + (a_3 - a_4) + \ldots + (a_{2n-1} - a_{2n}).$$

Each bracket is positive so that s_{2n} increases steadily, or

$$0 < s_2 < s_4 < \ldots .$$

Also
$$s_{2n} = a_1 - \{(a_2 - a_3) + (a_4 - a_5) + \ldots + (a_{2n-1} - a_{2n-1}) + a_{2n}\}$$
$$< a_1 \quad \text{since all the brackets are again positive.}$$

Hence s_{2n} is monotonic increasing, bounded above and therefore tends to a limit.

Also $\qquad s_{2n+1} = s_{2n} + a_{2n+1} \to s_{2n} \quad \text{as} \quad n \to \infty$

since $a_{2n+1} \to 0$, and thus s_{2n+1} converges. Hence the sequence $\{s_n\}$ converges and so does the alternating series. For example, the series $1 - \frac{1}{2} + \frac{1}{3} - \frac{1}{4} + \ldots$ converges since $a_n \to 0$ as $n \to \infty$ and $a_1 > a_2 > a_3 > \ldots > a_n > 0$ for all values of n.

3.6 POWER SERIES

A series of the form
$$a_0 + a_1 x + a_2 x^2 + \ldots = \sum_{n=0}^{\infty} a_n x^n$$

is called a *power series* in the variable x, where $a_0, a_1, a_2, \ldots$ are constants. If it converges it defines a function $f(x)$ given by $f(x) = a_0 + a_1 x + a_2 x^2 + \ldots$.

The series always converges when $x = 0$. It may or may not converge for any other value of x.

Example (i) The geometric series $1 + x + x^2 + \ldots$ converges to $1/(1 - x)$ provided $|x| < 1$ and diverges for all other values of x.

Example (ii) The series
$$1 + x + \frac{x^2}{2!} + \frac{x^3}{3!} + \ldots$$

converges for all values of x since by the ratio test of section 3.5.2
$$\left| \frac{x^{n+1}}{(n+1)!} \cdot \frac{n!}{x^n} \right| = \frac{|x|}{n+1} \to 0 \quad \text{as} \quad n \to \infty$$

for all values of x.

Example (iii) Consider the series
$$x - \frac{x^2}{2} + \frac{x^3}{3} - \frac{x^4}{4} + \ldots,$$

then $\qquad \left| \frac{x^{n+1}}{(n+1)} \cdot \frac{n}{x^n} \right| = \frac{|x|}{(1 + 1/n)} \to |x| \quad \text{as} \quad n \to \infty.$

Hence the series converges absolutely when $|x| < 1$ and diverges when $|x| > 1$. When $x = 1$ the series becomes $1 - \frac{1}{2} + \frac{1}{3} - \frac{1}{4} + \ldots$ which converges (see section 3.5.3). When $x = -1$ the series $-1 - \frac{1}{2} - \frac{1}{3} - \frac{1}{4} - \ldots$ diverges by comparison with the harmonic series (see section 3.3.2).

3.6.1 Radius of convergence

Consider the power series

$$a_0 + a_1 x + a_2 x^2 + \ldots.$$

The number R, such that the power series converges absolutely when $|x| < R$ and diverges when $|x| > R$, is called the *radius of convergence* of the series. The series may or may not converge when $|x| = R$. See example (iii) above. When the series is convergent everywhere, i.e. for all values of x, R is infinite.

If $\qquad f(x) = a_0 + a_1 x + a_2 x^2 + \ldots + a_n x^n + \ldots,$

then $\qquad f'(x) = a_1 + 2a_2 x + \ldots + n a_n x^{n-1} + \ldots$

for every value of x *inside* the range $|x| < R$ where R is the radius of convergence of the first series. That is, it is possible to differentiate a power series, whose sum is $f(x)$, term by term within the interval defined by its radius of convergence, and the differentiated series has as its sum the derivative of $f(x)$. Both series have the same radius of convergence.

Hence, a power series with radius of convergence R may be differentiated or integrated term by term within its radius of convergence.

The existence of R and the statements above require proofs that are beyond the scope of this book.

Example (i) Show that the series

$$\sum_{n=1}^{\infty} \frac{x^n}{n} = x + \frac{x^2}{2} + \frac{x^3}{3} + \ldots$$

converges for all values of x in the interval $-1 \leqslant x < 1$.

Solution Using the ratio test,

$$\left| \frac{a_{n+1}}{a_n} \right| = |x| \frac{n}{n+1}$$

$$= \frac{|x|}{(1 + 1/n)} \to |x| \quad \text{as} \quad n \to \infty.$$

Hence the series is convergent provided $|x| < 1$ and is divergent when $|x| > 1$. When $x = 1$, the series is $1 + \frac{1}{2} + \frac{1}{3} + \ldots$ which is divergent (see section 3.3.2). When $x = -1$ the series is

$$-1 + \tfrac{1}{2} - \tfrac{1}{3} + \tfrac{1}{4} - + \ldots$$

which is conditionally convergent by comparison with

$$1 - \tfrac{1}{2} + \tfrac{1}{3} - \tfrac{1}{4} + - \ldots.$$

Thus the given series converges in the interval $-1 \leqslant x < 1$ and diverges everywhere else. Its radius of convergence is $R = 1$.

The result of differentiating this series term by term is

$$1 + x + x^2 + \ldots,$$

which is the geometric series which converges for $-1 < x < 1$. The differentiated series converges at all interior points of the original region of convergence but not at the end point $x = -1$.

On the other hand, integration of the geometric series term by term gives a series which converges at one new point, $x = -1$, at which the original (geometric) series diverged.

Example (ii) The series

$$\sum_{n=0}^{\infty} \frac{x^n}{n!} = 1 + x + \frac{x^2}{2!} + \frac{x^3}{3!} + \cdots$$

converges for all values of x. See example (ii) of section 3.6. Differentiating this series term by term gives

$$1 + x + \frac{x^2}{2!} + \frac{x^3}{3!} + \cdots$$

which is the same series. Hence the function defined by this series is equal to its derivative and converges for all values of x.

EXERCISE 3

1 Do the following sequences have limits? If so, state these limits. Give reasons for your answer in each case.

   (i) $0, 1, 0, -\frac{1}{2}, 0, \frac{1}{4}, 0, -\frac{1}{8}, \ldots, 0, 1/2^{2n}, 0, -1/2^{2n+1}, 0, \ldots$

   (ii) $1, \frac{1}{2}, 2, \frac{1}{3}, 3, \ldots, \frac{1}{n}, n, \ldots$

2 For each of the following a_n determine whether the sequence $\{a_n\}$, so defined, has a limit, and find that limit when it exists.

   (i) $\dfrac{2n + 1}{1 - 3n}$ (vi) $\sin n$

   (ii) $\dfrac{n^2 - n}{2n^2 + n}$ (vii) $\dfrac{(-1)^n n}{n^2 + 1}$

   (iii) $1 + \dfrac{(-1)^n}{n}$ (viii) $\dfrac{n^2 + 2}{n^4 + 2}$

   (iv) $\dfrac{1 + (-1)^n}{n}$ (ix) $\dfrac{3^n + 1}{2^n + 1}$

   (v) $1 + (-1)^n$ (x) $\dfrac{1 + 3^n}{1 - n}$.

3 Find an expression for the nth partial sum of each of the following series. Find the sum of the series if it converges.

   (i) $2 + \dfrac{2}{3} + \dfrac{2}{9} + \dfrac{2}{27} + \cdots + \dfrac{2}{3^{n-1}} + \cdots$

   (ii) $5 - \dfrac{5}{2} + \dfrac{5}{4} - \dfrac{5}{8} + \cdots + \dfrac{(-1)^{n-1}5}{2^{n-1}} + \cdots$

(iii) $3 + 6 + 12 + \ldots + 3(2^{n-1}) + \ldots$

4 State the value of each of the following:

(i) $\sum_{r=1}^{n} (-1)^{r+1} x^r$;   (ii) $\sum_{r=0}^{n} \binom{n}{r} x^{r+2}$

5 Discuss the convergence or divergence of the following series.

(i) $x - \frac{1}{2}x^2 + \frac{1}{3}x^3 - \frac{1}{4}x^4 + \ldots + (-1)^{n+1} x^n/n + \ldots$

(ii) $\sum_{1}^{\infty} a_n$ where $a_n = \dfrac{(n!)^2}{(2n)!}$

(iii) $\sum_{1}^{\infty} a_n$ where $a_n = \dfrac{1}{2^n \sqrt{\{n(n+3)\}}}$

6 Discuss the convergence of the series whose general terms are

(i) $(-1)^n x^n n^{-\frac{1}{4}}$;   (ii) $(\sin nx)/n^2$;   (iii) $(-1)^n x^{2n}/(2n)!$.

7 Determine whether the following series converge.

(i) $\sum_{n=1}^{\infty} (n + \sqrt{n})^{-1}$;   (ii) $\sum_{n=1}^{\infty} \dfrac{n}{3^n}$;   (iii) $\sum_{n=1}^{\infty} \dfrac{n^2}{2^n}$.

8 Prove that, if the series $\sum_{}^{\infty} a_n$ converges, then $\underset{n \to \infty}{\text{Lim}} \ a_n = 0$, and show by an example that the converse is not necessarily true. Determine whether the following series converge or diverge

(i) $\sum_{n=1}^{\infty} (-1)^n \dfrac{n}{n+1}$;   (ii) $\sum_{n=1}^{\infty} \dfrac{1}{\sqrt{\{n(2n+1)\}}}$.

9 Find the radius of convergence of each of the following power series.

(i) $\sum_{n=0}^{\infty} \dfrac{(n!)^3 x^n}{(3n)!}$;   (ii) $\sum_{n=1}^{\infty} \dfrac{x^n}{n}$;   (iii) $\sum_{n=0}^{\infty} \dfrac{x^n}{n!}$.

10 Prove that if $\sum_{n=1}^{\infty} a_n$ converges absolutely then $\sum_{n=1}^{\infty} a_n \sin nx$ converges for all real values of x.

11 Prove that if a power series $a_0 + a_1 x + a_2 x^2 + \ldots$ converges when $x = x_0$ then it converges whenever $|x| < |x_0|$. For what values of x do the following series converge?

(i) $\sum_{n=0}^{\infty} \dfrac{1}{n+1} x^n$;   (ii) $\sum_{n=1}^{\infty} n^n x^n$.

4
Exponential, Logarithmic, and Hyperbolic Functions

4.1 THE EXPONENTIAL FUNCTION

The general exponential functions may be defined in one of several ways. Their fundamental property is that the derivative of the function is proportional to the value of the function. That is they all satisfy equations, called *differential* equations, of the type

$$\frac{dy}{dx} = cy$$

where c is a constant.

In particular the standard exponential function is defined to be that function which satisfies the differential equation

$$\frac{dy}{dx} = y \tag{1}$$

and such that $y = 1$ when $x = 0$. (2)

In example (ii) of section 3.6.1 the function $\sum_{n=0}^{\infty} \frac{x^n}{n!}$ was shown to be a function which satisfies (1) and when $x = 0$ its value is 1. Let this function be called $E(x)$, i.e.

$$E(x) = 1 + x + \frac{x^2}{2!} + \frac{x^3}{3!} + \ldots = \sum_{n=0}^{\infty} \frac{x^n}{n!} \tag{3}$$

From equations (1) and (2)

$$E'(x) = E(x) \tag{4}$$
$$E(0) = 1 \tag{5}$$

4.1.1 Properties of $E(x)$

$E(x)$ satisfies the functional relation

$$E(\alpha)E(\beta) = E(\alpha + \beta). \tag{6}$$

To prove this consider the function $F(x)$ defined by

$$F(x) = E(c - x)E(x) \tag{7}$$

where c is some constant. Differentiating equation (7) gives

$$\begin{aligned}
F'(x) &= -E'(c - x)E(x) + E(c - x)E'(x) \\
&= -E(c - x)E(x) + E(c - x)E(x) \text{ using (4)} \\
&= 0.
\end{aligned}$$

Hence $\qquad F(x) =$ constant, since its derivative is zero

or $\qquad\qquad F(x) = k$ say, for all values of x

or $\qquad\qquad E(c - x)E(x) = k$ substituting from (7).

Now, writing $x = 0$,

$$E(c)E(0) = k,$$

or $\qquad\qquad E(c) = k$ since $E(0) = 1$ from (5).

But $\qquad\qquad k = E(c - x)E(x),$

hence $\qquad\qquad E(c - x)E(x) = E(c),$

and writing $c - x = \alpha,\ x = \beta$ so that $c = \alpha + \beta$ gives

$$E(\alpha)E(\beta) = E(\alpha + \beta)$$

which is equation (6).

Replacing α by x and β by $-x$ in equation (6),

$$E(x)E(-x) = E(0) = 1 \text{ from (5)}$$

i.e. $\qquad\qquad\qquad E(-x) = \dfrac{1}{E(x)} \qquad\qquad\qquad (8)$

Repeated application of (6), setting $\alpha = \beta = \ldots = x$, gives

$$\{E(x)\}^n = E(x)E(x)E(x) \ldots E(x) \text{ to } n \text{ factors}$$
$$= E(x + x + x + \ldots + x) \ (n \text{ terms})$$

i.e. $\qquad \{E(x)\}^n = E(nx) \qquad n = 1, 2, 3, \ldots . \qquad (9)$

Hence if p and q are positive integers, setting $x = p/q,\ n = q$ in (9)

$$\{E(p/q)\}^q = E(p) = \{E(1)\}^p \qquad\qquad (10)$$

hence $\qquad\qquad E(p/q) = \{E(1)\}^{p/q} \qquad\qquad\qquad (11)$

and $\qquad\qquad E(x) = \{E(1)\}^x \quad$ whenever $x = p/q.$ $\qquad (12)$

Since a rational number is any number which can be written in the form p/q where p and q are integers and $q \neq 0$, equation (12) is true whenever x is a positive rational number. Writing $-p$ for p in equation (11) shows that equation (12) is true when x is any rational number, positive or negative.

4.1.2 The number e

This is chosen to be $E(1)$, the value of $E(x)$ when $x = 1$, i.e.

$$e = E(1) = 1 + 1 + \frac{1}{2!} + \frac{1}{3!} + \ldots = \sum_{n=0}^{\infty} \frac{1}{n!}$$

and its value may be calculated to any degree of accuracy. For example, if thirteen terms of the series are added, $e = 2{\cdot}718\ 281\ 8$ correct to seven decimal places. From (12)

$$\{E(1)\}^x = E(x),$$

so that

$$e^x = E(x),$$

and the properties of $E(x)$ developed in section 4.1.1. are written in terms of e^x as follows:

$$\frac{d}{dx}(e^x) = e^x \quad \text{from (1),}$$

$$e^x = 1 + x + \frac{x^2}{2!} + \frac{x^3}{3!} + \ldots = \sum_{n=0}^{\infty} \frac{x^n}{n!} \quad \text{from (3),}$$

$$e^\alpha e^\beta = e^{\alpha + \beta} \quad \text{from (6),}$$

$$e^{-x} = \frac{1}{e^x} \quad \text{from (8).}$$

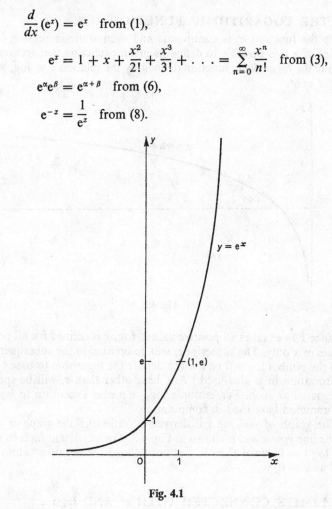

Fig. 4.1

4.1.3 The graph of $y = e^x$

When $x = 0$, $y = e^0 = 1$, $dy/dx = e^0 = 1$, so that the gradient of the curve at $(0, 1)$ is 1.

When x is positive, $e^x > 1 + x$ so that $e^x \to \infty$ as $x \to \infty$, and x increases steadily with x.

When x is negative, write $x = -t$, $t > 0$ then $e^x = e^{-t} = 1/e^t$ is always positive and as $x \to -\infty$, $t \to \infty$, $e^x \to 0$.

Hence as x increases, e^x increases monotonically and is always positive. The graph is shown in Fig. 4.1.

When the exponent is a lengthy expression it is sometimes convenient to write e^x as exp (x).

4.2 THE LOGARITHMIC FUNCTION

Since the function e^x is continuous and monotonic increasing for all values of x it is possible to define its inverse function (see section 1.2). This is the *logarithmic* function defined by the relation $y = \log_e x$ such that $x = e^y$.

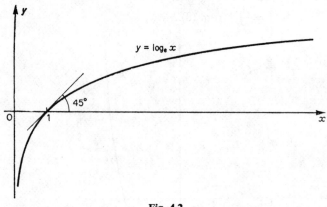

Fig. 4.2

Since $x = e^y$ takes all positive values, $\log_e x$ is defined for all positive values of x only. The subscript e, will be omitted in the subsequent text and the symbol log will be used to denote the logarithm to base e. (The abbreviation ln is also used.) Any base other than e, will be specially designated as such. For example $\log_{10} x$ means logarithm to base 10, the common base used in computation.

The graph of $y = \log x$ is derived by 'reflecting' the graph of $y = e^x$ in the line $y = x$ and is shown in Fig. 4.2. The important facts to notice are $\log 1 = 0$, since $e^0 = 1$; y is not defined for negative values of x; and as $x \to 0$, $y \to -\infty$.

4.3 LIMITS CONNECTED WITH e^x AND $\log x$

As $x \to \infty$, $e^x \to \infty$ more rapidly than any power of x, i.e.

$$\frac{e^x}{x^n} \to \infty \quad \text{as} \quad x \to \infty \tag{13}$$

for any fixed positive value of n.

The function e^x/x^n assumes the *indeterminate* form ∞/∞ as $x \to \infty$, but since

$$e^x = 1 + x + \frac{x^2}{2!} + \frac{x^3}{3!} + \cdots$$

then, with m any integer greater than n,

$$e^x > \frac{x^m}{m!}.$$

Therefore $\dfrac{e^x}{x^n} > \dfrac{x^{m-n}}{m!}$ which tends to infinity with x. Hence $e^x/x^n \to \infty$ as $x \to \infty$, or e^x 'grows' more rapidly than x^n.

Similarly it can be proved that e^{-x} tends to zero more rapidly than any power of x as $x \to \infty$, i.e.

$$x^n e^{-x} \to 0 \quad \text{as} \quad x \to \infty. \tag{14}$$

Similarly it may be proved that as $x \to \infty$, $\log x \to \infty$ more slowly than any power of x, or

$$\frac{\log x}{x^n} \to 0 \quad \text{as} \quad x \to \infty \tag{15}$$

4.4 DIFFERENTIATION OF e^x AND $\log x$

4.4.1 The derivative of e^x

From the definition of e^x,

$$\frac{d}{dx} e^x = e^x,$$

hence $\qquad \dfrac{d}{dx} e^{kx} = k e^{kx}$ where k is a constant,

using function of a function rule.

4.4.2 The derivative of $\log x$

If $\qquad\qquad\qquad\qquad y = \log x, \qquad x > 0$

then $\qquad\qquad\qquad\qquad x = e^y$

and $\qquad\qquad\qquad\qquad \dfrac{dx}{dy} = e^y$

or $\qquad\qquad\qquad\qquad \dfrac{dy}{dx} = \dfrac{1}{e^y} = \dfrac{1}{x}.$

Hence $\qquad\qquad\qquad \dfrac{d}{dx} \log x = \dfrac{1}{x}.$

As x increases, the derivative $1/x$ decreases, so that the gradient of the graph of $y = \log x$ decreases but is always positive (see Fig. 4.2).

More generally,

$$\frac{d}{dx} \log |x| = \frac{1}{x}, \qquad x \neq 0,$$

$|x|$ is always positive so that $\log |x|$ is always defined. When x is negative, $|x| = -x$, $\log |x| = \log (-x)$ so that

$$\frac{d}{dx} \log |x| = \frac{d}{dx} \log (-x)$$

$$= \frac{1}{(-x)} (-1)$$

$$= \frac{1}{x}.$$

Also $\dfrac{d}{dx} \log|f(x)| = \dfrac{f'(x)}{f(x)}$ provided $f(x) \neq 0$,

since, writing $f(x) = z$, $\dfrac{dz}{dx} = f'(x)$,

$$\frac{d}{dx} \log |f(x)| = \frac{d}{dz} \log |z| \frac{dz}{dx}$$

$$= \frac{1}{z} f'(x)$$

$$= \frac{f'(x)}{f(x)}.$$

4.4.3 The general exponential function

The general exponential function is the function a^x, $a > 0$. If $y = a^x$, then

$$\log y = x \log a$$

and $y = e^{x \log a}$

or $a^x = e^{x \log a}.$

Thus, instead of a^x the simpler function $e^{x \log a}$ can be used in calculations. In particular the simpler form is more useful in differentiation, for example,

$$\frac{d}{dx} a^x = \frac{d}{dx} e^{x \log a}$$

$$= e^{x \log a}. \log a \quad a > 0$$

$$= a^x \log a.$$

To differentiate the inverse function, $y = \log_a x$, write $x = a^y = e^{y \log a}$

$$\frac{dx}{dy} = e^{y \log a} \cdot \log a$$

$$= x \log a$$

and

$$\frac{dy}{dx} = \frac{1}{x \log a}$$

or

$$\frac{d}{dx} \log_a x = \frac{1}{x \log a}, \qquad a > 0.$$

4.4.4 Logarithmic differentiation

In the case of differentiation of a function which consists of a number of factors it is sometimes simpler to take the logarithm before differentiating. Thus if

$$y = f_1(x) f_2(x) f_3(x) \ldots f_r(x)$$

then for any value of x which gives y a non-zero value,

$$\ln y = \ln f_1(x) + \ln f_2(x) + \ln f_3(x) + \ldots + \ln f_r(x)$$

and differentiations with respect to x leads to

$$\frac{1}{y} \frac{dy}{dx} = \frac{f_1'(x)}{f_1(x)} + \frac{f_2'(x)}{f_2(x)} + \frac{f_3'(x)}{f_3(x)} + \ldots + \frac{f_r'(x)}{f_r(x)}.$$

Example (i) Differentiate x^α where α is any real number and $x > 0$.

Solution Now, $x^\alpha = e^{\alpha \log x}$ since $x = e^{\log x}$ and so

$$\frac{d}{dx}(x^\alpha) = \frac{d}{dx}(e^{\alpha \log x})$$

$$= e^{\alpha \log x} \frac{\alpha}{x}$$

$$= x^\alpha \frac{\alpha}{x}$$

$$= \alpha x^{\alpha - 1}.$$

Example (ii) Differentiate x^x, $x > 0$.

Solution Let $y = x^x = e^{x \log x}$

$$\frac{dy}{dx} = e^{x \log x} \left(\log x + x \cdot \frac{1}{x} \right)$$

$$= x^x (\log x + 1)$$

alternatively, let $y = x^x$

then
$$\log y = x \log x$$

and
$$\frac{1}{y}\frac{dy}{dx} = \log x + 1$$

$$\frac{dy}{dx} = y(\log x + 1)$$

$$= x^x(\log x + 1).$$

Example (iii) Differentiate $y = \log(\log x)$.

Solution

$$\frac{dy}{dx} = \frac{1}{\log x} \cdot \frac{1}{x}$$

$$= \frac{1}{x \log x}.$$

The function $\log(\log x)$ is defined for $(\log x) > 0$ and hence for $x > 1$.

Example (iv) Find $\frac{dy}{dx}$ if $y = \dfrac{e^{2x}(x^2 + 1)(x + 1)^3 \sin x}{(x^4 + 4)(3x - 2)}$.

Solution Write

$$\log y = 2x + \log(x^2 + 1) + 3\log(x + 1) + \log \sin x$$
$$- \log(x^4 + 4) - \log(3x - 2)$$

and $\dfrac{1}{y}\dfrac{dy}{dx} = 2 + \dfrac{2x}{x^2 + 1} + \dfrac{3}{x + 1} + \dfrac{\cos x}{\sin x} - \dfrac{4x^3}{x^4 + 4} - \dfrac{3}{3x - 2}$

for any value of x for which $y \neq 0$.

4.5 THE HYPERBOLIC FUNCTIONS

There are certain combinations of the exponential functions e^x and e^{-x} which are called 'hyperbolic functions'. They are used in solving certain engineering problems and are introduced now so that they may be used in the systematic study of integration in Chapter 6.

4.5.1 Definitions

The combinations $\frac{1}{2}(e^x + e^{-x})$ and $\frac{1}{2}(e^x - e^{-x})$ occur with such frequency that it is convenient to give special names to them. They have properties analogous to those of the circular functions $\cos x$ and $\sin x$, and are called *hyperbolic cosine* of x or *cosh x*, and *hyperbolic sine* of x, or *sinh x*, respectively, i.e.

$$\left. \begin{array}{l} \cosh x = \frac{1}{2}(e^x + e^{-x}) \\ \sinh x = \frac{1}{2}(e^x - e^{-x}) \end{array} \right\} \tag{16}$$

The remaining hyperbolic functions are *defined* in terms of cosh x and sinh x as follows:

$$\tanh x = \frac{\sinh x}{\cosh x} = \frac{e^x - e^{-x}}{e^x + e^{-x}}$$

$$\coth x = \frac{\cosh x}{\sinh x} = \frac{e^x + e^{-x}}{e^x - e^{-x}}$$

$$\operatorname{sech} x = \frac{1}{\cosh x} = \frac{2}{e^x + e^{-x}}$$

$$\operatorname{cosech} x = \frac{1}{\sinh x} = \frac{2}{e^x - e^{-x}}.$$

4.5.2 Identities

The basic identity of hyperbolic functions is

$$\cosh^2 x - \sinh^2 x = 1 \qquad (17)$$

which is proved as follows:

$$\cosh x + \sinh x = \tfrac{1}{2}(e^x + e^{-x}) + \tfrac{1}{2}(e^x - e^{-x}) = e^x,$$
$$\cosh x - \sinh x = \tfrac{1}{2}(e^x + e^{-x}) - \tfrac{1}{2}(e^x - e^{-x}) = e^{-x}.$$

Hence the left side of equation (17) is

$$\cosh^2 x - \sinh^2 x = (\cosh x + \sinh x)(\cosh x - \sinh x)$$
$$= e^x e^{-x}$$
$$= 1.$$

Dividing equation (17) by $\cosh^2 x$ gives

$$1 - \tanh^2 x = \operatorname{sech}^2 x \qquad (17a)$$

and dividing equation (17) by $\sinh^2 x$ gives

$$\coth^2 x - 1 = \operatorname{cosech}^2 x. \qquad (17b)$$

4.5.3 Addition formulae

There are results analogous to those of the circular functions given by

$$\sinh (x \pm y) = \sinh x \cosh y \pm \cosh x \sinh y$$
$$\cosh (x \pm y) = \cosh x \cosh y \pm \sinh x \sinh y \quad \text{(note signs)}$$

These follow from the definitions of section 4.5.1 and are easily proved, starting with the right side.

Also
$$\sinh 2x = 2 \sinh x \cosh x$$
$$\cosh 2x = \cosh^2 x + \sinh^2 x$$
but
$$1 = \cosh^2 x - \sinh^2 x$$

so that, adding and subtracting respectively,

$$\cosh^2 x = \tfrac{1}{2}(1 + \cosh 2x), \quad \sinh^2 x = \tfrac{1}{2}(\cosh 2x - 1).$$

4.5.4 Derivatives

From the definitions in section 4.5.1,

$$\frac{d}{dx}\sinh x = \frac{d}{dx}\tfrac{1}{2}(e^x - e^{-x})$$

$$= \tfrac{1}{2}(e^x + e^{-x})$$

$$= \cosh x.$$

Similarly
$$\frac{d}{dx}\cosh x = \sinh x,$$

$$\frac{d}{dx}\tanh x = \operatorname{sech}^2 x,$$

and
$$\frac{d}{dx}\coth x = -\operatorname{cosech}^2 x,$$

each of which may be checked by the reader.

4.5.5 Graphs of the hyperbolic functions

(i) The properties of $y = \sinh x = \tfrac{1}{2}(e^x - e^{-x})$ are

(a) $x > 0, \quad \sinh x > 0$

(b) $x = 0, \quad \sinh(0) = 0$

(c) $\dfrac{dy}{dx} = \cosh x = \tfrac{1}{2}(e^x + e^{-x}) > 0 \quad$ all x,

i.e. $\sinh x$ is an increasing function for all values of x. Also $\sinh x$ is an odd function since

$$\sinh(-x) = \tfrac{1}{2}(e^{-x} - e^x) = -\sinh x$$

(ii) The properties of $y = \cosh x = \tfrac{1}{2}(e^x + e^{-x})$ are

(a) $\cosh x > 0 \quad$ all x

(b) $\cosh(0) = 1$

(c) $\dfrac{dy}{dx} = \sinh x \begin{cases} > 0 & \text{when } x > 0 \\ < 0 & \text{when } x < 0. \end{cases}$

Hence when x is negative y decreases as x increases and when x is positive y increases as x increases. Also $\cosh x$ is an even function since

$$\cosh(-x) = \tfrac{1}{2}(e^{-x} + e^x) = \cosh x.$$

The graphs of $\sinh x$ and $\cosh x$ are shown in Fig. 4.3.

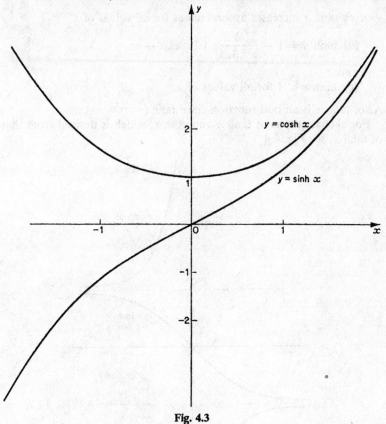

Fig. 4.3

(iii) From the expression $y = \tanh x$,

$$y = \frac{\sinh x}{\cosh x}$$

$$= \frac{e^x - e^{-x}}{e^x + e^{-x}}$$

$$= \frac{e^{2x} - 1}{e^{2x} + 1}$$

$$= 1 - \frac{2}{e^{2x} + 1}$$

it follows that

(a) $\tanh(0) = 0$,

(b) $\dfrac{dy}{dx} = \dfrac{1}{\cosh^2 x} > 0$ all x,

(c) tanh x increases as x increases for all values of x,

(d) $\tanh x = 1 - \dfrac{2}{e^{2x} + 1} \to 1$ as $x \to \infty$,

and

(e) tanh $x < 1$ for all values of x.

Also, tanh x is an odd function since $\tanh(-x) = -\tanh x$.

For sketch graphs of tanh x and coth x, which is derived from that of tanh x, see Fig. 4.4.

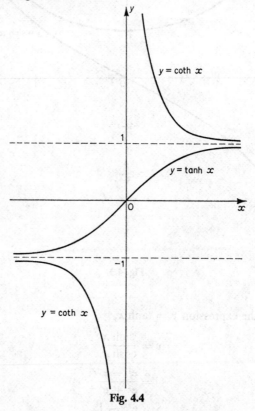

Fig. 4.4

4.6 THE INVERSE HYPERBOLIC FUNCTIONS

4.6.1 Range of the inverse hyperbolic functions

Since $x = \sinh y$ is a monotonically increasing function for all values of y it is possible to define the inverse function $y = \sinh^{-1} x$ (sometimes written $y = \operatorname{argsinh} x$) for all values of x and y. (*Note*, argsinh x but arcsin x.)

Similarly $y = \cosh^{-1} x$ (or argcosh x) is defined where $x = \cosh y$ is monotonic increasing, i.e. when $y \geqslant 0$, and $x \geqslant 1$.

The function $y = \tanh^{-1} x$ (or argtanh x) is defined for all values of y since $x = \tanh y$ is monotonic increasing in this range, and $|x| < 1$.

4.6.2 The logarithmic form of the inverse functions

(i) If $y = \sinh^{-1} x$, then

$$\sinh y = x = \tfrac{1}{2}(e^y - e^{-y}).$$

Multiplying by $2e^y$ and rearranging,

$$e^{2y} - 2xe^y - 1 = 0.$$

This is a quadratic equation in e^y whose roots are given by

$$e^y = x \pm \sqrt{(x^2 + 1)}.$$

Now e^y is positive so that $x \pm \sqrt{(x^2 + 1)}$ must be positive also. Since $\sqrt{(x^2 + 1)} > x$ the positive sign of the square root is chosen, and

$$e^y = x + \sqrt{(x^2 + 1)}$$

or $$y = \log [x + \sqrt{(x^2 + 1)}]$$

or $\quad \sinh^{-1} x = \log [x + \sqrt{(x^2 + 1)}]$ for all values of x. $\qquad$ (18)

(ii) Let $y = \cosh^{-1} x$, then $x = \cosh y$, $x \geqslant 1$, $y \geqslant 0$ and so $\sinh y \geqslant 0$. Using the definitions (16), together with $\cosh^2 y - \sinh^2 y = 1$,

$$e^y = \cosh y + \sinh y = x + \sqrt{(x^2 - 1)}.$$

Hence $\quad y = \cosh^{-1} x = \log [x + \sqrt{(x^2 - 1)}], \quad x \geqslant 1.$ $\qquad$ (19)

(iii) Let $y = \tanh^{-1} x$, then $x = \tanh y$, $|x| < 1$, and

$$x = \frac{e^y - e^{-y}}{e^y + e^{-y}}$$

$$= \frac{e^{2y} - 1}{e^{2y} + 1}$$

or $$e^{2y} = \frac{1 + x}{1 - x}.$$

Hence $\quad y = \tanh^{-1} x = \tfrac{1}{2} \log \dfrac{1 + x}{1 - x}, \quad |x| < 1.$ $\qquad$ (20)

Similarly $\quad \coth^{-1} x = \tfrac{1}{2} \log \dfrac{x + 1}{x - 1}, \quad |x| > 1.$ $\qquad$ (21)

4.6.3 Derivatives of the inverse hyperbolic functions

These are simple algebraic functions similar to the derivatives of the inverse circular functions.

(i) Let $y = \sinh^{-1} x$, then $x = \sinh y$

$$\frac{dx}{dy} = \cosh y$$

$$= \sqrt{(\sinh^2 y + 1)}$$

$$= \sqrt{(x^2 + 1)}$$

The positive square root is chosen since $\cosh y > 0$ for all values of y. Hence

$$\frac{dy}{dx} = \frac{1}{\sqrt{(x^2 + 1)}}$$

or
$$\frac{d}{dx} \sinh^{-1} x = \frac{1}{\sqrt{(x^2 + 1)}}, \quad \text{all } x.$$

Similarly

$$\frac{d}{dx} \cosh^{-1} x = \frac{1}{\sqrt{(x^2 - 1)}}, \quad x > 1$$

$$\frac{d}{dx} \tanh^{-1} x = \frac{1}{1 - x^2}, \quad |x| < 1$$

$$\frac{d}{dx} \coth^{-1} x = \frac{-1}{x^2 - 1}, \quad |x| > 1.$$

EXERCISE 4

1 Differentiate with respect to x the following functions.

  (i) $\exp(x^2)$;   (ii) $\exp(\sec x)$;   (iii) $x^3 \log(1 + x^3)$;

  (iv) $(1 + x^2) \log(1 + x^2)$;   (v) $x \tan^{-1} x - \log \sqrt{(1 + x^2)}$;

  (vi) $\log[x + \sqrt{(x^2 - 3)}]$;   (vii) $\sinh(2x^2)$;   (viii) $x \log x - x$.

2 Differentiate with respect to x the following functions:

  (i) x^x;   (ii) $(\sin x)^x$;   (iii) $(\sqrt{x})^x$;

  (iv) $x^{\sqrt{x}}$;   (v) $x^{\sin x}$;   (vi) $10^{\log \sin x}$.

3 Evaluate $f'(x)/f(x)$ when $f(x) = (1 + x)^x$ and by successive differentiation calculate the values of $f'(0)$, $f''(0)$, and $f'''(0)$.

4 Differentiate with respect to x the following functions:

  (i) $(\cosh x - \cos x)/(\sinh x - \sin x)$;   (ii) $\operatorname{sech} x$;

  (iii) $x \tanh^{-1} x + \log \sqrt{(1 - x^2)}$;   (iv) $\tan^{-1}(\sinh x)$;

  (v) $\log \tanh x$.

5 Sketch the curve $y = \tanh^2 x$.

6 Find $\partial u/\partial x$ and $\partial u/\partial t$ if a, b, and p are constants and

$$u = e^{-ax} \cosh(pt - bx).$$

7 Differentiate the following functions:

(i) 2^x;

(ii) $\exp(\sin^2 x)$;

(iii) $a^{\sqrt{x}}, a > 0$;

(iv) $\sinh 3x \cosh^3 x$;

(v) $(x^2 + 1)^3/(2x - 1)(3x + 2)^2$;

(vi) $\log[(1 - x)/(1 + x)]$;

(vii) $x^2 \exp(1/x)$;

(viii) $a^x x^a, a > 0$.

8 Given that $z = x^y$, calculate $\partial^2 z/\partial x \, \partial y$ and $\partial^2 z/\partial y^2$.

9 Determine the nature of the stationary points of the function $x^4 e^{-x}$, and show that for all values of $x \geqslant 0$, $x^4 e^{-x} \leqslant (4/e)^4$.

10 Determine the position and nature of the stationary point of the function $\cosh 2x - 4 \cosh x$.

11 Prove that, if $u = x \log(1 + x/y) + y \log(1 + y/x)$ then

$$x \frac{\partial u}{\partial x} + y \frac{\partial u}{\partial y} = u.$$

12 Differentiate (i) $\exp(x \sin x)$; (ii) $(\log x)^x$; (iii) $(x \log x)/(1 + x)$; (iv) $(\log x)^{-1}$; (v) $\log(\sec x + \tan x)$.

13 Prove that, if $y = \sin^{-1}(e^{-x})$ then

$$\cos^2 y \frac{d^2y}{dx^2} + \sin^2 y \frac{dy}{dx} = \sin y \cos y.$$

14 Show that, if A and B are constants and

$$y = A \cos(\log x) + B \sin(\log x),$$

then

$$x^2 \frac{d^2y}{dx^2} + x \frac{dy}{dx} + y = 0.$$

15 Show that $\sec x + \log(\cos x)$ is an increasing function of x in the range $0 \leqslant x < \frac{\pi}{2}$.

16 Prove that, if $y = \tan^{-1}(\sinh x)$, then

$$\frac{d^2y}{dx^2} + \tan y \left(\frac{dy}{dx}\right)^2 = 0.$$

17 Find dy/dx where $\log(x^2 + y^2) = \tan^{-1}(y/x)$, $x^2 + y^2 \neq 0$.

18 Use the definitions of $\cosh x$ and $\sinh x$ to solve the simultaneous equations

$$\cosh x + \cosh y = 4$$
$$\sinh x - \sinh y = 2.$$

Give your answers correct to three significant figures.

19 The temperature $y°$ of a body t seconds after it is placed in a certain medium is given by $y = 27 + 63e^{-t/6}$. Find correct to two places of decimals, after how many seconds the rate of cooling of the body will have fallen to less than $1°$ per second.

20 Solve the equation

$$16 \cosh x - 4 \sinh x = 19,$$

giving the answers correct to three places of decimals.

5
The Mean Value Theorem, Taylor Series, and Further Partial Differentiation

5.1 THE MEAN VALUE THEOREM

5.1.1 Introduction

It is geometrically intuitive that if a function $y = f(x)$ is continuous in the interval $a \leqslant x \leqslant b$ and has a derivative at each value of x in $a < x < b$ and the values of y at A where $x = a$ and B where $x = b$ are equal, then there is at least one point between $x = a$ and $x = b$ where the derivative is zero. Moreover, if the values of y at A and B are not equal and the other conditions hold, then there will be at least one point, $x = c$, between $x = a$ and $x = b$ where the tangent is parallel to the chord AB. The sketch of functions $y = f(x)$ in Fig. 5.1 demonstrates this property.

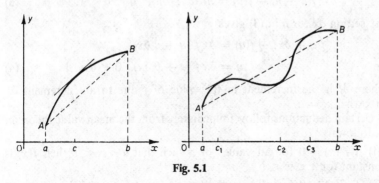

Fig. 5.1

The graph of $y = f(x)$ may have a vertical tangent at either or both of the end-points of the interval, so that for example, the function $\sqrt{(a^2 - x^2)}$, $-a \leqslant x \leqslant a$ which represents a semi-circle satisfies the requirements.

These two results may be stated precisely in the form of theorems as follows:

(1) *Rolle's theorem* Let $y = f(x)$ be (i) continuous for $a \leqslant x \leqslant b$, (ii) differentiable for $a < x < b$, and (iii) let $f(a) = f(b)$. Then there is at least one number c satisfying $a < c < b$ and such that $f'(c) = 0$.

(2) *The mean value theorem* Let $y = f(x)$ be (i) continuous for $a \leqslant x \leqslant b$, and (ii) differentiable for $a < x < b$, then there exists at least one number c such that

$$f(b) - f(a) = (b - a)f'(c) \quad \text{and} \quad a < c < b, \tag{1}$$

or
$$f'(c) = \frac{f(b) - f(a)}{b - a} \quad \text{and} \quad a < c < b. \tag{2}$$

The fraction $[f(b) - f(a)]/(b - a)$, which is the ratio of the increment in the value of the function to the increment in the value of x, measures the average rate of increase of the function in the interval $b - a$, and is represented by the gradient of the chord AB of Fig. 5.1. Hence the mean value theorem implies that the average or mean rate of increase in any interval is equal to the actual rate of increase at some point within the interval. For example, the average velocity of a moving point in any interval of time is equal to the actual velocity at some instant within the interval. Or, if the average speed of a moving vehicle is 50 km.p.h. over a given interval of time, then at some instant during that interval the speedometer registers 50 km.p.h.

Other ways of expressing equation (1) are as follows:

(i) writing $c = a + \theta(b - a)$ where θ is some number between 0 and 1, and writing $a + h$ for b so that $b - a = h$, then (1) becomes

$$f(a + h) - f(a) = hf'(a + \theta h), \quad 0 < \theta < 1. \tag{3}$$

(ii) setting δx for h in (3) gives

$$f(a + \delta x) - f(a) = \delta x f'(a + \theta \delta x)$$

or
$$\delta f = \delta x f'(a + \theta \delta x) \quad 0 < \theta < 1 \tag{4}$$

where δf is the increment in the value of f due to an increment δx in x when $x = a$.

Certain deductions follow immediately from the mean value theorem, as follows:

(i) if $f'(x) = 0$ for all values of x such that $a < x < b$ then $f(x)$ is constant for $a \leqslant x \leqslant b$.

(ii) if $f'(x) > 0$ for all x such that $a < x < b$, then $f(b) > f(a)$ and $f(x)$ is strictly increasing for $a \leqslant x \leqslant b$.

(iii) if $m \leqslant f'(x) \leqslant M$ for all x such that $a < x < b$, then

$$m(b - a) \leqslant f(b) - f(a) \leqslant M(b - a),$$

(iv) if $f'_1(x) = f'_2(x)$ for $a < x < b$ then

$$\frac{d}{dx}[f_1(x) - f_2(x)] = 0,$$

and $f_1(x) - f_2(x) = \text{constant}$ for $a \leqslant x \leqslant b$.

5.1.2 The generalized form of the mean value theorem

Suppose $f(x)$ and $g(x)$ are two functions which are continuous for $a \leqslant x \leqslant b$ and differentiable for $a < x < b$ and that $g(a) \neq g(b)$, then the function $\Phi(x)$ defined by

$$\Phi(x) = f(b) - f(x) - \frac{f(b) - f(a)}{g(b) - g(a)} \{g(b) - g(x)\},$$

satisfies the conditions of Rolle's theorem. Hence there is at least one number c between a and b such that $\Phi'(c) = 0$, i.e.

$$-f'(c) + \frac{f(b) - f(a)}{g(b) - g(a)} g'(c) = 0 \quad \text{and} \quad a < c < b.$$

Moreover if f' and g' are not simultaneously zero this may be written

$$\frac{f(b) - f(a)}{g(b) - g(a)} = \frac{f'(c)}{g'(c)}, \tag{5}$$

and this is the *generalized* or *Cauchy* form of the mean value theorem.

5.1.3 Application to small errors

Let $y = f(x)$ satisfy the conditions of the mean value theorem, then from equation (4)

$$\delta y = \delta x f'(a + \theta \delta x), \qquad 0 < \theta < 1. \tag{6}$$

This is an exact equation, and gives the value of δy when δx is known for some value $x = a$. If δx is *small* then (6) may be used to compute an approximate change in the value of y corresponding to a small change δx in the value of x when $x = a$. For if δx is small enough $f'(a)$ is an approximation for $f'(a + \theta \, \delta x)$ so that

$$\delta y \simeq \delta x f'(a) \tag{7}$$

(See equation (18) of section 2.4.)

Example (i) If $y = \sqrt{x}$, calculate the approximate error in the calculation of y when x is wrongly given as $1 \cdot 02$ instead of its correct value, 1.

Solution Here $f(x) = \sqrt{x}$, $f'(x) = \frac{1}{2}x^{-\frac{1}{2}}$, $f'(1) = \frac{1}{2}$, $\delta x = 0 \cdot 02$ and $\delta y \simeq 0 \cdot 01$.

Example (ii) The period T of a simple pendulum is given by

$$T = 2\pi \sqrt{(l/g)}$$

where l is the length of the pendulum. If an error of $\frac{1}{2}$ per cent is made in measuring l, and g is accurate, find the corresponding percentage error in the value of T.

Solution This question involves percentages and it is well to remember that this implies percentage *of* something.

Suppose the error in measuring l is positive, i.e. l is measured too large, then the error δl is $\frac{1}{2}$ per cent of l, or $\delta l/l = \frac{1}{2}$ per cent, and it is required to find $\delta T/T$.

From
$$T = 2\pi \sqrt{(l/g)}$$
$$\log T = \log 2\pi + \tfrac{1}{2}\log l - \tfrac{1}{2}\log g$$

and differentiating implicitly with respect to l leads to

$$\frac{1}{T}\frac{dT}{dl} = \frac{1}{2l}.$$

Using $\delta T \simeq (dT/dl)\,\delta l$ then gives

$$\frac{\delta T}{T} \simeq \frac{1}{T}\frac{dT}{dl}\,\delta l$$

$$= \frac{1}{2l}\,\delta l$$

$$= \tfrac{1}{2}\cdot\tfrac{1}{2}\text{ per cent,}$$

and the approximate percentage error in T is $\tfrac{1}{4}$.

If the measurement of l is too small, the error is negative and this simply means a change of sign of δl in the working.

5.1.4 l'Hôpital's rule

Suppose $f(x)$ and $g(x)$ are two functions which are continuous in $a \leqslant x \leqslant b$, and that $f'(x)$ and $g'(x)$ exist for each x in $a < x < b$, that $g'(x) \neq 0$ and that $f(a) = g(a) = 0$. Then l'Hôpital's rule states that

$$\operatorname*{Lim}_{x \to a} \frac{f(x)}{g(x)} = \operatorname*{Lim}_{x \to a} \frac{f'(x)}{g'(x)}, \tag{8}$$

provided the second limit exists. For, since $f(a) = g(a) = 0$,

$$\frac{f(x)}{g(x)} = \frac{f(x) - f(a)}{g(x) - g(a)} = \frac{f'(c)}{g'(c)} \quad \text{and} \quad a < c < x,$$

applying equation (5) of section 5.1.2. As $x \to a+$, $c \to a+$ and

$$\operatorname*{Lim}_{x \to a+} \frac{f(x)}{g(x)} = \operatorname*{Lim}_{c \to a+} \frac{f'(c)}{g'(c)} = \operatorname*{Lim}_{x \to a+} \frac{f'(x)}{g'(x)},$$

provided this limit exists.

This establishes the rule when x approaches a from above. The case when x approaches a from below is proved similarly and the two cases together give the result (8).

The rule (8) is used to determine $\operatorname*{Lim}_{x \to a}\{f(x)/g(x)\}$ where substitution of $x = a$ in the quotient leads to the expression $0/0$ which is *indeterminate*.

Example (i) $\operatorname*{Lim}_{x \to 0} \dfrac{\sin x}{x} \quad \left[= \dfrac{0}{0} \right]$

$$= \operatorname*{Lim}_{x \to 0} \frac{\cos x}{1} = 1.$$

Example (ii) $\displaystyle \lim_{x \to 1} \frac{\log x}{x - 1} \quad \left[= \frac{0}{0} \right]$

$$= \lim_{x \to 1} \frac{1/x}{1} = 1.$$

The rule was established on the assumption that $g'(x) \neq 0$ in the open interval $a < x < b$ and hence is valid when $g'(a) = 0$. Thus if $f'(a)$ is also zero then equation (8) gives another indeterminate form $0/0$ and applying l'Hôpital's rule again,

$$\lim_{x \to a} \frac{f'(x)}{g'(x)} = \lim_{x \to a} \frac{f''(x)}{g''(x)}$$

provided the second limit exists.

More generally, applying rule (8) m times

$$\lim_{x \to a} \frac{f(x)}{g(x)} = \lim_{x \to a} \frac{f^{(m)}(x)}{g^{(m)}(x)}$$

provided this limit exists. Conditions on f, g and their $m - 1$ derivatives may be deduced from considerations of the mean value theorem and conditions for l'Hôpital's rule.

In practice the method consists in differentiating numerator and denominator separately so long as the form $0/0$ appears at $x = a$. As soon as one or other of these derivatives is different from zero at $x = a$, the limit may be found.

Example (iii) $\displaystyle \lim_{x \to 0} \frac{\sqrt{(1 + x)} - (1 + \frac{1}{2}x)}{x^2} \quad \left[= \frac{0}{0} \right]$

$$= \lim_{x \to 0} \frac{\frac{1}{2}(1 + x)^{-\frac{1}{2}} - \frac{1}{2}}{2x} \quad \left[= \frac{0}{0} \right]$$

$$= \lim_{x \to 0} \frac{-\frac{1}{4}(1 + x)^{-\frac{3}{2}}}{2}$$

$$= -\frac{1}{8}.$$

It can be proved that if $f(x) \to \infty$ and $g(x) \to \infty$ as $x \to a$, then l'Hôpital's rule applies to the indeterminate form ∞/∞. Proof is beyond the scope of this book.

The indeterminate forms $\infty . 0$ and $\infty - \infty$ may be evaluated in some cases by reducing them to a form in which l'Hôpital's rule applies.

The forms $0^0, 1^\infty, \infty^0$, as limits, may also be evaluated if the expression

$$[f(x)]^{g(x)}$$

can be reduced to a form in which l'Hôpital's rule applies, for example by considering the logarithm of the expression.

Example (iv) Find the value of

$$\lim_{x \to 0} \left(\frac{1}{\sin x} - \frac{1}{x} \right) \quad \left[= \begin{array}{ll} \infty - \infty & \text{as } x \to 0+ \\ -\infty + \infty & \text{as } x \to 0- \end{array} \right]$$

Solution The expression may be re-written as

$$\lim_{x \to 0} \frac{x - \sin x}{x \sin x} \quad \left[= \frac{0}{0} \right],$$

and applying l'Hôpital's rule twice the limit is

$$\lim_{x \to 0} \frac{1 - \cos x}{\sin x + x \cos x} \quad \left[= \frac{0}{0} \right]$$

$$= \lim_{x \to 0} \frac{\sin x}{2 \cos x - x \sin x} = 0.$$

Example (v) Find the value of $\lim_{x \to 0+} x^x$.

Solution Writing $y = x^x$, then

$$\log y = x \log x$$

$$= \log x / (1/x),$$

and

$$\lim_{x \to 0+} \frac{\log x}{1/x} \quad \left[= \frac{\infty}{\infty} \right]$$

$$= \lim_{x \to 0+} \frac{1/x}{-1/x^2} = \lim_{x \to 0+} -x = 0.$$

Hence

$$\lim_{x \to 0+} \log y = 0$$

so that

$$\lim_{x \to 0+} y = e^0 = 1.$$

The assumption made in this last step is that

$$\lim_{x \to 0} (\log y) = \log \left(\lim_{x \to 0} y \right),$$

which can be proved since $\log y$ is continuous.

Example (vi) Evaluate $\lim_{x \to \infty} \dfrac{e^x}{x^3} \left[= \dfrac{\infty}{\infty} \right]$

Solution Applying l'Hôpital's rule, the required limit

$$\lim_{x \to \infty} \frac{e^x}{x^3} = \lim_{x \to \infty} \frac{e^x}{3x^2}$$

$$= \lim_{x \to \infty} \frac{e^x}{6x}$$

$$= \lim_{x \to \infty} \frac{e^x}{6}$$

and there is no 'limit', or the 'limit is infinity'. This shows that $e^x \to \infty$ more rapidly than does x^3. See section 4.3 and example 8 at the end of this chapter.

5.2 TAYLOR SERIES

The mean value theorem of section 5.1.1 states that if $f(x)$ is continuous for $a \leqslant x \leqslant b$ and has a derivative $f'(x)$ for $a < x < b$, then

$$f(b) - f(a) = (b - a)f'(c) \quad \text{where} \quad a < c < b.$$

Suppose now that $f'(x)$ is continuous for $a \leqslant x \leqslant b$ and that $f''(x)$ exists for $a < x < b$. Consider the function defined by

$$\Phi_2(x) - \frac{(b - x)^2}{(b - a)^2} \Phi_2(a) = f(b) - f(x) - (b - x)f'(x)$$
$$- \frac{(b - x)^2}{(b - a)^2}\{f(b) - f(a) - (b - a)f'(a)\}.$$

This function is zero when $x = a$ and when $x = b$ and satisfies the conditions of Rolle's theorem. Its derivative is

$$-f'(x) + f'(x) - (b - x)f''(x) + \frac{2(b - x)}{(b - a)^2}\{f(b) - f(a) - (b - a)f'(a)\},$$

i.e. $\quad \dfrac{2(b - x)}{(b - a)^2}\{f(b) - f(a) - (b - a)f'(a) - \tfrac{1}{2}(b - a)^2 f''(x)\}.$

Hence there exists a value c between a and b such that this derivative is zero, i.e.

$$f(b) = f(a) + (b - a)f'(a) + \tfrac{1}{2}(b - a)^2 f''(c) \quad \text{and} \quad a < c < b.$$

Extending this argument, suppose $f^{(n-1)}(x)$ is continuous for $a \leqslant x \leqslant b$ and $f^{(n)}(x)$ exists for $a < x < b$ and consider the function
$$\Phi_n(x) - \frac{(b - x)^n}{(b - a)^n} \Phi_n(a) \quad \text{where}$$

$$\Phi_n(x) = f(b) - f(x) - (b - x)f'(x) - \ldots - \frac{(b - x)^{n-1}}{(n - 1)!} f^{(n-1)}(x).$$

This function satisfies the conditions of Rolle's theorem and hence its derivative is zero for at least one value of $x = c$ between $x = a$ and $x = b$. This leads to the result known as *Taylor's theorem* that

$$f(b) = f(a) + (b - a)f'(a) + \frac{(b - a)^2}{2!} f''(a) + \ldots$$

$$+ \frac{(b - a)^{n-1}}{(n - 1)!} f^{(n-1)}(a) + \frac{(b - a)^n}{n!} f^{(n)}(c). \quad \text{and} \quad a < c < b. \quad (9)$$

This may be written more compactly, and slightly differently, by writing x for b and using the summation symbol, i.e.

$$f(x) = f(a) + \sum_{m=1}^{n-1} \frac{1}{m!} f^{(m)}(a)(x - a)^m + R_n(x) \quad (10)$$

where $R_n(x)$ is the *remainder* after n terms and

$$R_n(x) = \frac{1}{n!}(x - a)^n f^{(n)}(c), \quad a < c < x.$$

The special case $a = 0$ is known as *Maclaurin's* theorem, and writing $c = \theta x$ where θ is some number between 0 and 1,

$$f(x) = f(0) + \sum_{m=1}^{n-1} \frac{1}{m!} x^m f^{(m)}(0) + \frac{1}{n!} x^n f^{(n)}(\theta x). \tag{11}$$

The formulae (9), (10) and (11) are valid only if $f^{(n)}(x)$ is defined for all numbers x between a and b. This assumption is not true for all functions, so that not every function can be expanded in a Taylor series. It is true for most of the more common functions.

The form of the remainder $R_n(x)$ in (10) as $(x - a)^n f^{(n)}(c)/n!$ and in (11) as $x^n f^{(n)}(\theta x)/n!$ is known as the *Lagrange* form of the remainder. Even if $f^{(n)}(x)$ exists for all values of x for $a < x < b$ and the Taylor series can be found for $f(x)$, it is necessary to show that the difference between $f(x)$ and the sum of the first n terms of the series, or $R_n(x)$ is 'as small as we please' as $n \to \infty$.

5.3 POWER SERIES EXPANSIONS OF FUNCTIONS

The problem of representing a function by a power series is an important one, and can be divided into two parts. First it is necessary to develop a general formula to give a power series representation of a function of x, i.e. to give a series of the type

$$a_0 + a_1 x + a_2 x^2 + \ldots + a_n x^n + \ldots.$$

Secondly it is necessary to decide whether a given function can be expressed in this form and if so, for what range of values of x the series represents the function. The second part of the problem is by far the more difficult and is in general beyond the scope of this book.

5.3.1 The Maclaurin series

Let $y = f(x)$ be the given function and suppose the power series expansion of the function can be found, i.e.

$$f(x) = A_0 + A_1 x + A_2 x^2 + \ldots + A_n x^n + \ldots. \tag{12}$$

Let $\phi(x) = a_0 + a_1 x + a_2 x^2 + \ldots + a_r x^r + \ldots$ for $|x| < R \tag{13}$

(see section 3.6.1),

then $\phi'(x) = a_1 + 2a_2 x + 3a_3 x^2 + \ldots + ra_r x^{r-1} + \ldots$

$\phi''(x) = 2a_2 + 3 \cdot 2a_3 x + \ldots + r(r - 1)a_r x^{r-2} + \ldots$

and $\phi^{(n)}(x) = n! \, a_n +$ terms containing powers of x, for all x for $|x| < R$ and for all integer values of n. Setting $x = 0$ in these equations leads to

$$a_n = \frac{1}{n!} \phi^{(n)}(0). \tag{14}$$

Hence if $f(x)$ has an expansion (12) it must be of the form

$$f(0) + x f'(0) + \frac{x^2}{2!}f''(0) + \ldots + \frac{x^n}{n!}f^{(n)}(0) + \ldots \qquad (15)$$

which is Maclaurin's expansion (11) of section 5.2. This development shows no more than that a condition which must be satisfied by $f(x)$ is that its derivatives of all orders exist for $|x| < R$. By considering the sequence of remainders $R_1(x)$, $R_2(x)$, . . ., $R_n(x)$, . . ., it is possible to estimate the remainder after n terms of the expansion, and if the sequence tends to zero it is possible to show that the series (15) converges and for which values of x.

5.3.2 Some standard series
(i) For $f(x) = \log(1 + x)$, $f(0) = 0$, then

$$f^{(n)}(x) = (-1)^{n-1}(n-1)!(1+x)^{-n},$$

$$f^{(n)}(0) = (-1)^{n-1}(n-1)!,$$

and substituting in equation (15), gives

$$\log(1 + x) = x - \frac{x^2}{2} + \frac{x^3}{3} - + \ldots + \frac{x^r}{r!}(r-1)!(-1)^{r-1} + \ldots$$

$$= x - \frac{x^2}{2} + \frac{x^3}{3} - + \ldots + (-1)^{r-1}\frac{x^r}{r} + \ldots.$$

The remainder after n terms of (15) is R_n where

$$R_n(x) = (-1)^n \frac{x^n}{n!} \frac{(n-1)!}{(1+\theta x)^n}$$

and

$$|R_n(x)| = \frac{|x|^n}{n|1+\theta x|^n}, \qquad 0 < \theta < 1.$$

It is necessary to consider $\underset{n \to \infty}{\text{Lim}} \dfrac{|x|^n}{n|1+\theta x|^n}$. If $0 \leqslant x \leqslant 1$,

then

$$0 \leqslant \frac{x^n}{n(1+\theta x)^n} \leqslant \frac{1}{n}$$

and

$$\underset{n \to \infty}{\text{Lim}} |R_1(x)| = 0.$$

Hence $\log(1 + x) = x - \dfrac{x^2}{2} + \dfrac{x^3}{3} - + \ldots + (-1)^{n-1}\dfrac{x^n}{n} + \ldots$

$$\qquad (16)$$

is valid $0 \leqslant x \leqslant 1$.

If x is negative, it may be proved that the series is valid for values of x between 0 and -1 only, so that equation (16) is valid for $-1 < x \leqslant 1$.

(ii) For $f(x) = \sin x$, then
$$f^{(n)}(x) = \sin (x + \tfrac{1}{2}n\pi)$$
and
$$f^{(n)}(0) = \sin \tfrac{1}{2}n\pi.$$

When n is even, $n = 2k$, and $f^{(n)}(0) = \sin k\pi = 0$

When n is odd, $n = 2k + 1$, and $f^{(n)}(0) = \sin (2k + 1)\tfrac{1}{2}\pi = (-1)^k$.

The Maclaurin series for $\sin x$ is
$$x - \frac{x^3}{3!} + \frac{x^5}{5!} + \ldots + \frac{(-1)^{n-1}x^{2n-1}}{(2n-1)!} + \ldots$$

with
$$R_{2n+1}(x) = \frac{x^{2n+1}}{(2n+1)!} \sin (\theta x + \tfrac{1}{2}(2n+1)\pi)$$

and $0 < \theta < 1$. This is the remainder after $(2n + 1)$ terms of (11).

Since $|\sin y| \leqslant 1$, then
$$|R_{2n+1}(x)| \leqslant \frac{|x|^{2n+1}}{(2n+1)!}$$

for any fixed finite value of x, no matter how large, and hence
$$\operatorname*{Lim}_{n \to \infty} |R_{2n+1}(x)| = 0.$$

Thus $\sin x = x - \dfrac{x^3}{3!} + \dfrac{x^5}{5!} - + \ldots + \dfrac{(-1)^{n-1}x^{2n-1}}{(2n-1)!} + \ldots$

for all values of x.

(iii) Analagous methods may be used to show that

$$e^x = 1 + x + \frac{x^2}{2!} + \ldots + \frac{x^n}{n!} + \ldots, \quad -\infty < x < \infty$$

$$\cos x = 1 - \frac{x^2}{2!} + \frac{x^4}{4!} - + \ldots + \frac{(-1)^n x^{2n}}{(2n)!} + \ldots,$$
$$-\infty < x < \infty$$

$$(1 + x)^\alpha = 1 + \alpha x + \binom{\alpha}{2} x^2 + \ldots + \binom{\alpha}{n} x^n + \ldots, \quad |x| < 1$$

and α is any real number.

These standard series and their radii of convergence should be remembered and used. For example,

$$\log (1 - 2x) = (-2x) - \frac{(-2x)^2}{2} + \frac{(-2x)^3}{3} + \ldots$$
$$+ \frac{(-1)^{n-1}(-2x)^n}{n} + \ldots$$
$$= -2x - 2x^2 - \frac{8x^3}{3} - \ldots - \frac{2^n x^n}{n} - \ldots$$

is valid for $-\tfrac{1}{2} \leqslant x < \tfrac{1}{2}$.

(iv) To find the power series expansion of $\sin x$ about $x = \frac{1}{2}\pi$.

For $f(x) = \sin x$ then $f^{(n)}(x) = \sin(x + \frac{1}{2}n\pi)$ and $f(\frac{1}{2}\pi) = 1$, $f^{(n)}(\frac{1}{2}\pi) = \cos \frac{1}{2}n\pi$. If $n = 2k$ then $f^{(2k)}(\frac{1}{2}\pi) = \cos k\pi = (-1)^k$, and if $n = 2k + 1$, then $f^{(2k+1)}(\frac{1}{2}\pi) = \cos(k\pi + \frac{1}{2}\pi) = 0$. Using equation (10) of section 5.2 with $a = \frac{1}{2}\pi$, the Taylor series expansion is

$$1 - \frac{(x - \frac{1}{2}\pi)^2}{2!} + \frac{(x - \frac{1}{2}\pi)^4}{4!} - \ldots + \frac{(-1)^{n-1}(x - \frac{1}{2}\pi)^{2n-2}}{(2n - 2)!} + \ldots$$

and $$|R_{2n}(x)| \leqslant \frac{(x - \frac{1}{2}\pi)^{2n}}{(2n)!}$$

which tends to zero as n tends to infinity for all values of x.

5.4 TAYLOR SERIES FOR A FUNCTION OF TWO VARIABLES

Let $\omega = f(x, y)$ be a function of x and y such that ω and its partial derivatives up to a required order are continuous at (x_0, y_0). Consider the value of $f(x_0 + h, y_0 + k)$. Keeping the y coordinate $y_0 + k$ constant and using Taylor's theorem of section 5.2 with $a = x_0$, $b = x_0 + h$ so that $b - a = h$,

$$f(x_0 + h, y_0 + k) = f(x_0, y_0 + k) + hf_x(x_0, y_0 + k)$$
$$+ \frac{h^2}{2!}f_{xx}(x_0, y_0 + k) + \ldots \quad (17)$$

Again, applying Taylor's theorem to the terms on the right of (17),

$$f(x_0, y_0 + k) = f(x_0, y_0) + kf_y(x_0, y_0) + \frac{k^2}{2!}f_{yy}(x_0, y_0) + \ldots$$

$$f_x(x_0, y_0 + k) = f_x(x_0, y_0) + kf_{yx}(x_0, y_0) + \frac{k^2}{2!}f_{yyx}(x_0, y_0) + \ldots$$

$$f_{xx}(x_0, y_0 + k) = f_{xx}(x_0, y_0) + kf_{xyx}(x_0, y_0) + \ldots$$

Substituting these values in equation (17) as far as the second order terms gives

$$f(x_0 + h, y_0 + k) = f(x_0, y_0) + kf_y(x_0, y_0) + \frac{k^2}{2!}f_{yy}(x_0, y_0) + \ldots$$
$$+ hf_x(x_0, y_0) + hkf_{yx}(x_0, y_0) + \ldots$$
$$+ \frac{h^2}{2!}f_{xx}(x_0, y_0) + \ldots \quad (18)$$

This equation expresses the value of the function $f(x, y)$ at $x = x_0 + h$, $y = y_0 + k$ in terms of the values of the function and its partial derivatives up to the second order at (x_0, y_0), together with powers of $h = x - x_0$ and $k = y - y_0$. An expansion using higher derivatives is written more simply if the following 'shorthand' form is used. The expression $hf_x + kf_y$ is written $(h\,\partial/\partial x + k\,\partial/\partial y)f$ and may be thought of as the operator $(h\,\partial/\partial x + k\,\partial/\partial y)$ operating on f. Then

$$\left(h\frac{\partial}{\partial x} + k\frac{\partial}{\partial y}\right)^2 f$$

is a short form for $\left(h\dfrac{\partial}{\partial x} + k\dfrac{\partial}{\partial y}\right)$ operating on $h\dfrac{\partial f}{\partial x} + k\dfrac{\partial f}{\partial y}$.
The result is

$$h\frac{\partial}{\partial x}\left(h\frac{\partial f}{\partial x} + k\frac{\partial f}{\partial y}\right) + k\frac{\partial}{\partial y}\left(h\frac{\partial f}{\partial x} + k\frac{\partial f}{\partial y}\right)$$

or

$$h^2\frac{\partial^2 f}{\partial x^2} + 2hk\frac{\partial^2 f}{\partial x\,\partial y} + k^2\frac{\partial^2 f}{\partial y^2},$$

since the mixed partial derivatives are equal because they are continuous, and h, k are arbitrary constants. Similarly

$$\left(h\frac{\partial}{\partial x} + k\frac{\partial}{\partial y}\right)^3 f = h^3 f_{xxx} + 3h^2 k f_{xxy} + 3hk^2 f_{xyy} + k^3 f_{yyy}.$$

Hence the Taylor series expansion about the point (x_0, y_0) of the function $f(x, y)$ is

$$f(x_0 + h, y_0 + k) = f(x_0, y_0) + \sum_{r=1}^{n-1}\frac{1}{r!}\left(h\frac{\partial}{\partial x} + k\frac{\partial}{\partial y}\right)^r f(x_0, y_0) + R_n \tag{19}$$

remembering that the partial derivatives on the right are all evaluated at the point (x_0, y_0), and where

$$R_n = \frac{1}{n!}\left(h\frac{\partial}{\partial x} + k\frac{\partial}{\partial y}\right)^n f(x_0 + \theta_1 h, y_0 + \theta_2 k)$$

and $0 < \theta_1 < 1$, $0 < \theta_2 < 1$.

If $R_n \to 0$ as $n \to \infty$, the series represents the function $f(x, y)$ near (x_0, y_0).

5.5 APPLICATION TO MAXIMUM AND MINIMUM

Problems of the maximum and minimum of functions of two or more variables occur in many systems of physics, chemistry and engineering. In Chapter 2 conditions for the maximum and minimum of functions of one variable were developed. In this section the maximum and minimum of a function of two variables will be considered. Extension to functions of three or more variables may then be deduced.

5.5.1 Definitions

Consider the function $\omega = f(x, y)$ defined in some region R of the (xy) plane. The function $\omega = f(x, y)$ has its greatest value or *absolute* maximum at a point (x_0, y_0) within or on the boundary of R if the value of the function at this point is greater than the value of the function for *all* other points in the region, i.e. if $f(x_0, y_0) \geqslant f(x, y)$ for all (x, y) in R.

The function $f(x, y)$ is said to have a *local* or *relative* maximum, or simply a maximum at the point P (x_0, y_0) in R, if its value at P is greater than its value at any point in the *immediate neighbourhood* of P. In other words, the value of $\omega = f(x, y)$ at all points (x, y) *near to* (x_0, y_0) and in any direction from (x_0, y_0), is less than the value $\omega_0 = f(x_0, y_0)$. Thus there is a local maximum at (x_0, y_0) if

$$f(x_0 + h, y_0 + k) < f(x_0, y_0)$$

provided h and k are small enough.

An absolute minimum or least value and a local minimum value of f are similarly defined.

If the function $f(x, y)$ has either a local maximum or a local minimum value at (x_0, y_0) then it is said to have an *extreme* value at this point.

5.5.2 Analytic conditions for an extreme value

Suppose $\omega = f(x, y)$ has an extreme value at the point (x_0, y_0) in R. Keeping y constant and equal to $y_0, f(x, y_0)$ is a function of x only which has an extreme value at $x = x_0$ and hence its derivative is zero at $x = x_0$, i.e. $f_x(x_0, y_0) = 0$. Similarly, keeping x constant and equal to x_0, $f(x_0, y)$ is a function of y only which has an extreme value at $y = y_0$ and hence $f_y(x_0, y_0) = 0$.

Any point (x, y) at which $f_x = 0$ and $f_y = 0$ is called a *stationary point* of f, so that wherever f has an extreme value it has a stationary point. The converse is not true, since it is possible to find functions which have stationary points which do not give extreme values. For example, if $f(x, y) = x^2 y^2$, then f, f_x, and f_y are all zero at the point $(0, 0)$ but this point is not an extreme one because f_x and f_y are also zero at *all* points on both the x- and y-axes and f is positive everywhere else.

The condition $f_x = 0$, $f_y = 0$ determines those values of (x, y) at which there is an extreme value of f but this condition is not sufficient. Let A be the point (x_0, y_0) and B be $(x_0 + h, y_0 + k)$, where h and k are arbitrary, and let $f_x(x_0, y_0) = 0 = f_y(x_0, y_0)$. Then from the Taylor expansion near the point A,

$$f(x_0 + h, y_0 + k) - f(x_0, y_0) = \tfrac{1}{2}\{h^2 f_{xx} + 2hk f_{xy} + k^2 f_{yy}\} + R_3(h,k) \tag{20}$$

where the partial derivatives, f_{xx}, f_{yy}, and f_{xy} are evaluated at the point (x_0, y_0) and R_3 is the remainder. This term R_3 tends to zero with h and k and is a lower order of magnitude than any of the other terms of (20). It may be assumed for h and k sufficiently small, that (20) may be replaced by

$$f(x_0 + h, y_0 + k) - f(x_0, y_0) = ph^2 + 2qhk + rk^2$$

where $2p = f_{xx}(x_0, y_0)$; $2q = f_{xy}(x_0, y_0)$, and $2r = f_{yy}(x_0 \, y_0)$.

The quadratic form may be written

$$ph^2 + 2qhk + rk^2 = p\{h^2 + 2qhk/p + rk^2/p\}$$

$$= p\left\{\left(h + k\frac{q}{p}\right)^2 + \frac{k^2(pr - q^2)}{p^2}\right\}.$$

This is said to be positive definite for all values of h and k provided that $pr - q^2 > 0$ and $p > 0$ and is negative definite for all values of h and k provided that $pr - q^2 > 0$ and $p < 0$, and is zero only when $h = k = 0$.

Hence if the quadratic form on the right of equation (20) is positive definite for all values of h and k, $f(x_0 + h, y_0 + k) > f(x_0, y_0)$ and $f(x_0, y_0)$ is a minimum value of f; if the quadratic form is negative definite for all values of h and k, $f(x_0 + h, y_0 + k) < f(x_0, y_0)$ and $f(x_0, y_0)$ is a maximum value of f.

Moreover if h and k are small enough and the second order derivatives of f are continuous the condition

$$f_x = 0 = f_y, \quad f_{xx}f_{yy} - f_{xy}^2 > 0$$

at a point (x_0, y_0) means that f has an extreme value at that point. This is a maximum when $f_{xx} < 0$ and a minimum when $f_{xx} > 0$.

If at (x_0, y_0), $f_x = 0 = f_y$ and $f_{xx}f_{yy} - f_{xy}^2 < 0$, the binary quadratic on the right of equation (20) is said to be indefinite and its sign may be positive or negative according to the signs of h and k. This gives what is called a *saddle point* at (x_0, y_0).

If at (x_0, y_0), $f_x = 0 = f_y$ and $f_{xx}f_{yy} - f_{xy}^2 = 0$, the quadratic form is said to be semi-definite and the nature of the stationary point is undecided. Further terms of the Taylor series must then be considered, but in this book only conditions of second order terms are considered.

Example (i) Investigate the extreme values of the function

$$f(x, y) = x^2 + xy + y^2 + x - 4y + 5.$$

Solution The function $f(x, y)$ has partial derivatives

$$f_x = 2x + y + 1, \quad f_y = x + 2y - 4$$

and these are both zero at $(-2, 3)$.

The second order partial derivatives are all constant, $f_{xx} = 2$, $f_{yy} = 2, f_{xy} = 1$, and the expression which determines the nature of the stationary point at $(-2, 3)$ is

$$f_{xx}f_{yy} - f_{xy}^2 = 4 - 1 = 3 > 0.$$

The function therefore has a minimum at $(-2, 3)$. Moreover this is its least value, since

$$f(-2 + h, 3 + k) \geqslant f(-2, 3) \text{ for all } h, k.$$

Example (ii) Investigate the extreme values of the function

$$f(x, y) = (x^2 + y^2)^2 - 4(x^2 + 2y^2) + 13.$$

Solution The function $f(x, y)$ has partial derivatives

$$f_x = 4x(x^2 + y^2 - 2), \qquad f_y = 4y(x^2 + y^2 - 4),$$
$$f_{xx} = 4(3x^2 + y^2 - 2), \qquad f_{yy} = 4(x^2 + 3y^2 - 4),$$
$$f_{xy} = 8xy.$$

There are five stationary points whose coordinates are found by solving the equations $f_x = 0 = f_y$, and they are $(0, 0)$, $(0, \pm 2)$ and $(\pm\sqrt{2}, 0)$.

At $(0, 0)$, $\qquad f_{xx}f_{yy} - f_{xy}^2 = (-8)(-16) - 0 > 0$

$f_{xx} < 0$, hence f is a maximum at $(0, 0)$.

At $(0, \pm 2)$, $\qquad f_{xx}f_{yy} - f_{xy}^2 = (8)(32) - 0 > 0$

and $f_{xx} > 0$, hence f has minima at $(0, \pm 2)$.

At $(\pm\sqrt{2}, 0)$, $\qquad f_{xx}f_{yy} - f_{xy}^2 = (16)(-8) - 0 < 0$

and f has saddle points at $(\pm\sqrt{2}, 0)$.

5.6 FURTHER PARTIAL DIFFERENTIATION

5.6.1 The increment of a function of two variables

Consider now an approximate value of the increment of a function of two variables. Let $\omega = f(x, y)$ and let its partial derivatives f_x and f_y be continuous functions of x and y throughout a region R in the (xy) plane containing (x_0, y_0). Then the increment

$$\delta\omega = f(x_0 + \delta x, y_0 + \delta y) - f(x_0, y_0) \qquad (21)$$

is given by

$$\delta\omega = f(x_0 + \delta x, y_0 + \delta y) - f(x_0 + \delta x, y_0) + f(x_0 + \delta x, y_0) - f(x_0, y_0)$$
$$= \delta\omega_1 + \delta\omega_2$$

where

$$\delta\omega_1 = f(x_0 + \delta x, y_0 + \delta y) - f(x_0 + \delta x, y_0) \quad (x \text{ held constant})$$

and $\delta\omega_2 = f(x_0 + \delta x, y_0) - f(x_0, y_0) \qquad (y \text{ held constant})$.

Thus, holding y constant at $y = y_0$ and changing x from x_0 to $x_0 + \delta x$ and using the mean value theorem,

$$\frac{\delta\omega_2}{\delta x} = \frac{f(x_0 + \delta x, y_0) - f(x_0, y_0)}{\delta x}$$

is the value of $(\partial\omega_2/\partial x)_{(x_1, y_0)} = f_x(x_1, y_0)$ for some x_1 between x_0 and $x_0 + \delta x$, i.e.

$$\delta\omega_2 = f_x(x_1, y_0)\delta x \quad \text{and} \quad x_0 < x_1 < x_0 + \delta x. \qquad (22)$$

Since f_x is a continuous function of x and y, its value at (x_1, y_0) is nearly the same as its value at (x_0, y_0) when δx is small. Hence the difference denoted by ε_1 and given by

$$f_x(x_1, y_0) - f_x(x_0, y_0) = \varepsilon_1$$

approaches zero as δx approaches zero. Substituting for $f_x(x_1, y_0)$ in equation (22) gives

$$\delta \omega_2 = \{f_x(x_0, y_0) + \varepsilon_1\} \, \delta x, \tag{23}$$

and $\varepsilon_1 \to 0$ when $\delta x \to 0$.

Similarly $\delta \omega_1 = f_y\{x_0 + \delta x, y_0 + \theta \delta y\} \, \delta y = \{f_y(x_0, y_0) + \varepsilon_2\} \, \delta y$ (24)

where $0 < \theta < 1$ and $\varepsilon_2 \to 0$ when $\delta y \to 0$. Substituting from equations (23) and (24) in $\delta \omega = \delta \omega_1 + \delta \omega_2$ gives

$$\delta \omega = f_x(x_0, y_0)\delta x + f_y(x_0, y_0)\delta y + \varepsilon_1 \, \delta x + \varepsilon_2 \, \delta y \tag{25}$$

where ε_1 and ε_2 both approach zero when δx, δy approach zero.

An analagous result holds for a function of any finite number of independent variables. For the case of a function of three variables $\omega = f(x, y, z)$,

$$\delta \omega = f_x \, \delta x + f_y \, \delta y + f_z \, \delta z + \varepsilon_1 \, \delta x + \varepsilon_2 \, \delta y + \varepsilon_3 \, \delta z$$

where $\varepsilon_1, \varepsilon_2, \varepsilon_3 \to 0$ when $\delta x, \delta y, \delta z \to 0$.

5.6.2 Total differentiation

Let $\omega = f(x, y)$ be differentiable at all points along some curve C in a region R in the (xy) plane and let the equation of C in terms of the parameter t be $x = x(t)$, $y = y(t)$ where x and y are differentiable functions of t. Then ω is a differentiable function of t and

$$\frac{d\omega}{dt} = \frac{\partial \omega}{\partial x}\frac{dx}{dt} + \frac{\partial \omega}{\partial y}\frac{dy}{dt} \tag{26}$$

This follows from equation (25) where the parameter t has value t_0 at the point (x_0, y_0) and when $\delta t \to 0$, δx and δy both $\to 0$ and ε_1 and ε_2 both $\to 0$.

The formula (26) may be extended to give the total derivative of a function of n variables $f(x_1, x_2, x_3, \ldots, x_n)$ where $x_1, x_2, x_3, \ldots, x_n$ are differentiable functions of the parameter t. Then

$$\frac{df}{dt} = \frac{\partial f}{\partial x_1}\frac{dx_1}{dt} + \frac{\partial f}{\partial x_2}\frac{dx_2}{dt} + \frac{\partial f}{\partial x_3}\frac{dx_3}{dt} + \ldots + \frac{\partial f}{\partial x_n}\frac{dx_n}{dt}. \tag{27}$$

A particular case arises when $f(x, y)$ is a constant. The equation (26) then gives

$$0 = \frac{\partial f}{\partial x}\frac{dx}{dt} + \frac{\partial f}{\partial y}\frac{dy}{dt}$$

which leads to

$$\frac{dy}{dx} = \frac{dy/dt}{dx/dt} = -\frac{\partial f/\partial x}{\partial f/\partial y}$$

The latter result is the derivative of y with respect to x when x and y are related in the implicit form $f(x, y) =$ constant (see also section 2.2.11).

Example Find dy/dx given that $x^5 + 4x^2y^3 - 3y^5 = 2$. Use this result to find the equation of the tangent to the curve $x^5 + 4x^2y^3 - 3y^5 = 2$ at the point $(1, 1)$.

Solution Here $f(x, y) = x^5 + 4x^2y^3 - 3y^5$, and hence

$$\frac{dy}{dx} = -\frac{5x^4 + 8xy^3}{12x^2y^2 - 15y^4}.$$

The equation of the tangent to the curve at the point $(1, 1)$ is given by

$$(y - 1) = \left(\frac{dy}{dx}\right)_{1,1}(x - 1)$$

or

$$(y - 1) = \left(+\frac{13}{3}\right)(x - 1)$$

which is

$$3y - 13x + 10 = 0.$$

5.6.3 The chain rule

No essential complication is introduced by considering the derivative of the function $\omega = f(x, y)$ where x and y are functions of two parameters r and s, i.e. $x = x(r, s)$ and $y = y(r, s)$. Then when s is held constant and writing

$$\underset{\delta r \to 0}{\text{Lim}} \frac{\delta \omega}{\delta r} = \frac{\partial \omega}{\partial r} \quad \text{etc.}$$

the result is the two formulae

$$\left. \begin{aligned} \frac{\partial \omega}{\partial r} &= \frac{\partial \omega}{\partial x}\frac{\partial x}{\partial r} + \frac{\partial \omega}{\partial y}\frac{\partial y}{\partial r}, \\ \frac{\partial \omega}{\partial s} &= \frac{\partial \omega}{\partial x}\frac{\partial x}{\partial s} + \frac{\partial \omega}{\partial y}\frac{\partial y}{\partial s}. \end{aligned} \right\} \tag{28}$$

More generally suppose $\omega = f(y_1, y_2, y_3, \ldots, y_n)$ and $y_1, y_2, y_3,$

. . ., y_n are each functions of m variables $x_1, x_2, x_3, \ldots, x_m$, or
$y_k = y_k(x_1, x_2, x_3, \ldots, x_m)$, $k = 1, 2, 3, \ldots, n$, then

$$\frac{\partial \omega}{\partial x_r} = \sum_i \frac{\partial \omega}{\partial y_i} \frac{\partial y_i}{\partial x_r}. \tag{29}$$

This rule is the *chain rule* of differentiation and may be compared with the function of a function rule of section 2.2.9 of Chapter 2.

Example Given that $\omega = f(x, y)$ is transformed to $F(r, \theta)$ by substituting $x = r \cos \theta$ and $y = r \sin \theta$, find $\partial f/\partial x$ and $\partial f/\partial y$ in terms of r, θ, $\partial F/\partial r$ and $\partial F/\partial \theta$.

Solution First set of variables: x, y
          Second set of variables: r, θ

(a) $\partial f/\partial x$ and $\partial f/\partial y$ are to be found

(b) $\dfrac{\partial x}{\partial r} = \cos \theta$, $\dfrac{\partial y}{\partial r} = \sin \theta$, $\dfrac{\partial x}{\partial \theta} = -r \sin \theta$, $\dfrac{\partial y}{\partial \theta} = r \cos \theta$

(c) $\dfrac{\partial F}{\partial r} = \dfrac{\partial f}{\partial x} \dfrac{\partial x}{\partial r} + \dfrac{\partial f}{\partial y} \dfrac{\partial y}{\partial r}$

and $\dfrac{\partial F}{\partial \theta} = \dfrac{\partial f}{\partial x} \dfrac{\partial x}{\partial \theta} + \dfrac{\partial f}{\partial y} \dfrac{\partial y}{\partial \theta}$

or $\dfrac{\partial F}{\partial r} = \cos \theta \dfrac{\partial f}{\partial x} + \sin \theta \dfrac{\partial f}{\partial y}$ (i)

$\dfrac{\partial F}{\partial \theta} = -r \sin \theta \dfrac{\partial f}{\partial x} + r \cos \theta \dfrac{\partial f}{\partial y}$ (ii)

The question requires $\partial f/\partial x$ and $\partial f/\partial y$ which can be found in terms of r, θ, $\partial F/\partial r$, and $\partial F/\partial \theta$ by solving (i) and (ii) as simultaneous equations. Multiplying (i) by $r \cos \theta$ and (ii) by $\sin \theta$ and subtracting gives

$$\frac{\partial f}{\partial x} = \cos \theta \frac{\partial F}{\partial r} - \frac{1}{r} \sin \theta \frac{\partial F}{\partial \theta} \tag{30}$$

and multiplying (i) by $r \sin \theta$ and (ii) by $\cos \theta$ and adding gives

$$\frac{\partial f}{\partial y} = \sin \theta \frac{\partial F}{\partial r} + \frac{1}{r} \cos \theta \frac{\partial F}{\partial \theta}. \tag{31}$$

Notice that when $x = r \cos \theta$, $y = r \sin \theta$ are substituted in $f(x, y)$ a different function $F(r, \theta)$ is obtained.

Equations (30) and (31) may be written in operator form,

$$\frac{\partial}{\partial x} = \cos \theta \frac{\partial}{\partial r} - \frac{1}{r} \sin \theta \frac{\partial}{\partial \theta} \tag{32}$$

$$\frac{\partial}{\partial y} = \sin \theta \frac{\partial}{\partial r} + \frac{1}{r} \cos \theta \frac{\partial}{\partial \theta}. \tag{33}$$

Since $r^2 = x^2 + y^2$ and $\tan \theta = y/x$, it is possible to evaluate $\partial f/\partial x$ directly as

$$\frac{\partial f}{\partial x} = \frac{\partial F}{\partial r}\frac{\partial r}{\partial x} + \frac{\partial F}{\partial \theta}\frac{\partial \theta}{\partial x}$$

and this is left as a useful exercise for the reader.

5.6.4 Transformation of second order derivatives

Many applications involving second order partial derivatives are simplified by a change of variable. Suppose $\omega = f(u, v)$ becomes $F(x, y)$ on substitution of $u = u(x, y)$, $v = v(x, y)$ where x, y are independent variables and u, v are differentiable functions of x, y. Suppose also that the second order partial derivatives f_{uv} and f_{vu} are continuous and so equal. By equation (28)

$$\frac{\partial F}{\partial x} = \frac{\partial f}{\partial u}\frac{\partial u}{\partial x} + \frac{\partial f}{\partial v}\frac{\partial v}{\partial x} \tag{34}$$

or the operator

$$\frac{\partial}{\partial x} = \frac{\partial}{\partial u}\frac{\partial u}{\partial x} + \frac{\partial}{\partial v}\frac{\partial v}{\partial x}. \tag{35}$$

Using operator (35) on equation (34)

$$\frac{\partial}{\partial x}\left(\frac{\partial F}{\partial x}\right) = \frac{\partial}{\partial x}\left(\frac{\partial f}{\partial u}\frac{\partial u}{\partial x} + \frac{\partial f}{\partial v}\frac{\partial v}{\partial x}\right)$$

$$= \frac{\partial}{\partial x}(\partial f/\partial u)\frac{\partial u}{\partial x} + \frac{\partial f}{\partial u}\frac{\partial^2 u}{\partial x^2} + \frac{\partial}{\partial x}(\partial f/\partial v)\frac{\partial v}{\partial x} + \frac{\partial f}{\partial v}\frac{\partial^2 v}{\partial x^2}$$

$$= \frac{\partial f}{\partial u}\frac{\partial^2 u}{\partial x^2} + \frac{\partial f}{\partial v}\frac{\partial^2 v}{\partial x^2} + \frac{\partial u}{\partial x}\left\{\frac{\partial}{\partial u}(\partial f/\partial u)\frac{\partial u}{\partial x} + \frac{\partial}{\partial v}(\partial f/\partial u)\frac{\partial v}{\partial x}\right\}$$

$$+ \frac{\partial v}{\partial x}\left\{\frac{\partial}{\partial u}(\partial f/\partial v)\frac{\partial u}{\partial x} + \frac{\partial}{\partial v}(\partial f/\partial v)\frac{\partial v}{\partial x}\right\}$$

$$= \frac{\partial f}{\partial u}\frac{\partial^2 u}{\partial x^2} + \frac{\partial f}{\partial v}\frac{\partial^2 v}{\partial x^2} + \frac{\partial^2 f}{\partial u^2}\left(\frac{\partial u}{\partial x}\right)^2 + 2\frac{\partial^2 f}{\partial u \partial v}\frac{\partial u}{\partial x}\frac{\partial v}{\partial x} + \frac{\partial^2 f}{\partial v^2}\left(\frac{\partial v}{\partial x}\right)^2 \tag{36}$$

Similarly,

$$\frac{\partial^2 F}{\partial y^2} = \frac{\partial f}{\partial u}\frac{\partial^2 u}{\partial y^2} + \frac{\partial f}{\partial v}\frac{\partial^2 v}{\partial y^2} + \frac{\partial^2 f}{\partial u^2}\left(\frac{\partial u}{\partial y}\right)^2 + 2\frac{\partial^2 f}{\partial u \partial v}\frac{\partial u}{\partial y}\frac{\partial v}{\partial y} + \frac{\partial^2 f}{\partial v^2}\left(\frac{\partial v}{\partial y}\right)^2 \tag{37}$$

and

$$\frac{\partial^2 F}{\partial x \partial y} = \frac{\partial f}{\partial u}\frac{\partial^2 u}{\partial x \partial y} + \frac{\partial f}{\partial v}\frac{\partial^2 v}{\partial x \partial y} + \frac{\partial^2 f}{\partial u^2}\frac{\partial u}{\partial y}\frac{\partial u}{\partial x} + \frac{\partial^2 f}{\partial u \partial v}\left(\frac{\partial u}{\partial y}\frac{\partial v}{\partial x} + \frac{\partial u}{\partial x}\frac{\partial v}{\partial y}\right)$$

$$+ \frac{\partial^2 f}{\partial v^2}\frac{\partial v}{\partial y}\frac{\partial v}{\partial x} \tag{38}$$

The formulation of equations (37) and (38) are left as exercises for the reader.

Example Find $\partial^2 f/\partial x^2$ if $\omega = f(x, y) = F(r, \theta)$ where $x = r \cos \theta$ and $y = r \sin \theta$.

Solution Using operator (32) to find $\partial^2 f/\partial x^2$ gives

$$\frac{\partial}{\partial x}\left(\frac{\partial f}{\partial x}\right) = \left(\cos \theta \, \frac{\partial}{\partial r} - \frac{1}{r} \sin \theta \, \frac{\partial}{\partial \theta}\right)\left(\cos \theta \, \frac{\partial F}{\partial r} - \frac{1}{r} \sin \theta \, \frac{\partial F}{\partial \theta}\right)$$

$$= \cos \theta \, \frac{\partial}{\partial r}\left(\cos \theta \, \frac{\partial F}{\partial r} - \frac{1}{r} \sin \theta \, \frac{\partial F}{\partial \theta}\right)$$

$$- \frac{1}{r} \sin \theta \, \frac{\partial}{\partial \theta}\left(\cos \theta \, \frac{\partial F}{\partial r} - \frac{1}{r} \sin \theta \, \frac{\partial F}{\partial \theta}\right)$$

$$= \cos \theta \left(\cos \theta \, \frac{\partial^2 F}{\partial r^2} + \frac{1}{r^2} \sin \theta \, \frac{\partial F}{\partial \theta} - \frac{1}{r} \sin \theta \, \frac{\partial^2 F}{\partial r \partial \theta}\right)$$

$$- \frac{1}{r} \sin \theta \left(-\sin \theta \, \frac{\partial F}{\partial r} + \cos \theta \, \frac{\partial^2 F}{\partial \theta \partial r} - \frac{1}{r} \cos \theta \, \frac{\partial F}{\partial \theta}\right.$$

$$\left. - \frac{1}{r} \sin \theta \, \frac{\partial^2 F}{\partial \theta^2}\right)$$

$$= \cos^2 \theta \, \frac{\partial^2 F}{\partial r^2} - \frac{2}{r} \cos \theta \sin \theta \, \frac{\partial^2 F}{\partial r \partial \theta} + \frac{1}{r^2} \sin^2 \theta \, \frac{\partial^2 F}{\partial \theta^2}$$

$$+ \frac{2}{r^2} \cos \theta \sin \theta \, \frac{\partial F}{\partial \theta} + \frac{1}{r} \sin^2 \theta \, \frac{\partial F}{\partial r}.$$

Similarly f_{yy} and f_{xy} may be evaluated.

EXERCISE 5

1 By using the mean value theorem or otherwise, prove that for every positive number β
$$\beta e^{-\beta} < 1 - e^{-\beta} < \beta.$$

2 If in the observed values of the variables x, y, z there are percentage errors of $+1$, -2, $+1\cdot5$ respectively, find approximately the percentage error in the value of ω calculated from the formula
$$\omega = 27x^5 y^{3/4}/z^4.$$

3 The area Δ of a triangle of sides of lengths a, b, c is given by the formula $\Delta = \sqrt{\{s(s - a)(s - b)(s - c)\}}$ where $s = \frac{1}{2}(a + b + c)$. A plane triangular lamina of sides $a = 5$, $b = 6$, $c = 7$ units is to be cut from a plane piece of metal. If the side b is cut exactly, but small percentage errors α, γ are made in the sides a, c respectively, show that the percentage error in the area is approximately $(125\alpha + 49\gamma)/144$.

4 The volume of a cone is obtained by measuring its height h and its semi-vertical angle α and then using the formula $V = \frac{1}{3}\pi h^3 \tan^2 \alpha$.

If the measurements of h and α are subject to small percentage errors p_h and p_α respectively, find an approximate expression for the resulting percentage error in V if α is approximately $\frac{1}{4}\pi$.

5 Evaluate the following limits:

(i) $\underset{x \to 1}{\text{Lim}} \dfrac{x - 1}{x^2 - 1}$;   (ii) $\underset{x \to 0}{\text{Lim}} \dfrac{\sin x - \tan x}{x^3}$;   (iii) $\underset{x \to 0}{\text{Lim}} \dfrac{\log \sin x}{\log x}$;

(iv) $\underset{x \to 0}{\text{Lim}} \dfrac{x \sin^{-1} x}{\log \cos x}$;   (v) $\underset{x \to \infty}{\text{Lim}} \dfrac{x^2 + x}{2x^2 + 1}$;   (vi) $\underset{x \to 0}{\text{Lim}} \dfrac{\tan x - x}{x - \sin x}$

6 Evaluate the following limits:

(i) $\underset{x \to 1}{\text{Lim}} \dfrac{\log x - (x - 1) + \frac{1}{2}(x - 1)^2}{(x - 1)^3}$;   (ii) $\underset{x \to 0}{\text{Lim}} \dfrac{x - \sinh x}{1 + x^2 - \cosh x}$

7 Evaluate the following limits:

(i) $\underset{x \to 0}{\text{Lim}} \dfrac{\sinh x - x}{x^3}$;   (ii) $\underset{x \to \frac{1}{2}\pi}{\text{Lim}} \dfrac{\sec^2 x}{\tan^2 x}$

(iii) $\underset{x \to 0}{\text{Lim}} \dfrac{\sin (x^2)}{1 - \cos x}$;   (iv) $\underset{x \to \pi}{\text{Lim}} \dfrac{\log(x/\pi)}{\sin x}$;

(v) $\underset{x \to \infty}{\text{Lim}} x\{\sqrt{(x^2 + 1)} - x\}$.

8 Use l'Hôpital's rule to prove

$$\underset{x \to \infty}{\text{Lim}} \frac{x^n}{e^x} = 0 \quad \text{and} \quad \underset{x \to \infty}{\text{Lim}} \frac{\log x}{x^n} = 0$$

(see section 4.3).

9 The function $f(x)$ is defined by $f(x) = \dfrac{x}{e^x - 1}$ when $x \neq 0, f(0) = 1$.

Use the definition of the derivative of a function as a limit to evaluate $f'(0)$.

10 Derive a power series expansion for $\log (1 + x)$ stating its range of validity. Show that

$$\tanh^{-1} x = \tfrac{1}{2} \log [(1 + x)/(1 - x)], \quad |x| < 1.$$

Hence find the first three non-zero terms in the expansion of (i) $\tanh^{-1} x$, (ii) $\tanh^{-1} (\sin x)$.

11 Write down the series expansion of $\log (1 + x)$ and of $\cos x$ in ascending powers of x and show that, when x is small,

$$\log \cos x = -\tfrac{1}{2}x^2 - x^4/12 \ldots ..$$

12 Write down the Taylor expansion about $x = 0$ (Maclaurin expansion) for the following functions, stating the general term and the range of values of x for which your series is valid:

(i) $\sin x$;   (ii) $\sin (\theta + x)$;   (iii) e^{cx};
(iv) $(1 - x)^{-1}$;   (v) $(1 - x)^\alpha$ when α is not a positive integer.

13 Write down the Maclaurin series for the function $\log(1 + \frac{1}{2}x^2)$ stating the general term. For what values of x is the series valid?

14 The Taylor series for a function of two variables about the point (a, b) may be expressed in the form

$$f(a + h, b + k) = f(a, b) + \sum_{n=1}^{\infty} \frac{1}{n!}\left\{h\frac{\partial}{\partial x} + k\frac{\partial}{\partial y}\right\}^n f(a,b).$$

Write $a + h = x$ and $b + k = y$ and obtain the Taylor series about the point $(0, 0)$ for the function $e^x \cos y$ as far as the terms of the second degree in x and y.

15 Test the following functions for maxima, minima and saddle points:

(i) $x^2 + y^2 - 2x + 4y + 8$; (ii) $x^2 - y^2 - 2x + 4y + 8$;
(iii) $5 + 2x + 2y - 2x^2 - 2xy - y^2$.

16 Show that the function $f(x, y) = x^2 + xy + y^2 + x - 4y + 9$ has an absolute minimum at the point $(-2, 3)$.

17 Show that the function $2x^4 - 3x^2y^2 + y^4 + 8x^2 + 3y^2$ has stationary values at the points $(0, 0)$, $(5, \pm6)$, $(-5, \pm6)$. Investigate the second order properties of the function at each of these points.

18 Determine the maxima and minima of the function

$$f(x, y) = (ax^2 + by^2) \exp[-(x^2 + y^2)]$$

where a, b are constants and $0 < a < b$.

19 Determine the position and nature of maxima, minima, and saddle points on the surface

$$z = x^2 + y^2 - \tfrac{1}{2}x^4.$$

20 Show that the surface

$$z = x^2 - xy + y^2 + 2x + 2y - 4$$

has a minimum at the point $(-2, -2, -8)$ and no other maximum or minimum.

21 Determine the minimum value of the function

$$3(x + y)^3 + 2(x - y)^2 - 8x.$$

22 Show that the function $f = (x^2 + y^2)^2 + 2a^2(x^2 - y^2)$ where a is a constant, has one saddle point and two minima.

23 Show that the function $f(x, y) = x^3 + 2x^2y - y^2$ has a saddle point at $(-\tfrac{3}{4}, \tfrac{9}{16})$.

24 Show that the function

$$f(x, y) = (x + y)^3 + (x - y)^2 - 12(x + y)$$

has a minimum at the point $(1, 1, -16)$.

25 Prove that, if $x = r \cos \theta$, $y = r \sin \theta$ and $F(r, \theta) \equiv f(x, y)$, then

$$F_{rr} + \frac{1}{r} F_r + \frac{1}{r^2} F_{\theta\theta} = f_{xx} + f_{yy}.$$

Hence, or otherwise, show that $f(x, y) = \tan^{-1}(y/x) + \log(x^2 + y^2)$ satisfies the equation $f_{xx} + f_{yy} = 0$.

26 Change the variables from x, y to u, v where $u = x + y$ and $v = x/(x + y)$ and show that $x \log(x + y) - x + \tan^{-1}\{x/(x + y)\}$ satisfies the equation

$$x f_x + y f_y = x \log(x + y).$$

27 By changing the independent variables from x, y, z to u, v, w where $u = x$, $v = x + y/x$, $w = x + yz$ show that the equation

$$xy f_x + y(y - x^2) f_y = (x + yz - x^2 z) f_z$$

transforms to $xy F_u = 0$ where $f(x, y, z) \equiv F(u, v, w)$.

28 If z is a function of u and v where $u = x^2 - y^2$, $v = 2xy$, show that (i) $x z_x + y z_y = 2(u z_u + v z_v)$, (ii) $y z_x - x z_y = 2(v z_u - u z_v)$. Show also that if

$$z_{uu} + z_{vv} = 0,$$

then

$$z_{xx} + z_{yy} = 0.$$

29 Find dy/dx at the point $(-1, 1)$ given that $x^3 - 2y^3 + 3xy^2 + 6 = 0$.

30 Show that the tangent to the curve $2x + x \log 2x + y e^y = 1$ at the point $(\frac{1}{2}, 0)$ has the equation $2y + 6x = 3$.

6
Integration

6.1 THE INDEFINITE INTEGRAL

The problem is to find a function $y = F(x)$ such that its derivative in some given range, say $a < x < b$ is a given function $f(x)$. In symbols this may be written

given
$$\frac{dy}{dx} = f(x), \tag{1}$$

find
$$y = F(x). \tag{2}$$

For example, if $f(x) = 2x$ then one solution is $y = x^2$. On the other hand $y = x^2 + 2$, $y = x^2 - 3\pi$ are also valid solutions, and so also is $y = x^2 + C$ where C is any constant.

An equation such as (1) which specifies the derivative as a function of x (or as a function of x and y) is called a *differential equation*. Differential equations of more general types will be considered in Chapter 11.

A function $y = F(x)$ is called a *solution* of the differential equation (1) if, over the range $a < x < b$, $F(x)$ is differentiable and

$$F'(x) = f(x). \tag{3}$$

In these circumstances $F(x)$ is called an *integral* or a *primitive* of $f(x)$ with respect to x.

If $F(x)$ is an integral of $f(x)$ then $F(x) + C$ is also an integral where C is any constant. For if equation (3) is satisfied, $[F(x) + C]' = F'(x) = f(x)$. Moreover all integrals of $f(x)$ may be expressed in the form $F(x) + C$. For let $F_1(x)$ and $F_2(x)$ be two functions which satisfy equation (1), and using deduction (iv) of section 5.1.1, then $F_1(x)$ and $F_2(x)$ differ by a constant. The precise value of C cannot be determined without further data and hence C is called an *arbitrary constant*. This is indicated by writing

$$\int f(x)\, dx = F(x) + C \tag{4}$$

where $F(x)$ is any primitive of $f(x)$ and C is an arbitrary constant. The symbol $\int$ is called an integral sign and $\int \ldots dx$ is interpreted as the inverse of the symbol d/dx.

6.2 STANDARD FORMS

Integration as defined above requires the ability to guess the answer. The first requirement for skill in integration is a thorough knowledge of differentiation. The following list contains formulae for derivatives together with the corresponding indefinite integral. Arbitrary constants of integration are omitted, a and n are constants and only the simplest form of integral is given.

$$\frac{d}{dx} x^n = nx^{n-1} \qquad \int x^n \, dx = \frac{x^{n+1}}{n+1}, \quad n \neq -1 \tag{5}$$

$$\frac{d}{dx} \log |x| = \frac{1}{x} \qquad \int \frac{dx}{x} = \log |x| \tag{6}$$

$$\frac{d}{dx} e^x = e^x \qquad \int e^x \, dx = e^x \tag{7}$$

$$\frac{d}{dx} a^x = a^x \log a \qquad \int a^x \, dx = \frac{a^x}{\log a}, \quad a > 0, a \neq 1 \tag{8}$$

$$\frac{d}{dx} \sin x = \cos x \qquad \int \cos x \, dx = \sin x \tag{9}$$

$$\frac{d}{dx} \cos x = -\sin x \qquad \int \sin x \, dx = -\cos x \tag{10}$$

$$\frac{d}{dx} \tan x = \sec^2 x \qquad \int \sec^2 x \, dx = \tan x \tag{11}$$

$$\frac{d}{dx} \cot x = -\operatorname{cosec}^2 x \qquad \int \operatorname{cosec}^2 x \, dx = -\cot x \tag{12}$$

$$\frac{d}{dx} \sin^{-1} x = \frac{1}{\sqrt{(1-x^2)}} \qquad \int \frac{dx}{\sqrt{(1-x^2)}} = \sin^{-1} x, \quad |x| < 1 \tag{13}$$

$$\frac{d}{dx} \tan^{-1} x = \frac{1}{1+x^2} \qquad \int \frac{dx}{1+x^2} = \tan^{-1} x \tag{14}$$

$$\frac{d}{dx} \cosh x = \sinh x \qquad \int \sinh x \, dx = \cosh x \tag{15}$$

$$\frac{d}{dx} \sinh x = \cosh x \qquad \int \cosh x \, dx = \sinh x \tag{16}$$

$$\frac{d}{dx} \tanh x = \operatorname{sech}^2 x \qquad \int \operatorname{sech}^2 x \, dx = \tanh x \tag{17}$$

$$\frac{d}{dx} \coth x = -\operatorname{cosech}^2 x \qquad \int \operatorname{cosech}^2 x \, dx = -\coth x \tag{18}$$

$$\frac{d}{dx} \sinh^{-1} x = \frac{1}{\sqrt{(1+x^2)}} \qquad \int \frac{dx}{\sqrt{(1+x^2)}} = \sinh^{-1} x$$
$$= \log |x + \sqrt{(1+x^2)}| \tag{19}$$

$$\frac{d}{dx} \cosh^{-1} x = \frac{1}{\sqrt{(x^2-1)}} \qquad \int \frac{dx}{\sqrt{(x^2-1)}} = \cosh^{-1} x$$
$$= \log |x + \sqrt{(x^2-1)}|, \, x \geq 1 \tag{20}$$

$$\frac{d}{dx} \tanh^{-1} x = \frac{1}{1 - x^2} \qquad \int \frac{dx}{1 - x^2} = \tanh^{-1} x$$

$$= \tfrac{1}{2} \log \left| \frac{1 + x}{1 - x} \right|, \quad x^2 < 1 \quad (21)$$

$$\frac{d}{dx} \coth^{-1} x = \frac{-1}{x^2 - 1} \qquad \int \frac{dx}{x^2 - 1} = -\coth^{-1} x$$

$$= \tfrac{1}{2} \log \left| \frac{x - 1}{x + 1} \right|, \quad x^2 > 1 \quad (22)$$

6.2.1 Rules

The following rules help to reduce the amount of guesswork in integration in many cases and lead to further standard forms. Arbitrary constants of integration are omitted.

(a) It follows from section 2.2.5 that

$$\int \{af(x) + bg(x)\} \, dx = a \int f(x) \, dx + b \int g(x) \, dx \qquad (23)$$

where a and b are given constants. This rule may be extended to the integral of the sum of any finite number of functions.

(b) From equation (9) of section 2.2.9,

$$\int f(ax + b) \, dx = \frac{1}{a} F(ax + b), \quad a \neq 0 \qquad (24)$$

where a and b are known constants and $F'(x) = f(x)$.

Example (i) $\int \sin (3x + 2) \, dx = -\tfrac{1}{3} \cos (3x + 2).$

Example (ii) $\int (5x + 3)^{-\frac{1}{2}} \, dx = \tfrac{2}{5} \sqrt{(5x + 3)}.$

These results follow from equation (24) directly and compare with examples (i) and (ii) of section 2.2.9.

The following are some standard forms which follow from equation (24), a, b and n are constants and $a \neq 0$.

$$\int (ax + b)^n \, dx = \frac{1}{n + 1} \frac{1}{a} (ax + b)^{n+1}, \quad n \neq -1 \qquad (25)$$

$$\int \frac{dx}{ax + b} = \frac{1}{a} \log |ax + b|, \quad ax + b \neq 0 \qquad (26)$$

$$\int e^{ax} \, dx = \frac{1}{a} e^{ax} \qquad (27)$$

$$\int \cos (ax + b) \, dx = \frac{1}{a} \sin (ax + b) \qquad (28)$$

$$\int \frac{dx}{\sqrt{(a^2 - x^2)}} = \sin^{-1}\left(\frac{x}{a}\right), \quad \left|\frac{x}{a}\right| < 1 \tag{29}$$

$$\int \frac{dx}{a^2 + x^2} = \frac{1}{a} \tan^{-1}\left(\frac{x}{a}\right) \tag{30}$$

$$\int \frac{dx}{\sqrt{(a^2 + x^2)}} = \sinh^{-1}\left(\frac{x}{a}\right) = \log |x + \sqrt{(a^2 + x^2)}| \tag{31}$$

$$\int \frac{dx}{\sqrt{(x^2 - a^2)}} = \cosh^{-1}\left(\frac{x}{a}\right) = \log |x + \sqrt{(x^2 - a^2)}|,$$
$$\left|\frac{x}{a}\right| > 1. \tag{32}$$

(c) Since

$$\frac{d}{dx}\{f(x)\}^{n+1} = (n + 1)\{f(x)\}^n f'(x),$$

using the rule for differentiating the function of a function, then, when $n \neq -1$,

$$\int \{f(x)\}^n f'(x)\, dx = \frac{1}{n + 1}\{f(x)\}^{n+1}, \tag{33}$$

and when $n = -1$,

$$\int \frac{f'(x)}{f(x)}\, dx = \log |f(x)| \tag{34}$$

provided $f(x) \neq 0$.

Example (i) $\displaystyle \int \frac{\cos 2x}{\sin^5 2x}\, dx = \frac{1}{2}\int (\sin 2x)^{-5}\,(2\cos 2x)\, dx$

$$= \frac{1}{2}\int (\sin 2x)^{-5}\,(\sin 2x)'\, dx$$

$$= \frac{1}{2}(-\tfrac{1}{4})(\sin 2x)^{-4}.$$

Example (ii) $\displaystyle \int \tan dx = \int \frac{\sin x}{\cos x}\, dx$

$$= -\int \frac{(\cos x)'}{\cos x}\, dx$$

$$= -\log |\cos x|$$

i.e. $\displaystyle \int \tan x\, dx = \log |\sec x|. \tag{35}$

Example (iii) $\displaystyle \int \csc x\, dx = \int \frac{dx}{\sin x}$

$$= \int \frac{dx}{2 \sin \tfrac{1}{2}x \cos \tfrac{1}{2}x}.$$

Multiplying numerator and denominator of the integrand by $\frac{1}{2} \sec^2 \frac{1}{2}x$ makes this integrable in the form of equation (34), so that

$$\int \csc x \, dx = \int \frac{\frac{1}{2} \sec^2 \frac{1}{2}x}{\tan \frac{1}{2}x} \, dx$$

$$= \log |\tan \tfrac{1}{2}x|. \tag{36}$$

Example (iv) $\int \sec x \, dx$ may be evaluated in one of two ways.

First, using equation (36) and the identity $\cos x = \sin (x + \frac{1}{2}\pi)$,

$$\int \sec x \, dx = \int \csc (x + \tfrac{1}{2}\pi) \, dx = \log |\tan (\tfrac{1}{2}x + \tfrac{1}{4}\pi)| \tag{37}$$

Second, writing

$$\sec x = \frac{\sec x \, (\tan x + \sec x)}{(\tan x + \sec x)} = \frac{(\tan x + \sec x)'}{(\tan x + \sec x)}$$

and using equation (34),

$$\int \sec x \, dx = \log |\tan x + \sec x| \tag{38}$$

Example (v) $\int \cos^3 x \, dx = \int \cos^2 x \cos x \, dx = \int (1 - \sin^2 x) \cos x \, dx$

$$= \int \cos x \, dx - \int (\sin x)^2 (\sin x)' \, dx$$

$$= \sin x - \tfrac{1}{3} \sin^3 x.$$

This method applies whenever an odd power of $\sin x$ or $\cos x$ is to be integrated.

Example (vi) $\int \cos^m x \sin^n x \, dx, \quad m,n$ integers.

If one of m or n is an odd integer the method of example (v) may be used to evaluate the integral.

If both m and n are even integers or if one is an even integer and the other is zero, it is necessary to use the identities $\cos^2 A = \frac{1}{2}(1 + \cos 2A)$ and $\sin^2 A = \frac{1}{2}(1 - \cos 2A)$.

For example, if $m = 0, n = 2$,

$$\int \sin^2 x \, dx = \int \tfrac{1}{2}(1 - \cos 2x) \, dx = \tfrac{1}{2}x - \tfrac{1}{4} \sin 2x,$$

and if $m = 2, n = 4$,

$$\cos^2 x \sin^4 x = (2 - \cos 2x - 2 \cos 4x + \cos 6x)/32$$

and the integral is $\frac{1}{16}x - \frac{1}{64} \sin 2x - \frac{1}{64} \sin 4x + \frac{1}{192} \sin 6x.$

Example (vii) To integrate

$$\sin mx \cos nx, \sin mx \sin nx, \text{ and } \cos mx \cos nx,$$

where m and n are integers it is necessary to use the trigonometrical formulae

$$\sin A \cos B = \tfrac{1}{2}[\sin (A - B) + \sin (A + B)],$$
$$\sin A \sin B = \tfrac{1}{2}[\cos (A - B) - \cos (A + B)],$$
$$\cos A \cos B = \tfrac{1}{2}[\cos (A - B) + \cos (A + B)].$$

Thus for example, if $m \neq n$,

$$\int \cos mx \cos nx \, dx = \int \tfrac{1}{2}[\cos (m - n)x + \cos (m + n)x] \, dx$$
$$= \tfrac{1}{2}\left[\frac{\sin (m - n)x}{m - n} + \frac{\sin (m + n)x}{m + n}\right],$$

and if $m = n$,

$$\int \cos^2 nx \, dx = \int \tfrac{1}{2}(1 + \cos 2nx) \, dx = \tfrac{1}{2}[x + (\sin 2nx)/(2n)].$$

6.3 THE DEFINITE INTEGRAL

Let $f(x)$ be defined in the interval $a < x < b$ and let $F(x)$ be continuous in $a \leqslant x \leqslant b$ and such that $F'(x) = f(x)$ in $a < x < b$. Then from the mean value theorem (1) of section 5.1.1,

$$F(b) - F(a) = (b - a)F'(x) = (b - a)f(x) \quad \text{and } a < x < b.$$

This is written

$$\int_a^b f(x) \, dx = F(b) - F(a) \qquad (39)$$

and is defined to be the *definite* integral of $f(x)$ in the interval from $x = a$ to $x = b$.

6.3.1 Geometrical interpretation of the definite integral

Suppose the function $f(x)$ is positive and continuous in $a \leqslant x \leqslant b$. Using an intuitive idea of area, consider the area enclosed by the curve $y = f(x)$, the x-axis and the ordinates at a and x, where x is some point between a and b. This area is a function of x and may be written $A(x)$. The problem is to show that $A'(x) = f(x)$, $a \leqslant x \leqslant b$.

Using Fig. 6.1, let Q be the point between a and b on the x-axis whose x-coordinate is $x + h$, then the ordinate QR is $f(x + h)$. The area between the curve, the x-axis, and the ordinates at a and $x + h$ is then $A(x + h)$ and the area, shaded in Fig. 6.1 is $A(x + h) - A(x)$.

Let M be the greatest and m the least value of $f(x)$ in the range from x to $x + h$. Then, using the lettering of Fig. 6.1, area of rectangle $PQRS \leqslant A(x + h) - A(x) \leqslant$ area of rectangle $PQTU$

i.e. $\qquad mh \leqslant A(x + h) - A(x) \leqslant Mh$

or $\qquad m \leqslant \dfrac{A(x + h) - A(x)}{h} \leqslant M.$

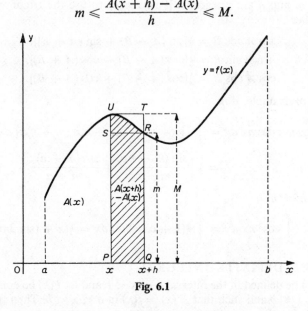

Fig. 6.1

Now, as $h \to 0$, x being held fixed, both m and $M \to f(x)$, since f is continuous, and hence

$$\underset{h \to 0}{\text{Lim}} \frac{A(x + h) - A(x)}{h} = f(x),$$

or $\qquad A'(x) = f(x). \qquad (40)$

Let $F(x)$ be any indefinite integral of $f(x)$ so that $F'(x) = f(x)$. Since $F(x)$ and $A(x)$ have the same derivative they differ by a constant, i.e. $A(x) - F(x) = k$, where k is some constant. When $x = a$, $A(a) = 0$, hence $k = -F(a)$ and

$$A(x) = F(x) - F(a),$$

or $\qquad A(x) = \int_a^x f(x)\, dx = F(x) - F(a), \qquad (41)$

using the notation of (39).

It is usual to denote $F(x) - F(a)$ as $[F(x)]_a^x$ where the square bracket with its limits of integration a and x, is an instruction to substitute and subtract.

The total area enclosed by the curve $y = f(x)$, the x-axis and the ordinates at $x = a$ and $x = b$ is hence $F(b) - F(a)$ or $\int_a^b f(x)\, dx$ and is usually called 'the area under the curve' from $x = a$ to $x = b$. Moreover it follows that

$$\int_a^b f(x)\,dx = \int_a^c f(x)\,dx + \int_c^b f(x)\,dx \quad \text{where } a < c < b. \quad (42)$$

6.3.2 The 'sign' of an area

When $f(x)$ is negative and $b > a > 0$, since

$$\int_a^b -f(x)\,dx = F(a) - F(b) = -\int_a^b f(x)\,dx,$$

then the integral is negative. This is explained geometrically by attaching a negative sign to any area below the x-axis.

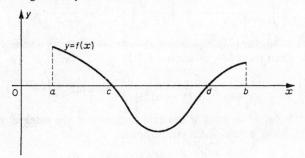

Fig. 6.2

More generally if $y = f(x)$ takes positive and negative values in the interval $a \leqslant x \leqslant b$ as does the function represented by the curve in Fig. 6.2 where the function is negative for $c < x < d$ and positive elsewhere in the interval, from (42),

$$\int_a^b f(x)\,dx = \int_a^c f(x)\,dx + \int_c^d f(x)\,dx + \int_d^b f(x)\,dx$$
$$= \text{the algebraic sum of the separate integrals.}$$

Hence the area enclosed between the curve and the x-axis and the ordinates at $x = a$, $x = b$ is the sum of the moduli of the separate integrals, i.e.

$$\text{area} = \left| \int_a^c f(x)\,dx \right| + \left| \int_c^d f(x)\,dx \right| + \left| \int_d^b f(x)\,dx \right|.$$

Example $\displaystyle\int_0^\pi \cos x \, dx = \Big[\sin x \Big]_0^\pi = \sin \pi - \sin 0 = 0,$

but the area enclosed between the curve $y = \cos x$, the x-axis, and the ordinates at $x = 0$ and $x = \pi$ is

$$\int_0^{\frac{1}{2}\pi} \cos x \, dx - \int_{\frac{1}{2}\pi}^\pi \cos x \, dx = 2 \int_0^{\frac{1}{2}\pi} \cos x \, dx = 2.$$

6.4 INTEGRATION OF THE RECIPROCAL OF A QUADRATIC

6.4.1 The integrand of the form $1/(x^2 + px + q)$

Consider the integral

$$\int \frac{dx}{x^2 + px + q}, \quad p, q \text{ given constants.} \tag{43}$$

Since $x^2 + px + q = (x + \frac{1}{2}p)^2 + q - \frac{1}{4}p^2$, the form of the integral depends on the value of $q - \frac{1}{4}p^2$ and there are three cases to consider:

(a) If $p^2 = 4q$, $x^2 + px + q = (x + \frac{1}{2}p)^2$ and

$$\int \frac{dx}{x^2 + px + q} = \int \frac{dx}{(x + \frac{1}{2}p)^2} = -\frac{1}{(x + \frac{1}{2}p)};$$

(b) if $p^2 < 4q$, $(x + \frac{1}{2}p)^2 + q - \frac{1}{4}p^2 = (x - \alpha)^2 + \beta^2$ where α and β are real. Then from (30) of section 6.2.1

$$\int \frac{dx}{x^2 + px + q} = \int \frac{dx}{(x - \alpha)^2 + \beta^2} = \frac{1}{\beta} \tan^{-1} \left(\frac{x - \alpha}{\beta} \right); \tag{44}$$

(c) if $p^2 > 4q$, $x^2 + px + q$ has real factors and the integral may be evaluated using *partial fractions*. Suppose

$$x^2 + px + q = (x - \alpha)(x - \beta)$$

where α and β are real, then

$$\frac{1}{(x - \alpha)(x - \beta)} = \frac{1/(\alpha - \beta)}{x - \alpha} - \frac{1/(\alpha - \beta)}{x - \beta},$$

and
$$\int \frac{dx}{x^2 + px + q} = \int \frac{dx}{(x - \alpha)(x - \beta)} = \frac{1}{\alpha - \beta} \log \left| \frac{x - \alpha}{x - \beta} \right|. \tag{45}$$

6.4.2 The integrand of the form $(ax + b)/(x^2 + px + q)$

Consider the integral

$$\int \frac{(ax + b)\, dx}{x^2 + px + q}, \quad a, b, p, q \text{ given real constants.}$$

(a) It is possible to choose constants λ, μ such that

$$ax + b = \lambda(2x + p) + \mu,$$

i.e. λ (derivative of the denominator) $+ \mu$. Hence $\lambda = \frac{1}{2}a, \mu = b - \frac{1}{2}ap$, then

$$\int \frac{(ax + b)\, dx}{x^2 + px + q} = \lambda \int \frac{(2x + p)\, dx}{x^2 + px + q} + \mu \int \frac{dx}{x^2 + px + q}. \tag{46}$$

The first integral is $\lambda \log |x^2 + px + q|$ and the second has already been dealt with in section 6.4.1.

(b) When the denominator $x^2 + px + q$ can be resolved into real, distinct factors it is easier to write $(ax + b)/(x^2 + px + q)$ in its partial fractions immediately. Suppose $x^2 + px + q = (x - \alpha)(x - \beta)$, α, β real, then write

$$\frac{ax + b}{(x - \alpha)(x - \beta)} \equiv \frac{A}{x - \alpha} + \frac{B}{x - \beta} \tag{47}$$

where A, B are constants to be found. The constants may be found by one of three methods, (i) the 'cover-up' method; (ii) the 'identity' method; (iii) the 'comparison' method, which are explained as follows:

(i) The 'cover-up' method

Multiply both sides of (47) by $x - \alpha$ to give

$$\frac{ax + b}{x - \beta} \equiv A + \frac{B(x - \alpha)}{x - \beta},$$

and set $x = \alpha$ so that $A = (a\alpha + b)/(\alpha - \beta)$. This is equivalent to 'covering-up' the factor $x - \alpha$ on the left of (47) and writing $x = \alpha$ in the remaining fraction to give the value of A. Similarly B is found to be $(a\beta + b)/(\beta - \alpha)$ when the factor $x - \beta$ is covered up on the left of (47) and β substituted for x in the remaining fraction.

This method is very quick and always works when the factors in the denominator are simple distinct linear factors.

(ii) The 'identity' method

Consider the identity, supposed true for all values of x,

$$ax + b \equiv A(x - \beta) + B(x - \alpha) \tag{48}$$

obtained from (47) by multiplying by $(x - \alpha)(x - \beta)$. In (48) set $x = \alpha$ and $A = (a\alpha + b)/(\alpha - \beta)$, set $x = \beta$ and $B = (a\beta + b)/(\beta - \alpha)$.

(iii) The 'comparison' method

Write (48) in the form

$$ax + b \equiv (A + B)x + (-A\beta - B\alpha),$$

and equate coefficients of x and constant terms to give two simultaneous equations, $a = A + B$ and $b = -A\beta - B\alpha$ from which A and B are found.

6.4.3 The integrand of the form $f(x)/(x^2 + px + q)$

When the function $f(x)$ is a polynomial of degree two or more it is necessary to divide by the denominator until the integral is of the form $(ax + b)/(x^2 + px + q)$. The integral is then evaluated using one of the methods outlined above.

Example (i) $\displaystyle\int \frac{dx}{2 - x - x^2} = \int \frac{dx}{(1 - x)(2 + x)}$

$$= \int \left\{ \frac{\frac{1}{3}}{1 - x} + \frac{\frac{1}{3}}{2 + x} \right\} dx$$

$$= \tfrac{1}{3} \log \left| \frac{x + 2}{1 - x} \right|.$$

Example (ii) $\displaystyle\int \frac{dx}{4x^2 - 4x + 1} = \int \frac{dx}{(2x - 1)^2} = \frac{-1}{2(2x - 1)}.$

Example (iii) $\displaystyle\int \frac{dx}{1 - x + x^2} = \int \frac{dx}{(x - \frac{1}{2})^2 + \frac{3}{4}}$

$$= \frac{2}{\sqrt{3}} \tan^{-1} \left(\frac{2x - 1}{\sqrt{3}} \right).$$

Example (iv)

$$\int \frac{(1 - x)\, dx}{1 - x + x^2} = \int \frac{\frac{1}{2} - \frac{1}{2}(2x - 1)}{1 - x + x^2}\, dx$$

$$= \frac{1}{\sqrt{3}} \tan^{-1} \left(\frac{2x - 1}{\sqrt{3}} \right) - \tfrac{1}{2} \log (1 - x + x^2).$$

Example (v) $\displaystyle\int \frac{(2x - 3)\, dx}{(x - 2)(x + 1)} = \int \left\{ \frac{\frac{1}{3}}{x - 2} + \frac{5/3}{x + 1} \right\} dx$

$$= \tfrac{1}{3} \log |x - 2| + \tfrac{5}{3} \log |x + 1|.$$

Example (vi)

$$\int \frac{x^2\, dx}{x^2 + 2x + 3} = \int \left\{ 1 - \frac{2x + 3}{x^2 + 2x + 3} \right\} dx$$

$$= \int \left\{ 1 - \frac{2x + 2}{x^2 + 2x + 3} - \frac{1}{(x + 1)^2 + 2} \right\} dx$$

$$= x - \log (x^2 + 2x + 3) - \frac{1}{\sqrt{2}} \tan^{-1} \left(\frac{x + 1}{\sqrt{2}} \right).$$

6.5 INTEGRATION OF RATIONAL FUNCTIONS

Consider the integration of the rational function

$$\frac{P(x)}{R(x)} \tag{49}$$

where $P(x)$ and $R(x)$ are polynomials in x, $R(x)$ is not necessarily a quadratic and the factors of $R(x)$ are known. The procedure is as follows:

(a) if the degree of $P(x)$ is not less than the degree of $R(x)$, divide the denominator into the numerator until the remainder is of the form

$$\frac{f(x)}{R(x)} \qquad (50)$$

where the degree of $f(x)$ is less than the degree of $R(x)$;

(b) separate the remainder (50) into its partial fractions. Let $R(x) = (x - \alpha)(x - \beta)^m(x^2 + px + q)^n$, where m and n are positive integers, so that (50) becomes

$$\frac{f(x)}{(x - \alpha)(x - \beta)^m(x^2 + px + q)^n}$$

and the degree of $f(x)$ is less than $m + 2n + 1$. Then the following rules apply:

(i) to each distinct linear factor such as $x - \alpha$, there corresponds a fraction $A/(x - \alpha)$ where A is a constant to be found by one of the methods of section 6.4.2 (b);

(ii) to each repeated factor such as $(x - \beta)^m$ there corresponds m fractions of the form

$$\frac{B_1}{x - \beta} + \frac{B_2}{(x - \beta)^2} + \frac{B_3}{(x - \beta)^3} + \cdots + \frac{B_m}{(x - \beta)^m},$$

and B_i, $i = 1, 2, 3, \ldots, m$ are constants to be found. The 'cover-up' method may be used to find the value of B_m;

(iii) to each quadratic factor such as $(x^2 + px + q)^n$ there correspond fractions of the form

$$\frac{C_1x + D_1}{x^2 + px + q} + \frac{C_2x + D_2}{(x^2 + px + q)^2} + \cdots + \frac{C_nx + D_n}{(x^2 + px + q)^n},$$

where C_i, D_i, $i = 1, 2, 3, \ldots, n$ are constants to be found.

Example Express

$$f(x) = \frac{x^4 - x^2 - 1}{(x - 2)(x^2 + 2x + 3)}$$

as a sum of partial fractions and hence evaluate $\int f(x)\,dx$.

Solution Since the degree of the numerator is greater than the degree of the denominator, it is necessary to divide so that

$$f(x) = x + \frac{6x - 1}{(x - 2)(x^2 + 2x + 3)}.$$

Since $x^2 + 2x + 3$ will not factorize into real factors, partial fractions are given by

$$\frac{6x - 1}{(x - 2)(x^2 + 2x + 3)} \equiv \frac{A}{x - 2} + \frac{Bx + C}{x^2 + 2x + 3}.$$

Then $A = 1$ by the 'cover-up' method, and the 'comparison' method leads to values $B = -1$, $C = 2$ and

$$f(x) = x + \frac{1}{x - 2} - \frac{x - 2}{x^2 + 2x + 3}.$$

Hence

$$\int f(x)\, dx = \int \left\{ x + \frac{1}{x - 2} - \frac{\frac{1}{2}(2x + 2)}{x^2 + 2x + 3} + \frac{3}{(x + 1)^2 + 2} \right\} dx$$

$$= \tfrac{1}{2}x^2 + \log |x - 2| - \tfrac{1}{2} \log (x^2 + 2x + 3)$$

$$+ \frac{3\sqrt{2}}{2} \tan^{-1} \left(\frac{x + 1}{\sqrt{2}} \right)$$

6.6 INTEGRATION OF FUNCTIONS WITH IRRATIONAL DENOMINATOR

Consider the integration of functions of the type

$$\frac{ax + b}{\sqrt{(x^2 + px + q)}}, \quad a, b, p, q \text{ real constants.}$$

6.6.1 Numerator unity

The method of evaluating

$$\int \frac{dx}{\sqrt{(x^2 + px + q)}}$$

is very similar to that of section 6.4.1, that is, by completing the square of the quadratic $x^2 + px + q$ and using one of the standard forms indicated by (29), (31) or (32).

Example Evaluate

$$I = \int_5^8 \frac{dx}{\sqrt{(x^2 - 4x + 13)}}.$$

Solution The quadratic $x^2 - 4x + 13 = 9 + (x - 2)^2$ and hence

$$I = \int_5^8 \frac{dx}{\sqrt{\{9 + (x - 2)^2\}}}$$

$$= \left[\sinh^{-1} \left(\frac{x - 2}{3} \right) \right]_5^8$$

$$= \sinh^{-1} 2 - \sinh^{-1} 1$$

$$= \log \{2 + \sqrt{(2^2 + 1)}\} - \log \{1 + \sqrt{(1^2 + 1)}\}$$

$$= \log \frac{2 + \sqrt{5}}{1 + \sqrt{2}}.$$

6.6.2 Numerator in linear form

The method of evaluating

$$\int \frac{(ax + b)}{\sqrt{(x^2 + px + q)}}\, dx, \quad a, b, p, q \text{ given, real constants,}$$

is very similar to that of section 6.4.2 (a).

Example Evaluate

$$I = \int \frac{(x - 2)\, dx}{\sqrt{(5 - 4x - x^2)}}.$$

Solution The derivative of $5 - 4x - x^2$ is $-4 - 2x$ and the numerator of the integrand is written $x - 2 = -\frac{1}{2}(-4 - 2x) - 4$, so that

$$I = -\frac{1}{2} \int \frac{(-4 - 2x)}{\sqrt{(5 - 4x - x^2)}}\, dx - 4 \int \frac{dx}{\sqrt{\{9 - (x + 2)^2\}}}$$

$$= -\sqrt{(5 - 4x - x^2)} - 4 \sin^{-1}\{\tfrac{1}{3}(x + 2)\}.$$

6.7 INTEGRATION BY PARTS

The method of integration called *integration by parts* depends on the formula for the differentiation of a product, namely that

$$\frac{d(uv)}{dx} = u\frac{dv}{dx} + v\frac{du}{dx},$$

where u and v are differentiable functions of x.

Suppose f and g are differentiable functions of x and that $G(x) = \int g(x)\, dx$, then the formula for integration by parts is

$$\int fg\, dx = fG - \int Gf'\, dx \qquad (51)$$

The derivative of the right side of this equation is $fg + f'G - Gf'$ which is fg, the derivative of the left side and this proves (51).

For this method to be of use it is necessary that the integral on the right of (51) should be easier to evaluate than the given integral. Note that of the two terms f and g the latter is integrated in *each* term on the right of (51).

Example (i) Evaluate $I = \displaystyle\int_0^1 x e^{3x}\, dx$.

Solution Both functions x and e^{3x} have simple integrals. If x is chosen as the function to be integrated, the power of x increases and the resulting integral is no simpler. If x is chosen as the function to be differentiated the resulting integral is simpler than the original. Hence, choosing $f = x$, $g = e^{3x}$, then $G = \frac{1}{3}e^{3x}$ and

$$I = \int_0^1 xe^{3x}\,dx$$

$$= \left[x\frac{1}{3}e^{3x} \right]_0^1 - \int_0^1 \frac{1}{3}e^{3x}\,dx$$

$$= \left[\frac{1}{3}xe^{3x} - e^{3x}/9 \right]_0^1$$

$$= (2e^3 + 1)/9.$$

This method of integration involves the product of two functions and it is necessary to integrate one of the functions and differentiate the other. At first the choice of function to be integrated may appear difficult. Once one of the functions is chosen to be integrated then this function must continue to be integrated should it be necessary to perform further integration by parts.

Example (ii) Evaluate $I = \int e^{2x} \sin 3x\,dx$.

Solution Choose $g = e^{2x}$ then $f = \sin 3x$ and

$$I = \frac{1}{2}e^{2x} \sin 3x - \int \frac{1}{2}e^{2x} (3 \cos 3x)\,dx.$$

Using formula (51) again for $\int e^{2x} \cos 3x\,dx$ and again choosing $g = e^{2x}$, leads to

$$\int e^{2x} \cos 3x\,dx = \frac{1}{2}e^{2x} \cos 3x - \int \frac{1}{2}e^{2x}(-3 \sin 3x)\,dx.$$

Thus $I = \frac{1}{2}e^{2x} \sin 3x - \frac{3}{4}e^{2x} \cos 3x - 9I/4$

and transposing the term containing I on the right to the left gives

$$13I/4 = \frac{1}{2}e^{2x} \sin 3x - \frac{3}{4}e^{2x} \cos 3x,$$

and $I = e^{2x}(2 \sin 3x - 3 \cos 3x)/13 +$ an arbitrary constant.

It is recommended that the reader evaluates this integral by choosing $g = \sin 3x$ and $f = e^{2x}$.

Formula (51) is used to integrate functions whose derivative is known and whose integral is not of standard form, such as $\log x$, $\tan^{-1} x$ etc.

Example (iii) Evaluate $\int \log x\,dx$.

Solution The integrand is considered as the product of two functions, $f = \log x$ and $g = 1$. Then

$$\int \log x \, dx = x \log x - \int x(1/x) \, dx$$

$$= x \log x - x + \text{an arbitrary constant.}$$

Example (iv) Evaluate $\displaystyle\int_0^1 \tan^{-1} x \, dx$.

Solution Let $f = \tan^{-1} x$, $g = 1$, then

$$\int_0^1 \tan^{-1} x \, dx = \left[x \tan^{-1} x \right]_0^1 - \int_0^1 \{ x/(1 + x^2) \} \, dx$$

$$= \left[x \tan^{-1} x - \tfrac{1}{2} \log (1 + x^2) \right]_0^1$$

$$= \tan^{-1} 1 - \tfrac{1}{2} \log 2$$

$$= \tfrac{1}{4}\pi - \tfrac{1}{2} \log 2.$$

6.8 REDUCTION FORMULAE

The technique of integrating by parts repeatedly is often necessary to evaluate an integral such as

$$I_n = \int e^x x^n \, dx. \tag{52}$$

Integrating by parts gives

$$I_n = e^x x^n - \int e^x n x^{n-1} \, dx \tag{53}$$

and the integral on the right is the same integral as the original of equation (52) with $n - 1$ in place of n. Hence it is written I_{n-1} and integral (53) becomes

$$I_n = e^x x^n - n I_{n-1}. \tag{54}$$

This is the *reduction formula* for the given integral which may be used for any value of $n \geqslant 0$. For example, if $n = 3$, $I_3 = e^x x^3 - 3I_2$, I_2 is evaluated by writing $n = 2$ in formula (54) as $I_2 = e^x x^2 - 2I_1$, and similarly $I_1 = e^x x - I_0$, $I_0 = e^x$ so that finally

$$I_3 = x^3 e^x - 3x^2 e^x + 6x e^x - 6 e^x + C$$

where C is an arbitrary constant.

Simple results are often achieved if the integrals to be reduced are definite integrals.

Example Obtain a reduction formula for

$$I_n = \int_0^{\frac{1}{2}\pi} \sin^n x \, dx, \qquad n \geqslant 0.$$

Solution The function $\sin^n x$ may be written as the product of $\sin^{n-1} x$ and $\sin x$ so that integrating $\sin x$ and differentiating $\sin^{n-1} x$ and integrating by parts,

$$I_n = \left[-\cos x \sin^{n-1} x \right]_0^{\frac{1}{2}\pi} - \int_0^{\frac{1}{2}\pi} (-\cos x)(n-1) \sin^{n-2} x \cos x \, dx$$

or

$$I_n = (n-1) \int_0^{\frac{1}{2}\pi} \sin^{n-2} x \cos^2 x \, dx. \tag{55}$$

Now

$$I_{n-2} = \int_0^{\frac{1}{2}\pi} \sin^{n-2} x \, dx.$$

Hence it is necessary to transform the integral in equation (55) using the identity $\cos^2 x + \sin^2 x = 1$, so that

$$I_n = (n-1) \int_0^{\frac{1}{2}\pi} \sin^{n-2} x (1 - \sin^2 x) \, dx,$$

or $I_n = (n-1)I_{n-2} - (n-1)I_n.$

Solving for I_n gives

$$nI_n = (n-1)I_{n-2}$$

or

$$I_n = \frac{n-1}{n} I_{n-2} \tag{56}$$

which is the required reduction formula.

When n is a positive integer the formula may be applied repeatedly until the remaining integral is either I_1, when n is odd, or I_0, when n is even, where

$$I_1 = \int_0^{\frac{1}{2}\pi} \sin x \, dx = 1 \quad \text{and} \quad I_0 = \int_0^{\frac{1}{2}\pi} dx = \tfrac{1}{2}\pi.$$

Writing $n - 2$ for n in formula (56) gives

$$I_{n-2} = \frac{n-3}{n-2} I_{n-4},$$

and writing $n - 4$ for n in formula (56) gives

$$I_{n-4} = \frac{n-5}{n-4} I_{n-6}$$

and so on, so that finally,

$$I_n = \frac{n-1}{n} \frac{n-3}{n-2} \frac{n-5}{n-4} \ldots I_0 \text{ or } I_1.$$

When n is odd, $I_n = \dfrac{(n-1)(n-3)(n-5)\ldots 4 . 2}{n(n-2)(n-4)\ldots 5 . 3},$

when n is even, $\quad I_n = \dfrac{(n-1)(n-3)(n-5)\ldots 3 \cdot 1}{n(n-2)(n-4)\ldots 4 \cdot 2} \cdot \dfrac{\pi}{2}.$

6.9 FURTHER SUBSTITUTIONS

Rule (c) of section 6.2.1 is a substitution rule, but more generally let $F'(x) = f(x)$, then

$$[F\{\phi(x)\}]' = f\{\phi(x)\}\phi'(x)$$

and $\quad \displaystyle\int f\{\phi(x)\}\phi'(x)\,dx = \int f\{\phi(x)\}d\phi(x) = \int f(u)\,du \qquad (57)$

where $u = \phi(x)$ and after integration u is replaced by $\phi(x)$. For example,

$$\int \{\phi(x)\}^n \phi'(x)\,dx = \int u^n\,du \quad \text{where } u = \phi(x),$$

$$= \frac{u^{n+1}}{n+1}$$

$$= \frac{1}{n+1}\{\phi(x)\}^{n+1}, \quad n \neq -1.$$

Sometimes it is necessary to write $x = g(u)$ say instead of $u = \phi(x)$ in the given integral. Then the function $g(u)$ is the inverse of $\phi(x)$ and hence must be chosen to be monotonic in the range of integration.

Consider the definite integral

$$\int_a^b f(x)\,dx = F(b) - F(a).$$

Suppose that the substitution $x = g(u)$ is made such that as x varies monotonically from a to b, u varies monotonically from α to β where $a = g(\alpha)$, $b = g(\beta)$, then

$$\int_a^b f(x)\,dx = F\{g(\beta)\} - F\{g(\alpha)\}.$$

Some general substitutions follow, where a and b are given constants:

if $\sqrt{(ax+b)}$ occurs, substitute $ax + b = u^2$; $\qquad (58)$

if $\sqrt{(a^2 - x^2)}$ occurs, substitute $x = a\sin\theta,\ -\tfrac{1}{2}\pi \leqslant \theta \leqslant \tfrac{1}{2}\pi$; $\qquad (59)$

if $a^2 + x^2$ occurs, substitute $x = a\tan\theta,\ -\tfrac{1}{2}\pi < \theta < \tfrac{1}{2}\pi$; $\qquad (60)$

if $\sqrt{(x^2 + a^2)}$ occurs, substitute $x = a\tan\theta,\ -\tfrac{1}{2}\pi < \theta < \tfrac{1}{2}\pi$ $\qquad (61)$

or $\qquad\qquad\qquad x = a\sinh u$; $\qquad (62)$

if $\sqrt{(x^2 - a^2)}$ occurs, substitute $x = a\cosh u,\ u > 0$. $\qquad (63)$

Example (i) Evaluate

$$I = \int_0^1 \frac{dx}{\{\sqrt{(1 + x^2)}\}^3}.$$

Solution Substitute $x = \tan \theta$, $-\frac{1}{2}\pi < \theta < \frac{1}{2}\pi$, then $dx = \sec^2 \theta \, d\theta$; when $x = 0$, $\theta = 0$; when $x = 1$, $\theta = \frac{1}{4}\pi$, and $\{\sqrt{(1 + x^2)}\}^3 = \sec^3 \theta$. Then

$$I = \int_0^{\frac{1}{4}\pi} \frac{\sec^2 \theta \, d\theta}{\sec^3 \theta} = \int_0^{\frac{1}{4}\pi} \cos \theta \, d\theta = \frac{1}{\sqrt{2}}.$$

Example (ii) Evaluate

$$I = \int \frac{x^2 \, dx}{\sqrt{(x^2 + 1)}}.$$

Solution The substitution $x = \tan \theta$ transforms the integral into $\int \tan^2 \theta \sec \theta \, d\theta$ which is no simpler than the original. Hence substitute $x = \sinh u$, $dx = \cosh u \, du$ and

$$\begin{aligned}
I &= \int \sinh^2 u \, du \\
&= \tfrac{1}{2} \int (\cosh 2u - 1) \, du \\
&= \tfrac{1}{4} \sinh 2u - \tfrac{1}{2}u \\
&= \tfrac{1}{2} \sinh u \cosh u - \tfrac{1}{2}u \\
&= \tfrac{1}{2}x\sqrt{(x^2 + 1)} - \tfrac{1}{2} \sinh^{-1} x.
\end{aligned}$$

6.9.1 Integration using $t = \tan \frac{1}{2}x$

Rational functions of $\sin x$ and $\cos x$ may be reduced to rational functions of t by the substitution $t = \tan \frac{1}{2}x$, $-\frac{1}{2}\pi < \frac{1}{2}x < \frac{1}{2}\pi$. In particular this substitution is of use in integrating functions containing expressions $a \cos x + b \sin x$, where a and b are constants. The method uses the following results:

$$\cos x = \frac{1 - t^2}{1 + t^2}, \quad \sin x = \frac{2t}{1 + t^2} \quad \text{and} \quad \tfrac{1}{2} \sec^2 \tfrac{1}{2}x \, dx = dt$$

so that $dx = 2dt/(1 + t^2)$.

The substitution is cumbersome and is used only when the simpler methods of earlier sections have failed.

Example Evaluate $\int_0^{\frac{1}{2}\pi} \dfrac{dx}{\sin x + \cos x + 1}.$

Solution Substitute $t = \tan \frac{1}{2}x$ and use the results above and the integral becomes

$$\int_0^1 \frac{dt}{1 + t} = \Big[\log (1 + t) \Big]_0^1 = \log 2.$$

6.10 INFINITE INTEGRALS

Infinite integrals are those in which the limits of integration or the integrand or both become infinite. In section 6.3 the integral of $f(x)$ was defined in the range $a < x < b$, on the assumptions that $f(x)$ is defined at each point in the range and that a and b are finite. This definition is now extended to the case where a or b or both become infinite, and to the case where $f(x)$ is undefined for a finite number of points in the range of integration.

6.10.1 The limits of integration become infinite

Let $F'(x) = f(x)$ for all values of $x > a$, where a is a finite real number, then $\int_a^\infty f(x)\,dx$ is defined by

$$\int_a^\infty f(x)\,dx = \lim_{X \to \infty} \int_a^X f(x)\,dx = \lim_{X \to \infty} F(X) - F(a) \qquad (64)$$

when this limit exists.

Similarly $\int_{-\infty}^b f(x)\,dx$ is defined as

$$\lim_{X \to -\infty} \int_X^b f(x)\,dx = F(b) - \lim_{X \to -\infty} F(X) \qquad (65)$$

when the limit exists.

If the integral as a limit exists the integral is said to be *convergent*. Otherwise it is *divergent*.

Example (i) Evaluate $\int_1^\infty \dfrac{dx}{x}$.

Solution Consider $\int_1^X \dfrac{dx}{x} = \log X - \log 1$

$$= \log X \to \infty \quad \text{as } X \to \infty.$$

Hence the integral is divergent.

Example (ii) Evaluate $\int_0^\infty e^{-kx}\,dx$, k a positive constant.

Solution Using (64),

$$\int_0^\infty e^{-kx}\,dx = \lim_{X \to \infty} \int_0^X e^{-kx}\,dx$$

$$= \lim_{X \to \infty} \frac{1}{k}(1 - e^{-kX})$$

$$= \frac{1}{k}.$$

6.10.2 Substitution in infinite integrals

It is possible to transform an integral with infinite limits into one with finite limits by a substitution, as in the following example.

Example Evaluate $\displaystyle\int_0^\infty \frac{dx}{1 + x^2}$.

Solution Consider $\displaystyle\int_0^X \frac{dx}{1 + x^2}$ and make the substitution $x = \tan \theta$ so that the integral becomes

$$\int_0^{\tan^{-1} X} d\theta = \tan^{-1} X \quad \text{which} \rightarrow \tfrac{1}{2}\pi \text{ as } X \rightarrow \infty.$$

Hence it is possible to evaluate the integral directly as

$$\int_0^\infty \frac{dx}{1 + x^2} = \int_0^{\frac{1}{2}\pi} d\theta = \tfrac{1}{2}\pi.$$

6.10.3 The integrand becomes infinite

Let the integrand $f(x)$ become infinite at the end points of the interval $a < x < b$. Strictly the integral does not exist since the integrand does not exist at these points. But a meaning may be attached to the integral, called an *improper* integral.

Let $F'(x) = f(x)$ for $a < x < b$. If both $F(a + 0) \rightarrow$ a finite limit and $F(b - 0) \rightarrow$ a finite limit then $\displaystyle\int_a^b f(x)\, dx$ is defined to be

$$F(b - 0) - F(a + 0). \tag{66}$$

Example (i) Evaluate $\displaystyle\int_0^2 \frac{dx}{\sqrt{(4 - x^2)}}$.

Solution The integrand becomes infinite when $x = 2$. Consider

$$\int_0^{2-\delta} \frac{dx}{\sqrt{(4 - x^2)}} = \left[\sin^{-1} \left(\frac{x}{2} \right) \right]_0^{2-\delta}$$
$$= \sin^{-1} (1 - \tfrac{1}{2}\delta).$$

As $\delta \rightarrow 0$, $\sin^{-1} (1 - \tfrac{1}{2}\delta) \rightarrow \tfrac{1}{2}\pi$ and hence this is the value of the integral.

Example (ii) Evaluate $\displaystyle\int_0^1 \log x\, dx$.

Solution The integrand becomes infinite (and negative) at $x = 0$. Consider

$$\int_\delta^1 \log x\, dx = \left[x \log x - x \right]_\delta^1$$
$$= \delta - 1 - \delta \log \delta.$$

Now $-\delta \log \delta = (\log X)/X$ where $X = 1/\delta$ and

$$\operatorname*{Lim}_{\delta \rightarrow 0} (-\delta \log \delta) = \operatorname*{Lim}_{X \rightarrow \infty} \{(\log X)/X\} = 0.$$

Hence
$$\int_0^1 \log x \, dx = -1.$$

The definition (66) may be extended to define integration of functions which are undefined at a finite number of points within the range. For if $f(x)$ becomes infinite at $x = c$, $a < c < b$, then using (42) of section 6.3.1 the integral may be written

$$\int_a^b f(x) \, dx = \int_a^c f(x) \, dx + \int_c^b f(x) \, dx.$$

If $F(c - 0)$ and $F(c + 0)$ both approach finite limits then the integrals on the right may be evaluated using (66).

6.11 OTHER METHODS OF INTEGRATION

The methods of integration of elementary functions given in this chapter are not exhaustive. There are many elementary functions which are not exactly integrable, such as $\cos(x^2)$, $1/\sqrt{(1 - x^4)}$, $(\sin x)/x$, $\exp(-x^2)$, and other methods of integration do exist to deal with these. Numerical methods of integration are discussed in section 14.8 of chapter 14. The integrals of some important functions are tabulated. Further methods are beyond the scope of this book.

EXERCISE 6

In the following examples a and b are given constants. Find the indefinite integrals of the functions 1–42:

1 $3a^2 x^6$,   **2** $(3x^2 + 4x + 2)$,

3 $(ax^3 + b)^2$,   **4** $\sqrt{(2ax)}$,

5 $\sqrt{(e^x)}$,   **6** $[\sqrt{(2 + x^2)} - \sqrt{(2 - x^2)}]/\sqrt{(4 - x^4)}$,

7 $\coth^2 x$,   **8** $2^x e^x$,

9 $(5 - 4x)^{-\frac{1}{2}}$,

10 $a/(a - x)$,   **11** $(3x + 2)/(3x + 1)$,

12 $(x^2 + 1)/(x - 1)$,   **13** $1/(1 + 9x^2)$,

14 $\sinh x/(1 + \cosh x)$,   **15** $x/\sqrt{(x^2 + 1)}$,

16 $\sec x \operatorname{cosec} x$,   **17** $(\sin x - \cos x)/(\sin x + \cos x)$,

18 $(1 - \sin x)/(x + \cos x)$,

19 $\cos^3 x$,   **20** $\sin^5 x$,

21 $\sin^2 x \cos^3 x$,   **22** $\sin^3 \frac{1}{2}x \cos^5 \frac{1}{2}x$,

23 $\cos^5 x \operatorname{cosec}^7 x$,   **24** $\sin^2 x \cos^4 x$,

25 $\operatorname{cosec}^4 x$,   **26** $\tan^2 5x$,

27 $\sin 3x \cos 5x$,   **28** $(2x - 1)/(x^2 + 4x + 4)$,

29 $(2x - 1)/(x^2 + 4x)$,   **30** $(2x - 1)/(x^2 + 4x + 8)$,

31 $(2x - 1)\sqrt{(x^2 + 4x + 8)}$,   **32** $1/\sqrt{(1 - x - x^2)}$,

33 $(x - 5)/\sqrt{(4x - x^2)}$,   **34** $1/\sqrt{\{x(1 - x)\}}$,

35 $e^x/(1 + e^{2x})$,   **36** $1/(x^2 + 4x + 7)$,

37 $1/x(1 - x^2)$,   **38** $(5x - 3)/(x + 1)(x - 3)$,

39 $1/x(x + 1)^2$,   **40** $1/(1 - x^4)$,

41 $(x + 1)/(x^2 + 4x - 5)$,   **42** $(x^4 - x^3 - x - 1)/(x^3 - x)$.

Evaluate the integrals **43–49**:

43 $\displaystyle\int_0^{\frac{1}{2}\pi} \tan x \, dx,$ **44** $\displaystyle\int_0^{\frac{1}{2}\pi} \sin^5 x \cos x \, dx,$

45 $\displaystyle\int_e^{e^2} \frac{dx}{x \log x},$ **46** $\displaystyle\int_1^e \frac{\log x}{x} \, dx,$

47 $\displaystyle\int_{-2}^{-1} \frac{dx}{(2x+1)^2},$ **48** $\displaystyle\int_0^1 \frac{dx}{x^2 - x + 1}.$

49 $\displaystyle\int_0^1 \frac{1-x^2}{1+x^2} \, dx.$

Find the indefinite integrals of the functions **50–54** by integrating by parts:

50 $\sin^{-1} x,$ **51** $x^2 \tan^{-1} x,$
52 $xe^{x/a},$ **53** $x \sin x,$
54 $x^n \log x, \quad n \neq -1.$

Evaluate the integrals **55–58**:

55 $\displaystyle\int_1^2 x^{-2} \log(1+x) \, dx,$ **56** $\displaystyle\int_0^{\frac{1}{2}\pi} x \cos x \, dx,$

57 $\displaystyle\int_0^{\frac{1}{2}\pi} x^2 \sin x \, dx,$ **58** $\displaystyle\int_0^{\frac{1}{2}\pi} e^{-2x} \cos 3x \, dx.$

59 Find a reduction formula for I_n in terms of I_{n-1} given that

$$I_n = \int e^x x^n \, dx.$$

Hence find the value of I_3.

60 Prove that if $I_n = \displaystyle\int \tan^n x \, dx$, then

$$I_n + I_{n-2} = (\tan^{n-1} x)/(n-1), \quad n \neq 1,$$

and hence evaluate $\displaystyle\int_0^{\frac{1}{2}\pi} \tan^6 x \, dx.$

61 Given that $I_n = \displaystyle\int_0^{\pi} e^x \sin^n x \, dx$ prove that, for $n > 1$,

$$(n^2 + 1)I_n = n(n-1)I_{n-2}.$$

Hence or otherwise evaluate I_4 and I_5.

62 Show that, if $I_n = \displaystyle\int_0^{\frac{1}{2}\pi} x \cos^n x \, dx$ then

$$I_n = \frac{n-1}{n} I_{n-2} - \frac{1}{n^2}, \quad n \geqslant 2,$$

and evaluate I_4 and I_5.

63 Find the value of $\int_0^{2\pi} \cos 5x \cos nx \, dx$ (i) when $n = 5$, (ii) when $n \neq 5$.

Find the indefinite integrals of the functions **64–68** using the substitution given:

64 $1/\{x\sqrt{(x^2 - 2)}\}$, $(x = 1/t)$, **65** $1/(e^x + 1)$, $(x = -\log u)$,

66 $x(3x^2 - 2)^9$, $(3x^2 - 2 = t)$, **67** $x/\sqrt{(x + 1)}$, $(\sqrt{(x + 1)} = u)$,

68 $\cos x/\sqrt{(1 + \sin^2 x)}$, $(u = \sin x)$.

Examine for convergence, and in cases where the integral converges evaluate, the definite integrals **69–76**:

69 $\int_{-\infty}^0 e^x \sin x \, dx$, **70** $\int_0^\infty \dfrac{dx}{(x^2 + a^2)^3}$,

71 $\int_1^\infty \dfrac{\log x}{x} \, dx$, **72** $\int_0^\infty \dfrac{dx}{(x + 1)(x + 2)}$,

73 $\int_0^1 \dfrac{dx}{\sqrt{(1 - x)}}$, **74** $\int_0^\infty \dfrac{dx}{\sqrt{(1 + x^2)}}$,

75 $\int_0^3 \dfrac{dx}{(x - 1)^2}$, **76** $\int_{-1}^1 \sqrt{\left(\dfrac{1 + x}{1 - x}\right)} \, dx$.

By writing $t = \tan \tfrac{1}{2}x$ find the indefinite integrals of the functions **77–79**:

77 $1/(3 + 5 \cos x)$, **78** $\sin x/(1 - \sin x)$,

79 $(1 + \tan x)/(1 - \tan x)$.

80 Evaluate

$$\int_0^{\frac{1}{2}\pi} \dfrac{dx}{2 + \sin x}.$$

7
Applications of Integration

7.0 INTRODUCTION

In this chapter the problem of integration will be explained briefly without rigour, with reference to geometrical and physical applications.

7.1 INTEGRATION AS THE LIMIT OF A SUM

7.1.1 Area as a limit

The problem is to consider the area A included between a continuous curve $y = f(x)$, the x-axis between $x = a$ and $x = b$, and the ordinates $x = a$ and $x = b$ as in Fig. 7.1.

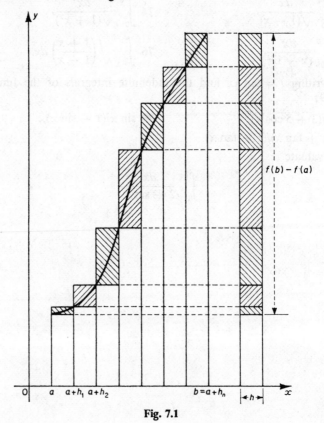

Fig. 7.1

The method is to construct a set of rectangles which lie underneath the curve whose total area is S_1 and another set which lie above it whose total area is S_2, and investigate as to whether they can be made to approach each other.

Let $y = f(x)$ be continuous, increasing, and positive in $a < x < b$ and let it be represented by the curve sketched in Fig. 7.1. Divide the range $b - a$ into a series of subdivisions with abscissae a, $a + h_1$, $a + h_2$, . . ., $a + h_n = b$ and complete the rectangles on bases h_1, $h_2 - h_1$, $h_3 - h_2$, . . ., $h_n - h_{n-1}$. The areas S_1 and S_2 are given by

$$S_1 = f(a)h_1 + f(a + h_1)(h_2 - h_1) + f(a + h_2)(h_3 - h_2) + \ldots$$
$$+ f(a + h_{n-1})(h_n - h_{n-1})$$

$$S_2 = f(a + h_1)h_1 + f(a + h_2)(h_2 - h_1) + f(a + h_3)(h_3 - h_2) + \ldots$$
$$+ f(a + h_n)(h_n - h_{n-1})$$

then $$S_1 < A < S_2.$$

As the number of subdivisions is increased, and the largest of the intervals $h_k - h_{k-1}$ is made 'as small as we please', then the difference between S_1 and S_2 approaches zero, i.e. S_1 and S_2 tend to a common limit A. If, for all possible values of S_1 and S_2, A is a fixed number, A is *defined* as the required area. The difference $S_2 - S_1$ is given by

$$S_2 - S_1 = \{f(a + h_1) - f(a)\}h_1 + \{f(a + h_2) - f(a + h_1)\}(h_2 - h_1)$$
$$+ \ldots + \{f(a + h_n) - f(a + h_{n-1})\}(h_n - h_{n-1}).$$

Let h be the greatest of the numbers h_1, $(h_2 - h_1)$, $(h_3 - h_2)$, . . ., $(h_n - h_{n-1})$. Then since $f(x)$ is increasing, the numbers $\{f(a + h_1) - f(a)\}$ $\{f(a + h_2) - f(a + h_1)\}$ etc. are all positive or zero. Hence

$$S_2 - S_1 \leqslant \{f(a + h_1) - f(a)\}h + \{f(a + h_2) - f(a + h_1)\}h$$
$$+ \ldots + \{f(a + h_n) - f(a + h_{n-1})\}h$$
$$= \{f(a + h_n) - f(a)\}h$$
$$= \{f(b) - f(a)\}h.$$

Thus, given a number c say, however small, it is possible to choose h such that

$$h < \frac{c}{f(b) - f(a)}$$

then $$S_2 - S_1 < c,$$

i.e. by taking the subdivisions sufficiently small, the difference $S_2 - S_1$ can be made 'as small as we please', and the area A is a definite number.

Let x_k be a value of x in the kth sub-interval. Then if

$$\sum_{k=1}^{n} f(x_k)(h_k - h_{k-1}) \to l$$

as the largest sub-interval $(h_k - h_{k-1}) \rightarrow 0$ where l is fixed and finite, then $f(x)$ is said to have a definite integral defined as

$$\operatorname*{Lim}_{\delta x \to 0} \sum_{x=a}^{x=b} f(x_k)\, \delta x_k, \quad \text{where } h_k - h_{k-1} = \delta x_k \tag{1}$$

Note that Fig. 7.1 shows the case of n equal subdivisions of $(b - a)$.

7.1.2 Area as integral

Suppose now that $F'(x) = f(x)$ also. Then, using the mean value theorem,

$$F(a + h_k) - F(a + h_{k-1}) = (h_k - h_{k-1})f(x_k) \tag{2}$$

since x_k is a value of x in the kth sub-interval. Summing over all values of k,

$$F(b) - F(a) = \sum_{k=1}^{n} (h_k - h_{k-1})f(x_k). \tag{3}$$

If the sum on the right of equation (3) approaches a limit as the largest of the intervals $h_k - h_{k-1} = \delta x_k$ approaches zero, then that limit is the same as in (1). When this limit exists it is the definite integral of the function $f(x)$ between the limits $x = a$ and $x = b$, i.e.

$$\operatorname*{Lim} \sum_{x=a}^{x=b} f(x)\, \delta x = \text{area} = \int_{a}^{b} f(x)\, dx \tag{4}$$

the idea of area being omitted as necessary.

7.1.3 Functions not necessarily continuous increasing and positive

In section 7.1.1, $y = f(x)$ was chosen to be continuous, increasing, and positive throughout the whole range $a < x < b$. Similar arrangements to those of sections 7.1.1 and 7.1.2 hold if the function decreases throughout the range, or if it is negative anywhere in the range. A function which decreases and increases through the range may be divided into intervals in which the function either decreases or increases, and the total area defined as the sum of the separate areas.

A function which becomes negative in the range may be divided into intervals in which the function is either positive or negative, and the total area defined as the sum of the separate areas. Since the integral is negative when the function is negative the area here is the modulus of the integral (see section 6.3.2).

Moreover, the argument applies to all increasing (or decreasing) functions, not only to those which are continuous. For example, it is possible to integrate the following function which is increasing but not continuous:

$$f(x) = \tfrac{1}{2} \quad \text{in} \quad 0 < x < \tfrac{1}{2},$$
$$f(x) = \tfrac{3}{4} \quad \text{in} \quad \tfrac{1}{2} < x < \tfrac{3}{4},$$
$$f(x) = \tfrac{7}{8} \quad \text{in} \quad \tfrac{3}{4} < x < \tfrac{7}{8},$$
$$\text{etc.}$$
$$f(x) = 1 \quad \text{when} \quad x = 1.$$

See Fig. 7.2 for a sketch of $y = f(x)$.

This function has an infinite number of discontinuities; nevertheless the definition of integral as the limit sum applies.

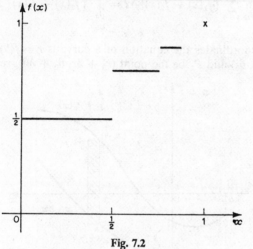

Fig. 7.2

7.1.4 Rules for definite integrals

(1)
$$\int_a^b f(x)\, dx = -\int_b^a f(x)\, dx$$

since
$$F(b) - F(a) = -\{F(a) - F(b)\}$$

(2)
$$\int_a^b f(x)\, dx = \int_a^c f(x)\, dx + \int_c^b f(x)\, dx$$

where c is some value of x between a and b.

7.2 APPLICATIONS OF INTEGRATION

7.2.1 Plane areas

The area enclosed by the curve $y = f(x)$, the x-axis from $x = a$ to $x = b$ and the ordinates at $x = a$, $x = b$ is, from (4) of section 7.1.2

$$\lim_{\delta x \to 0} \sum_{x=a}^{x=b} f(x)\, \delta x = \int_a^b f(x)\, dx \tag{5}$$

If $x = x(t)$, $y = y(t)$, $\alpha \leqslant t \leqslant \beta$ where t is a parameter,

$$\text{area} = \int_\alpha^\beta y(t)\, \frac{dx}{dt}\, dt \tag{6}$$

Suppose $y_1 = f_1(x)$ and $y_2 = f_2(x)$ are two functions of x which are continuous for $a \leqslant x \leqslant b$ and suppose $f_2(x) \leqslant f_1(x)$ for $a \leqslant x \leqslant b$.

Then the area bounded by the two curves and the ordinates at $x = a$, $x = b$ is given by

$$\underset{\delta x \to 0}{\text{Lim}} \sum_{x=a}^{x=b} \{f_1(x) - f_2(x)\} \, \delta x = \int_a^b \{f_1(x) - f_2(x)\} \, dx \qquad (7)$$

See Fig. 7.3.

In polar coordinates the equation of a curve is $r = f(\theta)$. Let P_0 be the point (r_0, θ_0) and P_1 be the point $(r_0 + \delta r, \theta_0 + \delta\theta)$, see Fig. 7.4.

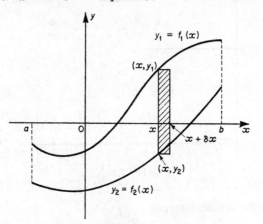

Fig. 7.3

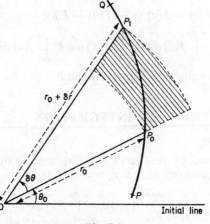

Fig. 7.4

Then the elementary area swept out by the radius OP lies between $\frac{1}{2}r_0^2 \, \delta\theta$ and $\frac{1}{2}(r_0 + \delta r)^2 \, \delta\theta$, i.e.

$$\tfrac{1}{2}r_0^2 \leqslant \frac{\delta A}{\delta\theta} \leqslant \tfrac{1}{2}(r_0 + \delta r)^2,$$

and thus, in the limit $dA/d\theta = \frac{1}{2}r^2$. Summing elementary areas, the area swept out by the radius OP as θ increases from α to β is

$$\underset{\delta\theta \to 0}{\text{Lim}} \sum_{\theta=\alpha}^{\theta=\beta} \tfrac{1}{2}r^2 \, \delta\theta = \int_{\alpha}^{\beta} \tfrac{1}{2}r^2 \, d\theta \qquad (8)$$

In terms of the corresponding Cartesian coordinates, $x = r \cos \theta$, $y = r \sin \theta$, $\tan \theta = y/x$ and $\sec^2 \theta \, d\theta = (x \, dy - y \, dx)/x^2$. Also since $r = x \sec \theta$, $r^2 \, d\theta = x \, dy - y \, dx$ and the area may be written

$$\int_{P}^{Q} \tfrac{1}{2}(x \, dy - y \, dx) \qquad (9)$$

which is the area swept out by the radius as P moves from P to Q along the curve.

Example (i) Find the area bounded by the parabola $y = 2 - x^2$ and the straight line $y = x$.

Solution The required area is sketched in Fig. 7.5.

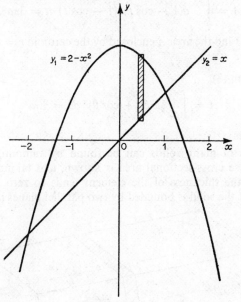

Fig. 7.5

The curves intersect when $x = 1$ and $x = -2$, and for all values of x between -2 and 1 the parabola is above the line by an amount $y_1 - y_2$ where $y_1 = 2 - x^2$ and $y_2 = x$. Hence the total area A between the curves is given by

$$A = \underset{\delta x \to 0}{\text{Lim}} \sum (y_1 - y_2) \, \delta x = \int_{-2}^{1} (2 - x^2 - x) \, dx = 4\tfrac{1}{2}.$$

Example (ii) Find the area between the *x*-axis and the cycloid $x = a(t - \sin t), y = a(1 - \cos t), 0 \leqslant t \leqslant 2\pi$, a sketch of which is given in Fig. 7.6.

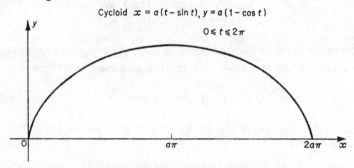

Cycloid $x = a(t - \sin t), y = a(1 - \cos t)$

$0 \leqslant t \leqslant 2\pi$

Fig. 7.6

Solution Using (6) the required area A is given by

$$A = \int_0^{2\pi} a(1 - \cos t)a(1 - \cos t)\, dt = 3\pi a^2.$$

Example (iii) Find the area A enclosed by the cardioid $r = a(1 + \cos \theta)$, $-\pi \leqslant 0 \leqslant \pi$.

Solution From (8),

$$A = \int_{-\pi}^{\pi} \tfrac{1}{2}a^2(1 + \cos \theta)^2\, d\theta = \tfrac{3}{2}\pi a^2.$$

7.2.2 Volumes

The volumes of many solids can be found by summing elementary volumes whose cross-sectional area is known, and taking the limit of this sum as the thickness of the element tends to zero. Suppose for example, that the solid is bounded by two parallel planes perpendicular

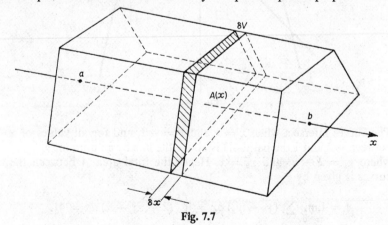

Fig. 7.7

to the x-axis at $x = a$ and $x = b$. Let δV be the volume of a representative elementary volume, or 'slice', whose area of cross-section is $A(x)$ and whose thickness is δx, as sketched in Fig. 7.7.

Then the required volume is the limit of the sum of such elementary volumes as the thickness $\delta x \to 0$. The total volume V is given by

$$V = \lim_{\delta x \to 0} \sum A(x)\, \delta x = \int_a^b A(x)\, dx. \tag{10}$$

In particular, when the solid is a *solid of revolution* formed by rotating a plane area about the x-axis or y-axis, a simple result may be obtained. Let the plane area be that bounded by the arc of the curve $y = f(x)$, the x-axis and the ordinates $x = a$, $x = b$, and suppose that this area is rotated about the x-axis to form a solid of revolution.

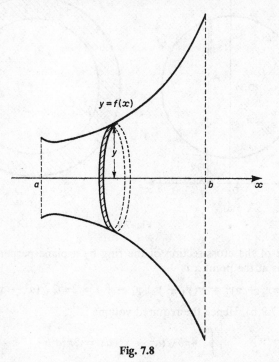

Fig. 7.8

The representative element of volume is then a disc of radius $y = f(x)$, and hence $A(x) = \pi y^2$, (see Fig. 7.8). Then from (10) the volume of the solid V is given by

$$V = \lim_{\delta x \to 0} \sum_{x=a}^{x=b} \pi y^2\, \delta x = \int_a^b \pi y^2\, dx \quad \text{or} \quad \int_a^b \pi \{f(x)\}^2\, dx. \tag{11}$$

Example (i) Find the volume of the ellipsoid formed by rotating the area enclosed by the ellipse $x^2/a^2 + y^2/b^2 = 1$, $y \geqslant 0$ about the x-axis.

Solution $y^2 = b^2(1 - x^2/a^2)$ $-a \leqslant x \leqslant a$

$$\text{Volume} = 2 \int_0^a \pi b^2(1 - x^2/a^2)\, dx = 4\pi ab^2/3.$$

Writing $b = a$ gives the volume of a sphere of radius a as $4\pi a^3/3$.

Example (ii) Find the volume of the ring generated when the circle $x^2 + (y - b)^2 = a^2$, $a < b$, is rotated about the x-axis.

Solution See Fig. 7.9(a). For each value of x between $-a$ and a there are two values of y, say y_1 and y_2 where $y_1 = b + \sqrt{(a^2 - x^2)}$ and $y_2 = b - \sqrt{(a^2 - x^2)}$.

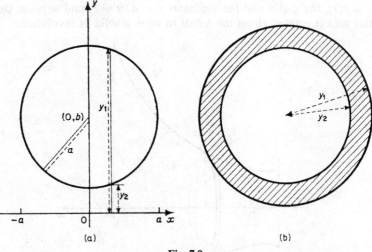

Fig. 7.9

The area of the cross-section of the ring by a plane perpendicular to the x-axis at the point x is

$$\pi y_1^2 - \pi y_2^2 = \pi(y_1 + y_2)(y_1 - y_2) = 2\pi b 2\sqrt{(a^2 - x^2)}.$$

See Fig. 7.9(b). Hence the required volume is

$$\int_{-a}^a 4\pi b \sqrt{(a^2 - x^2)}\, dx = 2\pi^2 a^2 b$$

(substitute $x = a \sin \theta$ to evaluate the integral).

It is of interest to note that the volume is the product of πa^2, the area of the generating circle, and $2\pi b$ which is the length of the path of the centre of that circle as it rotates through one revolution about Ox. This is investigated further in section 7.2.6.

7.2.3 Centre of mass, centroid

Let masses m_1, m_2, m_3, . . ., m_n be concentrated at points P_1, P_2, P_3, . . ., P_n at distances from a fixed plane $x = 0$ of x_1, x_2, x_3, . . ., x_n.

Their moment about the plane is defined to be

$$x_1 m_1 + x_2 m_2 + x_3 m_3 + \ldots + x_n m_n = \sum_{k=1}^{n} x_k m_k, \qquad (12)$$

known sometimes as the *first moment* of the system about the plane $x = 0$.

If the total mass $\sum_{k=1}^{n} m_k = M$ say, then the *centre of mass* G of the system is the point at which a concentrated mass M would be situated in order to have the same moment about $x = 0$. Let G have coordinate $\bar{x}$, then the position of G is given by

$$\bar{x} M \quad \text{or} \quad \bar{x} \sum_{k=1}^{n} m_k = \sum_{k=1}^{n} x_k m_k. \qquad (13)$$

If the masses are located at points whose coordinates are (x_1, y_1, z_1), $(x_2, y_2, z_2), \ldots, (x_n, y_n, z_n)$ with reference to fixed axes, it follows that G has coordinates $(\bar{x}, \bar{y}, \bar{z})$ given by

$$M\bar{x} = \sum_{k=1}^{n} x_k m_k, \quad M\bar{y} = \sum_{k=1}^{n} y_k m_k, \quad M\bar{z} = \sum_{k=1}^{n} z_k m_k. \qquad (14)$$

The *centroid* of the points $P_1, P_2, P_3, \ldots, P_n$ is the geometric point defined by equations (14) when the masses m_k are taken equal so that the centroid or *mean centre* of the system of points is the point whose coordinates are

$$\bar{x} = \frac{1}{n} \sum_{k=1}^{n} x_k, \quad \bar{y} = \frac{1}{n} \sum_{k=1}^{n} y_k, \quad \bar{z} = \frac{1}{n} \sum_{k=1}^{n} z_k \qquad (15)$$

In order to find the centre of mass of most physical objects it is necessary to make certain simplifying assumptions. One is that matter in a given solid is continuously distributed throughout the solid. Another is that if P is a point in the solid and δV is an element of volume containing P whose mass is δm, then the ratio $\delta m / \delta V$ tends to a definite limit called the density of the solid at P, and denoted by $\rho(P)$.

7.2.4 Centre of mass of a plane lamina

Consider the plane lamina formed by the area A enclosed between the curve $y = f(x)$, the x-axis from $x = a$ to $x = b$ and the ordinates $x = a, x = b$, see Fig. 7.10.

Let the coordinates of the centre of mass of this area be $(\bar{x}, \bar{y})$. The elementary area is $y \, \delta x$ so that the elementary mass $\delta m = \rho y \, \delta x$ and its centre of mass is approximately the point $G\,(x, \frac{1}{2}y)$. Taking moments about the x- and y-axes,

$$\bar{x} \sum y \delta x \, \rho = \sum xy \, \delta x \, \rho$$
$$\bar{y} \sum y \delta x \, \rho = \sum \tfrac{1}{2} y^2 \, \delta x \, \rho. \qquad (16)$$

If ρ is constant, mass is proportional to area and, in the limit (16) is equivalent to

$$A\bar{x} = \int_{a}^{b} yx \, dx$$

$$A\bar{y} = \int_a^b \tfrac{1}{2}y^2 \, dx \qquad (17)$$

i.e. $(\bar{x}, \bar{y})$ are the coordinates of the centroid of the area.

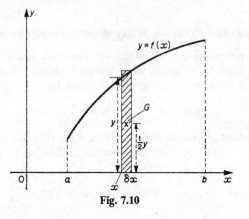

Fig. 7.10

Example (i) Find the coordinates of the centroid of a plane semi-circular lamina of radius a (Fig. 7.11).

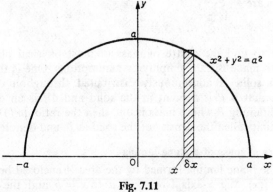

Fig. 7.11

Solution This is the area under the curve

$$y = \sqrt{(a^2 - x^2)}, \qquad -a \leqslant x \leqslant a$$

Total area $A = \tfrac{1}{2}\pi a^2$.

Let the coordinates of the centroid G be $(\bar{x}, \bar{y},)$, then $\bar{x} = 0$ from symmetry, and

$$A\bar{y} = \int_{-a}^a \tfrac{1}{2}y^2 \, dx$$

$$= \int_{-a}^a \tfrac{1}{2}(a^2 - x^2) \, dx$$

$$= \int_0^a (a^2 - x^2) \, dx = \tfrac{2}{3}a^3,$$

then
$$\bar{y} = \frac{4a}{3\pi}.$$

Example (ii) Find the coordinates of the centroid of the area contained between the x-axis and the cycloid $x = a(t - \sin t)$, $y = a(1 - \cos t)$, $0 \leqslant t \leqslant 2\pi$.

Solution From example (ii) of section 7.2.1 the area A enclosed is $3\pi a^2$. Let the centroid G have coordinates $(\bar{x}, \bar{y})$, then from symmetry $\bar{x} = a$, or alternatively $\bar{x}$ may be found from

$$A\bar{x} = \int_0^{2\pi} a(t - \sin t)a(1 - \cos t)a(1 - \cos t) \, dt$$

since $\quad dx = a(1 - \cos t) \, dt$.

Also, $\quad A\bar{y} = \int_0^{2\pi} \tfrac{1}{2}a^2(1 - \cos t)^2 a(1 - \cos t) \, dt$

$$= \tfrac{1}{2}a^3 \int_0^{2\pi} (\tfrac{5}{2} - 4 \cos t + \tfrac{3}{2} \cos 2t + \sin^2 t \cos t) \, dt$$

$$= \tfrac{1}{2}a^3 \left[\tfrac{5}{2}t - 4 \sin t + \tfrac{3}{4} \sin 2t + \tfrac{1}{3} \sin^3 t \right]_0^{2\pi}$$

$$= \tfrac{5}{2}\pi a^3.$$

Hence $\quad \bar{y} = \tfrac{5}{6}a$ and G is $(a, \tfrac{5}{6}a)$.

7.2.5 The centroid of the volume of a solid of revolution

Let the solid be generated by rotating the plane area under the curve $y = f(x)$ from $x = a$ to $x = b$ about the x-axis. Using (11) of section 7.2.2 the total volume V is given by

$$V = \int_a^b \pi y^2 \, dx.$$

The elementary volume is $\pi y^2 \delta x$ and its centroid is at $(x, 0)$ approximately. Hence $\bar{y} = 0$ and taking moments about the y-axis, summing and proceeding to the limits,

$$V\bar{x} = \int_a^b \pi y^2 x \, dx. \tag{18}$$

Example (i) Find the coordinates of the centroid of a solid hemisphere with radius a.

Solution Let the hemisphere be formed by rotating that part of the area of the circle bounded by $x^2 + y^2 = a^2$ for which $x \geqslant 0$, $y \geqslant 0$ about the x-axis (see Fig. 7.12).
Let the coordinates of the centroid G be $(\bar{x}, \bar{y})$, then $\bar{y} = 0$ and $V = \tfrac{2}{3}\pi a^3$. From equation (18),

$$V\bar{x} = \int_0^a \pi y^2 x \, dx$$

$$= \pi \int_0^a x(a^2 - x^2) \, dx$$
$$= \tfrac{1}{4}\pi a^4$$

and $\bar{x} = \tfrac{3}{8}a$. Hence, G is at $(\tfrac{3}{8}a, 0)$.

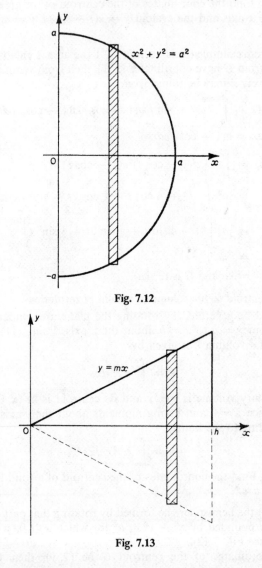

Fig. 7.12

Fig. 7.13

Example (ii) Find the coordinates of the centre of mass of a right circular cone of height h the density of which varies as the square of the distance along the axis from the vertex.

Solution Choose the vertex of the cone as the origin of the coordinates and the axis of the cone as the x-axis. Let the cone be generated by rotating about the x-axis the area enclosed by $y = mx$, $0 \leqslant x \leqslant h$ the x-axis between $x = 0$ and $x = h$ and the ordinate $x = h$. Let the density be kx^2 where k is a constant. Let the coordinates of the centre of mass G be $(\bar{x}, \bar{y})$ then $\bar{y} = 0$ by symmetry, see Fig. 7.13.

$$\text{Total mass} = \text{limit of } \sum \pi y^2 \delta x \, kx^2$$

$$= k\pi \int_0^h m^2 x^4 \, dx$$

$$= \tfrac{1}{5} k\pi m^2 h^5.$$

Taking moments about Oy,

$$\tfrac{1}{5} k\pi m^2 h^5 \bar{x} = \int_0^h (\pi y^2 \, dx)(kx^2)x$$

$$= k\pi \int_0^h m^2 x^5 \, dx$$

$$= \tfrac{1}{6} k\pi m^2 h^6$$

and $\bar{x} = \tfrac{5}{6}h$ and G is $(\tfrac{5}{6}h, 0)$.

The reader is left to check that if the density is constant, so that mass is proportional to volume, then G is the point $(\tfrac{3}{4}h, 0)$.

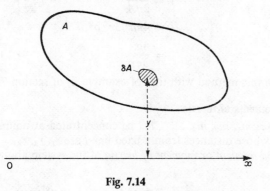

Fig. 7.14

7.2.6 Pappus' theorem

When a plane area A, as in Fig. 7.14, is rotated about an axis in its plane which does not intersect the area, there is a relation between the centroid of the area and the volume of the solid of revolution formed by rotating the area about the axis. This relation is a formula in the form of the theorem below, of use in calculating the volume when the centroid is known, or in calculating the centroid when the volume is known.

For, let the axis of revolution be the x-axis, and δA an element of area distant y from the x-axis. If V is the volume obtained by rotating A about Ox, δV the corresponding element of volume obtained by rotating δA about Ox, and $\bar{y}$ the coordinate of the centroid of A, then

$$\delta V \simeq 2\pi y\, \delta A.$$

Hence
$$
\begin{aligned}
V &= \text{Lim} \sum 2\pi y\, \delta A \\
&= 2\pi\, \text{Lim} \sum y\, \delta A \\
&= 2\pi\ (\text{moment of area about } Ox) \\
&= 2\pi(A\bar{y})
\end{aligned}
$$

or
$$V = 2\pi\bar{y}A, \tag{19}$$

and *Pappus' theorem* states that if a plane closed area is completely rotated about an axis in its own plane which does not intersect the area, the volume of the solid formed is equal to the product of the area and the length of the path traced out by the centroid of this area. Note that the axis may touch the area.

Example Find the centroid of a plane semicircular lamina of radius a.

Solution The semicircular lamina, as in Fig. 7.11, is chosen to be that bounded by the curve $x^2 + y^2 = a^2, y \geqslant 0$ and the x-axis, $-a \leqslant x \leqslant a$. Hence $A = \frac{1}{2}\pi a^2$, $V = 4\pi a^3/3$,

and
$$
\begin{aligned}
\bar{y} &= \frac{V}{2\pi A} \\
&= 4\pi a^3/3\pi^2\, a^2 \\
&= \frac{4a}{3\pi}.
\end{aligned}
$$

Compare this method with that of example (i) of section 7.2.4.

7.2.7 Moments of inertia

Let masses $m_1, m_2, m_3, \ldots, m_n$ be concentrated at points $P_1, P_2, P_3, \ldots, P_n$ whose distances from a fixed line l are $r_1, r_2, r_3, \ldots, r_n$. Then the *second moment* or *moment of inertia* of the system about l is defined as

$$I_l = m_1 r_1^2 + m_2 r_2^2 + m_3 r_3^2 + \ldots + m_n r_n^2 = \sum_{i=1}^{n} m_i r_i^2 \tag{20}$$

Let M be the total mass of the system, then the equation

$$Mk_l^2 = I_l \tag{21}$$

defines k_l, the *radius of gyration* of the system about l.

If instead of a system of discrete masses there is a continuous distribution of mass, the assumption is made that the mass may be divided into small elements of mass δm such that, if r represents the distance of one point of the element δm from an axis, then all points of that element will be within distances $r \pm c$ of the axis where $c \rightarrow 0$ as the largest value of $\delta m \rightarrow 0$. Then for each element of mass, its moment of inertia lies between $(r - c)^2 \delta m$ and $(r + c)^2 \delta m$ and hence differs from $r^2 \delta m$ by a quantity which is 'as small as we please' as $\delta m \rightarrow 0$, i.e.

$$\underset{\delta m \rightarrow 0}{\text{Lim}} \sum r^2 \delta m = \int r^2 \, dm = I. \tag{22}$$

Example (i) Find the moment of inertia of a thin uniform rod of mass M and length l about an axis which is the perpendicular bisector of the rod.

Solution Choose the origin as the mid-point of the rod and the x-axis along the rod. Then (22) gives

$$I = \int_{-\frac{1}{2}l}^{\frac{1}{2}l} x^2 \frac{M}{l} \, dx$$

$$= \frac{Ml^2}{12}.$$

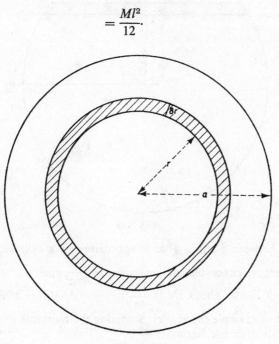

Fig. 7.15

Example (ii) Find the moment of inertia of a uniform disc of mass M, radius a about an axis through its centre perpendicular to the plane of the disc.

Solution Consider the disc split into concentric circles so that at distance r from the centre the width of an element of area is δr, see Fig. 7.15.

Then $\delta m \simeq 2\pi r \, \delta r (M/\pi a^2) = (2M/a^2) r \, \delta r$ and

$$I = \int_0^a \frac{2M}{a^2} r^3 \, dr = \tfrac{1}{2} M a^2.$$

Example (iii) Find the moment of inertia of a solid homogeneous sphere of total mass M and radius a about a diameter.

Solution Choose axes Ox, Oy, Oz through the centre of the sphere. Let the sphere be cut into elementary discs by planes perpendicular to the z-axis at intervals δz as in Fig. 7.16.

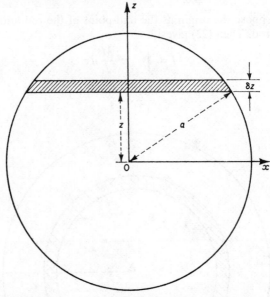

Fig. 7.16

The disc between z and $z + \delta z$ is approximately a cylinder of radius $\sqrt{(a^2 - z^2)}$, thickness δz and mass $\delta m = \dfrac{3M}{4\pi a^3} \pi (a^2 - z^2)\delta z$, and its moment of inertia about Oz is $\dfrac{3M}{4a^3} (a^2 - z^2)\delta z \, \tfrac{1}{2}(a^2 - z^2)$, (applying the result of example (ii) above). Summing the moments of inertia for the separate discs and letting $\delta z \to 0$

$$I_{Oz} = \int_{-a}^a \frac{3M}{8a^3} (a^2 - z^2)^2 \, dz$$
$$= \tfrac{2}{5} M a^2.$$

The moment of inertia of a rigid body about a line is important when the rotation of the body is being considered. Moments of inertia of only the simplest bodies can be found by integration. There are two theorems which may be used to calculate moments of inertia.

The *parallel axis theorem* relates the moment of inertia of a body about an axis through its centre of mass G to the moment of inertia about a parallel axis l distant d from G (see Fig. 7.17).

Let I_G be the moment of inertia of a body of mass M about an axis Gz through the centre of mass G, and let I_l be the moment of inertia about

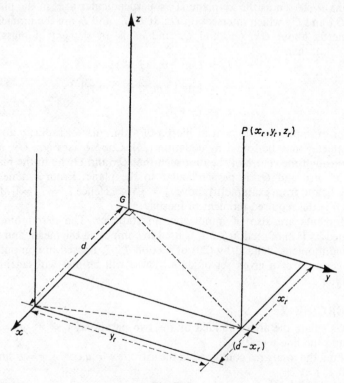

Fig. 7.17

a line l parallel to Gz so that l is distant d from Gz. Let m_r at $P(x_r, y_r, z_r)$ be an element of mass of the body. Then, from formula (20),

$$I_l = \text{Lim} \sum_r m_r\{y_r^2 + (d - x_r)^2\}$$

$$= \text{Lim} \left[\sum_r m_r(x_r^2 + y_r^2) - 2d \sum_r m_r x_r + d^2 \sum m_r\right],$$

i.e. $$I_l = I_G + d^2 M. \tag{23}$$

The term $\sum m_r x_r$ is zero since G is the centre of mass.

For example, using the result of example (i) of section 7.2.7 and the parallel axis theorem, the moment of inertia of a thin uniform rod of length l about an axis through one end perpendicular to the rod is $Ml^2/12 + (\tfrac{1}{2}l)^2 M$ which is $\tfrac{1}{3}Ml^2$.

A second theorem, for plane laminas only, which is sometimes called the *perpendicular axis theorem*, states that the moment of inertia of a plane lamina about an axis perpendicular to its plane is equal to the sum of the moments of inertia about any two perpendicular axes in the plane which intersect on the first axis. For suppose the plane of the lamina is chosen as the xy-plane. Two perpendicular axes in the plane are Ox and Oy which intersect on Oz. If I_1, I_2, and I_3 are the moments of inertia about Ox, Oy, and Oz and m_r is an element of mass at $(x_r, y_r, 0)$, then

$$I_3 = \text{Lim} \sum m_r(x_r^2 + y_r^2)$$
$$= \text{Lim} \left[\sum m_r x_r^2 + \sum m_r y_r^2 \right]$$

or $\qquad\qquad I_3 = I_2 + I_1.$ \hfill (24)

For example, the moment of inertia of a thin disc of radius a about a diameter may be found by equation (24). Choose axes Ox, Oy, and Oz through the centre of the disc such that Ox and Oy lie in the plane of the disc and Oz is perpendicular to the plane. From symmetry, $I_1 = I_2$, and from example (ii) above $I_3 = \frac{1}{2}Ma^2$. Hence $I_1 = I_2 = \frac{1}{4}Ma^2$, which is the required moment of inertia.

Moments are also of importance in statistics. The *first* moment, defined by (12) of section 7.2.3, is used in computing the mean, and the *second* moment, defined by (20) of section 7.2.7, is used in computing the variance of a given set of data. These will be met with again in Chapter 15.

EXERCISE 7

1 Calculate the area bounded by the two parabolas $y = x^2$, $y = \frac{1}{2}x^2$ and the line $y = 2x$.

2 Find the total area contained in the astroid $x = a \cos^3 t$, $y = b \sin^3 t$, $0 \leqslant t \leqslant 2\pi$.

3 Find the area bounded by the cardioid $x = a(2 \cos t - \cos 2t)$, $y = a(2 \sin t - \sin 2t)$, $0 \leqslant t \leqslant 2\pi$.

4 Find the total area inside the circle $r = 2a \sin \theta$.

5 Find the volume of the solid generated by rotating about the x-axis that part of the parabola $y = 2x - x^2$ which lies above the x-axis.

6 Prove that the volume of a spherical cap of height h cut from a sphere of radius a is $\frac{1}{3}\pi h^2(3a - h)$.

7 The centres of two spheres of radius 3 cm and 4 cm respectively are at a distance 5 cm apart. Prove that the volume common to both spheres is $92\pi/15$ cm^3.

8 Sketch the curve $y = x \cos x$ for $0 \leqslant x \leqslant \frac{1}{2}\pi$, and find the volume generated when the area contained between this curve and the x-axis is rotated about the x-axis.

9 Find the area of the region enclosed by the curves $x^2 = ay$, $ay^2 = x^3$ and find also the volume generated when this region is rotated about the x-axis.

10 Show that the curve $x(x^2 + y^2) = y^2$ lies entirely in the region $0 \leqslant x \leqslant 1$, and make a rough sketch of the curve. Find the area of the region in the first quadrant lying between the curve and the line $x = 1$ and find the x coordinate of the centroid of this region. (Hint: Use the substitution $x = \sin^2 \theta$ to evaluate the integral.)

11 Find the area bounded by one loop of the curve $y^2 = x^2(4 - x^2)$ and the coordinates of the centroid of this area. Hence, or otherwise find the volume generated when this area is rotated about the y-axis.

12 Use the substitution $x = 2 \sin^2 \theta$ to evaluate $\displaystyle\int_0^2 x^n (2 - x)^{\frac{1}{2}} \, dx$ where n is a positive integer. Find the area enclosed by the loop of the curve $y^2 = x^4(2 - x)$ and the coordinates of the centroid of that part of this area which lies in the first quadrant.

13 Sketch the curve $x = \cos^3 t$, $y = \sin t$, $0 \leqslant t \leqslant 2\pi$. Find the area enclosed by the curve and the volume of the solid formed when the upper half of the area is rotated about the x-axis.

14 Find the points of intersection of the curves $r = a(1 + \cos \theta)$ and $r = 3a(1 - \cos \theta)$ and show that the area enclosed between these curves is $(11\pi/2 - 9\sqrt{3})a^2$.

15 Find the centroid of the area bounded by the lines $y = 0$, $x = at^2$ and that part of the parabola $y^2 = 4ax$ from the origin to the point $(at^2, 2at)$, where a and t are both positive. Find the volumes generated when this area is rotated (i) about the x-axis, (ii) about the y-axis. Find the value of t such that these two volumes are equal.

16 The area bounded by the parabola $y = x^2$ and the line $y = 4$ is rotated about the y-axis to generate a solid. Find the moment of inertia of this solid about the y-axis, if the total mass of the solid is M.

17 Find the centroid of the sector of a circle of radius a which subtends an angle 2α at the centre. Hence, or otherwise find the volume of the solid generated when the sector is rotated through one revolution about a bounding radius.

18 Find the moment of inertia of a homogeneous solid sphere of mass M and radius a about an axis which is a tangent to the sphere.

19 A right circular cone has a base radius a and altitude $2a$ and mass M. Find the moment of inertia of the volume of the cone about its axis.

20 Find the coordinates of the centroid of the area in the first quadrant bounded by two concentric circles and the coordinate axes, if the circles have radii a and b, $b > a > 0$, and the origin is their common centre. Find also the limiting value of these coordinates as a approaches b and discuss the meaning of this result.

21 The area bounded by the curve $y^2 = 4ax$, the line $x = a$ and the x-axis is rotated (i) about the x-axis, (ii) about the line $x = a$, (iii) about the y-axis. Find the volume generated in each case.

8
Matrices and Determinants

8.0 INTRODUCTION
The linear algebraic equation of the type

$$y_1 = 3x_1 + 7x_2 + 6x_3$$

is of common occurrence in engineering. For example, the x's could be the electrical currents flowing in the branches of a resistive network as in Fig. 8.1 and the y_1 the voltage drop across the network.

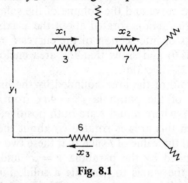

Fig. 8.1

Again, y_1 could be the weekly output of a factory which uses three different kinds of machine making the same product; the x's are the quantities of each machine installed and the weekly outputs of the machines are 3, 7, and 6 respectively.

It is usual to have to consider a set of two or more such linear algebraic equations and it is straightforward to handle up to three such equations with integer coefficients without using other than simple algebra. However, if the coefficients are decimal or if there are many more than three equations it becomes necessary to employ a systematic algebraic process which will permit the important features of the set of equations to be examined and manipulated, and the information contained in the set to be stored for reference without necessarily having to solve the equations.

The most important item of information contained within a set of equations is that provided by the coefficients of the algebraic terms. For example, the pair of equations

$$y_1 = 3x_1 + 7x_2 + 6x_3$$
$$y_2 = 5x_1 - 2x_2 + 3x_3$$

may be used to calculate the y's from given values of the x's. No matter how the x's are changed the y's are immediately determinable and are controlled by the coefficients of the right side, that is by the values of the numbers in the *ordered array*

$$\begin{pmatrix} 3 & 7 & 6 \\ 5 & -2 & 3 \end{pmatrix}.$$

Any rectangular array which stores definite information is called a *matrix*. A matrix has no numerical value, it is simply an assembly of terms which as a carrier of information may be manipulated according to certain algebraic rules. It is necessary in any such manipulation to know the exact nature of the information, for example the matrix

$$\begin{pmatrix} 6 & 2 & 5 \\ 7 & 5 & 1 \\ 12 & 0 & 1 \end{pmatrix}$$

becomes meaningful only when the 'key' is given as in

	W	L	D
Newcastle United	6	2	5
Norwich City	7	5	1
Newtown Wanderers	12	0	1

8.1 DEFINITION AND NOTATION

A matrix A of m rows and n columns is said to be of *order* $m \times n$. The components of the matrix are called *elements*. A double suffix notation is used, the element of the ith row and the jth column of the matrix A being written a_{ij}. The array is usually enclosed in curved brackets ( ). Other notations in common use are square brackets [ ] and double bars $\| \ \|$. When there is no ambiguity or doubt the matrix may be written A, (a_{ij}), or simply a_{ij}. Thus

$$A = (a_{ij}) = a_{ij} = \begin{pmatrix} a_{11} & a_{12} & a_{13} & \cdots & a_{1n} \\ a_{21} & a_{22} & a_{23} & \cdots & a_{2n} \\ a_{31} & a_{32} & a_{33} & \cdots & a_{3n} \\ \cdot & \cdot & \cdot & & \cdot \\ \cdot & \cdot & \cdot & & \cdot \\ \cdot & \cdot & \cdot & & \cdot \\ a_{m1} & a_{m2} & a_{m3} & \cdots & a_{mn} \end{pmatrix} \qquad (1)$$

An important matrix is the square matrix of order $n \times n$, i.e. one of n rows and n columns. Such a matrix occurs for instance when dealing with a linear transformation from one set of space variables to another. In the two-dimensional Fig. 8.2 the point P has coordinates (a, b) when referred to the rectangular Cartesian system Oxy, and coordinates (A, B) when referred to the system OXY.

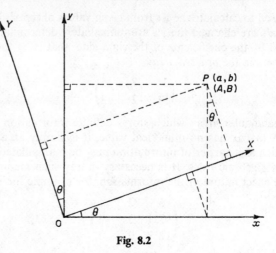

Fig. 8.2

The transformation is given by

$$A = a \cos \theta + b \sin \theta \\ B = -a \sin \theta + b \cos \theta \tag{2}$$

and the *transformation matrix* is the square matrix of order 2

$$\begin{pmatrix} \cos \theta & \sin \theta \\ -\sin \theta & \cos \theta \end{pmatrix}. \tag{3}$$

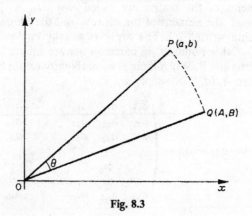

Fig. 8.3

The matrix (3) can also be interpreted as representing a rotation of the coordinate axes Oxy through an angle θ to the position of OXY. Alternatively if the coordinate frame Oxy is fixed as in Fig. 8.3 the matrix (3) can be interpreted as representing the rotation of the line OP, where P has coordinates (a, b), through an angle θ to take up the position OQ, where Q has coordinates (A, B). Changing the sense of rotation would then mean that the matrix

$$\begin{pmatrix} \cos \theta & -\sin \theta \\ \sin \theta & \cos \theta \end{pmatrix} \tag{4}$$

can represent the rotation of the line OQ in the plane of Oxy through an angle θ in the anticlockwise sense.

A matrix with $m = 1$

$$(a_{11} \quad a_{12} \quad a_{13} \ldots a_{1n})$$

is called a row matrix of order n, and a matrix with $n = 1$

$$\begin{pmatrix} a_{11} \\ a_{21} \\ a_{31} \\ \cdot \\ \cdot \\ \cdot \\ a_{m1} \end{pmatrix}$$

is called a column matrix of order m.

A general linear transformation of m variables y_i, $i = 1, 2, 3, \ldots, m$ and n variables x_j, $j = 1, 2, 3, \ldots, n$ has the form

$$\left. \begin{aligned} y_1 &= a_{11}x_1 + a_{12}x_2 + a_{13}x_3 + \ldots + a_{1n}x_n \\ y_2 &= a_{21}x_1 + a_{22}x_2 + a_{23}x_3 + \ldots + a_{2n}x_n \\ &\quad \cdot \qquad \cdot \qquad \cdot \qquad \cdot \qquad \cdot \\ &\quad \cdot \qquad \cdot \qquad \cdot \qquad \cdot \qquad \cdot \\ &\quad \cdot \qquad \cdot \qquad \cdot \qquad \cdot \qquad \cdot \\ y_m &= a_{m1}x_1 + a_{m2}x_2 + a_{m3}x_3 + \ldots + a_{mn}x_n. \end{aligned} \right\} \tag{5}$$

In (5) the variables y_i and x_j may be represented by two column matrices

$$y = \begin{pmatrix} y_1 \\ y_2 \\ \cdot \\ \cdot \\ \cdot \\ y_m \end{pmatrix} \qquad x = \begin{pmatrix} x_1 \\ x_2 \\ \cdot \\ \cdot \\ \cdot \\ x_n \end{pmatrix}$$

and the transformation matrix by $A = (a_{ij})$. The equations may be written in compact notation in the form

$$y_i = \sum_{j=1}^{n} a_{ij}x_j \qquad i = 1, 2, 3, \ldots, m$$

or simply $\qquad\qquad\qquad y_i = a_{ij}x_j \tag{6}$

where the repeated index j indicates a summation from $j = 1$ to $j = n$. It will be seen in section 8.2.6 that relation (6) is simply a matrix multiplication and that (5) can be written in matrix form

$$y = Ax.$$

If i, j take only the values 1, 2, 3 then equation (6) can be considered as a relationship connecting together the pair of rectangular Cartesian coordinates (x_1, x_2, x_3) and (y_1, y_2, y_3). Suppose P is the point (x_1, x_2, x_3) and Q is the point (y_1, y_2, y_3) such that $OP = OQ$, then OQ can be obtained by rotating OP through the angle POQ in the plane of OP and OQ. The matrix a_{ij} of equation (6) can then be thought of as representing this rotation and must carry information concerning both the magnitude of the angle and the location of the plane in which it lies. This representation of a matrix is investigated further in section 9.3.7.

8.2 MATRIX ALGEBRA
Consider now the basic algebraical manipulations of addition, subtraction, and division as applied to matrix arrays.

8.2.1 Equality of matrices
Two matrices are equal if and only if they are of the same order and the corresponding elements are equal.

8.2.2 Addition
Consider the linear transformation

$$\begin{aligned} y_1 &= a_{11}x_1 + a_{12}x_2 \\ y_2 &= a_{21}x_1 + a_{22}x_2 \\ y_3 &= a_{31}x_1 + a_{32}x_2 \end{aligned} \tag{7}$$

having the transformation matrix

$$A = \begin{pmatrix} a_{11} & a_{12} \\ a_{21} & a_{22} \\ a_{31} & a_{32} \end{pmatrix} \tag{8}$$

and a second transformation

$$\begin{aligned} z_1 &= b_{11}x_1 + b_{12}x_2 \\ z_2 &= b_{21}x_1 + b_{22}x_2 \\ z_3 &= b_{31}x_1 + b_{32}x_2 \end{aligned} \tag{9}$$

having the transformation matrix

$$B = \begin{pmatrix} b_{11} & b_{12} \\ b_{21} & b_{22} \\ b_{31} & b_{32} \end{pmatrix}. \tag{10}$$

If new variables w_1, w_2, and w_3 are constructed by adding the respective y and z variables, then equations

$$\begin{aligned} w_1 &= y_1 + z_1 = (a_{11} + b_{11})x_1 + (a_{12} + b_{12})x_2 \\ w_2 &= y_2 + z_2 = (a_{21} + b_{21})x_1 + (a_{22} + b_{22})x_2 \\ w_3 &= y_3 + z_3 = (a_{31} + b_{31})x_1 + (a_{32} + b_{32})x_2 \end{aligned} \tag{11}$$

form a linear transformation having a matrix

$$\begin{pmatrix} a_{11} + b_{11} & a_{12} + b_{12} \\ a_{21} + b_{21} & a_{22} + b_{22} \\ a_{31} + b_{31} & a_{32} + b_{32} \end{pmatrix}. \tag{12}$$

This process may be extended and leads to the rule of *matrix addition*.

Two matrices A and B, each of order $m \times n$, when added together form another matrix of order $m \times n$ whose elements are the sums of the associated elements of A and B, i.e.

$$A + B = (a_{ij}) + (b_{ij}) = (a_{ij} + b_{ij}) \tag{13}$$

Note 1 No meaning can be attached to the addition of matrices of different orders.

Note 2 The result (13) may be extended to any finite number of matrices, i.e.

$$A + B + C = (a_{ij}) + (b_{ij}) + (c_{ij}) = (a_{ij} + b_{ij} + c_{ij}) \tag{14}$$

and so on.

Note 3 The *commutative law* and the *associative law* are true, i.e.

$$A + B = B + A \quad \text{and} \quad (A + B) + C = A + (B + C).$$

Example

$$\begin{pmatrix} 3 & 6 \\ 7 & 2 \\ 5 & 0 \end{pmatrix} + \begin{pmatrix} 0 & 2 \\ -1 & 5 \\ 3 & -2 \end{pmatrix} = \begin{pmatrix} 3 & 8 \\ 6 & 7 \\ 8 & -2 \end{pmatrix}.$$

8.2.3 Multiplication by a scalar

Suppose new variables

$$p_1 = \lambda y_1, \quad p_2 = \lambda y_2, \quad \text{and} \quad p_3 = \lambda y_3$$

are defined, where λ is a numerical magnitude usually called a *scalar*, and used with equations (7), then

$$\left.\begin{aligned} p_1 &= \lambda a_{11} x_1 + \lambda a_{12} x_2 \\ p_2 &= \lambda a_{21} x_1 + \lambda a_{22} x_2 \\ p_3 &= \lambda a_{31} x_1 + \lambda a_{32} x_2 \end{aligned}\right\} \tag{15}$$

is a linear transformation with a matrix (λa_{ij}). This suggests the following rule.

Scalar multiplication

If a matrix is multiplied by a scalar λ then every element is multiplied by λ, i.e.

$$\lambda A = \lambda(a_{ij}) = (\lambda a_{ij}).$$

Example (i) $3\begin{pmatrix} 2 & 3 & 1 \\ 4 & 3 & 0 \end{pmatrix} = \begin{pmatrix} 6 & 9 & 3 \\ 12 & 9 & 0 \end{pmatrix}.$

Example (ii) $-1\begin{pmatrix} 3 & -2 \\ -1 & 5 \end{pmatrix} = \begin{pmatrix} -3 & 2 \\ 1 & -5 \end{pmatrix}.$

Conversely, a matrix has a factor λ if each element contains the factor λ.

8.2.4 Linear combination of matrices

The rules of sections 8.2.2 and 8.2.3 lead to the combination rule in the form

$$\lambda A + \mu B + \nu C = (\lambda a_{ij} + \mu b_{ij} + \nu c_{ij}) \tag{16}$$

where λ, μ, and ν are scalar quantities. In particular if $\nu = 0$, $\lambda = 1$, and $\mu = -1$, then

$$A - B = (a_{ij} - b_{ij}). \tag{17}$$

8.2.5 Zero matrix

The subtraction of two equal matrices of order $m \times n$ gives a zero matrix of order $m \times n$, that is one whose every element is zero.

Example

$$2\begin{pmatrix} 3 & 2 & 1 \\ 1 & 2 & -3 \end{pmatrix} + 3\begin{pmatrix} 4 & 3 & 2 \\ 2 & 3 & -2 \end{pmatrix} - \begin{pmatrix} 18 & 13 & 8 \\ 8 & 13 & -12 \end{pmatrix}$$

$$= \begin{pmatrix} 6 & 4 & 2 \\ 2 & 4 & -6 \end{pmatrix} + \begin{pmatrix} 12 & 9 & 6 \\ 6 & 9 & -6 \end{pmatrix} - \begin{pmatrix} 18 & 13 & 8 \\ 8 & 13 & -12 \end{pmatrix}$$

$$= \begin{pmatrix} 18 & 13 & 8 \\ 8 & 13 & -12 \end{pmatrix} - \begin{pmatrix} 18 & 13 & 8 \\ 8 & 13 & -12 \end{pmatrix}$$

$$= \begin{pmatrix} 0 & 0 & 0 \\ 0 & 0 & 0 \end{pmatrix}.$$

8.2.6 Multiplication of matrices

Consider the successive linear transformations

$$\begin{aligned} y_1 &= a_{11}x_1 + a_{12}x_2 \\ y_2 &= a_{21}x_1 + a_{22}x_2 \\ y_3 &= a_{31}x_1 + a_{32}x_2 \end{aligned} \quad \text{and} \quad \begin{aligned} x_1 &= b_{11}z_1 + b_{12}z_2 \\ x_2 &= b_{21}z_1 + b_{22}z_2 \end{aligned}$$

with coefficient matrices

$$A = \begin{pmatrix} a_{11} & a_{12} \\ a_{21} & a_{22} \\ a_{31} & a_{32} \end{pmatrix} \quad \text{and} \quad B = \begin{pmatrix} b_{11} & b_{12} \\ b_{21} & b_{22} \end{pmatrix}$$

respectively.

They are equivalent to a single linear transformation

$$y_1 = (a_{11}b_{11} + a_{12}b_{21})z_1 + (a_{11}b_{12} + a_{12}b_{22})z_2$$

$$y_2 = (a_{21}b_{11} + a_{22}b_{21})z_1 + (a_{21}b_{12} + a_{22}b_{22})z_2$$

$$y_3 = (a_{31}b_{11} + a_{32}b_{21})z_1 + (a_{31}b_{12} + a_{32}b_{22})z_2$$

with a coefficient matrix

$$C = \begin{pmatrix} a_{11}b_{11} + a_{12}b_{21} & a_{11}b_{12} + a_{12}b_{22} \\ a_{21}b_{11} + a_{22}b_{21} & a_{21}b_{12} + a_{22}b_{22} \\ a_{31}b_{11} + a_{32}b_{21} & a_{31}b_{12} + a_{32}b_{22} \end{pmatrix}.$$

It is seen that the elements of C are obtained from those of A and B by summing the products of the elements of the rows of A with those of the columns of B. For example, the element of the third row and the first column of C is obtained from the third row of A and the first column of B.

Matrix multiplication

Given a matrix A of order $m \times n$ and a second matrix B of order $n \times p$, then the product AB is a third matrix C of order $m \times p$, such that the element of the ith row and the kth column of C is the sum of the products of the corresponding elements of the ith row of A and the kth column of B. In a more concise notation, if $A = (a_{ij})$ and $B = (b_{jk})$ where $i = 1, 2, 3, \ldots, m; j = 1, 2, 3, \ldots, n;$ and $k = 1, 2, 3, \ldots, p$, then $C = (c_{ik})$ where

$$c_{ik} = a_{ij}b_{jk} \quad \text{(summed over } j) \tag{18}$$
$$= a_{i1}b_{1k} + a_{i2}b_{2k} + a_{i3}b_{3k} + \ldots + a_{in}b_{nk}.$$

Note that for the product AB to exist it is essential that there are the same number of *columns* in A as there are *rows* in B.

Example (i)

$$\begin{pmatrix} 2 & 3 & 1 \\ -1 & 2 & 4 \end{pmatrix} \begin{pmatrix} 5 & 3 \\ 4 & 0 \\ 3 & 1 \end{pmatrix} = \begin{pmatrix} 10 + 12 + 3 & 6 + 0 + 1 \\ -5 + 8 + 12 & -3 + 0 + 4 \end{pmatrix} = \begin{pmatrix} 25 & 7 \\ 15 & 1 \end{pmatrix}$$

$\quad (2 \times 3) \qquad (3 \times 2) \qquad\qquad\qquad\qquad\qquad\qquad\qquad (2 \times 2)$

Example (ii)

$$\begin{pmatrix} 5 & 3 \\ 4 & 0 \\ 3 & 1 \end{pmatrix} \begin{pmatrix} 2 & 3 & 1 \\ -1 & 2 & 4 \end{pmatrix} = \begin{pmatrix} 10 - 3 & 15 + 6 & 5 + 12 \\ 8 + 0 & 12 + 0 & 4 + 0 \\ 6 - 1 & 9 + 2 & 3 + 4 \end{pmatrix} = \begin{pmatrix} 7 & 21 & 17 \\ 8 & 12 & 4 \\ 5 & 11 & 7 \end{pmatrix}$$

$(3 \times 2) \qquad\quad (2 \times 3) \qquad\qquad\qquad\qquad\qquad\qquad\qquad\qquad (3 \times 3)$

Example (iii)

$$(4 \quad 5 \quad 6 \quad -1 \quad 0 \quad 2) \begin{pmatrix} 2 & -1 \\ 3 & 0 \\ 0 & -4 \\ 5 & 3 \\ -2 & 1 \\ 4 & 1 \end{pmatrix}$$

$$\text{(1 × 6)} \qquad \text{(6 × 2)}$$
$$= (8 + 15 + 0 - 5 + 0 + 8 \quad -4 + 0 - 24 - 3 + 0 + 2)$$
$$= (26 \quad -29)$$
$$\text{(1 × 2)}$$

Example (iv)

$$\begin{pmatrix} 2 & 4 \\ 3 & 1 \end{pmatrix} (5 \quad 6 \quad 7) \quad \text{has no meaning.}$$

$$\text{(2 × 2)} \quad \text{(1 × 3)}$$

From the definition of matrix multiplication it is clear that if A is of order $m \times n$ then AB and BA can both exist if and only if B is of order $n \times m$. The product AB is then of order $m \times m$ and the product BA is of order $n \times n$. These two products are of different order if $m \neq n$, and it may be concluded that the product of A and B is *non-commutative*, i.e. $AB \neq BA$. This is illustrated in the first two examples above. Even when $n = m$ and the two products are of the same order they are generally unequal.

Example (v)

$$\begin{pmatrix} 1 & 1 \\ 1 & 1 \end{pmatrix} \begin{pmatrix} 2 & 0 \\ 1 & 0 \end{pmatrix} = \begin{pmatrix} 3 & 0 \\ 3 & 0 \end{pmatrix} \quad \text{and} \quad \begin{pmatrix} 2 & 0 \\ 1 & 0 \end{pmatrix} \begin{pmatrix} 1 & 1 \\ 1 & 1 \end{pmatrix} = \begin{pmatrix} 2 & 2 \\ 1 & 1 \end{pmatrix}.$$

A particular case of the equality $AB = BA$ is discussed in section 8.2.9.

It is possible to multiply AB by a further matrix C providing the matrix order is correct. Given three matrices $A = (a_{ij})$, $B = (b_{jk})$, and $C = (c_{kl})$ of orders $m \times n$, $n \times p$, and $p \times q$ respectively the product AB, of order $m \times p$, is given by

$$AB = (a_{ij}b_{jk}),$$

summed over $j = 1, 2, 3, \ldots, n$. The product of AB and C, i.e. $(AB)C$, of order $m \times q$, is given by

$$(AB)C = (a_{ij}b_{jk}c_{kl})$$

in which each element is a double summation taken over $j = 1, 2, 3, \ldots, n$ and $k = 1, 2, 3, \ldots, p$. The order of the summation is immaterial and the same result would be obtained by forming the product

$A(BC)$. The product of A, B, and C is *associative*. It should be clear that the product is also *distributive*, i.e.

$$(R + S)C = RC + SC$$

where R and S are each of order $r \times p$.

Example (vi) If A and B are arbitrary square matrices of equal order then

$$(A + B)^2 = (A + B)(A + B) = AA + AB + BA + BB.$$

Example (vii) If A is any square matrix then

$$AAA = AA^2 = A^2A = A^3$$

where A^2 means AA.

Example (viii) Verify the associative law of multiplication of matrices by forming the product ABC in two different ways, where

$$A = \begin{pmatrix} 2 & 3 \\ 1 & 0 \\ 1 & 2 \end{pmatrix}, \quad B = \begin{pmatrix} 4 & 1 & 0 & 3 \\ 2 & -1 & -2 & 4 \end{pmatrix}, \quad C = \begin{pmatrix} 0 \\ 1 \\ 2 \\ 1 \end{pmatrix}.$$

Solution

$$ABC = (AB)C = \begin{pmatrix} 14 & -1 & -6 & 18 \\ 4 & 1 & 0 & 3 \\ 8 & -1 & -4 & 11 \end{pmatrix} \begin{pmatrix} 0 \\ 1 \\ 2 \\ 1 \end{pmatrix} = \begin{pmatrix} 5 \\ 4 \\ 2 \end{pmatrix}$$

and

$$ABC = A(BC) = \begin{pmatrix} 2 & 3 \\ 1 & 0 \\ 1 & 2 \end{pmatrix} \begin{pmatrix} 4 \\ -1 \end{pmatrix} = \begin{pmatrix} 5 \\ 4 \\ 2 \end{pmatrix}.$$

Example (ix) The general linear transformation equations (5) of section 8.1 may be written

$$\begin{pmatrix} y_1 \\ y_2 \\ \cdot \\ \cdot \\ \cdot \\ y_m \end{pmatrix} = \begin{pmatrix} a_{11} & a_{12} & \cdots & a_{1n} \\ a_{21} & a_{22} & \cdots & a_{2n} \\ \cdot & \cdot & & \cdot \\ \cdot & \cdot & & \cdot \\ \cdot & \cdot & & \cdot \\ a_{m1} & a_{m2} & \cdots & a_{mn} \end{pmatrix} \begin{pmatrix} x_1 \\ x_2 \\ \cdot \\ \cdot \\ \cdot \\ x_n \end{pmatrix}$$

or simply

$$y = Ax.$$

Example (x) Solve for x and y given

$$\begin{pmatrix} 21 & 6 \\ 2 & -5 \end{pmatrix} \begin{pmatrix} x \\ y \end{pmatrix} = \begin{pmatrix} 24 \\ 19 \end{pmatrix}.$$

Solution Multiply the left side

$$\begin{pmatrix} 21x + 6y \\ 2x - 5y \end{pmatrix} = \begin{pmatrix} 24 \\ 19 \end{pmatrix}$$

and hence, by section 8.2.1, it follows that

$$21x + 6y = 24$$
$$2x - 5y = 19.$$

The pair of equations has the solution $x = 2$, $y = -3$.

Example (xi) Find the points gained by each team shown in the football league table in the section 8.0, on the basis of 3 points for a win (W), 1 point for a draw (D), and 0 points for a lose (L).

Solution The points awarded are as follows:

$$\begin{pmatrix} \text{Newcastle United} \\ \text{Norwich City} \\ \text{Newtown Wanderers} \end{pmatrix} = \begin{pmatrix} 6 & 2 & 5 \\ 7 & 5 & 1 \\ 12 & 0 & 1 \end{pmatrix} \begin{pmatrix} 3 \\ 0 \\ 1 \end{pmatrix}$$

$$= \begin{pmatrix} 18 + 0 + 5 \\ 21 + 0 + 1 \\ 36 + 0 + 1 \end{pmatrix}$$

$$= \begin{pmatrix} 23 \\ 22 \\ 37 \end{pmatrix}.$$

8.2.7 The unit matrix

Suppose an arbitrary rectangular matrix $A = (a_{ij})$ of order $m \times n$ is unchanged after multiplication by a square matrix $B = (b_{jk})$ of order $n \times n$, so that $AB = A$. The elements b_{jk} must then satisfy the equation

$$a_{ik} = a_{ij}b_{jk},$$

(summed over $j = 1, 2, 3, \ldots, n$) with $i = 1, 2, 3, \ldots, m$ and $k = 1, 2, 3, \ldots, n$. This means that the b_{jk} must satisfy a set of equations

$$a_{11} = a_{11}b_{11} + a_{12}b_{21} + \ldots + a_{1n}b_{n1}$$
$$a_{12} = a_{11}b_{12} + a_{12}b_{22} + \ldots + a_{1n}b_{n2}$$
$$a_{13} = a_{11}b_{13} + a_{12}b_{23} + \ldots + a_{1n}b_{n3}$$
$$\text{etc.}$$

Since the $a_{11}, a_{12} \ldots$ are arbitrary, the set of equations leads to

$$b_{11} = b_{22} = b_{33} = \ldots = b_{nn} = 1$$

with all other $b_{jk} = 0$.

The matrix B has a similarity to unity in scalar algebra and is called the *unit matrix* of order n. It is denoted by I_n. For example,

$$I_3 = \begin{pmatrix} 1 & 0 & 0 \\ 0 & 1 & 0 \\ 0 & 0 & 1 \end{pmatrix} \quad \text{and} \quad I_4 = \begin{pmatrix} 1 & 0 & 0 & 0 \\ 0 & 1 & 0 & 0 \\ 0 & 0 & 1 & 0 \\ 0 & 0 & 0 & 1 \end{pmatrix}.$$

Given the matrix A of order $m \times n$ it is easily shown that not only is $AI_n = A$, but also $I_m A = A$, and for this reason the unit matrix is sometimes called the *identity matrix*.

When the order is either understood or immaterial the suffix may be dropped and the unit matrix written I.

Example (i) Evaluate IA and AI when $A = \begin{pmatrix} 3 & 7 & 2 \\ 1 & -4 & 3 \end{pmatrix}$.

Solution

$$IA = \begin{pmatrix} 1 & 0 \\ 0 & 1 \end{pmatrix}\begin{pmatrix} 3 & 7 & 2 \\ 1 & -4 & 3 \end{pmatrix} = \begin{pmatrix} 3 & 7 & 2 \\ 1 & -4 & 3 \end{pmatrix}$$

and $\quad AI = \begin{pmatrix} 3 & 7 & 2 \\ 1 & -4 & 3 \end{pmatrix}\begin{pmatrix} 1 & 0 & 0 \\ 0 & 1 & 0 \\ 0 & 0 & 1 \end{pmatrix} = \begin{pmatrix} 3 & 7 & 2 \\ 1 & -4 & 3 \end{pmatrix}.$

Example (ii) If

$$A = \begin{pmatrix} 2 & 1 & 1 \\ 1 & 2 & 1 \\ 1 & 1 & 2 \end{pmatrix} \quad \text{show that } A^2 - 5A + 4I = 0.$$

Solution

$$A^2 = \begin{pmatrix} 2 & 1 & 1 \\ 1 & 2 & 1 \\ 1 & 1 & 2 \end{pmatrix}\begin{pmatrix} 2 & 1 & 1 \\ 1 & 2 & 1 \\ 1 & 1 & 2 \end{pmatrix} = \begin{pmatrix} 6 & 5 & 5 \\ 5 & 6 & 5 \\ 5 & 5 & 6 \end{pmatrix}$$

$$-5A = \begin{pmatrix} -10 & -5 & -5 \\ -5 & -10 & -5 \\ -5 & -5 & -10 \end{pmatrix}$$

$$4I = \begin{pmatrix} 4 & 0 & 0 \\ 0 & 4 & 0 \\ 0 & 0 & 4 \end{pmatrix}$$

and $A^2 - 5A + 4I = 0$.

A matrix of the form λI where λ is a scalar is called a *scalar matrix*. The latter is itself a particular case of a matrix with non-zero diagonal terms and all other terms zero. A matrix such as

$$\begin{pmatrix} a_{11} & 0 & 0 & \ldots & 0 \\ 0 & a_{22} & 0 & \ldots & 0 \\ 0 & 0 & a_{33} & \ldots & 0 \\ \cdot & \cdot & \cdot & & \cdot \\ \cdot & \cdot & \cdot & & \cdot \\ \cdot & \cdot & \cdot & & \cdot \\ 0 & 0 & 0 & \ldots & a_{nn} \end{pmatrix}$$

is called a *diagonal matrix*.

8.2.8 The transposed matrix

Given an $m \times n$ matrix $A = (a_{ij})$, the transpose of A is written A' or A^T and is that matrix of order $n \times m$ obtained from A by interchanging the rows and columns. For example,

$$A = \begin{pmatrix} 6 & 2 & 4 & 1 \\ 7 & 3 & 2 & 8 \end{pmatrix} \quad \text{and} \quad A' = A^T = \begin{pmatrix} 6 & 7 \\ 2 & 3 \\ 4 & 2 \\ 1 & 8 \end{pmatrix}.$$

8.2.9 The inverse matrix

The next algebraic operation is that corresponding to division. Multiplication and division are inverse processes and it is to be expected that 'matrix division' will be dependent on the multiplication process. In scalar algebra division by 3, for example, can be considered as the multiplication by the reciprocal of 3. The latter is that number which when multiplied by 3 gives unity. In matrix algebra, given a matrix A the *reciprocal* or *inverse matrix* of A is defined as the matrix B such that $AB = BA = I$. Such a statement requires, of necessity, that both A and B are square matrices. When A is square and B exists it will be denoted by A^{-1}, i.e.

$$AA^{-1} = A^{-1}A = I \tag{19}$$

Note 1 It is possible that A^{-1} does not exist. This possibility is discussed in section 8.5.

Note 2 The condition that the matrices A and B commute in a general manner, i.e. that $AB = BA \neq I$, cannot be dealt with here. It may be shown that such matrices can be found by considering the products of

$$A = \begin{pmatrix} 3 & -1 \\ 2 & 1 \end{pmatrix} \quad \text{and} \quad B = \begin{pmatrix} 11 & -3 \\ 6 & 5 \end{pmatrix}.$$

Example (i) Verify that

$$\begin{pmatrix} 1 & 1 & -2 \\ -2 & -1 & 4 \\ 0 & -1 & 1 \end{pmatrix} \text{ is the inverse of } \begin{pmatrix} 3 & 1 & 2 \\ 2 & 1 & 0 \\ 2 & 1 & 1 \end{pmatrix}.$$

Solution

$$\begin{pmatrix} 3 & 1 & 2 \\ 2 & 1 & 0 \\ 2 & 1 & 1 \end{pmatrix}\begin{pmatrix} 1 & 1 & -2 \\ -2 & -1 & 4 \\ 0 & -1 & 1 \end{pmatrix} = \begin{pmatrix} 1 & 0 & 0 \\ 0 & 1 & 0 \\ 0 & 0 & 1 \end{pmatrix} = I$$

and $$\begin{pmatrix} 1 & 1 & -2 \\ -2 & -1 & 4 \\ 0 & -1 & 1 \end{pmatrix}\begin{pmatrix} 3 & 1 & 2 \\ 2 & 1 & 0 \\ 2 & 1 & 1 \end{pmatrix} = \begin{pmatrix} 1 & 0 & 0 \\ 0 & 1 & 0 \\ 0 & 0 & 1 \end{pmatrix} = I.$$

The fuller understanding and the evaluation of an inverse matrix is dependent on the use of *determinants* to be defined and discussed in the next section. However, one application of the inverse matrix is illustrated in the following example.

Example (ii) Solve the system of equations

$$3x_1 + x_2 + 2x_3 = 5$$
$$2x_1 + x_2 \quad\quad = -3$$
$$2x_1 + x_2 + x_3 = 3.$$

Solution In matrix form, the system of equations is written $Ax = B$ where

$$A = \begin{pmatrix} 3 & 1 & 2 \\ 2 & 1 & 0 \\ 2 & 1 & 1 \end{pmatrix}, \quad x = \begin{pmatrix} x_1 \\ x_2 \\ x_3 \end{pmatrix}, \quad \text{and} \quad B = \begin{pmatrix} 5 \\ -3 \\ 3 \end{pmatrix}.$$

Pre-multiply by the inverse matrix A^{-1}, then

$$A^{-1}Ax = A^{-1}B$$

or $$Ix = A^{-1}B$$

or $$x = A^{-1}B.$$

Using the result of the previous example

$$x = \begin{pmatrix} 1 & 1 & -2 \\ -2 & -1 & 4 \\ 0 & -1 & 1 \end{pmatrix}\begin{pmatrix} 5 \\ -3 \\ 3 \end{pmatrix} = \begin{pmatrix} -4 \\ 5 \\ 6 \end{pmatrix}$$

and hence $x_1 = -4$, $x_2 = 5$, and $x_3 = 6$.

8.3 DETERMINANTS

The solution of the pair of equations

$$\left.\begin{array}{l} a_{11}x_1 + a_{12}x_2 + b_1 = 0 \\ a_{21}x_1 + a_{22}x_2 + b_2 = 0 \end{array}\right\} \tag{20}$$

is

$$\left.\begin{array}{l} x_1 = (a_{12}b_2 - a_{22}b_1)/(a_{11}a_{22} - a_{12}a_{21}) \\ x_2 = -(a_{11}b_2 - a_{21}b_1)/(a_{11}a_{22} - a_{12}a_{21}) \end{array}\right\} \tag{21}$$

provided that the denominator is non-zero. The three expressions in brackets arise from the cross products of the pairs of elements

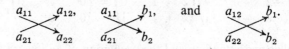

Such expressions are called *determinants* of the *second order*. The *augmented coefficient matrix* of (20) is the 2 × 3 matrix

$$\begin{pmatrix} a_{11} & a_{12} & b_1 \\ a_{21} & a_{22} & b_2 \end{pmatrix} \tag{22}$$

and each determinant of (21) is an evaluation associated with the three 2 × 2 matrices available in (22). For example $(a_{12}b_2 - a_{22}b_1)$ is the evaluation of the determinant associated with the matrix

$$\begin{pmatrix} a_{12} & b_1 \\ a_{22} & b_2 \end{pmatrix}.$$

This determinant is written

$$\begin{vmatrix} a_{12} & b_1 \\ a_{22} & b_2 \end{vmatrix}.$$

The solution (21) may be written in determinant form

$$\frac{x_1}{\begin{vmatrix} a_{12} & b_1 \\ a_{22} & b_2 \end{vmatrix}} = \frac{-x_2}{\begin{vmatrix} a_{11} & b_1 \\ a_{21} & b_2 \end{vmatrix}} = \frac{1}{\begin{vmatrix} a_{11} & a_{12} \\ a_{21} & a_{22} \end{vmatrix}}. \tag{23}$$

Example The matrix $\begin{pmatrix} 3 & -7 & -1 \\ 2 & 1 & 0 \end{pmatrix}$ contains the three determinants of the second order

$$\begin{vmatrix} 3 & -7 \\ 2 & 1 \end{vmatrix}, \quad \begin{vmatrix} 3 & -1 \\ 2 & 0 \end{vmatrix}, \quad \text{and} \quad \begin{vmatrix} -7 & -1 \\ 1 & 0 \end{vmatrix}$$

whose values are 17, 2, and 1 respectively.

Likewise, if three equations in three unknowns

$$\left.\begin{array}{l} a_{11}x_1 + a_{12}x_2 + a_{13}x_3 + b_1 = 0 \\ a_{21}x_1 + a_{22}x_2 + a_{23}x_3 + b_2 = 0 \\ a_{31}x_1 + a_{32}x_2 + a_{33}x_3 + b_3 = 0 \end{array}\right\} \qquad (24)$$

have a solution, each x_i is found as the ratio of two six-termed expressions. The denominator is common to each x_i and has the value

$$a_{11}a_{22}a_{33} - a_{11}a_{23}a_{32} + a_{12}a_{23}a_{31} - a_{12}a_{21}a_{33} + a_{13}a_{21}a_{32}$$
$$- a_{13}a_{22}a_{31} \qquad (25)$$

The reader should verify this expression by solving equations (24) for x_1 by eliminating x_2 and x_3.

An expression of the type (25) associated, in this case, with a 3×3 matrix whose elements are the coefficients of the x_i of (24) is called a determinant of the *third order* and is written

$$\begin{vmatrix} a_{11} & a_{12} & a_{13} \\ a_{21} & a_{22} & a_{23} \\ a_{31} & a_{32} & a_{33} \end{vmatrix}.$$

The other three third order determinants in the solution of (24) are associated with the further three 3×3 matrices obtained from the augmented coefficient matrix of (24). Each determinant contains six terms, three prefixed with a $+$ sign and three with a $-$ sign.

Again, if four equations in four unknowns x_i have a solution, each x_i is found as the ratio of two 24-termed expressions. The denominator term is common to each x_i and has the value

$$a_{11}a_{22}a_{33}a_{44} - a_{11}a_{22}a_{34}a_{43} + a_{11}a_{23}a_{34}a_{42} + \ldots \qquad (26)$$

to 24 terms. The expression (26) is associated with the 4×4 coefficient matrix (a_{ij}) and is the evaluation of the *fourth order* determinant written

$$\begin{vmatrix} a_{11} & a_{12} & a_{13} & a_{14} \\ a_{21} & a_{22} & a_{23} & a_{24} \\ a_{31} & a_{32} & a_{33} & a_{34} \\ a_{41} & a_{42} & a_{43} & a_{44} \end{vmatrix}. \qquad (27)$$

The 24 terms in the evaluation of (27) (in which 12 are prefixed with the $-$ sign) are each of the type $a_{1\alpha}a_{2\beta}a_{3\gamma}a_{4\delta}$ where α, β, γ, δ are the numbers representing the *columns* from which the element is taken and $\alpha\beta\gamma\delta$ is some arrangement, or permutation, of the natural order 1234. The first term with the $+$ sign is $a_{11}a_{22}a_{33}a_{44}$ and is called the *leading term*. The number 24 is simply the number of ways in which the natural order 1234 can be arranged. Six of these arrangements are 1234, 1243, 1324, 1342, 1432, 1423. The reader should satisfy himself that there are 4!, i.e. 24 terms in all.

8.3.1 The determinant of the nth order

An $n \times n$ matrix A or (a_{ij}) has associated with it a determinant of the nth order, i.e. n rows and n columns, written

$$|A| = |a_{ij}| = \begin{vmatrix} a_{11} & a_{12} & \cdots & a_{1n} \\ a_{21} & a_{22} & \cdots & a_{2n} \\ \cdot & \cdot & & \cdot \\ \cdot & \cdot & & \cdot \\ \cdot & \cdot & & \cdot \\ a_{n1} & a_{n2} & \cdots & a_{nn} \end{vmatrix}. \tag{28}$$

The determinant (28), unlike the matrix A which remains a manipulative array of elements, has a definite value and consists of the sum of $n!$ terms each of which is of the form $\pm a_{1\alpha}a_{2\beta}a_{3\gamma} \ldots a_{n\lambda}$ where $\alpha\beta\gamma \ldots \lambda$ is some arrangement of the natural order $123 \ldots n$ and the sign of the term depends on the arrangement.

The formal mathematical approach is to define the value of (28) by this sum and to derive the properties of determinants using the term by term evaluation. This approach will not be dealt with here but instead the evaluation and properties will be deduced from (25) for third order determinants, verified as necessary for fourth order determinants and then assumed to be true for the general determinant of order n.

8.3.2 Minors and cofactors

Consider the third order determinant

$$|A| = \begin{vmatrix} a_{11} & a_{12} & a_{13} \\ a_{21} & a_{22} & a_{23} \\ a_{31} & a_{32} & a_{33} \end{vmatrix}$$

$$= a_{11}a_{22}a_{33} - a_{11}a_{23}a_{32} + a_{12}a_{23}a_{31}$$
$$- a_{12}a_{21}a_{33} + a_{13}a_{21}a_{32} - a_{13}a_{22}a_{31}. \tag{29}$$

Each element a_{ij} of the determinantal array (29) has associated with it a determinant of the second order obtained by striking out the row and column containing the element a_{ij}. This determinant is called the *minor* of a_{ij} and written M_{ij}. For example, the minor of a_{12} is the second order determinant

$$M_{12} = \begin{vmatrix} a_{21} & a_{23} \\ a_{31} & a_{33} \end{vmatrix}$$

having the value $(a_{21}a_{33} - a_{31}a_{23})$. A re-arrangement of the right side of (29) shows that

$$|A| = a_{11}(a_{22}a_{33} - a_{32}a_{23}) - a_{12}(a_{21}a_{33} - a_{31}a_{23})$$
$$+ a_{13}(a_{21}a_{32} - a_{31}a_{22})$$
$$= a_{11}M_{11} - a_{12}M_{12} + a_{13}M_{13}. \tag{30}$$

The re-arrangement of (29) can be performed in five other ways using the elements of the two other rows and three columns. For example, two further ways are given by

$$|A| = -a_{12}M_{12} + a_{22}M_{22} - a_{32}M_{32}$$
$$= a_{31}M_{31} - a_{32}M_{32} + a_{33}M_{33}. \qquad (31)$$

It is more convenient to introduce the *cofactor* of the element a_{ij}. This is usually written A_{ij} and is the signed minor of a_{ij} so that $A_{ij} = \pm M_{ij}$. The sign for the leading element is $+$ and the sign of any other cofactor is taken from the array of signs

$$\begin{vmatrix} + & - & + \\ - & + & - \\ + & - & + \end{vmatrix}. \qquad (32)$$

For example,

$$A_{23} = -M_{23} = - \begin{vmatrix} a_{11} & a_{12} \\ a_{31} & a_{32} \end{vmatrix} \quad \text{and} \quad A_{31} = +M_{31}.$$

The determinant (29) may be written

$a_{11}A_{11} + a_{12}A_{12} + a_{13}A_{13}$, evaluating along the first row,

or $a_{11}A_{11} + a_{21}A_{21} + a_{31}A_{31}$, evaluating along the first column,

or $a_{12}A_{12} + a_{22}A_{22} + a_{32}A_{32}$, evaluating along the second column,

or in three other ways. The six possible ways of evaluating the third order determinants are

$$\left. \begin{aligned} |A| &= a_{i1}A_{i1} + a_{i2}A_{i2} + a_{i3}A_{i3}, \text{ evaluating along the } i\text{th row} \\ &= a_{1j}A_{1j} + a_{2j}A_{2j} + a_{3j}A_{3j}, \text{ evaluating along the } j\text{th column} \end{aligned} \right\}. \quad (33)$$

Extending this process to the fourth order determinant and using the elements of the first row, say, (27) can be written

$$a_{11}A_{11} + a_{12}A_{12} + a_{13}A_{13} + a_{14}A_{14} \qquad (34)$$

in which each of the cofactors A_{1j} is a third order determinant that can be evaluated using (33). Note that (34) becomes

$$a_{11} \begin{vmatrix} a_{22} & a_{23} & a_{24} \\ a_{32} & a_{33} & a_{34} \\ a_{42} & a_{43} & a_{44} \end{vmatrix} + \ldots \text{ to 4 terms,}$$

$$\text{or } a_{11} \left\{ a_{22} \begin{vmatrix} a_{33} & a_{34} \\ a_{43} & a_{44} \end{vmatrix} - a_{23} \begin{vmatrix} a_{32} & a_{34} \\ a_{42} & a_{44} \end{vmatrix} + a_{24} \begin{vmatrix} a_{32} & a_{33} \\ a_{42} & a_{43} \end{vmatrix} \right\}$$

$$+ \ldots \text{ to 12 terms,}$$

or $a_{11}a_{22}a_{33}a_{44} - a_{11}a_{22}a_{34}a_{43} + a_{11}a_{23}a_{34}a_{42} + \ldots$ to 24 terms. When written out in full the last expression may be shown to be identical with the expression (26).

The evaluation of the nth order determinant is now defined by the extension of (33). Thus,

$$
\begin{aligned}
|A| &= a_{i1}A_{i1} + a_{i2}A_{i2} + \ldots + a_{in}A_{in} \quad (i\text{th row}) \\
&= a_{1j}A_{1j} + a_{2j}A_{2j} + \ldots + a_{nj}A_{nj} \quad (j\text{th column})
\end{aligned}
\Bigg\}
\tag{35}
$$

in which A_{ij} is the cofactor of a_{ij} and is that determinant of $(n-1)$th order obtained by striking out the elements of the ith row and the jth column of (28) prefixed with the sign taken from the array

$$
\begin{vmatrix}
+ & - & + & \ldots & (-1)^{n-1} \\
- & + & - & \ldots & \\
+ & - & + & \ldots & \\
\cdot & \cdot & \cdot & & \\
\cdot & \cdot & \cdot & & \\
\cdot & \cdot & \cdot & & + & - \\
(-1)^{n-1} & & & \ldots & - & +
\end{vmatrix}.
$$

Repeated application of (35) will reduce any nth order determinant to $n!/6$ determinants of the third order or $n!/2$ determinants of the second order prior to evaluation. In practice determinants of large order may be simplified by using the properties of the next section, evaluated by numerical methods, or may be avoided altogether.

Example (i) Evaluate
$$
\begin{vmatrix}
2 & 1 & 1 & 1 \\
0 & 2 & 1 & 0 \\
3 & 0 & 1 & 2 \\
1 & -1 & 2 & 1
\end{vmatrix}.
$$

Solution Using the elements of the second row, the evaluation is

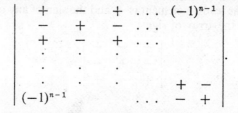

Further evaluation along the row and column indicated leads to

$$
2\left\{ 2\begin{vmatrix} 1 & 2 \\ 2 & 1 \end{vmatrix} - 1\begin{vmatrix} 3 & 2 \\ 1 & 1 \end{vmatrix} + 1\begin{vmatrix} 3 & 1 \\ 1 & 2 \end{vmatrix} \right\}
$$

$$
\qquad - 1\left\{ -1\begin{vmatrix} 3 & 2 \\ 1 & 1 \end{vmatrix} + 0 - (-1)\begin{vmatrix} 2 & 1 \\ 3 & 2 \end{vmatrix} \right\}
$$

$$
= 2\{2(-3) - 1 + 5\} - \{-1(+1) + 0 + 1\}
$$

$$
= 2\{-2\} - 0
$$

$$
= -4.
$$

A useful method of evaluation for a third order determinant *only* is given by the *Rule of Sarrus*. Repeat the first two columns of the determinant as shown below and then form the two groups of triple products one prefixed with the $+$ sign and one with the $-$ sign. It will be seen that the two groups consist only of the terms of (25)

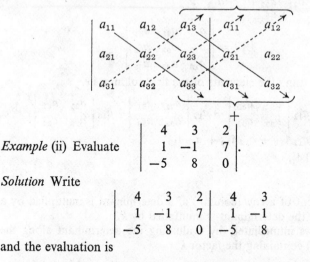

Example (ii) Evaluate $\begin{vmatrix} 4 & 3 & 2 \\ 1 & -1 & 7 \\ -5 & 8 & 0 \end{vmatrix}$.

Solution Write

$$\begin{array}{ccc|cc} 4 & 3 & 2 & 4 & 3 \\ 1 & -1 & 7 & 1 & -1 \\ -5 & 8 & 0 & -5 & 8 \end{array}$$

and the evaluation is

$+ \{(4)(-1)(0) + (3)(7)(-5) + (2)(1)(8)\}$
$\qquad\qquad - \{(-5)(-1)(2) + (8)(7)(4) + (0)(1)(3)\}$
$= \{0 - 105 + 16\} - \{10 + 224 + 0\}$
$= -89 - 234$
$= -323.$

8.3.3 Some properties of determinants

The results to be stated in this section are general results and may be applied to determinants of any order but illustrations are usually given by referring to the third order determinant

$$|A| = \begin{vmatrix} a_{11} & a_{12} & a_{13} \\ a_{21} & a_{22} & a_{23} \\ a_{31} & a_{32} & a_{33} \end{vmatrix}$$

with cofactors A_{ij}.

Property 1

If every element of a row of a determinant is zero then the value of the determinant is zero.

This follows immediately from (35) by evaluating along the row of zero elements.

Property 2

The determinant of any square matrix is equal to the determinant of the transposed matrix, i.e. to any property concerning the rows of a determinant there is an equivalent property concerning the columns.

This result is clearly true for a second order determinant because

$$\begin{vmatrix} a_{11} & a_{12} \\ a_{21} & a_{22} \end{vmatrix} = \begin{vmatrix} a_{11} & a_{21} \\ a_{12} & a_{22} \end{vmatrix} = a_{11}a_{22} - a_{21}a_{12}.$$

Consider

$$|A'| = \begin{vmatrix} a_{11} & a_{21} & a_{31} \\ a_{12} & a_{22} & a_{32} \\ a_{13} & a_{23} & a_{33} \end{vmatrix}$$

and evaluate using the elements of the first column, i.e.

$$\begin{aligned} |A'| &= a_{11}\begin{vmatrix} a_{22} & a_{32} \\ a_{23} & a_{33} \end{vmatrix} - a_{12}\begin{vmatrix} a_{21} & a_{31} \\ a_{23} & a_{33} \end{vmatrix} + a_{13}\begin{vmatrix} a_{21} & a_{31} \\ a_{22} & a_{32} \end{vmatrix} \\ &= a_{11}A_{11} + a_{12}A_{12} + a_{13}A_{13} \\ &= |A|. \end{aligned}$$

Property 3

If each element of a row (column) of a determinant is multiplied by a scalar k then the determinant is multiplied by k.

This follows immediately by evaluating the determinant along the row (column) containing the factor k.

Example (i)

$$\begin{vmatrix} a_{11} & a_{12} & a_{13} \\ ka_{21} & ka_{22} & ka_{23} \\ a_{31} & a_{32} & a_{33} \end{vmatrix} = k|A| = \begin{vmatrix} a_{11} & ka_{12} & a_{13} \\ a_{21} & ka_{22} & a_{23} \\ a_{31} & ka_{32} & a_{33} \end{vmatrix}.$$

Property 4

If any two rows (columns) of a determinant are interchanged, then the sign of the determinant is changed.

Consider

$$|B| = \begin{vmatrix} a_{31} & a_{32} & a_{33} \\ a_{21} & a_{22} & a_{23} \\ a_{11} & a_{12} & a_{13} \end{vmatrix}$$

obtained from $|A|$ by interchanging the first and third rows, and evaluate along the third row.

$$\begin{aligned} |B| &= a_{11}\begin{vmatrix} a_{32} & a_{33} \\ a_{22} & a_{23} \end{vmatrix} - a_{12}\begin{vmatrix} a_{31} & a_{33} \\ a_{21} & a_{23} \end{vmatrix} + a_{13}\begin{vmatrix} a_{31} & a_{32} \\ a_{21} & a_{22} \end{vmatrix} \\ &= a_{11}(-A_{11}) \quad - a_{12}(A_{12}) \quad + a_{13}(-A_{13}) \\ &= -|A|. \end{aligned}$$

Corollary to Property 4

If any two rows (columns) of a determinant are identical then the determinant is zero because interchange of the rows (columns) would lead to a change of sign by this property and yet must at the same time leave the determinant value unchanged.

Example (ii)

$$\begin{vmatrix} 3 & 2 & 1 \\ 1 & 3 & 2 \\ 6 & 4 & 2 \end{vmatrix} = 2 \begin{vmatrix} 3 & 2 & 1 \\ 1 & 3 & 2 \\ 3 & 2 & 1 \end{vmatrix} = 0.$$

Property 5

The sum of the products of the elements of any row (column) with the cofactors of any other row (column) is zero.

Consider $|B| = a_{11} A_{21} + a_{12} A_{22} + a_{13} A_{23}$ in which the elements of the first row of $|A|$ are multiplied by the cofactors of the second row.

$$\text{Then } |B| = -a_{11} \begin{vmatrix} a_{12} & a_{13} \\ a_{32} & a_{33} \end{vmatrix} + a_{12} \begin{vmatrix} a_{11} & a_{13} \\ a_{31} & a_{33} \end{vmatrix} - a_{13} \begin{vmatrix} a_{11} & a_{12} \\ a_{31} & a_{32} \end{vmatrix}$$

$$= - \begin{vmatrix} a_{11} & a_{12} & a_{13} \\ a_{11} & a_{12} & a_{13} \\ a_{31} & a_{32} & a_{33} \end{vmatrix}$$

$$= 0$$

by the corollary to Property 4.

This property may be generalized and, together with the cofactor evaluation (35) for the ith row, may be expressed in the single statement

$$\sum_{j=1}^{n} a_{ij} A_{kj} = \begin{cases} 0 & \text{if } i \neq k \\ |A| & \text{if } i = k. \end{cases}$$

Employing the summation convention for a repeated index this may be written

$$a_{ij} A_{kj} = \delta_{ik} |A| \tag{36}$$

where δ_{ik} is called the *Kronecker delta* and has the property that

$$\delta_{ik} = \begin{cases} 0 & \text{if } i \neq k \\ 1 & \text{if } i = k. \end{cases} \tag{37}$$

Using the elements of the jth column in (35) a result similar to (36) may be stated

$$a_{ij} A_{ik} = \delta_{jk} |A| \tag{38}$$

Property 6

A determinant is unchanged in value if to the elements of a row (column) are added a scalar multiple of the elements of any other row (column).

Consider

$$|B| = \begin{vmatrix} a_{11} + ka_{21} & a_{12} + ka_{22} & a_{13} + ka_{23} \\ a_{21} & a_{22} & a_{23} \\ a_{31} & a_{32} & a_{33} \end{vmatrix},$$

then

$$|B| = (a_{11} + ka_{21})A_{11} + (a_{12} + ka_{22})A_{12} + (a_{13} + ka_{23})A_{13}$$
$$= (a_{11}A_{11} + a_{12}A_{12} + a_{13}A_{13}) + k(a_{21}A_{11} + a_{22}A_{12} + a_{23}A_{13})$$
$$= |A|$$

the last line following from (36).

This property may be applied repeatedly to a determinant in order to reduce it to a simpler form before or during evaluation.

Example (iii) Evaluate

$$\begin{vmatrix} 1 & 5 & -3 \\ 2 & 7 & 0 \\ 3 & 9 & 3 \end{vmatrix} \begin{matrix} R_1 \\ R_2 \\ R_3 \end{matrix}.$$

Solution Perform the operations indicated

$$\begin{vmatrix} 1 & 5 & -3 \\ 1 & 2 & +3 \\ 1 & 2 & 3 \end{vmatrix} \begin{matrix} R_1 \\ R_2 - R_1 \\ R_3 - R_2 \end{matrix} = 0 \quad \text{by the corollary to Property 4.}$$

Example (iv) Evaluate

$$\begin{vmatrix} 1 & 1 & 1 \\ a & b & c \\ a^2 & b^2 & c^2 \end{vmatrix}.$$
$$\begin{matrix} C_1 & C_2 & C_3 \end{matrix}$$

Solution Perform the operations indicated

$$\begin{vmatrix} 1 & 0 & 0 \\ a & b-a & c-a \\ a^2 & b^2-a^2 & c^2-a^2 \end{vmatrix} = (b-a)(c-a) \begin{vmatrix} 1 & 0 & 0 \\ a & 1 & 1 \\ a^2 & b+a & c+a \end{vmatrix}$$
$$\begin{matrix} C_1 & C_2-C_1 & C_3-C_1 \end{matrix}$$
$$= (b-a)(c-a)(c-b).$$

8.4 THE ADJUGATE OR ADJOINT MATRIX

Before the inverse matrix introduced in section 8.2.9 can be finally defined it is necessary to examine a square matrix called the *adjugate* or *adjoint matrix* of $A = a_{ij}$ written adj A and defined by

$$\text{adj } A = \begin{pmatrix} A_{11} & A_{21} & A_{31} & \dots & A_{n1} \\ A_{12} & A_{22} & A_{32} & \dots & A_{n2} \\ A_{13} & A_{23} & A_{33} & \dots & A_{n3} \\ \cdot & \cdot & \cdot & & \cdot \\ \cdot & \cdot & \cdot & & \cdot \\ \cdot & \cdot & \cdot & & \cdot \\ A_{1n} & A_{2n} & A_{3n} & \dots & A_{nn} \end{pmatrix} = (A_{ji}) \qquad (39)$$

where A_{ij} is the cofactor of the element a_{ij} of the ith row and jth column of the determinant $|A|$. Note that adj A may be obtained from A', the transpose of A, by replacing the elements of A' by their cofactors in $|A'|$.

Example (i) If

$$A = \begin{pmatrix} 4 & 2 & 1 \\ 3 & 0 & 1 \\ 0 & 1 & 2 \end{pmatrix} \quad \text{then} \quad A' = \begin{pmatrix} 4 & 3 & 0 \\ 2 & 0 & 1 \\ 1 & 1 & 2 \end{pmatrix}$$

and hence

$$\text{adj } A = \begin{pmatrix} -1 & -3 & 2 \\ -6 & 8 & -1 \\ 3 & -4 & -6 \end{pmatrix}.$$

Let B be the matrix product of A and adj A then

$$B = (b_{ik}) = \begin{pmatrix} a_{11} & a_{12} & a_{13} & \ldots & a_{1n} \\ a_{21} & a_{22} & a_{23} & \ldots & a_{2n} \\ \cdot & \cdot & \cdot & & \cdot \\ \cdot & \cdot & \cdot & & \cdot \\ \cdot & \cdot & \cdot & & \cdot \\ a_{n1} & a_{n2} & a_{n3} & \ldots & a_{nn} \end{pmatrix} \begin{pmatrix} A_{11} & A_{21} & A_{31} & \ldots & A_{n1} \\ A_{12} & A_{22} & A_{32} & \ldots & A_{n2} \\ \cdot & \cdot & \cdot & & \cdot \\ \cdot & \cdot & \cdot & & \cdot \\ \cdot & \cdot & \cdot & & \cdot \\ A_{1n} & A_{2n} & A_{3n} & \ldots & A_{nn} \end{pmatrix} \quad (40)$$

and from the definition of matrix multiplication

$$b_{ik} = a_{ij}A_{kj} \quad \text{summed over } j$$
$$= \delta_{ik}|A|$$

in the notation of (36), section 8.3. The property of δ_{ik} shows that B is a scalar matrix and

$$A \text{ adj } A = \begin{pmatrix} |A| & 0 & 0 & \ldots & 0 \\ 0 & |A| & 0 & \ldots & 0 \\ 0 & 0 & |A| & \ldots & 0 \\ \cdot & \cdot & \cdot & & \cdot \\ \cdot & \cdot & \cdot & & \cdot \\ \cdot & \cdot & \cdot & & \cdot \\ 0 & 0 & 0 & \ldots & |A| \end{pmatrix}$$

$$= |A|I. \quad (41)$$

Using (38) of section 8.3, an exactly similar result follows for (adj A)A so that

$$A (\text{adj } A) = |A|I = (\text{adj } A)A. \quad (42)$$

Example (ii) If

$$A = \begin{pmatrix} 2 & 3 \\ 1 & 4 \end{pmatrix}, \quad \text{adj } A = \begin{pmatrix} 4 & -3 \\ -1 & 2 \end{pmatrix}$$

and then

$$A(\text{adj } A) = \begin{pmatrix} 2 & 3 \\ 1 & 4 \end{pmatrix} \begin{pmatrix} 4 & -3 \\ -1 & 2 \end{pmatrix} = \begin{pmatrix} 5 & 0 \\ 0 & 5 \end{pmatrix} = 5I,$$

$$(\text{adj } A)A = \begin{pmatrix} 4 & -3 \\ -1 & 2 \end{pmatrix} \begin{pmatrix} 2 & 3 \\ 1 & 4 \end{pmatrix} = \begin{pmatrix} 5 & 0 \\ 0 & 5 \end{pmatrix} = 5I.$$

Example (iii) If

$$A = \begin{pmatrix} 2 & 3 & 1 \\ 1 & 2 & 0 \\ 0 & 1 & 2 \end{pmatrix} \quad \text{and} \quad B = \begin{pmatrix} 1 & 0 & 1 \\ 0 & 2 & 1 \\ 1 & 1 & 1 \end{pmatrix}$$

show that adj $(AB) = (\text{adj } B)(\text{adj } A)$.

Solution

$$AB = \begin{pmatrix} 3 & 7 & 6 \\ 1 & 4 & 3 \\ 2 & 4 & 3 \end{pmatrix} \quad \text{and} \quad \text{adj } AB = \begin{pmatrix} 0 & 3 & -3 \\ 3 & -3 & -3 \\ -4 & 2 & 5 \end{pmatrix}$$

$$(\text{adj } B)(\text{adj } A) = \begin{pmatrix} 1 & 1 & -2 \\ 1 & 0 & -1 \\ -2 & -1 & 2 \end{pmatrix} \begin{pmatrix} 4 & -5 & -2 \\ -2 & 4 & 1 \\ 1 & -2 & 1 \end{pmatrix} = \begin{pmatrix} 0 & 3 & -3 \\ 3 & -3 & -3 \\ -4 & 2 & 5 \end{pmatrix}.$$

8.5 THE RECIPROCAL OR INVERSE MATRIX

In section 8.2.9 the inverse matrix of a square matrix A was written in symbolic form as A^{-1} such that

$$AA^{-1} = A^{-1}A = I. \tag{43}$$

Comparison with (42) shows that $(\text{adj } A)/|A|$ is identical with A^{-1}, i.e.

$$A^{-1} = \frac{1}{|A|} \text{adj } A \tag{44}$$

provided that $|A| \neq 0$.

If $|A| = 0$, the matrix is called singular and the inverse A^{-1} does not exist.

Example (i) Given $A = \begin{pmatrix} 3 & 1 & 2 \\ 2 & 1 & 0 \\ 2 & 1 & 1 \end{pmatrix}$ find A^{-1}.

Solution Here $|A| = 2(0) + 1(1) = 1$ so that A is non-singular and

$$A^{-1} = \frac{1}{|A|} \text{adj } A = \begin{pmatrix} 1 & 1 & -2 \\ -2 & -1 & 4 \\ 0 & -1 & 1 \end{pmatrix}.$$

Example (ii) Given $A = \begin{pmatrix} 2 & 1 & 1 \\ 0 & 11 & 0 \\ 6 & 4 & 3 \end{pmatrix}$ find A^{-1}.

Solution Here $|A| = 0$, A is singular, and A^{-1} does not exist.

8.6 SOLUTION OF REGULAR LINEAR ALGEBRAIC EQUATIONS

Consider the system of n linear equations in n unknowns x_i

$$Ax = B \tag{45}$$

where A is the $n \times n$ matrix (a_{ij}), x is the column matrix (x_i) and B is the column matrix (b_i). If A is non-singular the system (45) is called *regular*. With A non-singular, A^{-1} exists and equation (45) may be written

$$A^{-1}Ax = A^{-1}B$$

i.e. $$Ix = A^{-1}B$$

or $$x = A^{-1}B \tag{46}$$

The equation (46) is then the formal solution of (45).

Example (i) To solve

$$2x_1 + x_2 - x_3 = 4$$
$$x_1 - 2x_2 + x_3 = -10$$
$$- 3x_1 - 2x_3 = 9$$

write

$$A = \begin{pmatrix} 2 & 1 & -1 \\ 1 & -2 & 1 \\ -3 & 0 & -2 \end{pmatrix}, \quad x = \begin{pmatrix} x_1 \\ x_2 \\ x_3 \end{pmatrix} \quad \text{and} \quad B = \begin{pmatrix} 4 \\ -10 \\ 9 \end{pmatrix}.$$

Since $|A| = (-3)(-1) + (-2)(-5) = 13$, then

$$A^{-1} = \frac{1}{13} \begin{pmatrix} 4 & 2 & -1 \\ -1 & -7 & -3 \\ -6 & -3 & -5 \end{pmatrix}$$

and

$$x = \frac{1}{13} \begin{pmatrix} 4 & 2 & -1 \\ -1 & -7 & -3 \\ -6 & -3 & -5 \end{pmatrix} \begin{pmatrix} 4 \\ -10 \\ 9 \end{pmatrix} = \frac{1}{13} \begin{pmatrix} -13 \\ 39 \\ -39 \end{pmatrix}$$

Hence, $x_1 = -1$, $x_2 = 3$ and $x_3 = -3$.

The solution (46) may be expressed in terms of determinants and cofactors. As an illustration consider the case $n = 3$, then

$$x = \frac{1}{|A|} (\text{adj } A)B = \frac{1}{|A|} \begin{pmatrix} A_{11} & A_{21} & A_{31} \\ A_{12} & A_{22} & A_{32} \\ A_{13} & A_{23} & A_{33} \end{pmatrix} \begin{pmatrix} b_1 \\ b_2 \\ b_3 \end{pmatrix}$$

$$= \frac{1}{|A|} \begin{pmatrix} b_1A_{11} + b_2A_{21} + b_3A_{31} \\ b_1A_{12} + b_2A_{22} + b_3A_{32} \\ b_1A_{13} + b_2A_{23} + b_3A_{33} \end{pmatrix}$$

where A_{ij} is the cofactor of the element a_{ij} in A. Hence,

$$\frac{x_1}{\Delta_1} = \frac{x_2}{\Delta_2} = \frac{x_3}{\Delta_3} = \frac{1}{|A|}$$

or

$$x_j = \frac{\Delta_j}{|A|} \qquad (47)$$

where $\Delta_j = b_i A_{ij}$ (summed over i).

In equation (47), $|A|$ is the determinant of the matrix $A = (a_{ij})$ and Δ_j is the determinant of that matrix obtained from A on replacing the jth column of A by the column (b_j). The solution of equation (45) expressed in the form (47) is called *Cramer's rule*. The result (47) may be extended immediately for the case $n > 3$.

Example (ii) Solving example (i) by this rule leads to

$$\frac{x_1}{\begin{vmatrix} 4 & 1 & -1 \\ -10 & -2 & 1 \\ 9 & 0 & -2 \end{vmatrix}} = \frac{x_2}{\begin{vmatrix} 2 & 4 & 1 \\ 1 & -10 & 1 \\ -3 & 9 & -2 \end{vmatrix}} = \frac{x_3}{\begin{vmatrix} 2 & 1 & 4 \\ 1 & -2 & -10 \\ -3 & 0 & 9 \end{vmatrix}} = \frac{1}{\begin{vmatrix} 2 & 1 & -1 \\ 1 & -2 & 1 \\ -3 & 0 & -2 \end{vmatrix}}$$

It is left to the reader to evaluate each determinant and verify the solution.

8.6.1 Solution by the elimination method

A general method of dealing with the system of equations (45) is to eliminate unknowns x_i until one equation contains only one unknown, solve for this unknown, and then work back through successive equations to find the other unknowns. A systematic method of procedure was suggested by Gauss. This method is sometimes laborious but being systematic, and of a routine nature, it is suitable for a computer program.

Consider as an example the system of equations

$$\left. \begin{array}{rcr} 2x_1 - x_2 + 3x_3 &=& 10 \\ x_1 + 3x_2 - 4x_3 &=& -11 \\ 4x_1 + x_2 - 2x_3 &=& 3 \end{array} \right\} \qquad (48)$$

or

$$\begin{pmatrix} 2 & -1 & 3 \\ 1 & 3 & -4 \\ 4 & 1 & -2 \end{pmatrix} \begin{pmatrix} x_1 \\ x_2 \\ x_3 \end{pmatrix} = \begin{pmatrix} 10 \\ -11 \\ 3 \end{pmatrix} \qquad (49)$$

$$A \qquad\quad x \ = \ B$$

The matrices of (49) may be written in *augmented matrix form* (A, B)

$$\left(\begin{array}{ccc|c} 2 & -1 & 3 & 10 \\ 1 & 3 & -4 & -11 \\ 4 & 1 & -2 & 3 \end{array} \right) \qquad (50)$$

and because of the equality signs in equations (48) certain row manipulations may be performed on (50) without affecting the final solution. The three important row transformations are:

  I Interchange of rows,
  II Multiplication of a row by any non-zero constant,
  III Addition of rows.

Apply these row transformations to (50) to make the matrix A into an *upper triangular matrix*, i.e. one with zero elements below the leading diagonal. The successive steps are indicated below.

Apply **I**,
$$\begin{pmatrix} 1 & 3 & -4 & | & -11 \\ 2 & -1 & 3 & | & 10 \\ 4 & 1 & -2 & | & 3 \end{pmatrix} \begin{matrix} R_1 \\ R_2 \\ R_3 \end{matrix}$$

and then apply **II** and **III**
$$\begin{pmatrix} 1 & 3 & -4 & | & -11 \\ 0 & -7 & 11 & | & 32 \\ 0 & -11 & 14 & | & 47 \end{pmatrix} \begin{matrix} R_1 \\ R_2' \\ R_3' \end{matrix}$$

where
$$R_2' = R_2 - 2R_1; \quad R_3' = R_3 - 4R_1$$

Apply **II** and **III** again
$$\begin{pmatrix} 1 & 3 & -4 & | & -11 \\ 0 & -7 & 11 & | & 32 \\ 0 & 0 & -23 & | & -23 \end{pmatrix} \begin{matrix} R_1 \\ R_2' \\ R_3'' \end{matrix}$$

where
$$R_3'' = 7R_3' - 11R_2'$$

and a final application of **II** leads to
$$\begin{pmatrix} 1 & 3 & -4 & | & -11 \\ 0 & 7 & -11 & | & -32 \\ 0 & 0 & 1 & | & 1 \end{pmatrix}.$$

The solution may be written down working from the last row, i.e. $x_3 = 1$, $7x_2 - 11x_3 = -32$, and $x_1 + 3x_2 - 4x_3 = -11$ or $x_3 = 1$, $x_2 = -3$ and $x_1 = 2$. In a practical problem the coefficients of a system of algebraic equations are usually decimals and it is convenient to use **II** to make the coefficients of the variable being eliminated equal to unity. The numerical details of this, together with other methods of solution, is given in Chapter 13.

8.6.2 Alternative method of determination of the inverse matrix A^{-1}

If A is a non-singular matrix then the equation

$$Ax = IB \tag{51}$$

has the formal solution

$$Ix = A^{-1}B \tag{52}$$

which suggests that if A is transformed into I by the row transforms **I**, **II**, and **III**, then I will be transformed into A^{-1} by the same processes.

The method is illustrated in the following example where $\sim$ is a symbol used to indicate that successive matrices are obtained by row transforms.

Example Find the inverse of

$$A = \begin{pmatrix} 1 & 2 & 2 \\ 1 & 3 & 4 \\ 1 & 3 & 3 \end{pmatrix}.$$

Solution Write

$$(A, I) = \begin{pmatrix} 1 & 2 & 2 & | & 1 & 0 & 0 \\ 1 & 3 & 4 & | & 0 & 1 & 0 \\ 1 & 3 & 3 & | & 0 & 0 & 1 \end{pmatrix} \begin{matrix} R_1 \\ R_2 \\ R_3 \end{matrix}$$

$$\sim \begin{pmatrix} 1 & 2 & 2 & | & 1 & 0 & 0 \\ 0 & 1 & 2 & | & -1 & 1 & 0 \\ 0 & 0 & -1 & | & 0 & -1 & 1 \end{pmatrix} \begin{matrix} R'_1 \\ R'_2 \\ R'_3 \end{matrix}$$

where $R'_3 = R_3 - R_2, R'_2 = R_2 - R_1; R'_1 = R_1$

$$\sim \begin{pmatrix} 1 & 0 & 0 & | & 3 & 0 & -2 \\ 0 & 1 & 0 & | & -1 & -1 & 2 \\ 0 & 0 & 1 & | & 0 & 1 & -1 \end{pmatrix} \begin{matrix} R''_1 \\ R''_2 \\ R''_3 \end{matrix} = (I, A^{-1}).$$

where $R''_1 = R'_1 - 2R'_2 - 2R'_3; R''_2 = R'_2 + 2R'_3; R''_3 = -R'_3$

Hence, $$A^{-1} = \begin{pmatrix} 3 & 0 & -2 \\ -1 & -1 & 2 \\ 0 & 1 & -1 \end{pmatrix}.$$

These steps are not the only ones by which (A, I) can be transformed into (I, A^{-1}) and the reader is invited to attempt the transformation using different steps and also to verify that the result obtained is the same as adj $A/|A|$.

8.7 FURTHER DISCUSSION OF THE SOLUTION OF n LINEAR EQUATIONS IN n UNKNOWNS

The discussion of the system of n equations in n unknowns

$$Ax = B \tag{53}$$

is given here without too much detail. The reader who requires a more formal treatment is advised to read this section and then to consult a text devoted exclusively to matrix theory.

8.7.1 Regular solution, A non-singular.

If the $n \times n$ matrix is non-singular, then $|A| \neq 0$ and it is evident that Cramer's rule,

$$x_j = \Delta_j/|A|, \quad j = 1, 2, 3, \ldots, n$$

leads to n distinct values of x_j, i.e. to a unique solution. In this case the n linear equations are said to be *linearly independent*. Some of the Δ_j may be zero to give zeros in the corresponding x_j, but the equations (53) cannot have a solution with all x_j zero and A non-singular unless each Δ_j is zero, and this implies that B is zero. The equation (53) then reduces to

$$Ax = 0 \qquad (54)$$

This equation is called a homogeneous equation (the equation $Ax = B$ being called non-homogeneous) and has a unique but zero solution if and only if A is non-singular. This solution of equation (54) is called the *trivial solution*.

8.7.2 Singular solution, A singular

Suppose the matrix A is singular. The $|A| = 0$ and the form given by Cramer shows that if the $\Delta_j \neq 0$ the x_j cannot be found, i.e. no solution exists and the system of equations is said to be *inconsistent*.

Example (i) The system

$$x_1 + x_2 = 3$$
$$2x_1 + 2x_2 = 7$$

is obviously inconsistent. Here $|A| = 0$, $\Delta_1 = -1$, and $\Delta_2 = 1$.

Example (ii) The system of equations

$$3x_1 + 2x_2 + x_3 = 10$$
$$2x_1 + 3x_2 - x_3 = 5$$
$$4x_1 + x_2 + 3x_3 = 8$$

has no obvious inconsistency but $|A| = 0$, $\Delta_1 = 35$, and Δ_2, Δ_3 are also non-zero.

A closer inspection of the two examples shows that in the left side of each system one of the expressions is dependent on the others, for example

$$(4x_1 + x_2 + 3x_3) = 2(3x_1 + 2x_2 + x_3) - (2x_1 + 3x_2 - x_3),$$

but that this is not the case for the right side, i.e. $8 \neq 2(10) - (5)$.

Suppose now that all the Δ_j are zero, then Cramer's solution $x_j = \Delta_j/|A|$ is of the form $0/0$ which is *indeterminate*.

Example (iii) The system

$$3x_1 + 2x_2 + x_3 = 10$$
$$2x_1 + 3x_2 - x_3 = 5$$
$$4x_1 + x_2 + 3x_3 = 15$$

has $|A| = 0$ and $\Delta_1 = \Delta_2 = \Delta_3 = 0$ and there is no unique solution. In this example the last equation is a linear combination of the other two and the system reduces to one of two equations and three unknowns

$$3x_1 + 2x_2 + x_3 = 10$$
$$2x_1 + 3x_2 - x_3 = 5.$$

The original system of three equations is said to be *linearly dependent*. One of the unknowns can be disposed of arbitrarily, say $x_3 = \alpha$, and the others expressed in terms of α, i.e.

$$x_1 = 4 - \alpha, \quad x_2 = \alpha - 1, \quad x_3 = \alpha.$$

Note that there is a solution to the system of equations although indeterminate to the extent that α is arbitrary. Any system that possesses a solution regular or indeterminate is called *consistent*.

To sum up, a system of equations is either inconsistent ($|A| = 0$, $\Delta_j \neq 0$) or consistent and only the latter possesses a solution. A consistent system may have a unique solution (i.e. one fixed set of values for x_j) in which case the solution is called regular or it may have an indeterminate solution (i.e. an infinite number of different sets of values x_j). An illustration in which one variable may be assigned arbitrary values is given by considering the single equation $2x_1 + 3x_2 = 6$ with two unknowns. Geometrically this is the equation of a straight line in the plane Ox_1x_2 and the solution $x_1 = 3 - \frac{3}{2}\alpha$, $x_2 = \alpha$ gives the coordinates of an arbitrary point on the line. Similarly, the pair of equations above

$$3x_1 + 2x_2 + x_3 = 10$$
$$2x_1 + 3x_2 - x_3 = 5$$

may be geometrically interpreted as the intersection of two planes in three dimensions (see section 9.7). The solution

$$x_1 = 4 - \alpha, \quad x_2 = \alpha - 1, \quad x_3 = \alpha$$

gives the coordinates of the general point on the common line of intersection of the planes. Different values of α simply give different points on this line, e.g. $(4, -1, 0)$, and $(10, -7, -6)$ are two such points.

It is necessary to consider the general case. Without proof, it can be stated that the system is consistent if and only if r of the equations are linearly independent where r is any non-zero positive integer such that $r \leqslant n$ and then $(n - r)$ of the unknowns can be disposed of arbitrarily. In order to investigate a system of equations the elimination method of section 8.6.1 may be used. Any inconsistency in the system will appear in the equivalent form of the augmented matrix (A, B). The following examples should illustrate the process.

Example (iv) Discuss the system of equations

$$2x_1 + x_2 + 5x_3 = 1$$
$$x_1 - 3x_2 + 6x_3 = 2$$
$$3x_1 + 5x_2 + 4x_3 = 0.$$

Solution Write

$$\begin{pmatrix} 2 & 1 & 5 \\ 1 & -3 & 6 \\ 3 & 5 & 4 \end{pmatrix}\begin{pmatrix} x_1 \\ x_2 \\ x_3 \end{pmatrix} = \begin{pmatrix} 1 \\ 2 \\ 0 \end{pmatrix};$$

and $(A, B) = \begin{pmatrix} 2 & 1 & 5 & | & 1 \\ 1 & -3 & 6 & | & 2 \\ 3 & 5 & 4 & | & 0 \end{pmatrix} \sim \begin{pmatrix} 1 & -3 & 6 & | & 2 \\ 2 & 1 & 5 & | & 1 \\ 3 & 5 & 4 & | & 0 \end{pmatrix} \begin{matrix} R_1 \\ R_2 \\ R_3 \end{matrix}$

$$\sim \begin{pmatrix} 1 & -3 & 6 & | & 2 \\ 0 & 7 & -7 & | & -3 \\ 0 & 14 & -14 & | & -6 \end{pmatrix} \sim \begin{pmatrix} 1 & -3 & 6 & | & 2 \\ 0 & 7 & -7 & | & -3 \\ 0 & 0 & 0 & | & 0 \end{pmatrix}.$$

$$R_2 - 2R_1; \; R_3 - 3R_1 \qquad R'_3 - 2R'_2$$

The last row of zeros shows that the system is consistent but that one of the equations is linearly dependent on the others. One unknown may be arbitrarily assigned, let $x_3 = \alpha$ then $7x_2 = 7\alpha - 3$ and $7x_1 = 5 - 21\alpha$. Note that here $n = 3$ and $r = 2$.

Example (v) Discuss the system

$$\begin{pmatrix} 2 & 1 & 5 \\ 1 & -3 & 6 \\ 3 & 5 & 4 \end{pmatrix}\begin{pmatrix} x_1 \\ x_2 \\ x_3 \end{pmatrix} = \begin{pmatrix} 1 \\ 2 \\ 1 \end{pmatrix}.$$

Solution In this case

$$(A, B) \sim \begin{pmatrix} 1 & -3 & 6 & | & 2 \\ 0 & 7 & -7 & | & -3 \\ 0 & 0 & 0 & | & 1 \end{pmatrix}$$

and the last row shows that the system is inconsistent. Assuming that x_1 and x_2 are non-zero the equation corresponding to the last row is $0x_3 = 1$ which is impossible for finite x_3. The system has no solution.

Finally when $B = 0$ the system is homogeneous, $Ax = 0$ and the system is always consistent. If $|A| \neq 0$ there is only one solution $x_j = 0$ (see section 8.7.1) called the trivial solution but if $|A| = 0$ then $(n - r)$ of the variables may be assigned arbitrary values where r is an integer such that $0 < r < n$, and the remaining r variables obtained in terms of these assigned values. Thus the condition that the equation $Ax = 0$ has a solution other than the trivial one is that $|A| = 0$.

Example (vi) Solve

$$\begin{aligned}
3x_1 + x_2 + 2x_3 - x_4 &= 0 \\
x_1 - x_2 - x_3 &= 0 \\
4x_1 + x_3 - x_4 &= 0 \\
9x_1 - x_2 + x_3 - 2x_4 &= 0.
\end{aligned}$$

Solution The matrix is transformed as follows

$$A = \begin{pmatrix} 3 & 1 & 2 & -1 \\ 1 & -1 & -1 & 0 \\ 4 & 0 & 1 & -1 \\ 9 & -1 & 1 & -2 \end{pmatrix} \sim \begin{pmatrix} 1 & -1 & -1 & 0 \\ 3 & 1 & 2 & -1 \\ 4 & 0 & 1 & -1 \\ 9 & -1 & 1 & -2 \end{pmatrix}$$

$$\sim \begin{pmatrix} 1 & -1 & -1 & 0 \\ 0 & 4 & 5 & -1 \\ 0 & 4 & 5 & -1 \\ 0 & 8 & 10 & -2 \end{pmatrix} \sim \begin{pmatrix} 1 & -1 & -1 & 0 \\ 0 & 4 & 5 & -1 \\ 0 & 0 & 0 & 0 \\ 0 & 0 & 0 & 0 \end{pmatrix}$$

and the last two rows show that the system has two equations that are linearly dependent. Two of the variables may be assigned arbitrary values, let $x_3 = \alpha$ and $x_4 = \beta$ then $4x_2 = \beta - 5\alpha$ and $4x_1 = \beta - \alpha$.

8.8 THE SOLUTION OF A GENERAL SYSTEM OF m LINEAR EQUATIONS IN n UNKNOWNS

When $m = n$ the system reduces to that already discussed in section 8.7.

Consider now the two cases $m < n$ and $m > n$.

8.8.1 Case $m < n$
Given the system

$$Ax = B \tag{55}$$

A is an $m \times n$ rectangular matrix, x is an $n \times 1$ column matrix and B is an $m \times 1$ column matrix and there are more unknowns x_i than there are equations. The system (55) is consistent if and only if r of the equations are linearly independent where r is a non-zero positive integer such that $r \leqslant m$. In the solution, $(n - r)$ of the unknowns may be assigned arbitrary values and the remaining r variables found in terms of these assigned variables.

Example (i) Examine the system

$$3x_1 + x_2 - x_3 + x_4 = 4$$
$$x_1 \qquad + 2x_3 + 2x_4 = 5$$
$$3x_2 + x_3 - x_4 = 3$$

in which $m = 3$ and $n = 4$.

Solution Proceeding by the elimination method

$$(A, B) = \begin{pmatrix} 3 & 1 & -1 & 1 & | & 4 \\ 1 & 0 & 2 & 2 & | & 5 \\ 0 & 3 & 1 & -1 & | & 3 \end{pmatrix} \sim \begin{pmatrix} 1 & 0 & 2 & 2 & | & 5 \\ 0 & 1 & -7 & -5 & | & -11 \\ 0 & 3 & 1 & -1 & | & 3 \end{pmatrix}$$

$$\sim \begin{pmatrix} 1 & 0 & 2 & 2 & | & 5 \\ 0 & 1 & -7 & -5 & | & -11 \\ 0 & 0 & 11 & 7 & | & 18 \end{pmatrix}$$

The equivalent form of (A, B) shows that the system has no inconsistency, the equations are linearly independent, and $r = m = 3$. One $(n - r = 1)$ of the variables may be arbitrarily assigned. Let $x_4 = \alpha$ then $11x_3 = 18 - 7\alpha$, $11x_2 = 5 + 6\alpha$ and $11x_1 = 19 - 8\alpha$.

Example (ii) Examine the system

$$
\begin{aligned}
x_1 + x_2 - x_3 - x_4 &= 1 \\
2x_1 - x_2 \quad\quad + x_4 &= 2 \\
3x_1 \quad\quad - x_3 \quad\quad &= 3
\end{aligned}
$$

in which $m = 3$ and $n = 4$.

Solution

$$
(A, B) = \begin{pmatrix} 1 & 1 & -1 & -1 & | & 1 \\ 2 & -1 & 0 & 1 & | & 2 \\ 3 & 0 & -1 & 0 & | & 3 \end{pmatrix} \sim \begin{pmatrix} 1 & 1 & -1 & -1 & | & 1 \\ 0 & -3 & 2 & 3 & | & 0 \\ 0 & -3 & 2 & 3 & | & 0 \end{pmatrix}
$$

$$
\sim \begin{pmatrix} 3 & 0 & -1 & 0 & | & 3 \\ 0 & -3 & 2 & 3 & | & 0 \\ 0 & 0 & 0 & 0 & | & 0 \end{pmatrix}
$$

$$
R_3 - R_2;\ 3R_1 + R_2
$$

which shows that the system is consistent but that one of the equations is linearly dependent on the other two, i.e. $r = 2$. Then two $(n - r = 2)$ of the unknowns may be assigned arbitrary values. Let $x_4 = \alpha$, $x_3 = \beta$, then $3x_2 = 2\beta + 3\alpha$ and $3x_1 = 3 + \beta$.

Example (iii) Examine the system

$$
\begin{aligned}
x_1 - 2x_2 - x_3 - x_4 &= -4 \\
3x_1 + x_2 + x_3 - 2x_4 &= 11 \\
x_1 + 12x_2 + 6x_3 + x_4 &= 42
\end{aligned}
$$

in which $m = 3$ and $n = 4$.

Solution

$$
(A, B) = \begin{pmatrix} 1 & -2 & -1 & -1 & | & -4 \\ 3 & 1 & 1 & -2 & | & 11 \\ 1 & 12 & 6 & 1 & | & 42 \end{pmatrix} \sim \begin{pmatrix} 1 & -2 & -1 & -1 & | & -4 \\ 0 & 7 & 4 & 1 & | & 23 \\ 0 & 14 & 7 & 2 & | & 46 \end{pmatrix}
$$

$$
\sim \begin{pmatrix} 1 & -2 & -1 & -1 & | & -4 \\ 0 & 7 & 4 & 1 & | & 23 \\ 0 & 0 & -1 & 0 & | & 0 \end{pmatrix}
$$

and the system is consistent, $r = m = 3$ and one $(n - r = 1)$ of the variables may be arbitrarily assigned. The last equation formed from the equivalent matrix shows that $x_3 = 0$, let $x_4 = \alpha$, then the solution is $x_4 = \alpha$; $x_3 = 0$; $7x_2 = 23 - \alpha$ and $7x_1 = 18 + 5\alpha$.

Example (iv) Examine the system

$$x_1 - 2x_2 - x_3 - x_4 = -4$$
$$3x_1 + x_2 + x_3 - 2x_4 = 11$$
$$x_1 + 12x_2 + 7x_3 + x_4 = 31$$

in which $m = 3$ and $n = 4$.

Solution

$$(A, B) = \begin{pmatrix} 1 & -2 & -1 & -1 & -4 \\ 3 & 1 & 1 & -2 & 11 \\ 1 & 12 & 7 & 1 & 31 \end{pmatrix} \sim \begin{pmatrix} 1 & -2 & -1 & -1 & -4 \\ 0 & 7 & 4 & 1 & 23 \\ 0 & 14 & 8 & 2 & 35 \end{pmatrix}$$

$$\sim \begin{pmatrix} 1 & -2 & -1 & -1 & -4 \\ 0 & 7 & 4 & 1 & 23 \\ 0 & 0 & 0 & 0 & -11 \end{pmatrix}$$

and the equivalent form of the matrix shows that the system is inconsistent.

8.8.2 Case $m > n$

In this case the system (55) has more equations than unknowns. The system is either inconsistent in which case there is no solution, or it is consistent and the equations are linearly dependent.

If the system is consistent then, as in section 8.8.1, r of the equations are linearly independent and $(n - r)$ of the unknowns may be assigned arbitrary values. The non-zero integer r is such that $r \leqslant n$.

Example (i) Examine the system

$$2x_1 + x_2 = 7$$
$$x_1 + x_2 = 5$$
$$4x_1 + x_2 = 11$$
$$5x_1 - x_2 = 7$$
$$3x_1 - 2x_2 = 0$$

in which $m = 5$ and $n = 2$.

Solution

$$(A, B) \sim \begin{pmatrix} 1 & 1 & 5 \\ 2 & 1 & 7 \\ 4 & 1 & 11 \\ 5 & -1 & 7 \\ 3 & -2 & 0 \end{pmatrix} \sim \begin{pmatrix} 1 & 1 & 5 \\ 0 & -1 & -3 \\ 0 & -3 & -9 \\ 0 & -6 & -18 \\ 0 & -5 & -15 \end{pmatrix}$$

$$\sim \begin{pmatrix} 1 & 1 & 5 \\ 0 & -1 & -3 \\ 0 & 0 & 0 \\ 0 & 0 & 0 \\ 0 & 0 & 0 \end{pmatrix} \sim \begin{pmatrix} 1 & 0 & 2 \\ 0 & 1 & 3 \\ 0 & 0 & 0 \\ 0 & 0 & 0 \\ 0 & 0 & 0 \end{pmatrix}$$

which shows that the system is consistent and that three equations are linearly dependent on the other two, i.e. $r = 2$, $n - r = 0$. Finally, $x_2 = 3$ and $x_1 = 2$.

Example (ii) Examine the system

$$2x_1 + x_2 = 5$$
$$3x_1 + 2x_2 = 3$$
$$x_1 + x_2 = 2$$

in which $m = 3$ and $n = 2$.

Solution

$$(A, B) = \begin{pmatrix} 2 & 1 & 5 \\ 3 & 2 & 3 \\ 1 & 1 & 2 \end{pmatrix} \sim \begin{pmatrix} 1 & 1 & 2 \\ 0 & -1 & 1 \\ 0 & 0 & -4 \end{pmatrix}$$

The last row shows that the system is inconsistent.

Example (iii) Examine the system

$$3x_1 - x_2 + x_3 = 1$$
$$x_1 + 2x_2 - x_3 = 2$$
$$7x_1 - 7x_2 + 5x_3 = -1$$
$$12x_1 + 3x_2 = 9$$
$$7x_1 + x_3 = 4$$

in which $m = 5$ and $n = 3$.

Solution

$$(A, B) = \begin{pmatrix} 3 & -1 & 1 & 1 \\ 1 & 2 & -1 & 2 \\ 7 & -7 & 5 & -1 \\ 12 & 3 & 0 & 9 \\ 7 & 0 & 1 & 4 \end{pmatrix} \sim \begin{pmatrix} 1 & 2 & -1 & 2 \\ 3 & -1 & 1 & 1 \\ 7 & -7 & 5 & -1 \\ 12 & 3 & 0 & 9 \\ 7 & 0 & 1 & 4 \end{pmatrix}$$

$$\sim \begin{pmatrix} 1 & 2 & -1 & 2 \\ 0 & -7 & 4 & -5 \\ 0 & -21 & 12 & -15 \\ 0 & 7 & -4 & 5 \\ 0 & 7 & -4 & 5 \end{pmatrix} \sim \begin{pmatrix} 1 & 2 & -1 & 2 \\ 0 & -7 & 4 & -5 \\ 0 & 0 & 0 & 0 \\ 0 & 0 & 0 & 0 \\ 0 & 0 & 0 & 0 \end{pmatrix}$$

At this stage it may be seen that the system is consistent, that three of the equations are redundant, i.e. $r = 2$, and that one $(n - r = 1)$ of the unknowns can be given an arbitrary value. Let $x_3 = \alpha$, then $7x_2 = 4\alpha + 5$ and $7x_1 = 4 - \alpha$.

Finally, the condition that a system of $(n + 1)$ equations with n unknowns is consistent may be expressed as a simple determinant.

Consider, for simplicity, the case of 3 equations with 2 unknowns, i.e.

$$a_{11}x_1 + a_{12}x_2 = b_1$$
$$a_{21}x_1 + a_{22}x_2 = b_2$$
$$a_{31}x_1 + a_{32}x_2 = b_3.$$

Replace b_i by tb_i and the system may then be considered as a homogeneous one of three equations with three unknowns, x_1, x_2, and t. From section 8.7.2 the homogeneous system has a non-trivial solution if and only if the determinant of the coefficients is zero, i.e.

$$\begin{vmatrix} a_{11} & a_{12} & b_1 \\ a_{21} & a_{22} & b_2 \\ a_{31} & a_{32} & b_3 \end{vmatrix} = 0. \tag{56}$$

Since $t = 1$, this must be the necessary condition for the original three equations in two unknowns to have a unique solution. The condition (56) can be immediately extended to the general case of n unknowns.

8.8.3 Summary

A system of linear equations $Ax = B$ in n unknowns may be tested for consistency by carrying out a process of transformation of the matrix (A, B). The method involves an elimination of successive unknowns from the equations of the system by an application of row manipulations to the matrix (A, B). The process concludes when the matrix (A, B) has been replaced by the equivalent matrix (C, K) where K is a column matrix and C is an upper triangular matrix, i.e. one whose elements below the leading diagonal are all zero. Any inconsistency in the system will be shown by a row of (C, K) having all its elements zero except the element of the column K. If the system is consistent then the number of rows of (C, K) whose elements are not all zero is the number of equations of the system that are independent. The difference between the number of unknowns and the number of independent equations determines the number of unknowns that may be assigned arbitrary values.

EXERCISE 8

1 $A = (a_{ij})$, $B = (b_{jk})$, and $C = (c_{kl})$ are matrices of order 7×3, 3×4, and 4×5 respectively. The product ABC is given by $(a_{ij}b_{jk}c_{kl})$. Write out in full the element of the 4th row and the 2nd column of ABC.

2 Given $A = \begin{pmatrix} 3 & 2 \\ 1 & 4 \\ -1 & 7 \end{pmatrix}$, $B = \begin{pmatrix} 4 & -1 \\ 2 & 0 \end{pmatrix}$, and $C = \begin{pmatrix} 0 & 1 \\ -1 & 2 \end{pmatrix}$:

    (i) evaluate $(AB)C$ and $A(BC)$,
    (ii) verify that $A(B + C) = AB + AC$,
    (iii) find the matrix M such that $B^2 + 2C^3 + 2M = 0$.

3 Given matrices $A = (a_{ij})$ and $B = (b_{jk})$ of orders $m \times n$ and $n \times p$ respectively, write down:

(i) the element of the ith row and the kth column of AB,

(ii) the element of the kth row and the ith column of $(AB)'$,

(iii) the elements of the kth row of B',

(iv) the elements of the ith column of A'.

Show that $(AB)' = B'A'$.

4 Verify $(AB)' = B'A'$ for $A = \begin{pmatrix} 2 & -1 \\ 0 & 3 \\ 1 & 0 \end{pmatrix}$, $B = \begin{pmatrix} 4 \\ 2 \end{pmatrix}$.

5 If $A = \begin{pmatrix} 1 & 0 & 0 & 0 \\ 1 & -1 & 0 & 0 \\ 1 & -2 & 1 & 0 \\ 1 & -3 & 3 & -1 \end{pmatrix}$

show that (i) $A^2 = I$

(ii) $PQ = QP$

where $P = AD_1A$, $Q = AD_2A$, and D_1, D_2 are arbitrary diagonal matrices.

6 Write down the cofactors of the elements f and k in the determinant

$$\begin{vmatrix} a & b & c \\ d & e & f \\ g & h & k \end{vmatrix}.$$

7 Evaluate $\begin{vmatrix} 2 & 3 & -1 \\ 1 & 0 & 2 \\ 4 & 2 & 1 \end{vmatrix}$

(i) using cofactors of the first row,

(ii) using cofactors of the second column,

(iii) by the rule of Sarrus.

8 Given that $\begin{vmatrix} 3 & 1 & -2 \\ 8 & -5 & 7 \\ 4 & 0 & 1 \end{vmatrix} = -35,$

state, without expansion, the value of the following determinants

(i) $\begin{vmatrix} 3 & 8 & 4 \\ 1 & -5 & 0 \\ -2 & 7 & 1 \end{vmatrix}$ (ii) $\begin{vmatrix} 3 & -2 & -2 \\ 8 & 7 & 10 \\ 4 & 1 & 0 \end{vmatrix}$ (iii) $\begin{vmatrix} 3 & 1 & -2 \\ 4 & -5 & 6 \\ 4 & 0 & 1 \end{vmatrix}$

(iv) $\begin{vmatrix} 11 & 1 & -2 \\ -20 & -5 & 7 \\ 0 & 0 & 1 \end{vmatrix}$ (v) $\begin{vmatrix} 3 & 3 & 10 \\ 8 & -15 & -35 \\ 4 & 0 & -5 \end{vmatrix}$

9 Evaluate
$$\begin{vmatrix} 2 & 1 & -1 & 1 \\ 1 & 0 & 2 & 3 \\ 1 & 2 & 1 & 3 \\ -1 & 4 & 0 & -1 \end{vmatrix}$$

(i) by first forming the equivalent determinant whose elements of the second row are $(1, 0, 0, 0)$,

(ii) by first forming the equivalent determinant whose elements of the third column are $(-1, 0, 0, 0)$.

10 Find the inverse of the matrix
$$A = \begin{pmatrix} 4 & 3 & 1 \\ 2 & 0 & -2 \\ 1 & -1 & 1 \end{pmatrix}$$

by two different methods.

11 Given the system of equations

(i) $\begin{aligned} 2x_1 + x_2 + x_3 &= 6 \\ x_1 + 2x_2 + x_3 &= 10 \\ 10x_1 - x_2 + 3x_3 &= 2 \end{aligned}$ (ii) $\begin{aligned} 2x_1 + x_2 + x_3 &= 6 \\ x_1 + 2x_2 + x_3 &= 10 \\ 7x_1 + 5x_2 + 4x_3 &= 7. \end{aligned}$

Show that one is inconsistent and find the solution of the other.

12 Solve, if possible, the systems

(i) $\begin{aligned} x_1 - 3x_2 + 4x_3 - x_4 &= 3 \\ x_1 \quad\quad - x_3 + 2x_4 &= 1 \\ x_1 - 8x_2 + 11x_3 - 7x_4 &= 7 \\ x_2 - 3x_3 \quad\quad &= 0 \end{aligned}$ (ii) $\begin{aligned} 3x_1 - 2x_2 + x_3 &= 0 \\ 2x_1 + x_2 + 4x_3 &= 28 \\ x_1 - 3x_2 + 2x_3 &= 7 \\ 4x_1 \quad\quad + x_3 &= 3 \end{aligned}$

(iii) $\begin{aligned} 2x_1 + 3x_2 &= 7 \\ 4x_1 + 5x_2 &= 1 \\ 2x_1 - x_2 &= -19 \\ 2x_1 + 4x_2 &= 5 \end{aligned}$ (iv) $\begin{aligned} x_1 + 2x_2 + x_3 - x_4 &= -2 \\ 3x_1 - x_2 + 2x_3 \quad &= 2 \\ 2x_1 + 5x_2 + x_3 + x_4 &= 4 \end{aligned}$

13 Find the condition that $ax^2 + bx + c = 0$ and $px^2 + qx + r = 0$ have a common root.

[Hint: Form 4 equations in 3 unknowns x^3, x^2, x and use the consistency condition.]

9
Vectors

9.1 SCALARS

Certain physical quantities are completely described by a magnitude, or a measurement in a given system of units, for example, a volume, a mass, a temperature, or a speed. Such quantities are called *scalars* and may be manipulated using the basic laws of algebra, which for scalars α, β, and μ are:

The law of commutation

$$\text{(i) } \alpha + \beta = \beta + \alpha$$
$$\text{(ii) } \alpha\beta = \beta\alpha$$

The law of association

$$\text{(i) } (\alpha + \beta) + \mu = \alpha + (\beta + \mu)$$
$$\text{(ii) } (\alpha\beta)\mu = \alpha(\beta\mu) = \alpha\beta\mu$$

The law of distribution

$$\alpha(\beta + \mu) = \alpha\beta + \alpha\mu.$$

9.2 DIRECTED MAGNITUDES

Other quantities are not completely defined by a magnitude but require also a precise indication of the direction in which the measurement is taken. Typical examples are, a displacement, a force, a velocity, or a finite rotation. In the latter case, the direction is indicated by the axis of rotation and the magnitude by the angle of rotation. Such quantities may be conveniently represented on a figure by a line segment in the requisite direction and of a suitably scaled length to represent the magnitude.

It is not necessary that these *directed magnitudes* have a prescribed point of application, for example, if a solid body moves without rotation in a straight line then every point of the body has the same velocity. Such quantities are said to be *free*. If a directed magnitude has its point of application prescribed as, for example, a given force acting at a point of a body then it is said to be *bound*.

9.2.1 Notation

When directed magnitudes are indicated on a diagram it is convenient to use letters to label the figure and a directed magnitude may then be

represented by a line segment $\overline{OP}$, $\overrightarrow{OP}$, or **OP** in Clarendon type.

The notation $\overline{OP}$ means that the direction of the measurement is from O to P and that the magnitude is equivalent to the length from O to P. The direction in which a vector acts should be emphasized by adding an arrow to the line segment as in Fig. 9.1.

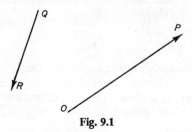

Fig. 9.1

9.2.2 Equality of directed magnitudes

In Fig. 9.2 the two directed magnitudes $\overline{OP}$ and $\overline{QR}$ are equal, i.e. $\overline{OP} = \overline{QR}$, if, and only if, the lengths OP and QR are equal and they are in the same direction (note this means that not only are the lines of action parallel but that the magnitudes are directed in the same sense). If two directed magnitudes have the same magnitude but are in opposite directions, as are $\overline{OP}$ and $\overline{ST}$ in Fig. 9.2, then this is indicated by writing $\overline{OP} = -\overline{ST}$. In particular, $\overline{OP} = -\overline{PO}$. A *null* or *zero directed magnitude* is one of zero magnitude in any direction.

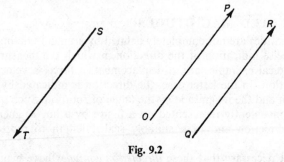

Fig. 9.2

9.2.3 Addition of directed magnitudes

Suppose a man walks 1 kilometre due East from a given point and then walks in a direction $N\ 30°\ E$ for 1 kilometre, his path may be represented by the displacement diagram Fig. 9.3 where O represents his starting point, Q his final position, and P his intermediate position.

His path represented by OPQ is equivalent to $\overline{OQ}$, i.e. walking $\sqrt{3}$ kilometre in a direction $N\ 60°\ E$ from his initial starting point. The equivalence of paths may be formally written

$$\overline{OQ} = \overline{OP} + \overline{PQ} \tag{1}$$

and called the *triangle of addition*.

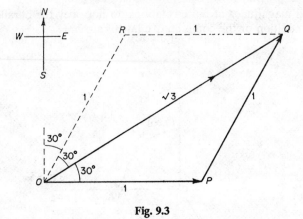

Fig. 9.3

It is immaterial in which order the directed magnitudes of displacement are taken because, assuming it to be physically possible, the man could have walked on a path represented in the figure first by the displacement $\overline{OR}$ and second by $\overline{RQ}$. From the definition of equality, $\overline{OR} = \overline{PQ}$ and $\overline{OP} = \overline{RQ}$ and it follows that

$$\overline{OQ} = \overline{OP} + \overline{PQ} = \overline{OR} + \overline{RQ} = \overline{PQ} + \overline{OP} = \overline{OR} + \overline{OP} \quad (2)$$

When the parallelogram $OPQR$ is completed, equations (2) state that the sum of $\overline{OP}$ and $\overline{OR}$ is given by the diagonal $\overline{OQ}$. This is called the *parallelogram law of addition*. Equations (2) also show that displacements obey the commutative law of addition. In the example given the various paths taken all lie on the surface of the Earth. The distances involved are so small compared to the radius of the Earth that the curvature of the latter may be neglected. The points represented by O, P, Q, and R then all lie on a plane. In comparison see example 2 of Exercise 9.

However, all directed magnitudes do not combine according to the triangle or parallelogram law of addition. Consider for example the addition of two finite rotations as follows. A point P (see Fig. 9.4) is situated at the position whose Cartesian coordinates are $(1, 0, 0)$ and the line OP is subjected to a rotation (R_1) through an angle of $\pi/2$ about the axis of Ox_3 and to a rotation (R_2) through an angle of $\pi/2$ about the axis of Ox_1. The two rotations in the sense indicated in the figure can be represented by directed magnitudes $\overline{OB}$ and $\overline{OA}$ (where the lengths OA and OB are both $\pi/2$). The rotation (R_1) followed by the rotation (R_2) makes P move, via Q, into S and is therefore equivalent to a single rotation about the axis Ox_2.

If, however, the order is reversed so that rotation (R_2) is followed by rotation (R_1) it is seen that P moves to Q. The conclusion is that in this example the sum $\overline{OA} + \overline{OB}$ is not the same as the sum $\overline{OB} + \overline{OA}$; the directed magnitudes of finite rotations do not obey the parallelogram law of addition or the commutative law of algebra.

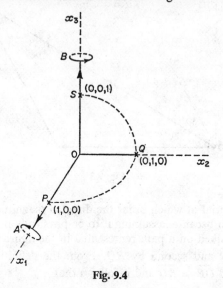

Fig. 9.4

9.3 VECTORS

A directed magnitude that obeys the parallelogram law of addition so that, in Fig. 9.3,

$$\overline{OQ} = \overline{OP} + \overline{OR} = \overline{OR} + \overline{OP}$$

is called a *vector*. Examples of vectors are displacement, force, velocity, acceleration, and angular velocity. From the definition it follows that a finite rotation is not a vector. In addition to the notation of section 9.2.1, a vector is frequently denoted by a single letter such as **a** or **b** printed in Clarendon type. The non-negative scalar magnitude of a vector **a** is then denoted by $|\mathbf{a}|$ or simply a in italic type.

9.3.1 Equality of vectors

Two vectors **a** and **b** are equal if, and only if, $a = b$ and the directions are the same. If $a = b$ but the directions are opposite then $\mathbf{a} = -\mathbf{b}$.

9.3.2 Zero vector and unit vector

A *null* or *zero vector* is of zero length and has no particular direction; it is frequently denoted by **0**. A *unit vector* is one whose magnitude is unity.

9.3.3 Addition and subtraction of vectors

If more than two vectors are added together then the addition law is applied successively as illustrated in Fig. 9.5 until a single vector (the *resultant*) is obtained.

It may be easily shown, using the properties of parallelograms, that the order of continued addition is immaterial and that

$$R = (a + b) + c = a + (b + c)$$

so that vectors obey the associative law of addition. The vectors of Fig. 9.5 do not necessarily lie in one plane.

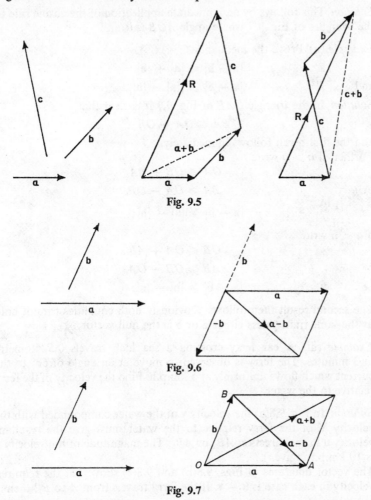

Fig. 9.5

Fig. 9.6

Fig. 9.7

The difference of two vectors **a** and **b** is simply the sum of the vector **a** and the vector (−**b**). In Fig. 9.6, **a** is added to (−**b**) using the parallelogram rule.

If (**a** − **b**) is a free vector then it may be moved to any parallel position and in Fig. 9.7 the sum and difference of two vectors **a** and **b** are represented by the diagonals of the parallelogram with sides **a** and **b**.

(Note that the vector $\mathbf{a} - \mathbf{b}$ is in the direction from the end of $\mathbf{b}$ to the end of $\mathbf{a}$.)

Example (i) If the vectors $\mathbf{a}$ and $\mathbf{b}$ are inclined at an angle θ, then

$$|\mathbf{a} - \mathbf{b}| = \sqrt{(a^2 + b^2 - 2ab \cos \theta)}.$$

and

$$|\mathbf{a} + \mathbf{b}| = \sqrt{(a^2 + b^2 + 2ab \cos \theta)}.$$

Solution This follows by an immediate application of the cosine rule to the triangles of Fig. 9.7 where angle AOB is θ.

Example (ii) Prove the inequalities

$$|\mathbf{a} - \mathbf{b}| \leqslant |\mathbf{a}| + |\mathbf{b}|$$

and

$$|\mathbf{a} - \mathbf{b}| \geqslant ||\mathbf{a}| - |\mathbf{b}||.$$

Solution In the triangle OAB of Fig. 9.7 it is clear that

$$BA \leqslant OA + OB$$

and the first result follows immediately.
Further, if $a > b$ write

$$OA \leqslant OB + BA$$

or

$$BA \geqslant OA - OB,$$

i.e.

$$|\mathbf{a} - \mathbf{b}| \geqslant |\mathbf{a}| - |\mathbf{b}|;$$

if $a < b$ write

$$OB \leqslant OA + AB$$

or

$$AB \geqslant OB - OA,$$

i.e.

$$|\mathbf{a} - \mathbf{b}| \geqslant |\mathbf{b}| - |\mathbf{a}|.$$

The second result then follows. Obviously both equalities cannot hold at the same time unless either $\mathbf{a}$ or $\mathbf{b}$ is the null vector.

Example (iii) A car ferry crossing a sea loch travels 0·5 kilometre in 3 minutes. The ferry is observed to move at an angle of 60° to the current which flows uniformly at 5 km.p.h. Find the velocity of the ferry relative to the water.

Solution In Fig. 9.8(a) the velocity $\mathbf{v}$ of the water compounded with the velocity $\mathbf{V}$ of the ferry relative to the water must give the resultant velocity $\mathbf{u}$ in the direction AB_1 or AB_2. The magnitude of the velocity $\mathbf{u}$ is 10 km.p.h. (given).
The vector diagrams in Figs. 9.8(b) and 9.8(c) show that the required velocity in each case is $\mathbf{u} - \mathbf{v}$. If the ferry travels from A to B_1, then

$$V = V_1 = \sqrt{(5^2 + 10^2 + 2 \cdot 5 \cdot 10 \cos 60°)}$$
$$= 5\sqrt{7} \text{ km.p.h.}$$

and

$$\tan \theta_1 = 10 \sin 60° / (5 + 10 \cos 60°)$$
$$= \sqrt{3}/2$$

i.e.

$$\theta_1 = 40° \, 54'.$$

If the ferry travels from A to B_2 then

$$V = V_2 = \sqrt{(5^2 + 10^2 - 2 \cdot 5 \cdot 10 \cos 60°)}$$
$$= 5\sqrt{3} \text{ km.p.h.}$$

and $\qquad \tan \theta_2 = 10 \sin 60°/(5 - 10 \cos 60°)$

i.e. $\qquad \theta_2 = 90°.$

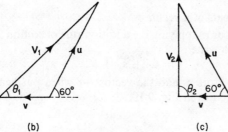

(a)

(b) $\qquad\qquad\qquad\qquad$ (c)

Fig. 9.8

9.3.4 Multiplication of a vector by a scalar

If λ is a scalar quantity then $\lambda \mathbf{a}$ is a vector of magnitude $|\lambda \mathbf{a}|$, i.e. $|\lambda| a$ whose direction is that of $\mathbf{a}$ if λ is positive and that of $-\mathbf{a}$ if λ is negative. The Fig. 9.9 shows a free vector $\mathbf{a}$ together with three vectors $2\mathbf{a}$, $-3\mathbf{a}$, and $\frac{1}{2}\mathbf{a}$.

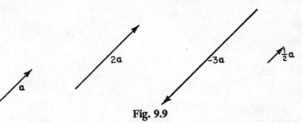

Fig. 9.9

If $\mathbf{a}$ is an arbitrary vector then $\mathbf{a}/|\mathbf{a}|$ is a unit vector in the direction of $\mathbf{a}$. A notation in common use is to reserve $\hat{\mathbf{a}}$ for this vector. Thus

$$\mathbf{a} = a\hat{\mathbf{a}} \qquad \text{or} \qquad \mathbf{a} = |\mathbf{a}|\hat{\mathbf{a}}. \qquad (3)$$

Using the properties of similar triangles and simple proportion it may be shown that if λ and μ are two scalars and $\mathbf{a}$ and $\mathbf{b}$ are two vectors then

$$\mu(\lambda\mathbf{a}) = \mu\lambda\mathbf{a} = \lambda(\mu\mathbf{a}) \qquad (4)$$
$$(\lambda + \mu)\mathbf{a} = \lambda\mathbf{a} + \mu\mathbf{a} \qquad (5)$$
$$\lambda(\mathbf{a} + \mathbf{b}) = \lambda\mathbf{a} + \lambda\mathbf{b} \qquad (6)$$

Equation (4) is the associative law and equations (5) and (6) form the distributive law for multiplication of a vector by a scalar.

Example Express in vector notation that (a) P is the mid-point of OA, (b) CD is parallel to AB, three times as long as AB, and in the opposite sense to AB.

Solution (a) $OP = \frac{1}{2}OA$, and vector $\overline{OP}$ has the same line of action as $\overline{OA}$. Hence, $\overline{OP} = \frac{1}{2}\overline{OA}$.

(b) $CD = 3AB$ and hence $\overline{CD} = -3\overline{AB}$, the opposite direction being indicated by the negative sign.

9.3.5 Components of a vector

Given any vector $\mathbf{r}$, applying the addition rule of section 9.3.3 in reverse it is possible to write

$$\mathbf{r} = \mathbf{a}_1 + \mathbf{a}_2 + \ldots + \mathbf{a}_n$$

where the n vectors $\mathbf{a}_i$ and the vector $-\mathbf{r}$ when put head to tail form a closed polygon as in Fig. 9.10.

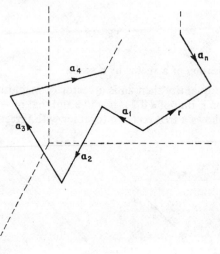

Fig. 9.10

This polygon is, in general, a three-dimensional one. The vectors $\mathbf{a}_1, \mathbf{a}_2, \ldots, \mathbf{a}_n$ are called the *component vectors* of $\mathbf{r}$. Any $(n-1)$ of the vectors $\mathbf{a}_i$ may be chosen arbitrarily but the nth vector must close the polygon.

A case of particular interest arises when the vector **r** is expressed as the sum of three vectors parallel to three given non-zero, non-coplanar vectors **a**, **b**, and **c**. In Fig. 9.11, let $\overline{OR}$ represent the vector **r** and construct the parallelepiped with OR as diagonal and the sides OA, OB, and OC parallel to the given vectors **a**, **b**, and **c** respectively. This

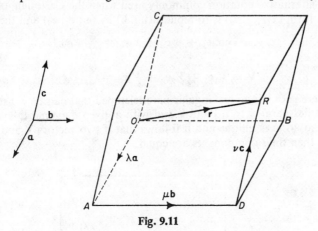

Fig. 9.11

construction is unique, that is, only one such parallelepiped can be drawn and

$$\mathbf{r} = \overline{OR} = \overline{OA} + \overline{AD} + \overline{DR}$$
$$= \overline{OA} + \overline{OB} + \overline{OC}$$
$$= \lambda\mathbf{a} + \mu\mathbf{b} + \nu\mathbf{c}$$

where λ, μ, and ν are determinable scalars. In particular if **a**, **b**, and **c** are unit vectors then λ, μ, and ν are the number of such units in OA, OB, and OC respectively. The component vectors of **r** are $\lambda\mathbf{a}$, $\mu\mathbf{b}$, and $\nu\mathbf{c}$ but the scalar quantities λ, μ, and ν are frequently referred to as the *scalar components* of the vector **r** or, when no ambiguity arises, simply as the *components* of **r**.

The case in which the vectors **a**, **b**, and **c** are mutually orthogonal is of special importance because the three directions of these vectors can be identified with those of the orthogonal Cartesian coordinate axes Ox, Oy, and Oz. Unit vectors in these directions are usually indicated by **i**, **j**, and **k** respectively. If **r** is a displacement vector then using Fig. 9.12, **r** may be written

$$\mathbf{r} = x\mathbf{i} + y\mathbf{j} + z\mathbf{k}$$

where (x, y, z) are the coordinates of the point R relative to the point O. If O is the origin of the coordinates then the displacement **r** is called the *position vector* of the point R.

If **V** is any vector it may be written in the form

$$\mathbf{V} = V_1\mathbf{i} + V_2\mathbf{j} + V_3\mathbf{k}$$

where V_1, V_2, and V_3 are the orthogonal (scalar) projections of the vector **V** on to the coordinate axes.

An alternative notation commonly used when the Cartesian axes are in the form (x_1, x_2, x_3) is to replace $(\mathbf{i}, \mathbf{j}, \mathbf{k})$ by $(\mathbf{e}_1, \mathbf{e}_2, \mathbf{e}_3)$ and then

$$\mathbf{r} = x_1\mathbf{e}_1 + x_2\mathbf{e}_2 + x_3\mathbf{e}_3 = \sum_{i=1}^{3} x_i\mathbf{e}_i$$

and

$$\mathbf{V} = V_1\mathbf{e}_1 + V_2\mathbf{e}_2 + V_3\mathbf{e}_3 = \sum_{i=1}^{3} V_i\mathbf{e}_i.$$

There is only one rectangular parallelepiped that can be constructed with sides parallel to **i**, **j** and **k** and with a given diagonal OR so that the resolution is unique and it follows that if two vectors **V** and **U** are equal then their components are equal.

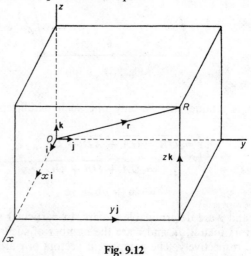

Fig. 9.12

Thus given that $\mathbf{V} = V_1\mathbf{i} + V_2\mathbf{j} + V_3\mathbf{k}$ and $\mathbf{U} = U_1\mathbf{i} + U_2\mathbf{j} + U_3\mathbf{k}$ are equal, so that $\mathbf{V} = \mathbf{U}$, then

$$V_1 = U_1, \quad V_2 = U_2, \quad \text{and} \quad V_3 = U_3.$$

Note 1 An even further abbreviated notation is to write $\mathbf{r} = x_i\mathbf{e}_i$, the convention being that a repeated index means a summation taken over all possible values of that index.

Note 2 A vector may be conveniently represented by its components in the form (x_1, x_2, x_3) or

$$\begin{pmatrix} x_1 \\ x_2 \\ x_3 \end{pmatrix}$$

providing that there is no ambiguity or confusion. The vector components then form a 1×3 row matrix or a 3×1 column matrix and may be used in conjunction with other matrices. For example if $\mathbf{x} = x_i \mathbf{e}_i$ and $\mathbf{y} = y_i \mathbf{e}_i$ and A is the 3×3 matrix a_{ij} then

$$\mathbf{y} = A\mathbf{x}$$

is the vector equivalent to the matrix equation

$$y = Ax$$

of example (ix) of section 8.2.6 where $y^T = y' = (y_1 \, y_2 \, y_3)$,

$$x^T = x' = (x_1 \, x_2 \, x_3) \text{ and}$$

$$A = \begin{pmatrix} a_{11} & a_{12} & a_{13} \\ a_{21} & a_{22} & a_{23} \\ a_{31} & a_{32} & a_{33} \end{pmatrix}.$$

9.3.6 Direction cosines

Suppose a general vector $\mathbf{V}$ is represented by $\overline{PQ}$ as in Fig. 9.13 and that $\mathbf{V} = V_1\mathbf{i} + V_2\mathbf{j} + V_3\mathbf{k}$, then by Pythagoras' theorem

$$V = |\mathbf{V}| = PQ = \sqrt{(PB^2 + BQ^2)}$$

$$= \sqrt{(PA^2 + AB^2 + BQ^2)}$$

$$= \sqrt{(V_1^2 + V_2^2 + V_3^2)} \tag{7}$$

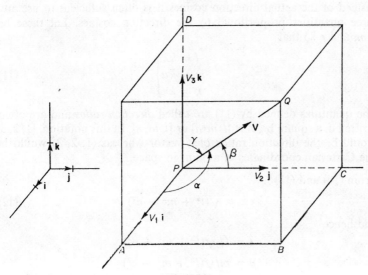

Fig. 9.13

The direction of $\overline{PQ}$ can be specified by the angles that $\overline{PQ}$ makes with the coordinate axes Ox, Oy, and Oz (or with PA, PC, and PD). These angles are usually designated α, β, and γ respectively and are called the *direction angles*. The angles are measured positively in the sense indicated in Fig. 9.13.

It is, however, the cosines of these angles that frequently occur and these are designated λ, μ, and ν where

$$\begin{aligned} \lambda &= \cos \alpha \\ \mu &= \cos \beta \\ \nu &= \cos \gamma \end{aligned} \tag{8}$$

and called the *direction cosines*. From the right-angled triangles of Fig. 9.13 it follows that $\cos \alpha = PA/PQ$, $\cos \beta = PC/PQ$, and $\cos \gamma = PD/PQ$ or

$$\left.\begin{aligned} \cos \alpha &= V_1/V \\ \cos \beta &= V_2/V \\ \cos \gamma &= V_3/V \end{aligned}\right\} \tag{9}$$

Since the position of Q relative to P is uniquely determined by only 3 quantities (viz. V_1, V_2, and V_3) and 4 quantities have been introduced (viz. λ, μ, ν, and V) there must exist some relationship between these quantities. Using equations (7), (8), and (9) leads to

$$\lambda^2 + \mu^2 + \nu^2 = 1$$

or $$\cos^2 \alpha + \cos^2 \beta + \cos^2 \gamma = 1 \tag{10}$$

Instead of the actual direction cosines it is often sufficient to use any three quantities proportional to the direction cosines. Let these be l, m, and n so that

$$\left.\begin{aligned} l &= k\lambda \\ m &= k\mu \\ n &= k\nu \end{aligned}\right\} \tag{11}$$

The quantities defined by (11) are called *direction ratios* and are often written in a square bracket $[l:m:n]$ or $[l, m, n]$. In this notation, $[1, 2, 3]$ would be the direction ratios of a vector whereas $(1, 2, 3)$ would be the Cartesian coordinates of a point in space.

From (10) and (11)

$$k = \sqrt{(l^2 + m^2 + n^2)} \tag{12}$$

and hence

$$\begin{aligned} \lambda &= l / \sqrt{(l^2 + m^2 + n^2)} \\ \mu &= m/\sqrt{(l^2 + m^2 + n^2)} \\ \nu &= n/\sqrt{(l^2 + m^2 + n^2)}. \end{aligned}$$

To obtain the direction cosines from the direction ratios it is therefore necessary to divide by $\sqrt{(l^2 + m^2 + n^2)}$. In particular, if

$$\mathbf{V} = V_1\mathbf{i} + V_2\mathbf{j} + V_3\mathbf{k}$$

then $[V_1 : V_2 : V_3]$ may be used as direction ratios.

Example (i) Find the sum of the two vectors $3\mathbf{i} + 7\mathbf{j} - 4\mathbf{k}$ and $6\mathbf{i} - 2\mathbf{j} + 12\mathbf{k}$ and calculate its magnitude and direction.

Solution Sum vector = $\mathbf{R} = 9\mathbf{i} + 5\mathbf{j} + 8\mathbf{k}$

$$\text{Magnitude} = R = \sqrt{(81 + 25 + 64)}$$
$$= \sqrt{170}.$$

Direction ratios are $[9:5:8]$ and the direction cosines are $(9/\sqrt{170}, 5/\sqrt{170}, 8/\sqrt{170})$.

Example (ii) Find the direction of the line PQ where P is the point $(3, -7, 8)$ and Q is the point $(2, 5, 5)$.

Solution If O is the origin then in vector form

$$\overline{OP} = 3\mathbf{i} - 7\mathbf{j} + 8\mathbf{k}$$

and

$$\overline{OQ} = 2\mathbf{i} + 5\mathbf{j} + 5\mathbf{k}$$

and hence

$$\overline{PQ} = \overline{OQ} - \overline{OP}$$
$$= -\mathbf{i} + 12\mathbf{j} - 3\mathbf{k}.$$

The direction cosines of the line PQ are then

$$(-1/\sqrt{154}, 12/\sqrt{154}, -3/\sqrt{154}).$$

Example (iii) Three forces $\mathbf{F}_1 = 2\mathbf{i} + 3\mathbf{j} + 2\mathbf{k}$, $\mathbf{F}_2 = 2\mathbf{j} - 3\mathbf{k}$, and $\mathbf{F}_3 = 4\mathbf{i} - 2\mathbf{j} - \mathbf{k}$ act on a point mass. Find the resultant force. Calculate the angles it makes with the coordinate axes.

Solution Let the forces act on the point mass at the point P in Fig. 9.14.

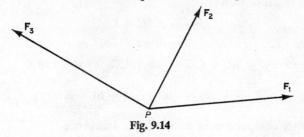

Fig. 9.14

The resultant $\mathbf{R} = \mathbf{F}_1 + \mathbf{F}_2 + \mathbf{F}_3$
$$= 6\mathbf{i} + 3\mathbf{j} - 2\mathbf{k}.$$

The magnitude of $\mathbf{R}$ is $\sqrt{(36 + 9 + 4)}$ or 7.

The direction cosines are (6/7, 3/7, −2/7) and the direction angles are

$$\alpha = \cos^{-1}(6/7) \quad = 31°,$$
$$\beta = \cos^{-1}(3/7) \quad = 64° \, 37',$$

and
$$\gamma = \cos^{-1}(-2/7) = 106° \, 36'.$$

(Note here the value of γ corresponding to the negative sign for $\cos \gamma$.)

Example (iv) Prove that the centroid of any triangle divides the median in the ratio 2:1.

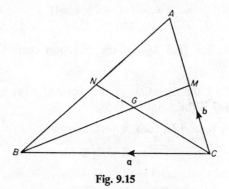

Fig. 9.15

Solution In Fig. 9.15 let M and N be the mid-points of AC and AB respectively of triangle ABC and let G be the centroid.
Let $\mathbf{CA} = \mathbf{b}$ and $\mathbf{CB} = \mathbf{a}$, then $\mathbf{BA} = \mathbf{b} - \mathbf{a}$, $\mathbf{BN} = \frac{1}{2}(\mathbf{b} - \mathbf{a})$, $\mathbf{CM} = \frac{1}{2}\mathbf{b}$.
Let $\mathbf{CG} = \alpha\mathbf{CN}$ where α is a scalar, then

$$\mathbf{CG} = \alpha(\mathbf{CB} + \mathbf{BN})$$
$$= \alpha(\mathbf{a} + \tfrac{1}{2}(\mathbf{b} - \mathbf{a}))$$
$$= \frac{\alpha}{2}(\mathbf{a} + \mathbf{b}).$$

Let $\mathbf{BG} = \beta\mathbf{BM}$, then

$$\mathbf{BG} = \beta(\mathbf{BC} + \mathbf{CM})$$
$$= \beta(-\mathbf{a} + \tfrac{1}{2}\mathbf{b}).$$

Now since $\mathbf{BG} = \mathbf{BC} + \mathbf{CG} = -\mathbf{a} + \mathbf{CG}$, it follows that

$$\beta(-\mathbf{a} + \tfrac{1}{2}\mathbf{b}) = -\mathbf{a} + \frac{\alpha}{2}(\mathbf{a} + \mathbf{b}) = \mathbf{R} \text{ (say)}.$$

The vector components of $\mathbf{R}$ are unique and hence

$$-\beta = -1 + \frac{\alpha}{2}$$

and
$$\frac{\beta}{2} = \frac{\alpha}{2}$$

from which it follows that

$$\alpha = \beta = \tfrac{2}{3}.$$

The result then follows immediately.

9.3.7 The transformation matrix

The transformation matrix of section 8.1 can now be given a geometrical significance. Suppose that P is a point that has Cartesian coordinates (b_1, b_2, b_3) when referred to a frame of axes $Ox_1x_2x_3$ and has coordinates (B_1, B_2, B_3) when referred to a second frame $Oy_1y_2y_3$ obtained from the first by a rotation about O as illustrated in Fig. 9.16.

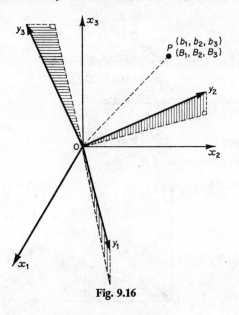

Fig. 9.16

From sections 9.3.5 and 9.3.6 it follows that

$$B_1 = \text{sum of projections of } b_1, b_2, \text{ and } b_3 \text{ on to } Oy_1$$
$$= b_1 \cos (x_1 O y_1) + b_2 \cos (x_2 O y_1) + b_3 \cos (x_3 O y_1)$$
$$= \lambda_1 b_1 + \mu_1 b_2 + \nu_1 b_3$$

where $[\lambda_1, \mu_1, \nu_1]$ are the direction cosines of Oy_1 referred to the frame $Ox_1x_2x_3$. Likewise

$$B_2 = \lambda_2 b_1 + \mu_2 b_2 + \nu_2 b_3$$

and

$$B_3 = \lambda_3 b_1 + \mu_3 b_2 + \nu_3 b_3$$

where $[\lambda_2, \mu_2, \nu_2]$ and $[\lambda_3, \mu_3, \nu_3]$ are the direction cosines of Oy_2 and Oy_3 respectively.

In matrix notation

$$\begin{pmatrix} B_1 \\ B_2 \\ B_3 \end{pmatrix} = \begin{pmatrix} \lambda_1 & \mu_1 & \nu_1 \\ \lambda_2 & \mu_2 & \nu_2 \\ \lambda_3 & \mu_3 & \nu_3 \end{pmatrix} \begin{pmatrix} b_1 \\ b_2 \\ b_3 \end{pmatrix}. \tag{13}$$

In an exactly similar manner a relationship can be derived in the form

$$\begin{pmatrix} b_1 \\ b_2 \\ b_3 \end{pmatrix} = \begin{pmatrix} \lambda_1 & \lambda_2 & \lambda_3 \\ \mu_1 & \mu_2 & \mu_3 \\ \nu_1 & \nu_2 & \nu_3 \end{pmatrix} \begin{pmatrix} B_1 \\ B_2 \\ B_3 \end{pmatrix} \tag{14}$$

where $[\lambda_1, \lambda_2, \lambda_3]$ are the direction cosines of Ox_1 referred to the frame $Oy_1y_2y_3$. In a more concise notation the equation (13) may be written

$$B_i = a_{ij}b_j$$

or
$$\mathbf{B} = A\mathbf{b}$$

where $A = (a_{ij})$ is the transformation matrix and

$$a_{ij} = \cos (x_j O y_i).$$

Similarly equation (14) may be written

$$b_i = a_{ji}B_j$$

or
$$\mathbf{b} = A'\mathbf{B}$$

where A' is the transpose of A.

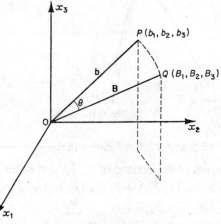

Fig. 9.17

Thus the transformation matrix (a_{ij}) represents the rotation of the frame of reference $Ox_1x_2x_3$ into the frame of reference $Oy_1y_2y_3$ and has elements which are direction cosines of one frame with respect to the other, i.e. it specifies the magnitude and direction of the rotation. Alternatively, if the frame $Ox_1x_2x_3$ is fixed then (a_{ij}) represents the rotation in space of the line OP through an angle θ, to take up a position OQ where P is the point whose coordinates are (b_1, b_2, b_3) and Q is the point whose coordinates are (B_1, B_2, B_3) and $OP = OQ$. This is illustrated in Fig. 9.17.

Now let x be a vector in the direction of OP, i.e. $x = t\,b$, and let y be a vector in the direction of OQ, i.e. $y = sB$, where t, s are scalars, then

$$y = Cx$$

where

$$C = (c_{ij})$$

$$= \frac{s}{t}(a_{ij}).$$

The matrix C contains all the information necessary to give the relative positions and magnitudes of the vectors x and y.

Note that solving equation (13) for b leads to $b = A^{-1}B$ and hence it follows that

$$A' = A^{-1}$$

or

$$AA' = I.$$

Matrices such as A which satisfy this relationship are called *orthogonal* and have an important role in matrix application to engineering problems, particularly in connection with those involving vibration.

Note also that in suffix notation the equation $y = Cx$ may be written

$$y_i = c_{ij}x_j$$

where (x_j) and (y_i) are column matrices of order (3×1) and (c_{ij}) is *necessarily* a square matrix of order (3×3).

9.4 VECTOR EQUATION OF A STRAIGHT LINE

The direction of a straight line in space is determined by its direction cosines (or ratios) and its position is fixed in space if the coordinates of any point on the line are known.

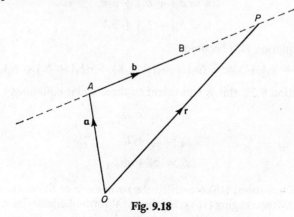

Fig. 9.18

Suppose a straight line has a direction specified by a constant vector $\mathbf{b}$ (i.e. in a Cartesian coordinate system by $\mathbf{b} = b_1\mathbf{i} + b_2\mathbf{j} + b_3\mathbf{k}$ so that the line has direction ratios $[b_1, b_2, b_3]$) and passes through a fixed point A in space whose position vector relative to an origin O is $\mathbf{a}$ (i.e. in Cartesian coordinates $\mathbf{a} = a_1\mathbf{i} + a_2\mathbf{j} + a_3\mathbf{k}$ so that A has coordinates (a_1, a_2, a_3)). In Fig. 9.18 the points A and B are given (for a given $\mathbf{a}$ and $\mathbf{b}$) and the straight line is obtained by producing the line segment AB in both directions.

The equation of the line is obtained by stating mathematically the geometrical condition that the variable point P is constrained to lie on AB and on AB (or BA) produced. Let $\mathbf{OP} = \mathbf{r}$ be the position vector of P, then in Fig. 9.18

$$\mathbf{OP} = \mathbf{r} = \mathbf{OA} + \mathbf{AP}. \tag{15}$$

But, from section 9.3.4

$$\mathbf{AP} = t\mathbf{AB}$$

$$= t\mathbf{b} \tag{16}$$

where t is some scalar.

For fixed $\mathbf{b}$, different values of t give different points along the line. From equations (15) and (16) the required equation is

$$\mathbf{r} = \mathbf{a} + t\mathbf{b}. \tag{17}$$

It is important to note that the vector equation (17) is independent of any system of coordinates. The scalar quantity t is a parameter that can range from $-\infty$ to $+\infty$. If $t > 1$, P lies in AB produced; if $t = 1$, P coincides with B; if $0 < t < 1$, P lies on AB; if $t = 0$, P coincides with A and if $t < 0$, P lies on BA produced.

In a rectangular Cartesian coordinate system

$$\mathbf{r} = x_1\mathbf{i} + x_2\mathbf{j} + x_3\mathbf{k},$$

$$\mathbf{a} = a_1\mathbf{i} + a_2\mathbf{j} + a_3\mathbf{k},$$

and

$$\mathbf{b} = b_1\mathbf{i} + b_2\mathbf{j} + b_3\mathbf{k}$$

so that equation (17) becomes

$$x_1\mathbf{i} + x_2\mathbf{j} + x_3\mathbf{k} = (a_1\mathbf{i} + a_2\mathbf{j} + a_3\mathbf{k}) + t(b_1\mathbf{i} + b_2\mathbf{j} + b_3\mathbf{k}).$$

From section 9.3.5 this is equivalent to three scalar equations

$$x_1 = a_1 + tb_1$$

$$x_2 = a_2 + tb_2 \tag{18}$$

$$x_3 = a_3 + tb_3.$$

The set of equations (18) are called the *parametric* or *freedom equations* of the line. Rearranging (18) gives the usual form of constraint equation of the line

$$\frac{x_1 - a_1}{b_1} = \frac{x_2 - a_2}{b_2} = \frac{x_3 - a_3}{b_3} = t \qquad (19)$$

It is worth noting that if the line lies in the two-dimensional plane Ox_1x_2 then $a_3 = 0$ and, since the line is at right angles to the x_3-axis, $b_3 = 0$ so that equation (19) reduces to

$$\frac{x_1 - a_1}{b_1} = \frac{x_2 - a_2}{b_2} = \frac{x_3}{0}. \qquad (20)$$

The equations (20) are equivalent to

$$\frac{x_1 - a_1}{b_1} = \frac{x_2 - a_2}{b_2}, \; x_3 = 0. \qquad (21)$$

If (x_1, x_2, x_3) is replaced by (x, y, z) the equation (21) becomes

$$y - a_2 = b_2(x - a_1)/b_1, z = 0$$

which is equivalent to the well-known form

$$y = mx + c, \quad z = 0.$$

Example (i) Find the equation of the line passing through the two points R and S whose Cartesian coordinates are $R(3, 2, +1)$ and $S(6, -3, 4)$.

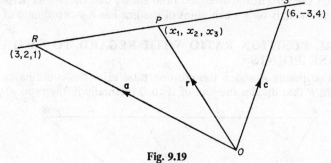

Fig. 9.19

Solution The direction of the line is specified by the join of the two points. In the notation of Fig. 9.19

$$\mathbf{a} = 3\mathbf{i} + 2\mathbf{j} + \mathbf{k},$$
$$\mathbf{r} = x_1\mathbf{i} + x_2\mathbf{j} + x_3\mathbf{k}$$

and
$$\mathbf{c} = 6\mathbf{i} - 3\mathbf{j} + 4\mathbf{k}$$

so that
$$\mathbf{RS} = \mathbf{b} = \mathbf{c} - \mathbf{a}$$
$$= 3\mathbf{i} - 5\mathbf{j} + 3\mathbf{k}.$$

The vector equation of the line is then in the form

$$\mathbf{r} = \mathbf{a} + t\mathbf{b}$$

which, using equations (17) and (19), leads to

$$\frac{x_1 - 3}{3} = \frac{x_2 - 2}{-5} = \frac{x_3 - 1}{3} (= t) \qquad (22)$$

Note that the vector equation could have been written using the point S instead of the point R, the equation would then have taken the form

$$r = c + sb$$

where s was the parameter. The equation equivalent to (22) would then have been

$$\frac{x_1 - 6}{3} = \frac{x_2 + 3}{-5} = \frac{x_3 - 4}{3} (= s) \tag{23}$$

That equation (23) is the same as equation (22) is immediately evident if s is replaced by $t - 1$.

Example (ii) Find the equation of the line through (0, 3, 2) and parallel to the join of (4, 3, 1) to (10, 3, −2).

Solution The join (4, 3, 1) to (10, 3, −2) has direction ratios [6:0:−3]. Write $a = 3j + 2k$ and $b = 6i − 3k$ then the vector equation is $r = a + tb$ which leads to the scalar equations

$$\frac{x}{6} = \frac{y - 3}{0} = \frac{z - 2}{-3} (=t).$$

Note that the zero in the direction ratio shows that the line is perpendicular to the axis of y. Each point of the line has a y-coordinate of 3.

9.5 THE POSITION RATIO WITH REGARD TO TWO BASE POINTS

Given two points A and B the position ratio gives the coordinates of the point P that divides the join of A to B internally in the ratio $p:q$.

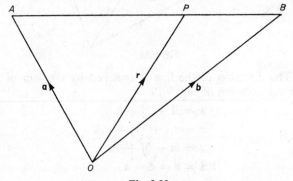

Fig. 9.20

In Fig. 9.20 let O be the origin and let $\mathbf{OA} = a$, $\mathbf{OB} = b$ be the position vectors of the given points A and B. Let P divide the join A to B in the ratio $p:q$ then

$$AP/PB = p/q$$
or $$AP/AB = p/(p + q) \tag{24}$$

In vector form,

$$\mathbf{AB} = \mathbf{b} - \mathbf{a}$$

and $$\mathbf{AP} = \{p/(p + q)\}(\mathbf{b} - \mathbf{a}).$$

The position of P is determined by the vector $\mathbf{OP}$ or $\mathbf{r}$ where

$$\mathbf{r} = \mathbf{OA} + \mathbf{AP} = \mathbf{a} + \{p/(p + q)\}(\mathbf{b} - \mathbf{a})$$
$$= (p\mathbf{b} + q\mathbf{a})/(p + q) \tag{25}$$

If $\mathbf{a} = x_1\mathbf{i} + y_1\mathbf{j} + z_1\mathbf{k}$; $\mathbf{b} = x_2\mathbf{i} + y_2\mathbf{j} + z_2\mathbf{k}$; and $\mathbf{r} = x\mathbf{i} + y\mathbf{j} + z\mathbf{k}$ then equation (25) leads to three scalar equations for the coordinates of P, namely,

$$x = \frac{px_2 + qx_1}{p + q}, \quad y = \frac{py_2 + qy_1}{p + q}, \quad z = \frac{pz_2 + qz_1}{p + q} \tag{26}$$

If the join A to B is divided externally in the ratio $p:q$ the equation (24) is replaced by

$$AP/AB = p/(p - q)$$

and equation (26) is replaced by a similar equation with $(-q)$ instead of q.

Example Find the coordinates of the point dividing the join of $(1, 2, 3)$ to $(5, 7, 8)$, (a) internally in the ratio $3:2$, and (b) externally in the ratio $5:7$.

Solution
(a) With $(x_1, y_1, z_1) \equiv (1, 2, 3)$ and $(x_2, y_2, z_2) \equiv (5, 7, 8)$ the point has coordinates

$$x = \frac{3 \cdot 5 + 2 \cdot 1}{3 + 2} = \frac{17}{5}$$

$$y = \frac{3 \cdot 7 + 2 \cdot 2}{3 + 2} = 5$$

$$z = \frac{3 \cdot 8 + 2 \cdot 3}{3 + 2} = 6$$

(b) Using $p:q = 5:-7$ gives the coordinates $(-9, -21/2, -19/2)$.

9.6 PRODUCT OF TWO VECTORS

So far only the addition of vectors and the multiplication by a scalar have been considered, but any vector algebra is lacking unless some meaning can be attached to the product of vector quantities. It is evident that the ordinary ideas of (scalar) multiplication cannot be sufficient because of the directional properties possessed by vectors. In order to give a motivation for the subsequent definition of the product of two vectors, consider now the multiplication of a force vector $\mathbf{F}$ and a displacement vector $\mathbf{d}$. Suppose for simplicity that the force $\mathbf{F}$ is constant (i.e. constant in both magnitude and direction) and that it acts at a point P as in Fig. 9.21. Consider the work done by the force as the point P moves to a point Q where the displacement vector $\overline{PQ}$ is denoted by **d**. Let the angle between **F** and **d** be θ, then the work done by **F** is defined as the product of the displacement and the force component in the direction of the displacement, i.e. $(F \cos \theta)d$. This scalar quantity $Fd \cos \theta$ has been arrived at by forming, in a particular way, the 'product' of the two vectors **F** and **d**.

As a second illustration of the product of two vectors, suppose that the force **F** acts at one end Q of a rigid rod PQ whose other end P is freely pivoted. Consider the moment of the force about the point P. In Fig. 9.21 the magnitude of this moment is defined as the product of the force and the length of the perpendicular from P to the line of action of **F**, i.e. $F(d \sin \theta)$. The moment of the force has, however, a direction associated with it as well as a magnitude and is a vector quantity. The direction of the moment of **F** about P is perpendicular to the plane of

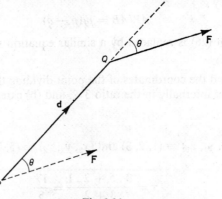

Fig. 9.21

F and **d** and in the sense of the longitudinal movement of a right-handed screw laid in that direction and rotated in the direction that **F** is attempting to turn PQ about P. In Fig. 9.21 if **F** and **d** are in the plane of the paper then the moment about P is perpendicular to the paper and directed away from the reader. Hence it is possible to arrive at a vector quantity by forming, in a particular way, the 'product' of two vectors **F** and **d**.

9.6.1 The scalar product

Given two vectors **a** and **b** the *scalar product* (sometimes called the *dot product*) is defined as the scalar quantity $ab \cos \theta$, where θ is the angle between the two vectors and is usually written **a . b**, i.e.

$$\begin{aligned}
\mathbf{a} \cdot \mathbf{b} &= ab \cos \theta \\
&= a(b \cos \theta) \\
&= b(a \cos \theta) \\
&= \mathbf{b} \cdot \mathbf{a}. \qquad (27)
\end{aligned}$$

Note that if **a** and **b** do not lie in the same plane then θ is the angle between two vectors parallel to **a** and **b** respectively drawn through any given point. The scalar product may be positive or negative depending on the angle θ. If $\theta = 0$ the vectors are parallel and **a** . **b** = ab, in particular **a** . **a** = $\mathbf{a}^2 = a^2$.

The scalar product may be used to find the projection of a vector in any given direction. In Fig. 9.22 the scalar projection of the vector **b** in the direction of **a** is PN, and from equation (27)

$$PN = b \cos \theta = \frac{ab \cos \theta}{a} = \frac{\mathbf{a} \cdot \mathbf{b}}{a}. \tag{28}$$

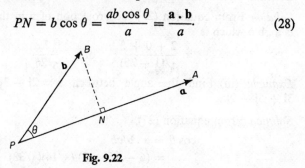

Fig. 9.22

Likewise the projection of **b** in any direction **n** is $\dfrac{\mathbf{b} \cdot \mathbf{n}}{n}$. Note particularly that **b** and **n** do not have to be coplanar vectors. The scalar product obeys the following laws of algebra:

The commutative law, $\quad \mathbf{a} \cdot \mathbf{b} = \mathbf{b} \cdot \mathbf{a}$ \hfill (29)

The associative law, $\quad \alpha(\mathbf{a} \cdot \mathbf{b}) = \alpha \mathbf{a} \cdot \mathbf{b} = \mathbf{a} \cdot \alpha \mathbf{b}$ \hfill (30)

The distributive law, $\quad \mathbf{a} \cdot (\mathbf{b} + \mathbf{c}) = \mathbf{a} \cdot \mathbf{b} + \mathbf{a} \cdot \mathbf{c}.$ \hfill (31)

If equation (31) is divided by a then the distributive law states that the component of $(\mathbf{b} + \mathbf{c})$ in the direction of **a** is equal to the sum of the component of **b** in the direction of **a** and the component of **c** in the direction of **a**. The distributive law may be extended in the form

$$(\mathbf{a} + \mathbf{b}) \cdot (\mathbf{c} + \mathbf{d}) = \mathbf{a} \cdot \mathbf{c} + \mathbf{a} \cdot \mathbf{d} + \mathbf{b} \cdot \mathbf{c} + \mathbf{b} \cdot \mathbf{d}. \tag{32}$$

From equation (27) it is easily seen that the scalar product **a** . **b** vanishes not only when either **a** or **b** is zero but also when $\cos \theta$ is zero. Thus, two non-zero vectors whose scalar product vanishes must be perpendicular. In particular if **i**, **j**, and **k** are the unit vectors along the orthogonal Cartesian coordinate axes then

$$\mathbf{i} \cdot \mathbf{i} = \mathbf{j} \cdot \mathbf{j} = \mathbf{k} \cdot \mathbf{k} = 1 \tag{33}$$

and $\qquad\qquad \mathbf{i} \cdot \mathbf{j} = \mathbf{i} \cdot \mathbf{k} = \mathbf{j} \cdot \mathbf{k} = 0.$ \hfill (34)

It follows immediately, using equations (32), (33), and (34), that if $\mathbf{a} = a_1\mathbf{i} + a_2\mathbf{j} + a_3\mathbf{k}$ and $\mathbf{b} = b_1\mathbf{i} + b_2\mathbf{j} + b_3\mathbf{k}$, then

$$\begin{aligned}
\mathbf{a} \cdot \mathbf{b} &= (a_1\mathbf{i} + a_2\mathbf{j} + a_3\mathbf{k}) \cdot (b_1\mathbf{i} + b_2\mathbf{j} + b_3\mathbf{k}) \\
&= a_1b_1 + a_2b_2 + a_3b_3.
\end{aligned} \tag{35}$$

Example (i) Find the scalar product of $\mathbf{a} = 2\mathbf{i} - 3\mathbf{j} + 7\mathbf{k}$ and $\mathbf{b} = 3\mathbf{i} + \mathbf{j} - 5\mathbf{k}$.

Solution Here,

$$\mathbf{a} \cdot \mathbf{b} = 6 - 3 - 35$$

using equation (35), i.e.

$$\mathbf{a} \cdot \mathbf{b} = -32.$$

Example (ii) Find the projection of $\mathbf{a} = 2\mathbf{i} - 3\mathbf{j} + \mathbf{k}$ on to $\mathbf{b} = \mathbf{i} + 5\mathbf{k}$.

Solution From equation (28) the projection of $\mathbf{a}$ in the direction of $\mathbf{b}$ is $\mathbf{a} \cdot \mathbf{b}/b$ which is

$$\frac{2 + 0 + 5}{\sqrt{(1 + 25)}} \quad \text{or} \quad \frac{7}{\sqrt{26}}.$$

Example (iii) Find the angle between $\mathbf{a} = 2\mathbf{i} - 3\mathbf{j} + \mathbf{k}$ and $\mathbf{b} = 3\mathbf{i} + 5\mathbf{j} - 2\mathbf{k}$.

Solution From equation (27),

$$\cos \theta = \mathbf{a} \cdot \mathbf{b}/ab$$
$$= (6 - 15 - 2)/((\sqrt{14})(\sqrt{38})$$
$$= -0.4768$$

and
$$\theta = 118° \, 29' = 180° - 61° \, 31'.$$

Note that in any given problem it may be sufficient to find the acute angle between the two vectors, i.e. to ignore any negative sign. Strictly the angle of 61° 31' in the example above is the angle between $\mathbf{a}$ and $-\mathbf{b}$.

Example (iv) The point of application of the force $10\mathbf{i} + 7\mathbf{j} + 8\mathbf{k}$ newton moves a distance of 3 metre in the direction of (a) the vector $2\mathbf{i} + 5\mathbf{j} + 3\mathbf{k}$, and (b) the vector $\mathbf{i} + 2\mathbf{j} - 3\mathbf{k}$. Find the work done by the force in each case.

Solution Let the force be $\mathbf{F}$ and the displacement be $\mathbf{d}$ then the work done in each case is $\mathbf{F} \cdot \mathbf{d}$.
(a) The unit vector in the direction of $2\mathbf{i} + 5\mathbf{j} + 3\mathbf{k}$ is $(2\mathbf{i} + 5\mathbf{j} + 3\mathbf{k})/\sqrt{38}$, therefore

$$\mathbf{d} = 3(2\mathbf{i} + 5\mathbf{j} + 3\mathbf{k})/\sqrt{38}$$

and the work done by the force is

$$W = \mathbf{F} \cdot \mathbf{d} = (10\mathbf{i} + 7\mathbf{j} + 8\mathbf{k}) \cdot (2\mathbf{i} + 5\mathbf{j} + 3\mathbf{k})(3/\sqrt{38})$$
$$= (20 + 35 + 24)3/\sqrt{38}$$
$$= 237/\sqrt{38} \text{ joule}$$

(b) The displacement vector in this case is

$$\mathbf{d} = 3(\mathbf{i} + 2\mathbf{j} - 3\mathbf{k})/\sqrt{14}$$

and the work done is

$$W = \mathbf{F} \cdot \mathbf{d} = (10 + 14 - 24)(3/\sqrt{14})$$
$$= 0 \text{ joule.}$$

No work is done by the force when it moves in a direction at right angles to its line of action.

Example (v) Show that the angle in a semi-circle is a right angle.

Solution It is required to show that in Fig. 9.23 AP and BP are at right angles for any position P on the given circle centre O, diameter AB.

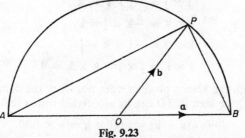

Fig. 9.23

Write $\mathbf{OB} = \mathbf{a}$ and $\mathbf{OP} = \mathbf{b}$, then $\mathbf{AO} = \mathbf{a}$ and $\mathbf{AP} = \mathbf{AO} + \mathbf{OP}$ $= \mathbf{a} + \mathbf{b}$, and $\mathbf{BP} = \mathbf{BO} + \mathbf{OP} = -\mathbf{a} + \mathbf{b}$.

The scalar product

$$\mathbf{AP} \cdot \mathbf{BP} = (\mathbf{a} + \mathbf{b}) \cdot (-\mathbf{a} + \mathbf{b})$$
$$= -\mathbf{a} \cdot \mathbf{a} + \mathbf{b} \cdot \mathbf{b}$$
$$= b^2 - a^2$$
$$= 0, \text{ since } OP = OB.$$

The scalar product is zero and since $\mathbf{AP}$ and $\mathbf{BP}$ are non-zero they are at right angles.

9.6.2 The vector product

Given two vectors $\mathbf{a}$ and $\mathbf{b}$ the *vector product* (sometimes called the *cross product*) is another vector $\mathbf{c}$ written $\mathbf{c} = \mathbf{a} \times \mathbf{b}$ such that

(i) $\mathbf{c}$ is at right angles to both $\mathbf{a}$ and $\mathbf{b}$,
(ii) $|\mathbf{c}| = ab \sin \theta$, where θ is that angle between $\mathbf{a}$ and $\mathbf{b}$ such that $0 \leqslant \theta \leqslant \pi$,
(iii) a right-handed rotation about $\mathbf{c}$ through an angle θ would move a into the direction of $\mathbf{b}$.

The vector product may also be written

$$\mathbf{a} \times \mathbf{b} = (ab \sin \theta)\mathbf{n} \qquad (36)$$

where $\mathbf{n}$ is a unit vector, $|\mathbf{n}| = 1$, in the direction of $\mathbf{c}$ (Fig. 9.24). These properties imply that the vector $\mathbf{b} \times \mathbf{a}$ is identical with $-\mathbf{c}$, or

$$\mathbf{b} \times \mathbf{a} = -(\mathbf{a} \times \mathbf{b}). \qquad (37)$$

The vector product does not obey the commutative law of algebra and it is essential in any manipulation of vectors involving the cross product that the order of the vectors is strictly maintained.

The vector product $\mathbf{a} \times \mathbf{b}$ is zero when either of $\mathbf{a}$, $\mathbf{b}$ is zero or when $\sin \theta = 0$. Thus two non-zero vectors have a zero vector product when they are parallel.

If $\mathbf{a}$ and $\mathbf{b}$ are at right angles then $\mathbf{a} \times \mathbf{b} = ab\mathbf{n}$. The base unit vectors $\mathbf{i}$, $\mathbf{j}$, $\mathbf{k}$, of a right-handed Cartesian coordinate system are mutually orthogonal and have the following properties

$$\mathbf{i} \times \mathbf{j} = -\mathbf{j} \times \mathbf{i} = \mathbf{k} \tag{38}$$

$$\mathbf{j} \times \mathbf{k} = -\mathbf{k} \times \mathbf{j} = \mathbf{i} \tag{39}$$

$$\mathbf{k} \times \mathbf{i} = -\mathbf{i} \times \mathbf{k} = \mathbf{j} \tag{40}$$

$$\mathbf{i} \times \mathbf{i} = \mathbf{j} \times \mathbf{j} = \mathbf{k} \times \mathbf{k} = 0. \tag{41}$$

Although the vector product does not obey the commutative law it may readily be seen that it can be associated with a scalar α in the form

$$\alpha(\mathbf{a} \times \mathbf{b}) = (\alpha\mathbf{a}) \times \mathbf{b} = \mathbf{a} \times (\alpha\mathbf{b}). \tag{42}$$

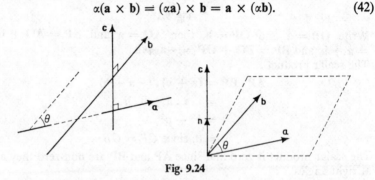

Fig. 9.24

The vector product satisfies the distributive law in the form

$$\mathbf{a} \times (\mathbf{b} + \mathbf{c}) = \mathbf{a} \times \mathbf{b} + \mathbf{a} \times \mathbf{c} \tag{43}$$

but the proof of this law is not obvious. (For a geometrical proof the reader is advised to consult a book devoted to vector analysis.)

The Cartesian form of $\mathbf{a} \times \mathbf{b}$ may be derived as follows. Suppose $\mathbf{c}$ of Fig. 9.24 represents $\mathbf{a} \times \mathbf{b}$ and the Cartesian forms of the vectors are

$$\mathbf{a} = a_1\mathbf{i} + a_2\mathbf{j} + a_3\mathbf{k}$$

$$\mathbf{b} = b_1\mathbf{i} + b_2\mathbf{j} + b_3\mathbf{k}$$

with $\qquad \mathbf{c} = \mathbf{a} \times \mathbf{b} = ab \sin \theta \mathbf{n} = c_1\mathbf{i} + c_2\mathbf{j} + c_3\mathbf{k}. \tag{44}$

Thus $\qquad \mathbf{c} = (a_1\mathbf{i} + a_2\mathbf{j} + a_3\mathbf{k}) \times (b_1\mathbf{i} + b_2\mathbf{j} + b_3\mathbf{k})$

$$\left.\begin{aligned} = {} & a_1\mathbf{i} \times (b_1\mathbf{i} + b_2\mathbf{j} + b_3\mathbf{k}) \\ & + a_2\mathbf{j} \times (b_1\mathbf{i} + b_2\mathbf{j} + b_3\mathbf{k}) \\ & + a_3\mathbf{k} \times (b_1\mathbf{i} + b_2\mathbf{j} + b_3\mathbf{k}) \end{aligned}\right\} \tag{45}$$

using the distributive law (43).

A further application of (43) and the use of equations (38)–(41) leads to

$$c = a \times b = (a_2b_3 - a_3b_2)i + (a_3b_1 - a_1b_3)j + (a_1b_2 - a_2b_1)k \quad (46)$$

or, in determinant form,

$$a \times b = \begin{vmatrix} i & j & k \\ a_1 & a_2 & a_3 \\ b_1 & b_2 & b_3 \end{vmatrix}. \quad (47)$$

Example (i) Find the vector product of $a = 3i + 2j - k$ and $b = i - 4j - 2k$.

Solution Here,

$$a \times b = \begin{vmatrix} i & j & k \\ 3 & 2 & -1 \\ 1 & -4 & -2 \end{vmatrix} = -8i + 5j - 14k.$$

Example (ii) Find the equation of the straight line passing through the point $(3, 4, 5)$ having a direction perpendicular to the lines whose direction ratios are $[1:0:2]$ and $[-1:2:3]$.

Solution The direction of the line is to be perpendicular to the vectors $a = i + 2k$ and $b = -i + 2j + 3k$. This direction is parallel to the vector

$$a \times b = \begin{vmatrix} i & j & k \\ 1 & 0 & 2 \\ -1 & 2 & 3 \end{vmatrix} = -4i - 5j + 2k.$$

The equation of the required line is given in vector form by

$$r = (3i + 4j + 5k) + t(-4i - 5j + 2k)$$

where $r = xi + yj + zk$, or in scalar form by the equations

$$\frac{x - 3}{-4} = \frac{y - 4}{-5} = \frac{z - 5}{2}.$$

9.6.3 The matrix representation of $a \times b$

Let
$$a \times b = c = c_i e_i$$

then the components of a, b, and c all form (1×3) row matrices. It is of interest to note that the vector product may be expressed in a matrix notation by

$$(c_1\ c_2\ c_3) = (a_1\ a_2\ a_3) \begin{pmatrix} 0 & -b_3 & b_2 \\ b_3 & 0 & -b_1 \\ -b_2 & b_1 & 0 \end{pmatrix}.$$

This may be further reduced into a concise matrix notation given by

$$c_i = e_{ijk}\, a_j b_k$$

in which i, j, and k are suffixes that take all possible values 1, 2, and 3 and e_{ijk} is defined as follows,

$$e_{ijk} = 0, \quad \text{if any two suffixes have the same value,}$$
$$= +1, \text{ if the suffixes have the cyclic order 123, 231, or 312,}$$
$$= -1, \text{ if the suffixes have the acyclic order 132, 321, or 213.}$$

For example,

$$c_2 = e_{2jk}\, a_j b_k \quad (\text{summed over all values of } j, k)$$
$$= e_{231}\, a_3 b_1 + e_{213}\, a_1 b_3 + \text{terms in which } e_{2jk} \text{ is zero}$$
$$= (+1)a_3 b_1 + (-1)a_1 b_3$$
$$= a_3 b_1 - a_1 b_3.$$

Note that the suffixes i, j, and k are simply symbols that take the values 1, 2, or 3 and have no connection with unit vectors $\mathbf{i}$, $\mathbf{j}$, and $\mathbf{k}$.

9.7 THE PLANE AND THE STRAIGHT LINE

A plane is a surface in space for which the join of any pair of points of the surface is a straight line lying entirely in the surface and perpendicular to a given direction known as the *normal* to the plane.

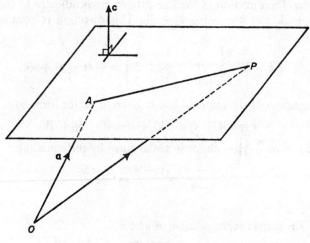

Fig. 9.25

Suppose, as in Fig. 9.25, that A lies in the plane and that the position vector of A is $\mathbf{OA} = \mathbf{a}$. Let P be a variable point such that $\mathbf{OP} = \mathbf{r}$. Let $\mathbf{c}$ be the vector giving the normal direction, then the join of A to P lies entirely in the plane passing through A if AP is perpendicular to $\mathbf{c}$, i.e.

$$\mathbf{AP} \cdot \mathbf{c} = 0$$

or

$$(\mathbf{r} - \mathbf{a}) \cdot \mathbf{c} = 0. \tag{48}$$

Equation (48) is then the equation of the plane and is a scalar expression.

In Cartesian form, if

$$a = a_1 i + a_2 j + a_3 k,$$
$$c = li + mj + nk,$$

and $$r = xi + yj + zk,$$

then the plane equation has the form

$$l(x - a_1) + m(y - a_2) + n(z - a_3) = 0 \qquad (49)$$

From equation (48), the equation of any plane whose normal is in the direction of c has the form

$$r . c = \beta \qquad (50)$$

or $$lx + my + nz = \beta. \qquad (51)$$

In another notation the plane equation has the form

$$a_{11}x_1 + a_{12}x_2 + a_{13}x_3 = b_1$$

and is a linear, algebraic equation of the type discussed in Chapter 8. Geometrically, a plane in space will usually meet a second plane in a common line of intersection. This line will then meet a third plane in a single point. Thus three planes will in general meet in a single point of intersection. The discussion of section 8.7 concerning the solution of n linear, algebraic equations in n unknowns has a simple geometrical interpretation when $n = 3$. The regular solution of section 8.7.1 is equivalent to the case of three planes meeting in a unique point of intersection. The equation (54) of section 8.7.1 is simply the case of three planes through the origin. The inconsistent case of section 8.7.2 is geometrically equivalent to the case of three planes whose three lines of intersection form the edges of a triangular prism, i.e. the lines are parallel. The indeterminate case of section 8.7.2 is equivalent to the three planes having a common line of intersection.

Many problems concerning the straight line and the plane can be dealt with using vectors and a few problems are solved below.

Example (i) Find the equation of the family of planes whose normals are in the direction of the vector $2i - 3j + k$. Find that member of the family passing through the point $(3, -1, 2)$.

Solution From equation (51) the family equation is

$$2x - 3y + z = \beta$$

The particular plane through the point $(3, -1, 2)$ has the value of β given by

$$2(3) - 3(-1) + (2) = \beta$$

that is $$\beta = 11.$$

The required equation is then

$$2x - 3y + z = 11.$$

Example (ii) Find the equation of the plane containing the line

$$\frac{x-3}{2} = \frac{y-1}{1} = \frac{z+1}{3},$$

and parallel to the line

$$\frac{x-1}{1} = \frac{y+2}{3} = \frac{z+4}{-1}.$$

Solution The plane must pass through the point $(3, 1, -1)$ and have its normal perpendicular to the directions $[2:1:3]$ and $[1:3:-1]$. If the normal vector is $\mathbf{c}$ then

$$\mathbf{c} = \begin{vmatrix} \mathbf{i} & \mathbf{j} & \mathbf{k} \\ 2 & 1 & 3 \\ 1 & 3 & -1 \end{vmatrix} = -10\mathbf{i} + 5\mathbf{j} + 5\mathbf{k}$$

and the equation of the plane is

$$-10(x-3) + 5(y-1) + 5(z+1) = 0$$

or

$$2x - y - z = 6.$$

(Note that the point $(1, -2, -4)$ is not used in determining the equation.)

Example (iii) Find the coordinates of the point of intersection of the line $\mathbf{r} = \mathbf{a} + t\mathbf{b}$ with the plane $\mathbf{r} \cdot \mathbf{c} = \beta$.

Solution A straight line meets a plane in one point, in general. The point will be determined when the line and plane have a common position vector, i.e. when

$$(\mathbf{a} + t\mathbf{b}) \cdot \mathbf{c} = \beta$$

or

$$\mathbf{a} \cdot \mathbf{c} + t(\mathbf{b} \cdot \mathbf{c}) = \beta.$$

The value of t is then given by

$$t = \frac{\beta - \mathbf{a} \cdot \mathbf{c}}{\mathbf{b} \cdot \mathbf{c}} \quad \text{if} \quad \mathbf{b} \cdot \mathbf{c} \neq 0,$$

and substitution into the equation $\mathbf{r} = \mathbf{a} + t\mathbf{b}$ gives the required position vector.

If $\mathbf{b} \cdot \mathbf{c} = 0$ the line is perpendicular to the normal, i.e. parallel to the plane, and there is no point of intersection (except at infinity) unless β also has the value $\mathbf{a} \cdot \mathbf{c}$ (which implies that $\mathbf{a}$ lies on the plane) and then t is indeterminate, the line lies in the plane, and every point of the line is a solution.

Example (iv) Show that the lines

$$\frac{x-5}{2} = \frac{y-1}{1} = \frac{z-2}{3} \quad \text{and} \quad \frac{x+9}{4} = \frac{y-6}{-2} = \frac{z+4}{1}$$

intersect and find the coordinates of the point of intersection.

Solution Usually two lines in space are skew (i.e. they do not intersect) and the condition for intersection is simply that they have a point in common.

Let the parameter of the first line be t and that of the second line be s, then the condition that there is a common point is the condition that the equations

$$5 + 2t = -9 + 4s$$
$$1 + t = 6 - 2s$$
$$2 + 3t = -4 + s$$

are consistent. The condition for consistency from equation (56) of section 8.8.2 is that

$$\begin{vmatrix} 14 & 2 & 4 \\ -5 & 1 & -2 \\ 6 & 3 & 1 \end{vmatrix} = 0.$$

The reader can easily verify that this is so and also that $t = -1$ (or $s = 3$). Using the first equation with $t = -1$, the intersection point is given by $x = 3$, $y = 0$, and $z = -1$.

Example (v) Find the shortest distance between the lines

$$\frac{x-3}{2} = \frac{y+1}{1} = \frac{z-4}{1} \quad \text{and} \quad \frac{x+4}{3} = \frac{y+2}{-1} = \frac{z-1}{-3}.$$

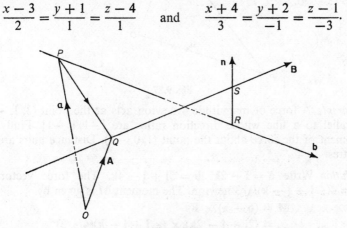

Fig. 9.26

Solution In Fig. 9.26 the shortest distance is given by RS the common perpendicular. The direction of the perpendicular is given by **n** where **n** = **b** × **B**, and **b** and **B** are vectors in the direction of the lines.

Let P and Q be points on the lines with position vectors $\mathbf{a}$ and $\mathbf{A}$ respectively, then the length RS is the projection of $\mathbf{PQ}$ on to $\mathbf{n}$. Since $\mathbf{PQ} = \mathbf{A} - \mathbf{a}$ then

$$RS = \left| \frac{(\mathbf{A} - \mathbf{a}) \cdot \mathbf{n}}{n} \right|.$$

Here, $\mathbf{a} = 3\mathbf{i} - \mathbf{j} + 4\mathbf{k}$, $\mathbf{b} = 2\mathbf{i} + \mathbf{j} + \mathbf{k}$, $\mathbf{A} = -4\mathbf{i} - 2\mathbf{j} + \mathbf{k}$, and $\mathbf{B} = 3\mathbf{i} - \mathbf{j} - 3\mathbf{k}$, hence

$$\mathbf{n} = \mathbf{b} \times \mathbf{B} = \begin{vmatrix} \mathbf{i} & \mathbf{j} & \mathbf{k} \\ 2 & 1 & 1 \\ 3 & -1 & -3 \end{vmatrix}$$
$$= -2\mathbf{i} + 9\mathbf{j} - 5\mathbf{k}.$$

The shortest distance is

$$\left| \frac{(-7\mathbf{i} - \mathbf{j} - 3\mathbf{k}) \cdot (-2\mathbf{i} + 9\mathbf{j} - 5\mathbf{k})}{\sqrt{(4 + 81 + 25)}} \right| = 20/\sqrt{110}.$$

9.8 MOMENT OF A VECTOR ABOUT A POINT

If a force acts on a particle at a point Q then the moment of the force about a point P is $\mathbf{d} \times \mathbf{F}$ where $\mathbf{d}$ is the position vector of Q relative to P as in Fig. 9.27. The direction of the moment is perpendicular to the plane of $\mathbf{d}$ and $\mathbf{F}$ and in the right-handed sense. The magnitude of the moment is $Fd \sin \theta$.

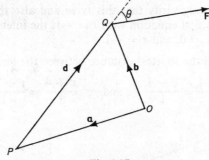

Fig. 9.27

Example A force of magnitude 4 newton acts at the point $(3, 1, -4)$ parallel to a line whose direction ratios are $[-1:1:-1]$. Find the moment of the force about the point $(1, 0, -2)$. Distance units are in metres.

Solution Write $\mathbf{a} = \mathbf{i} - 2\mathbf{k}$, $\mathbf{b} = 3\mathbf{i} + \mathbf{j} - 4\mathbf{k}$. The force vector is $\mathbf{F} = 4(-\mathbf{i} + \mathbf{j} - \mathbf{k})/\sqrt{3}$ newton. The moment $\mathbf{M}$ is given by

$$\mathbf{M} = (\mathbf{b} - \mathbf{a}) \times \mathbf{F}$$
$$= (2\mathbf{i} + \mathbf{j} - 2\mathbf{k}) \times (-\mathbf{i} + \mathbf{j} - \mathbf{k})(4/\sqrt{3})$$
$$= \frac{4}{\sqrt{3}} \begin{vmatrix} \mathbf{i} & \mathbf{j} & \mathbf{k} \\ 2 & 1 & -2 \\ -1 & 1 & -1 \end{vmatrix}$$
$$= 4(\mathbf{i} + 4\mathbf{j} + 3\mathbf{k})/\sqrt{3} \text{ newton metre.}$$

It may be noted that **M** has three components $(4/\sqrt{3})$, $(16/\sqrt{3})$, and $(12/\sqrt{3})$ respectively, in the directions of the coordinate axes.

The concept of the moment of a force about a point can be extended to any vector quantity although the extension may not always have any physical significance. For example, if **v** is the velocity vector of a particle of mass m at the point P and **r** is the position vector **OP** then m**v** is the linear momentum vector and **r** × m**v** is the *moment of momentum* (or *angular momentum*) of the particle about the origin O. This quantity is fundamental in mechanics.

9.9 THE PRODUCT OF THREE VECTORS

The product of two vectors is either a scalar or a vector quantity and in each case may be multiplied by a third vector.

Given two vectors **b** and **c**, the product **b . c** is a scalar, λ say, and may be multiplied by the third vector **a** to give (**b . c**)**a** or λ**a** which is simply a vector of length λa in the direction of **a**. The product of **a** with the vector product **b** × **c** is of more interest and has many applications in mechanics and other fields.

9.9.1 The triple scalar product

The product of **a** with **b** × **c** may be evaluated in the form **a . (b** × **c)** and the result, being a scalar, is called the *triple scalar product* of **a**, **b**, and **c**. Let **a**, **b**, and **c** be drawn at a point P and the parallelepiped constructed as in Fig. 9.28.

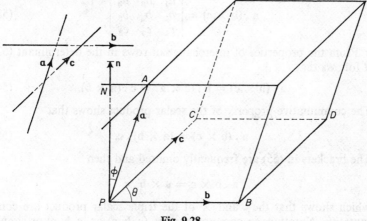

Fig. 9.28

Let θ be the angle between **b** and **c**, and ϕ the angle between **a** and the normal to the plane containing **b** and **c**. By definition

$$\mathbf{b} \times \mathbf{c} = (bc \sin \theta)\mathbf{n}$$

where **n** is a unit vector in the direction of the normal to the plane containing **b** and **c** in the right-handed sense with respect to **b** and **c**.

The triple scalar product is given by

$$\mathbf{a} \cdot (\mathbf{b} \times \mathbf{c}) = \mathbf{a} \cdot (bc \sin \theta)\mathbf{n}$$
$$= (bc \sin \theta)(\mathbf{a} \cdot \mathbf{n})$$
$$= (bc \sin \theta)(a \cos \phi)$$
$$= (bc \sin \theta)PN$$

where PN is the 'height' of the parallelepiped. Since $bc \sin \theta$ is the area of the 'base' parallelogram $PBDC$, the triple scalar product may be represented as the volume of the parallelepiped with sides a, b, and c. If

$$\mathbf{a} = a_1\mathbf{i} + a_2\mathbf{j} + a_3\mathbf{k}, \quad \mathbf{b} = b_1\mathbf{i} + b_2\mathbf{j} + b_3\mathbf{k} \text{ and } \mathbf{c} = c_1\mathbf{i} + c_2\mathbf{j} + c_3\mathbf{k},$$

then

$$\mathbf{b} \times \mathbf{c} = \begin{vmatrix} \mathbf{i} & \mathbf{j} & \mathbf{k} \\ b_1 & b_2 & b_3 \\ c_1 & c_2 & c_3 \end{vmatrix} \tag{52}$$

and

$$\mathbf{a} \cdot (\mathbf{b} \times \mathbf{c}) = (a_1\mathbf{i} + a_2\mathbf{j} + a_3\mathbf{k}) \cdot (A_1\mathbf{i} + A_2\mathbf{j} + A_3\mathbf{k})$$

where A_1, A_2, and A_3 are the cofactors of $\mathbf{i}, \mathbf{j},$ and $\mathbf{k}$ in determinant (52). Evaluating,

$$\mathbf{a} \cdot (\mathbf{b} \times \mathbf{c}) = \begin{vmatrix} a_1 & a_2 & a_3 \\ b_1 & b_2 & b_3 \\ c_1 & c_2 & c_3 \end{vmatrix}. \tag{53}$$

From the properties of interchange of rows in the determinant (53) it follows that

$$\mathbf{a} \cdot (\mathbf{b} \times \mathbf{c}) = \mathbf{b} \cdot (\mathbf{c} \times \mathbf{a}) = \mathbf{c} \cdot (\mathbf{a} \times \mathbf{b}). \tag{54}$$

The commutative property of the scalar product shows that

$$\mathbf{a} \cdot (\mathbf{b} \times \mathbf{c}) = (\mathbf{a} \times \mathbf{b}) \cdot \mathbf{c}. \tag{55}$$

The brackets in (55) are frequently omitted and then

$$\mathbf{a} \cdot \mathbf{b} \times \mathbf{c} = \mathbf{a} \times \mathbf{b} \cdot \mathbf{c}$$

which shows that the $.$ and $\times$ of the triple scalar product are commutative. Notations in common use are $(\mathbf{a}, \mathbf{b}, \mathbf{c})$ or $[\mathbf{a}, \mathbf{b}, \mathbf{c}]$ or simply **abc**. Although the brackets are frequently omitted, the reader is advised to retain brackets until confidence has been gained in vector manipulation. Note that a quantity such as $(\mathbf{a} \cdot \mathbf{b}) \times \mathbf{c}$ is meaningless because $(\mathbf{a} \cdot \mathbf{b})$ is a scalar quantity.

From the property of the determinant (53)

$$(\mathbf{a}, \mathbf{b}, \mathbf{c}) = -(\mathbf{a}, \mathbf{c}, \mathbf{b}).$$

The triple scalar product vanishes

(i) if any of the vectors is zero,

(ii) if, in Fig. 9.28, $\cos \phi = 0$, i.e. if $\phi = \pi/2$,

(iii) if any one vector is a scalar multiple of either of the other vectors. This follows immediately from the property concerning equal rows in the determinant (53).

From (iii) the triple scalar product vanishes if any two of the vectors are parallel. From (ii) the necessary condition that three non-parallel vectors are coplanar is that the triple scalar product vanishes.

Example (i) Find the triple scalar product of the vectors $2\mathbf{i} - 7\mathbf{j} + 3\mathbf{k}$, $\mathbf{i} - \mathbf{j}$, and $3\mathbf{i} + 4\mathbf{k}$.

Solution The triple scalar product is

$$\begin{vmatrix} 2 & -7 & 3 \\ 1 & -1 & 0 \\ 3 & 0 & 4 \end{vmatrix} = 3(3) + 4(5) = 29.$$

Example (ii) Find the condition that the two lines $\mathbf{r} = \mathbf{a} + t\mathbf{b}$ and $\mathbf{r} = \mathbf{A} + s\mathbf{B}$ intersect.

Solution An inspection of Fig. 9.26 shows that the lines will intersect if the vectors $(\mathbf{a} - \mathbf{A})$, $\mathbf{b}$, and $\mathbf{B}$ are coplanar. The condition for this is simply

$$(\mathbf{a} - \mathbf{A}) \cdot (\mathbf{b} \times \mathbf{B}) = 0$$

which is

$$\begin{vmatrix} a_1 - A_1 & b_1 & B_1 \\ a_2 - A_2 & b_2 & B_2 \\ a_3 - A_3 & b_3 & B_3 \end{vmatrix} = 0.$$

9.9.2 The triple vector product

The product of $\mathbf{a}$ with $\mathbf{b} \times \mathbf{c}$ may be evaluated in the form $\mathbf{a} \times (\mathbf{b} \times \mathbf{c})$ and the result, being a vector, is called the *triple vector product* of $\mathbf{a}$, $\mathbf{b}$, and $\mathbf{c}$. With reference to Fig. 9.28,

$$\mathbf{b} \times \mathbf{c} = (bc \sin \theta)\mathbf{n}$$

and
$$\mathbf{a} \times (\mathbf{b} \times \mathbf{c}) = (bc \sin \theta)(\mathbf{a} \times \mathbf{n})$$
$$= (bc \sin \theta)(a \sin \phi)\mathbf{m} \tag{56}$$

where $\mathbf{m}$ is a unit vector perpendicular to $\mathbf{a}$ and $\mathbf{n}$ in the right-handed sense. This vector $\mathbf{m}$, and hence $\mathbf{a} \times (\mathbf{b} \times \mathbf{c})$, is in the plane of $\mathbf{b}$ and $\mathbf{c}$ and from Fig. 9.11, with $\mathbf{a}$ a null vector and $\overline{OR}$ in the plane of $\mathbf{b}$ and $\mathbf{c}$, it follows that

$$\mathbf{a} \times (\mathbf{b} \times \mathbf{c}) = \mu\mathbf{b} + \nu\mathbf{c} \tag{57}$$

where μ and ν are scalars.

Likewise,

$$(\mathbf{a} \times \mathbf{b}) \times \mathbf{c} = -\mathbf{c} \times (\mathbf{a} \times \mathbf{b})$$
$$= \alpha\mathbf{a} + \beta\mathbf{b} \tag{58}$$

where α and β are scalars.

The results (57) and (58) show that, in general, $\mathbf{a} \times (\mathbf{b} \times \mathbf{c})$ and $(\mathbf{a} \times \mathbf{b}) \times \mathbf{c}$ are different vectors. The position of the bracket in a triple vector product is vital to the definition and must be retained. The quantity $\mathbf{a} \times \mathbf{b} \times \mathbf{c}$ is not fully defined.

To find the value of μ and ν of equation (57) first form the scalar product of both sides with the vector $\mathbf{a}$, i.e.

$$\mathbf{a} . \{\mathbf{a} \times (\mathbf{b} \times \mathbf{c})\} = \mu(\mathbf{a} . \mathbf{b}) + \nu(\mathbf{a} . \mathbf{c}) \tag{59}$$

The left side of (59) is a triple scalar product $[\mathbf{a}, \mathbf{a}, (\mathbf{b} \times \mathbf{c})]$ with two equal vectors and is therefore zero. Hence,

$$\frac{\mu}{\mathbf{a} . \mathbf{c}} = \frac{-\nu}{\mathbf{a} . \mathbf{b}} = t \text{ (say)} \tag{60}$$

and substitution in (57) gives

$$\mathbf{a} \times (\mathbf{b} \times \mathbf{c}) = t\{(\mathbf{a} . \mathbf{c})\mathbf{b} - (\mathbf{a} . \mathbf{b})\mathbf{c}\}. \tag{61}$$

The equation (61) is to be true for all vectors $\mathbf{a}$, $\mathbf{b}$, and $\mathbf{c}$. Suppose

$$\mathbf{a} = a_1\mathbf{i} + a_2\mathbf{j} + a_3\mathbf{k}$$
$$\mathbf{b} = b_1\mathbf{i} + b_2\mathbf{j} + b_3\mathbf{k}$$

and $$\mathbf{c} = c_1\mathbf{i} + c_2\mathbf{j} + c_3\mathbf{k},$$

then $\mathbf{b} \times \mathbf{c} = (b_2c_3 - b_3c_2)\mathbf{i} + (b_3c_1 - b_1c_3)\mathbf{j} + (b_1c_2 - b_2c_1)\mathbf{k}$

and the component of $\mathbf{a} \times (\mathbf{b} \times \mathbf{c})$ in the x-direction is

$$a_2(b_1c_2 - b_2c_1) - a_3(b_3c_1 - b_1c_3). \tag{62}$$

The component of $(\mathbf{a} . \mathbf{c})\mathbf{b} - (\mathbf{a} . \mathbf{b})\mathbf{c}$ in the x-direction is

$$(a_1c_1 + a_2c_2 + a_3c_3)b_1 - (a_1b_1 + a_2b_2 + a_3b_3)c_1 \tag{63}$$

Comparison of (62) and (63) shows that in (61)

$$t = 1$$

and that finally

$$\mathbf{a} \times (\mathbf{b} \times \mathbf{c}) = (\mathbf{a} . \mathbf{c})\mathbf{b} - (\mathbf{a} . \mathbf{b})\mathbf{c}. \tag{64}$$

Example (i) Verify the result (64) if $\mathbf{a} = 2\mathbf{i} - \mathbf{j} + \mathbf{k}$, $\mathbf{b} = \mathbf{i} - \mathbf{j}$, and $\mathbf{c} = \mathbf{i} + \mathbf{j} + \mathbf{k}$.

Solution Here,

$$\mathbf{b} \times \mathbf{c} = \begin{vmatrix} \mathbf{i} & \mathbf{j} & \mathbf{k} \\ 1 & -1 & 0 \\ 1 & 1 & 1 \end{vmatrix} = -\mathbf{i} - \mathbf{j} + 2\mathbf{k}.$$

The left side of (64) is

$$\mathbf{a} \times (\mathbf{b} \times \mathbf{c}) = \begin{vmatrix} \mathbf{i} & \mathbf{j} & \mathbf{k} \\ 2 & -1 & 1 \\ -1 & -1 & 2 \end{vmatrix} = -\mathbf{i} - 5\mathbf{j} - 3\mathbf{k}.$$

With $(\mathbf{a} \cdot \mathbf{c}) = 2$ and $(\mathbf{a} \cdot \mathbf{b}) = 3$, the right side of (64) is

$$(\mathbf{a} \cdot \mathbf{c})\mathbf{b} - (\mathbf{a} \cdot \mathbf{b})\mathbf{c} = 2(\mathbf{i} - \mathbf{j}) - 3(\mathbf{i} + \mathbf{j} + \mathbf{k})$$
$$= -\mathbf{i} - 5\mathbf{j} - 3\mathbf{k}$$

which is identical with the left side.

Example (ii) Given $\mathbf{x} \times \mathbf{a} = \mathbf{b}$ and $\mathbf{x} \cdot \mathbf{c} = \alpha$, find $\mathbf{x}$.

Solution The triple vector product of $\mathbf{c}$, $\mathbf{x}$, and $\mathbf{a}$ is, from (64)

$$\mathbf{c} \times (\mathbf{x} \times \mathbf{a}) = (\mathbf{c} \cdot \mathbf{a})\mathbf{x} - (\mathbf{c} \cdot \mathbf{x})\mathbf{a}$$
$$= (\mathbf{a} \cdot \mathbf{c})\mathbf{x} - \alpha\mathbf{a}.$$

Put $\mathbf{x} \times \mathbf{a} = \mathbf{b}$ then

$$\mathbf{c} \times \mathbf{b} = (\mathbf{a} \cdot \mathbf{c})\mathbf{x} - \alpha\mathbf{a}$$

and so
$$\mathbf{x} = \frac{(\mathbf{c} \times \mathbf{b}) + \alpha\mathbf{a}}{(\mathbf{a} \cdot \mathbf{c})}.$$

Example (iii) Prove that $(\mathbf{a} \times \mathbf{b}) \cdot (\mathbf{c} \times \mathbf{d}) = (\mathbf{a} \cdot \mathbf{c})(\mathbf{b} \cdot \mathbf{d}) - (\mathbf{a} \cdot \mathbf{d})(\mathbf{b} \cdot \mathbf{c})$.

Solution First consider $(\mathbf{c} \times \mathbf{d})$ as a single vector $\mathbf{V}$, say, then

$$(\mathbf{a} \times \mathbf{b}) \cdot (\mathbf{c} \times \mathbf{d}) = (\mathbf{a} \times \mathbf{b}) \cdot \mathbf{V}$$
$$= (\mathbf{a}, \mathbf{b}, \mathbf{V})$$
$$= \mathbf{a} \cdot (\mathbf{b} \times \mathbf{V})$$
$$= \mathbf{a} \cdot [\mathbf{b} \times (\mathbf{c} \times \mathbf{d})]$$
$$= \mathbf{a} \cdot [(\mathbf{b} \cdot \mathbf{d})\mathbf{c} - (\mathbf{b} \cdot \mathbf{c})\mathbf{d}]$$
$$= (\mathbf{a} \cdot \mathbf{c})(\mathbf{b} \cdot \mathbf{d}) - (\mathbf{a} \cdot \mathbf{d})(\mathbf{b} \cdot \mathbf{c}).$$

9.10 DIFFERENTIATION OF VECTORS

In most physical applications the vectors involved are not constants but are variable with respect to some scalar quantity. This scalar is frequently the time. For example the velocity $\mathbf{v}$ of a rocket being projected from the Earth is clearly a function of the time since blast-off.

9.10.1 Differentiation with respect to a scalar

Suppose $\mathbf{a}$ is a vector which is a function of a scalar u, this is written

$$\mathbf{a} = \mathbf{f}(u) \quad \text{or} \quad \mathbf{a}(u). \tag{65}$$

Let u change to $u + \delta u$ such that $\mathbf{a}$ changes to $\mathbf{a} + \delta\mathbf{a}$ where $\mathbf{a} + \delta\mathbf{a}$ is $\mathbf{a}(u + \delta u)$. The derivative of $\mathbf{a}$ with respect to u is defined in the same way as for the scalar functions of Chapter 2 by the limiting value of the ratio of δa to δu and written

$$\frac{da}{du} = \operatorname*{Lim}_{\delta u \to 0} \frac{\delta a}{\delta u} = \operatorname*{Lim}_{\delta u \to 0} \frac{a(u + \delta u) - a(u)}{\delta u} \tag{66}$$

The ratio of δa to δu is in the form of a scalar $(1/\delta u)$ multiplying a vector δa and is a vector in the direction of δa. In Fig. 9.29, **OP** represents the vector **a** and **OQ** the vector $\mathbf{a} + \delta \mathbf{a}$ so that **PQ** represents the vector $\delta \mathbf{a}$.

In general the vector $\delta \mathbf{a}$ is in a different direction from that of the vector **a**. Write $\mathbf{PQ} = PQ\mathbf{s}$ where **s** is a unit vector in the direction of **PQ**, then

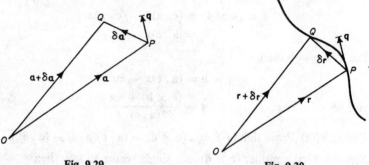

Fig. 9.29   Fig. 9.30

the limit process of (66) involves both the limit of the magnitude ratio $(PQ/\delta u)$ and the limit of the direction of the vector **s**. Let the latter be the unit vector **q** then

$$\frac{da}{du} = \left[\operatorname*{Lim}_{\delta u \to 0} \frac{PQ}{\delta u} \right] \mathbf{q} \tag{67}$$

that is, the derivative da/du is in the direction of **q**. Note again that this direction is not necessarily in the direction of **a**.

Consider the particular case when **a** is the position vector of a moving particle P. Replace **a** by **r** and u by the time t, then in Fig. 9.30

$$\mathbf{OP} = \mathbf{r} = \mathbf{r}(t)$$

and $\delta \mathbf{r}$ is the displacement **PQ** of two points on the path of the particle. The limit of $PQ/\delta t$ is the speed of the particle and the limiting direction specified by the unit vector **q** is that of the tangent to the space curve at P. But speed with an associated direction is simply the velocity; this is written

$$\frac{d\mathbf{r}}{dt} = \dot{\mathbf{r}} = \mathbf{v} = v\mathbf{q}. \tag{68}$$

Similarly, the derivative of the velocity may be evaluated to give the acceleration **f**, i.e.

$$\mathbf{f} = \frac{d\mathbf{v}}{dt} = \frac{d^2\mathbf{r}}{dt^2} = \dot{\mathbf{v}} = \ddot{\mathbf{r}}$$

$$= \underset{\delta t \to 0}{\text{Lim}} \frac{\mathbf{v}(t + \delta t) - \mathbf{v}(t)}{\delta t}. \tag{69}$$

Using Fig. 9.29 with **a** replaced by **v** it is evident that the acceleration vector is not necessarily in the direction of the velocity vector. In general the vectors of position, velocity, and acceleration are in three different directions.

The derivative of a vector has been defined in the same way as for a scalar function and the following results may be obtained

$$\frac{d}{du}(\mathbf{a} + \mathbf{b}) = \frac{d\mathbf{a}}{du} + \frac{d\mathbf{b}}{du} \tag{70}$$

$$\frac{d}{du}(\beta\mathbf{a}) = \frac{d\beta}{du}\mathbf{a} + \beta\frac{d\mathbf{a}}{du} \tag{71}$$

$$\frac{d}{du}(\mathbf{a} \cdot \mathbf{b}) = \frac{d\mathbf{a}}{du} \cdot \mathbf{b} + \mathbf{a} \cdot \frac{d\mathbf{b}}{du} \tag{72}$$

$$\frac{d}{du}(\mathbf{a}, \mathbf{b}, \mathbf{c}) = \left(\frac{d\mathbf{a}}{du}, \mathbf{b}, \mathbf{c}\right) + \left(\mathbf{a}, \frac{d\mathbf{b}}{du}, \mathbf{c}\right) + \left(\mathbf{a}, \mathbf{b}, \frac{d\mathbf{c}}{du}\right) \tag{73}$$

$$\frac{d}{du}(\mathbf{a} \times \mathbf{b}) = \frac{d\mathbf{a}}{du} \times \mathbf{b} + \mathbf{a} \times \frac{d\mathbf{b}}{du} \tag{74}$$

$$\frac{d}{du}\{\mathbf{a} \times (\mathbf{b} \times \mathbf{c})\} = \frac{d\mathbf{a}}{du} \times (\mathbf{b} \times \mathbf{c}) + \mathbf{a} \times \left(\frac{d\mathbf{b}}{du} \times \mathbf{c}\right) + \mathbf{a} \times \left(\mathbf{b} \times \frac{d\mathbf{c}}{du}\right) \tag{75}$$

It is important that the order of the vectors be unchanged whenever a vector product is involved.

Example (i) A particle moves so that its position at time t is given by $\mathbf{r} = 2 \sin t\mathbf{i} + 3 \cos t\mathbf{j} + t^2\mathbf{k}$ where **i**, **j**, and **k** are constant base vectors. Show that

$$\ddot{\mathbf{r}} + \mathbf{r} = (t^2 + 2)\mathbf{k}.$$

Solution If $\mathbf{r} = 2 \sin t\mathbf{i} + 3 \cos t\mathbf{j} + t^2\mathbf{k}$

then $\dot{\mathbf{r}} = 2 \cos t\mathbf{i} - 3 \sin t\mathbf{j} + 2t\mathbf{k}$

and $\ddot{\mathbf{r}} = -2 \sin t\mathbf{i} - 3 \cos t\mathbf{j} + 2\mathbf{k}$

$$= -(\mathbf{r} - t^2\mathbf{k}) + 2\mathbf{k}.$$

Hence $\ddot{\mathbf{r}} + \mathbf{r} = (t^2 + 2)\mathbf{k}.$

This result is a vector differential equation and represents three scalar equations of the type

$$\ddot{x} + x = 0, \quad \ddot{y} + y = 0, \quad \text{and} \quad \ddot{z} + z = (t^2 + 2).$$

Example (ii) What can be deduced about the acceleration of a particle that moves such that $\mathbf{r} \times \dot{\mathbf{r}}$ is constant?

Solution Let $\mathbf{c}$ be a constant vector such that

$$\mathbf{r} \times \dot{\mathbf{r}} = \mathbf{c}$$

then, differentiating and using (74),

$$\dot{\mathbf{r}} \times \dot{\mathbf{r}} + \mathbf{r} \times \ddot{\mathbf{r}} = 0.$$

Since $\dot{\mathbf{r}} \times \dot{\mathbf{r}}$ is zero, by definition, the acceleration is such that

$$\mathbf{r} \times \ddot{\mathbf{r}} = 0.$$

Either the acceleration is zero for all time or the acceleration vector is parallel to the position vector, that is the acceleration is directed towards (or away from) the origin. The particle is therefore moving under the action of a central force.

If, in equation (72), $\mathbf{a} = \mathbf{b}$ then

$$\frac{d}{du}(\mathbf{a} \cdot \mathbf{a}) = 2\mathbf{a} \cdot \frac{d\mathbf{a}}{du},$$

but $\mathbf{a} \cdot \mathbf{a} = a^2$ and it follows that

$$\mathbf{a} \cdot \frac{d\mathbf{a}}{du} = a\frac{da}{du} \tag{76}$$

Hence if $\mathbf{a}$ is a vector of constant magnitude, i.e. $|\mathbf{a}| = a = $ a constant, then $da/du = 0$ and equation (76) shows that

$$\mathbf{a} \cdot d\mathbf{a}/du = 0 \tag{77}$$

from which can be deduced the fact that $\mathbf{a}$ and $d\mathbf{a}/du$ are perpendicular. Note that this is all that can be deduced from (77), as in general $|d\mathbf{a}/du|$ is not zero. In particular, $da/du = 0$ but $|d\mathbf{a}/du| \neq 0$.

9.10.2 Differentiation of a unit vector in two-dimensions

Consider a unit vector $\mathbf{p}$ at the origin of coordinates O at time t. Suppose $\mathbf{p}$ makes an angle θ with some given direction and that in a small time interval δt the vector takes up a new position at angle $\theta + \delta\theta$ (Fig. 9.31).

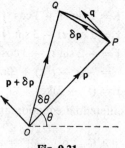

Fig. 9.31

Because $|\mathbf{p}| = |\mathbf{p} + \delta\mathbf{p}| = 1 = $ constant, then $d\mathbf{p}/dt$ is at right angles to $\mathbf{p}$ in the direction of the unit vector $\mathbf{q}$. The magnitude of $d\mathbf{p}/dt$ is

$$\left|\frac{d\mathbf{p}}{dt}\right| = \lim_{\delta t \to 0} \frac{PQ}{\delta t}$$

$$= \lim_{\delta t \to 0} \frac{2 \sin (\delta\theta/2)}{\delta t}$$

$$= \lim_{\delta t \to 0} \frac{\sin (\delta\theta/2)}{(\delta\theta/2)} \frac{\delta\theta}{\delta t}$$

$$= \frac{d\theta}{dt}$$

$$= \dot{\theta}$$

and so

$$\frac{d\mathbf{p}}{dt} = \dot{\theta}\mathbf{q}. \tag{78}$$

Any vector $\mathbf{a}$ may be written in the form $a\mathbf{p}$ where $\mathbf{p}$ is a unit vector in the direction of $\mathbf{a}$. Differentiating and applying (71) and (78) leads to

$$\frac{d\mathbf{a}}{dt} = \frac{d}{dt}(a\mathbf{p})$$

$$= \frac{da}{dt}\mathbf{p} + a\frac{d\mathbf{p}}{dt}$$

$$= \frac{da}{dt}\mathbf{p} + a\dot{\theta}\mathbf{q} \tag{79}$$

The equation (79) shows immediately that

$$\left|\frac{d\mathbf{a}}{dt}\right|^2 = \left(\frac{da}{dt}\right)^2 + a^2\dot{\theta}^2 \tag{80}$$

and that $|d\mathbf{a}/dt|$ and (da/dt) are different quantities, unless $\dot{\theta} = 0$, i.e. unless $\mathbf{a}$ is in a constant direction.

9.10.3 The velocity and acceleration components of a point moving in two-dimensions

Let $\mathbf{OP} = \mathbf{r}$ be the position vector of a moving point at time t. In many applications it is necessary to express the velocity and acceleration in Cartesian or polar coordinates.

In Cartesian coordinates

$$\left.\begin{array}{r} \mathbf{r} = x\mathbf{i} + y\mathbf{j} \\ \mathbf{v} = \dot{\mathbf{r}} = \dot{x}\mathbf{i} + \dot{y}\mathbf{j} \\ \mathbf{f} = \dot{\mathbf{v}} = \ddot{\mathbf{r}} = \ddot{x}\mathbf{i} + \ddot{y}\mathbf{j} \end{array}\right\} \tag{81}$$

and

In polar coordinates, write

$$\mathbf{r} = r\mathbf{p}$$

where $\mathbf{p}$ is a unit vector in the direction of $\mathbf{r}$ then from equation (79)

$$\mathbf{v} = \dot{\mathbf{r}} = \left(\frac{dr}{dt}\right)\mathbf{p} + (r\dot{\theta})\mathbf{q}$$

$$= \dot{r}\mathbf{p} + r\dot{\theta}\mathbf{q}. \tag{82}$$

The result (82) states that the velocity $\mathbf{v}$ may be resolved into two components at right angles, $\dot{r}$ in the radial direction and $r\dot{\theta}$ in the transverse or cross-radial direction. These components together with the Cartesian components are illustrated in Fig. 9.32(a).

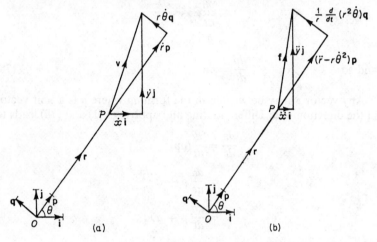

Fig. 9.32

Differentiating equation (82) again

$$\mathbf{f} = \dot{\mathbf{v}} = \ddot{\mathbf{r}} = \ddot{r}\mathbf{p} + \dot{r}\frac{d\mathbf{p}}{dt} + r\ddot{\theta}\mathbf{q} + r\dot{\theta}\frac{d\mathbf{q}}{dt} + \dot{r}\dot{\theta}\mathbf{q}$$

$$= \ddot{r}\mathbf{p} + \dot{r}\dot{\theta}\mathbf{q} + r\ddot{\theta}\mathbf{q} + \dot{r}\dot{\theta}\mathbf{q} + r\dot{\theta}^2(-\mathbf{p})$$

on noting that $d\mathbf{q}/dt$ has a magnitude of $\dot{\theta}$ and a direction at right angles to $\mathbf{p}$ in the sense of increasing θ, i.e. in the direction of $-\mathbf{p}$.
Finally,

$$\mathbf{f} = (\ddot{r} - r\dot{\theta}^2)\mathbf{p} + (2\dot{r}\dot{\theta} + r\ddot{\theta})\mathbf{q}. \tag{83}$$

The two components of the acceleration are then $\ddot{r} - r\dot{\theta}^2$ in the radial direction and $2\dot{r}\dot{\theta} + r\ddot{\theta}$ or $\dfrac{1}{r}\dfrac{d}{dt}(r^2\dot{\theta})$ in the transverse direction. These acceleration components are illustrated in Fig. 9.32(b).

If the point P moves in a circle of radius a so that $|\mathbf{r}| = r = a$ the polar acceleration components are given by $(-a\dot{\theta}^2)$ in the radial direction and $a\ddot{\theta}$ in the cross-radial direction. If further, the angular speed $\dot{\theta}$ is constant then the cross-radial acceleration component vanishes and the radial component of $(-a\dot\theta^2)$ remains. Thus a point moving round a circle with a constant angular speed experiences an acceleration directed towards the centre of the circle, the so-called centripetal acceleration.

9.11 EIGENVALUES AND EIGENVECTORS

In 9.3.7 it was observed that the relationship $\mathbf{y} = C\mathbf{x}$ could be considered as a transformation process, the vector $\mathbf{x}$ being transformed into the vector $\mathbf{y}$ by the matrix C. The two vectors are of the same order but are, in general, of different magnitudes and directions. The matrix C is necessarily square.

Suppose that the matrix C is given. Then of all the possible transformations there exists a vector $\mathbf{x}$ which will be transformed into a scalar multiple of itself, that is, such that

$$C\mathbf{x} = \lambda\mathbf{x} \tag{84}$$

where λ is a scalar.

A particular transformation such as (84) is of interest in the mathematical sense but it is also of interest to the engineer when it is found that equations of this form play an important role in engineering problems (see section 17.3.3 of Chapter 17).

The equation (84) may be written

$$(C - \lambda I)\mathbf{x} = 0 \tag{85}$$

where I is the unit matrix of the same order as C.

Suppose that $\mathbf{x}$ is represented as the column matrix $(x_1, x_2, x_3)^T$ and $C = (c_{ij})$, with $i, j = 1, 2, 3$ then (85) is equivalent to the set of equations

$$\left.\begin{array}{l}
(c_{11} - \lambda)x_1 + c_{12}x_2 + c_{13}x_3 = 0 \\
c_{21}x_1 + (c_{22} - \lambda)x_2 + c_{23}x_3 = 0 \\
c_{31}x_1 + c_{32}x_2 + (c_{33} - \lambda)x_3 = 0
\end{array}\right\} \tag{86}$$

The system (86) has a non-trivial solution if the determinant of the coefficients is zero [see the end of section 8.7.2].

i.e. $$|C - \lambda I| = 0 \tag{87}$$

or

$$\begin{vmatrix}
c_{11} - \lambda & c_{12} & c_{13} \\
c_{21} & c_{22} - \lambda & c_{23} \\
c_{31} & c_{32} & c_{33} - \lambda
\end{vmatrix} = 0 \tag{87a}$$

The equation (87) leads to a cubic equation in λ. This means that, in general, there are three different values of λ and hence three different vectors $\mathbf{x}$ for which (86) has a non-trivial solution, that is for which $\mathbf{x}$ is transformed into a scalar multiple of itself by the matrix C. The alert reader will observe that two of the roots of the cubic equation may be complex and will also note that it is possible that there are equal roots. It is assumed here that all the roots are real.

The values of λ obtained by solving (87) are called the *eigenvalues* of the matrix C (sometimes called *characteristic values* or *latent roots*), and the associated vectors $\mathbf{x}$ are called the *eigenvectors* (sometimes *characteristic functions*). The equation (87) is called the *characteristic equation*.

Example (i) Find the eigenvalues and associated eigenvectors if

$$C = \begin{pmatrix} 2 & -1 & 0 \\ -1 & 5 & -1 \\ 0 & -3 & 2 \end{pmatrix}$$

Solution The eigenvalues are the roots of the equation

$$\begin{vmatrix} 2-\lambda & -1 & 0 \\ -1 & 5-\lambda & -1 \\ 0 & -3 & 2-\lambda \end{vmatrix} = 0$$

that is, of

$$(2-\lambda)[(5-\lambda)(2-\lambda) - 3] + 1[-2 + \lambda] = 0$$

The equation reduces to $(2-\lambda)(\lambda-1)(\lambda-6) = 0$ and the three eigenvalues are $\lambda = 1, 2,$ and 6. The associated eigenvectors are the vectors $(x_1, x_2, x_3)^T$ given by solving the equations

$$\begin{aligned} (2-\lambda)x_1 - x_2 &= 0 \\ -x_1 + (5-\lambda)x_2 - x_3 &= 0 \\ -3x_2 + (2-\lambda)x_3 &= 0 \end{aligned}$$

for the appropriate eigenvalue.

If $\lambda = 1$, then $x_1 = x_2$ and $3x_2 = x_3$ and, choosing the value of x_2 arbitrarily leads to the eigenvector

$$\begin{pmatrix} \alpha \\ \alpha \\ 3\alpha \end{pmatrix}$$

If $\lambda = 2$, then

$$x_2 = 0 \text{ and } -x_1 + 3x_2 - x_3 = 0$$

which leads to $x_2 = 0$ and $x_1 = -x_3$. The associated eigenvector is then

$$\begin{pmatrix} \beta \\ 0 \\ -\beta \end{pmatrix}.$$

where β is arbitrary.

If $\lambda = 6$ then

$$\begin{aligned} -4x_1 - x_2 &= 0 \\ -x_1 - x_2 - x_3 &= 0 \\ -3x_2 - 4x_3 &= 0 \end{aligned}$$

which leads to $4x_1 = -x_2$, and $3x_2 = -4x_3$. The associated eigenvector (letting $x_1 = \gamma$) is

$$\begin{pmatrix} \gamma \\ -4\gamma \\ 3\gamma \end{pmatrix}$$

where γ is arbitrary.

It is common practice to remove the arbitrariness of an eigenvector by making it of unit magnitude. Thus in the example above the three vectors would become, respectively

$$\begin{pmatrix} 1/\sqrt{11} \\ 1/\sqrt{11} \\ 3/\sqrt{11} \end{pmatrix}, \quad \begin{pmatrix} 1/\sqrt{2} \\ 0 \\ -1/\sqrt{2} \end{pmatrix} \quad \text{and} \quad \begin{pmatrix} 1/\sqrt{26} \\ -4/\sqrt{26} \\ 3/\sqrt{26} \end{pmatrix}$$

The eigenvectors are then said to be *normalised*.

In a geometrical sense the above example shows that the transformation matrix

$$C = \begin{pmatrix} 2 & -1 & 0 \\ -1 & 5 & -1 \\ 1 & -3 & 2 \end{pmatrix}$$

will transform the vector $\mathbf{x}^{(1)} = \begin{pmatrix} 1/\sqrt{11} \\ 1/\sqrt{11} \\ 3/\sqrt{11} \end{pmatrix}$ into the vector $\mathbf{y}^{(1)}$ given by $\mathbf{y}^{(1)} = \mathbf{x}^{(1)}$

or   the vector $\mathbf{x}^{(2)} = \begin{pmatrix} 1/\sqrt{2} \\ 0 \\ -1/\sqrt{2} \end{pmatrix}$ into the vector $\mathbf{y}^{(2)}$ given by $\mathbf{y}^{(2)} = 2\mathbf{x}^{(2)}$

or   the vector $\mathbf{x}^{(3)} = \begin{pmatrix} 1/\sqrt{26} \\ -4/\sqrt{26} \\ 3/\sqrt{26} \end{pmatrix}$ into the vector $\mathbf{y}^{(3)}$ given by $\mathbf{y}^{(3)} = 6\mathbf{x}^{(3)}$

Example (ii) Find the eigenvalues of the matrix

$$C = \begin{pmatrix} -1 & 2 & 0 \\ 1 & 0 & 1 \\ 1 & -1 & 0 \end{pmatrix}$$

and the associated eigenvectors $\mathbf{x}$ such that $C\mathbf{x} = \lambda\mathbf{x}$.
Solution The characteristic equation is

$$\begin{vmatrix} -1-\lambda & 2 & 0 \\ 1 & -\lambda & 1 \\ 1 & -1 & -\lambda \end{vmatrix} = 0$$

which reduces to $(\lambda + 1)^2(\lambda - 1) = 0$. There are three values of λ but only two are different, i.e. $\lambda = 1$ and $\lambda = -1$ (twice).

If $\lambda = 1$, the eigenvector is given by solving

$$
\begin{aligned}
-2x_1 &+ 2x_2 &&= 0 \\
x_1 &- x_2 &+ x_3 &= 0 \\
x_1 &- x_2 &- x_3 &= 0
\end{aligned}
$$

which leads to $x_3 = 0$ and $x_1 = x_2$. The normalised eigenvector is

$$
\begin{pmatrix} 1/\sqrt{2} \\ 1/\sqrt{2} \\ 0 \end{pmatrix}
$$

If $\lambda = -1$, the equations for $\mathbf{x}$ reduce to

$$
\begin{aligned}
2x_2 &&&= 0 \\
x_1 &+ x_2 &+ x_3 &= 0 \\
x_1 &- x_2 &+ x_3 &= 0
\end{aligned}
$$

which give $x_2 = 0$ and $x_1 = -x_3$. There is only one eigenvector associated with the double root and this is

$$
\begin{pmatrix} 1/\sqrt{2} \\ 0 \\ -1/\sqrt{2} \end{pmatrix}
$$

in normalised form.

In many practical problems the matrix C is *symmetric* (i.e. $c_{ij} = + c_{ji}$) and it may be shown (but not here) that in this case there is a distinct eigenvector for each eigenvalue whether repeated or not.

Example (iii) Find the eigenvalues and the associated eigenvectors for the symmetric matrix

$$
C = \begin{pmatrix} -1 & 2 & 4 \\ 2 & 2 & -2 \\ 4 & -2 & -1 \end{pmatrix}
$$

Solution The equation for the eigenvalues is given by

$$
\begin{vmatrix} -1-\lambda & 2 & 4 \\ 2 & 2-\lambda & -2 \\ 4 & -2 & -1-\lambda \end{vmatrix} = 0
$$

which reduces to the cubic equation $(\lambda - 3)^2(\lambda + 6) = 0$.

If $\lambda = -6$, the eigenvector is given by

$$
\begin{aligned}
5x_1 &+ 2x_2 &+ 4x_3 &= 0 \\
2x_1 &+ 8x_2 &- 2x_3 &= 0 \\
4x_1 &- 2x_2 &+ 5x_3 &= 0
\end{aligned}
$$

Simple Gaussian elimination (see 8.6.1) leads to $-x_1 = 2x_2 = x_3$, and the eigenvector is

$$
\alpha \begin{pmatrix} 2 \\ -1 \\ -2 \end{pmatrix}
$$

where α is arbitrary, or

$$\begin{pmatrix} 2/3 \\ -1/3 \\ -2/3 \end{pmatrix}$$

in the normalised form.

If $\lambda = 3$, the eigenvector is given by solving

$$\begin{aligned} -4x_1 &+2x_2 &+4x_3 &= 0 \\ 2x_1 &- x_2 &-2x_3 &= 0 \\ 4x_1 &-2x_2 &-4x_3 &= 0 \end{aligned}$$

The three equations have a redundancy and are equivalent to the single equation $2x_1 - x_2 - 2x_3 = 0$, hence two of the variables, say x_1 and x_3, can be assigned arbitrary values β and γ, and the vector takes the form

$$\begin{pmatrix} \beta \\ 2\beta - 2\gamma \\ \gamma \end{pmatrix} \quad \text{or} \quad \beta \begin{pmatrix} 1 \\ 2 \\ 0 \end{pmatrix} + \gamma \begin{pmatrix} 0 \\ -2 \\ 1 \end{pmatrix}$$

In this form the eigenvector is a linear combination of the two eigenvectors

$$\begin{pmatrix} 1 \\ 2 \\ 0 \end{pmatrix} \quad \text{and} \quad \begin{pmatrix} 0 \\ -2 \\ 1 \end{pmatrix}$$

It is an additional property of a symmetric matrix that the eigenvectors are orthogonal, i.e. the scalar products of the vectors are zero. In Example (iii) the vectors $(1, 2, 0)^T$ and $(0, -2, 1)^T$ are orthogonal to the eigenvectors $(2, -1, -2)^T$ since

$$(1 \quad 2 \quad 0) \begin{pmatrix} 2 \\ -1 \\ 2 \end{pmatrix} = 0 \quad \text{and} \quad (0 - 2 \quad 1) \begin{pmatrix} 2 \\ -1 \\ 2 \end{pmatrix} = 0$$

It follows that β and γ may be chosen to give any pair of eigenvectors which are orthogonal to each other and to the third eigenvector. For example, suppose the vector with $\beta = 1$, $\gamma = 0$, i.e. $(1, 2, 0)^T$ is called **a** and the vector associated with $\lambda = 6$ is called **b** then the third vector perpendicular to the plane of these two may be called **c** and the three vectors are mutually orthogonal if $\mathbf{c} = \mathbf{a} \times \mathbf{b}$.

Thus **c** may be written

$$\mathbf{c} = \mathbf{a} \times \mathbf{b} = \begin{vmatrix} \mathbf{i} & \mathbf{j} & \mathbf{k} \\ 1 & 2 & 0 \\ 2 & -1 & -2 \end{vmatrix}$$

$$= -4\mathbf{i} + 2\mathbf{j} - 5\mathbf{k}$$

where **i**, **j**, and **k** are the usual orthogonal base vectors in the Cartesian frame $Ox_1x_2x_3$.

The vector $\mathbf{c}$ then has the form

$$\begin{pmatrix} 4 \\ -2 \\ 5 \end{pmatrix}$$

and it may be verified that

$$\begin{pmatrix} 2 \\ -1 \\ -2 \end{pmatrix}, \quad \begin{pmatrix} 1 \\ 2 \\ 0 \end{pmatrix}, \quad \text{and} \quad \begin{pmatrix} 4 \\ -2 \\ 5 \end{pmatrix}$$

are mutually orthogonal.

9.12 THE GRADIENT OF A SCALAR FIELD

A scalar function of position (and possibly time) $V(x, y, z, t)$ together with the region over which the function is defined is called a *scalar field*. Examples of such fields are the temperature T in a heated region or the electrical potential in an electrical conducting medium.

9.12.1 Two-dimensional scalar field

A two-dimensional scalar field may be conveniently represented geometrically by drawing *level curves*

$$V(x, y, t) = c = \text{constant}$$

For example, the Fig. 9.33 shows a possible set of equi-temperature curves across a rectangular region heated on two sides and held at

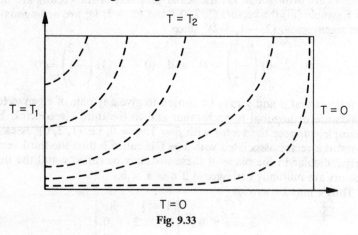

Fig. 9.33

zero temperature on the other two. If the temperatures T_1 and T_2 are variable in time then the picture of level curves will be different for different times. The scalar, i.e. the temperature, is then called *unsteady*. However if the scalar function is independent of time there is only one level curve picture and the field is called *steady*. If the rectangular region of Fig. 9.33 was initially at zero temperature and the two sides were suddenly raised to, and maintained at, temperatures T_1 and T_2 (constant with respect to time), then the temperature distribution over the region would be time dependent long enough for the heat to flow over the entire region. After a sufficiently long period of time the distribution would become steady and the level curve picture would be the same for all subsequent time.

Suppose that $V(x, y)$ is a steady two-dimensional scalar field and P is any point on the level curve (see Fig. 9.34) $V(x, y) = k = $ constant. Let N be an arbitrary point $(x + \delta x, y + \delta y)$ on an adjacent level curve corresponding to $V + \delta V$ and let PN be at angle θ to the axis of x. If s is a spatial distance measured in the direction PN such that PN = δs then the rate at which V is changing in the direction of PN is given by Limit $(\delta V/\delta s)$ as $\delta s \to 0$, i.e. by dV/ds. This is called the *directional*

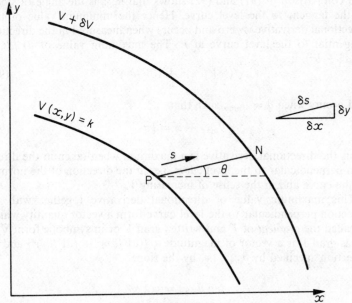

Fig. 9.34

derivative of V (in the direction PN). Using equation (26) of 5.6.2 and noting that V is a function of x and y, as well as a function of s,

$$\frac{dV}{ds} = \frac{\partial V}{\partial x}\frac{dx}{ds} + \frac{\partial V}{\partial y}\frac{dy}{ds}$$

$$= \frac{\partial V}{\partial x}\cos\theta + \frac{\partial V}{\partial y}\sin\theta \tag{88}$$

$$= a\cos\theta + b\sin\theta$$

$$= \sqrt{(a^2 + b^2)}\sin(\theta + \alpha) \tag{89}$$

in which, $\tan \alpha = a/b$, and $a = \partial V/\partial x$, $b = \partial V/\partial y$ are the rates of change of V in the directions of x and y respectively and are independent of θ.

From (89), it follows that the numerically least value of dV/ds is zero and that it occurs when $\theta = \theta_{min}$ such that

$$\theta_{min} = -\alpha \qquad (90)$$

i.e.
$$\tan \theta_{min} = -a/b = -\frac{\partial V/\partial x}{\partial V/\partial y}. \qquad (91)$$

But if $V(x, y) = k = $ constant, it follows (see section 5.6.2) that

$$\frac{dy}{dx} = -\frac{\partial V/\partial x}{\partial V/\partial y} \qquad (92)$$

and comparison of (91) and (92) shows that θ_{min} is the angle of slope of the tangent to the level curve. Hence the minimum value of the directional derivative is zero and occurs when measured in the direction tangential to the level curve at P. The maximum value of dV/ds is $\sqrt{(a^2 + b^2)}$, i.e.

$$\sqrt{(\{\partial V/\partial x\}^2 + \{\partial V/\partial y\}^2)} \qquad (93)$$

and occurs when $\theta = \theta_{max}$ such that

$$\theta_{max} + \alpha = \tfrac{1}{2}\pi \qquad (94)$$

Thus the directional derivative is a maximum when taken in the direction perpendicular to the level curve, i.e. in the direction of the normal to the curve and in the sense of increasing V.

This maximum value of directional derivative together with its direction perpendicular to the level curve form a vector quantity which is called the *gradient* of V and written grad V or in symbolic form ∇V. Thus grad V is a vector of magnitude $\sqrt{\{(\partial V/\partial x)^2 + (\partial V/\partial y)^2\}}$ and of direction specified by θ_{max}, i.e. by the slope

$$\tan \theta_{max} = \cot \alpha$$
$$= b/a$$
$$= \frac{\partial V/\partial y}{\partial V/\partial x}.$$

The vector grad V thus has components in the x and y directions given by $\partial V/\partial x$ and $\partial V/\partial y$ respectively. It follows immediately that

$$\text{grad } V = \nabla V = \frac{\partial V}{\partial x}\mathbf{i} + \frac{\partial V}{\partial y}\mathbf{j} \qquad (95)$$

where $\mathbf{i}$ and $\mathbf{j}$ are the usual unit vectors in the x and y directions respectively. In the alternative notation

$$\text{grad } V = \nabla V = \frac{\partial V}{\partial x_1} \mathbf{e}_1 + \frac{\partial V}{\partial x_2} \mathbf{e}_2$$

$$= \frac{\partial V}{\partial x_i} \mathbf{e}_i \qquad (96)$$

using the summation convention on the symbol i.

9.12.2 Three-dimensional scalar field

The definition of grad V in (95) and (96) may be extended immediately to a three-dimensional field, i.e.

$$\text{grad } V = \nabla V = \frac{\partial V}{\partial x} \mathbf{i} + \frac{\partial V}{\partial y} \mathbf{j} + \frac{\partial V}{\partial z} \mathbf{k} \qquad (97)$$

or alternatively

$$\text{grad } V = \frac{\partial V}{\partial x_i} \mathbf{e}_i \ i = 1, 2, 3. \qquad (98)$$

In this case the geometrical representation of the field requires a three-dimensional picture of level *surfaces* $V(x, y, z) = $ constant. The vector grad V is then of magnitude

$$\sqrt{\{(\partial V/\partial x)^2 + (\partial V/\partial y)^2 + (\partial V/\partial z)^2\}}$$

at the point P and in the direction of the normal to the surface (i.e. perpendicular to the tangent plane at P).

Note The symbol ∇ is pronounced *del* (sometimes *nabla*) and is a *vector operator*, i.e. an operator which when applied to V gives the vector grad V. It is often written

$$\nabla = \frac{\partial}{\partial x} \mathbf{i} + \frac{\partial}{\partial y} \mathbf{j} + \frac{\partial}{\partial z} \mathbf{k}$$

$$= \frac{\partial}{\partial x_i} \mathbf{e}_i$$

Example (i) Find grad V if $V = x^3 y + 3xy^2 z - 2x^2 z^2$

Solution Here

$$\text{grad } V = \frac{\partial V}{\partial x} \mathbf{i} + \frac{\partial V}{\partial y} \mathbf{j} + \frac{\partial V}{\partial z} \mathbf{k}$$

$$= (3x^2 y + 3y^2 z - 4xz^2)\mathbf{i} + (x^3 + 6xyz)\mathbf{j} + (3xy^2 - 4x^2 z)\mathbf{k}$$

Example (ii) Find the equation of the normal to the ellipsoid $2x^2 + 3y^2 + z^2 = 21$ at the point $(3, -1, 0)$.

Solution Let $V(x, y, z) \equiv 2x^2 + 3y^2 + z^2$, for all x, y, z, be a scalar field, then grad V is perpendicular to the level surface $V(x, y, z) = 21$, i.e. is in the direction of the normal to the surface of the ellipsoid. Now,

$$\text{grad } V = 4x\,\mathbf{i} + 6y\,\mathbf{j} + 2z\,\mathbf{k}$$
$$= 12\,\mathbf{i} - 6\,\mathbf{j} + 0\,\mathbf{k}$$

at the point $(3, -1, 0)$. The equation of the required normal is

$$\frac{x-3}{12} = \frac{y+1}{-6} = \frac{z}{0}$$

The normal is the line $2y + x = 1$, $z = 0$, i.e. the line $2y + x = 1$ in the (x, y) plane.

Example (iii) A particle at the point P (x, y, z) is subjected to a force **F** whose magnitude is inversely proportional to the square of the distance OP and whose direction is towards the origin 0 (the inverse square law). *Verify* that there exists a scalar potential function V which is inversely proportional to OP and such that $\mathbf{F} = \text{grad } V$.

Solution Let OP $= r$
then $V = k/r$

where k is a constant and $r = \sqrt{(x^2 + y^2 + z^2)}$.

Then $\text{grad } V = k\nabla(1/r)$
$$= k(-1/r^2)\nabla(r),$$

using the chain rule of differentiation

$$\frac{\partial}{\partial x}\left(\frac{1}{r}\right) = \frac{d}{dr}\left(\frac{1}{r}\right)\frac{\partial r}{\partial x} = \left(-\frac{1}{r^2}\right)\frac{\partial r}{\partial x}$$

and two similar results for differentiation with respect y and z.

Now $r^2 = x^2 + y^2 + z^2$

and so $2r\partial r/\partial x = 2x,$

i.e. $\partial r/\partial x = x/r$, with similar results $\partial r/\partial y = y/r$, and $\partial r/\partial z = z/r$. Finally,

$$\mathbf{F} = \text{grad } V = k(-1/r^2)\left(\frac{x}{r}\mathbf{i} + \frac{y}{r}\mathbf{j} + \frac{z}{r}\mathbf{k}\right)$$
$$= -(k/r^3)\mathbf{r}.$$

This last result shows that **F** is proportional to $(1/r^2)$ and that **F** is in the direction of $-\mathbf{r}$, i.e. towards O.

9.12.3 The directional derivative

Suppose **a** is a given vector, then the directional derivative of V in the direction of **a** is dV/ds where s is measured in the direction of **a**.

Now,

$$\frac{dV}{ds} = \text{Lim} \frac{\delta V}{\delta s} = \text{Lim} \frac{\delta V}{\delta n} \frac{\delta n}{\delta s}$$

where δn is measured in the direction of the normal to the level curve or surface (see Fig. 9.35).

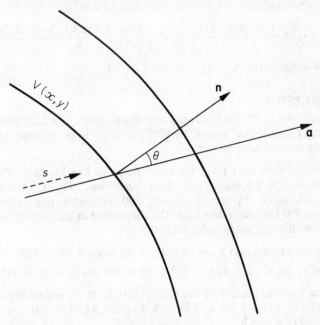

Fig. 9.35

Thus,

$$\frac{dV}{ds} = \text{Lim} \frac{\delta V}{\delta n} \text{Lim} \frac{\delta n}{\delta s}$$

$$= \frac{dV}{dn} \cos \theta$$

$$= \frac{dV}{dn} \frac{\mathbf{n} \cdot \mathbf{a}}{|\mathbf{a}|}$$

$$= \frac{(dV/dn)\mathbf{n} \cdot \mathbf{a}}{|\mathbf{a}|}$$

where $\mathbf{n}$ is the unit normal and θ is the angle between $\mathbf{n}$ and $\mathbf{a}$. Finally,

$$\frac{dV}{ds} = \frac{\nabla V \cdot \mathbf{a}}{|\mathbf{a}|} = \frac{\text{grad } V \cdot \mathbf{a}}{|\mathbf{a}|} \qquad (99)$$

Example Find the directional derivative of $V = 3xyz - x^2y + 2xz^2$ at the point $(1, 1, 1)$ in the direction specified by the join of $(2, 1, 0)$ to $(-1, 3, 4)$.

Solution The required direction is **a** and

$$\mathbf{a} = -3\mathbf{i} + 2\mathbf{j} + 4\mathbf{k}.$$

The directional derivative in the direction of **a** is

$$\frac{[(3yz - 2xy + 2z^2)\mathbf{i} + (3xz - x^2)\mathbf{j} + (3xy + 4xz)\mathbf{k}] \cdot [-3\mathbf{i} + 2\mathbf{j} + 4\mathbf{k}]}{\sqrt{(9 + 4 + 16)}}$$

which at the point $(1, 1, 1)$ becomes $23/\sqrt{29}$.

EXERCISE 9

1 State which of the following are scalar quantities and which are vector quantities: speed, weight, pressure, time, voltage, tension, momentum, temperature.

2 Two aircraft start from the same point on the Equator. The first travels 500 kilometre due East and then 500 kilometre due North while the second travels 500 kilometre due North and then 500 kilometre due East. Do they arrive at the same point? Are these directed magnitudes vectors?

3 Free vectors **a** and **b** are inclined at an angle θ, show that

$$|\mathbf{a}| - |\mathbf{b}| \leqslant |\mathbf{a} + \mathbf{b}| = \sqrt{(a^2 + b^2 + 2ab \cos \theta)} \leqslant |\mathbf{a}| + |\mathbf{b}|.$$

4 The position vectors of the four points A, B, C, and D are respectively, **a**, **b**, $3\mathbf{a} + 2\mathbf{b}$, and $2\mathbf{a} - \mathbf{b}$. Express **AC**, **DB**, **CB**, and **CA** in terms of **a** and **b**.

5 In any triangle ABC prove by vector methods that (i) the straight line joining the middle points of AB, AC is parallel to BC and half the length of BC, and (ii) if D, E, F are the mid-points of BC, CA, and AB respectively and O is any other point,

$$\mathbf{OD} + \mathbf{OE} + \mathbf{OF} = \mathbf{OA} + \mathbf{OB} + \mathbf{OC}.$$

6 Two forces act at the corner A of a quadrilateral $ABCD$ and two at the corner C. The forces are represented in magnitude and direction by **AB**, **AD**, **CB**, and **CD** respectively. Show that their resultant is represented by **4PQ** where P and Q are the mid-points of AC and BD respectively.

7 If **a** and **b** are the position vectors of two points A and B find the position vector of a point C on AB produced such that $AC = 3AB$.

8 Vectors from O to A, B, C, and D are represented by **a**, **b**, **c**, and **d** respectively. The joins CB and AD intersect at P. If $AP = \frac{1}{4}AD$ and $CP = \frac{1}{3}CB$ show that $\mathbf{d} - \mathbf{b} = 2(\mathbf{c} - \mathbf{a})$.

9 A, B, C, and G are the points $(2, 1, 3)$, $(3, -1, -2)$, $(-2, 3, 5)$, and $(1, 1, 2)$ respectively and P is any point. Show that $\overline{PA} + \overline{PB} + \overline{PC} = 3\overline{PG}$ and state what can be deduced about the point G.

10 Find the magnitude and direction of the resultant of the vectors (i) $7\mathbf{i} + 2\mathbf{j} - 5\mathbf{k}$, $3\mathbf{i} - \mathbf{j} - 2\mathbf{k}$, $-2\mathbf{i} - 2\mathbf{j} + \mathbf{k}$; (ii) of magnitudes 6, 5, and 3 in the directions $[2:1:2]$, $[4:3:0]$, and $[2:2:-1]$ respectively.

11 Find the centre of gravity of masses 3, 4, 5 at the points $(2, -3, 3)$, $(5, -3, -4)$, and $(2, -3, -1)$ respectively.

12 Forces of magnitudes 1, 2, and 3 act at one corner of a cube along the diagonals of the faces meeting at that corner, find the magnitude of their resultant and its inclination to the edges of the cube.

13 The position vector of a point A is $2\mathbf{i} - 3\mathbf{j} + \mathbf{k}$. Find the vector equation of the line through A in the direction of the vector $7\mathbf{j} + \mathbf{k}$. Find the coordinates of the point B on the line such that $AB = 6\sqrt{2}$.

14 Find the equation of the line passing through the points $(1, 1, 1)$ and $(-2, -1, 3)$.

15 A and B are the points whose position vectors are $3\mathbf{i} + 4\mathbf{j} - \mathbf{k}$ and $-3\mathbf{i} + 7\mathbf{j} + 2\mathbf{k}$ respectively. Find the position vector of the point that divides the join of AB (i) internally in the ratio $3:4$, and (ii) externally in the ratio $4:7$.

16 A force of 13 units acts at the point $(1, 3, -4)$ and its line of action is parallel to the join of the point $(1, 0, 3)$ to the point $(5, 1, 7)$. Find the vector form of the force.

17 Find the scalar projection of the vector $2\mathbf{i} - 3\mathbf{j} + 7\mathbf{k}$ on to the vector $6\mathbf{i} + 4\mathbf{j} - \mathbf{k}$.

18 Find the vector projection of $2\mathbf{i} + 3\mathbf{j}$ on to the vector $4\mathbf{j} + 5\mathbf{k}$.

19 Show that the perpendicular distance from the origin to the line $\mathbf{r} = \mathbf{a} + t\mathbf{b}$ is given by $|\mathbf{a} - (\mathbf{a} \cdot \mathbf{b})\mathbf{b}/b^2|$. Find the perpendicular distance from the origin to the line $3\mathbf{j} + 4\mathbf{k} + t(5\mathbf{i} + 5\mathbf{j})$.

20 Show that the perpendicular distance from the point whose position vector is $\mathbf{c}$ to the line $\mathbf{r} = \mathbf{a} + t\mathbf{b}$ is $|(\mathbf{c} - \mathbf{a}) \times \mathbf{b}|/b$. Find the perpendicular distance from the point $P(-1, 3, 4)$ to the join of $A(2, 3, 5)$ and $B(-6, 7, 9)$.

21 A particle moves in a straight line from a point $A(0, 0, 0)$ to a point $B(0, 1, 2)$ and then in a second straight line to a point $C(1, 0, 3)$ and finally in a third straight line back to A. It is acted on by a force $2\mathbf{i} + 4\mathbf{j} + 3\mathbf{k}$ while moving along AB, by a force $2\mathbf{i} + 3\mathbf{k}$ while moving along BC and by a force $-3\mathbf{k}$ while moving along CA. Find the total work done by these forces.

22 Prove that the vectors $\mathbf{a} = 2\mathbf{i} + \mathbf{j} + 2\mathbf{k}$, $\mathbf{b} = \mathbf{i} + 3\mathbf{j} - 7\mathbf{k}$, $\mathbf{c} = 3\mathbf{i} + 4\mathbf{j} - 5\mathbf{k}$ form a right angled triangle.

23 Find the area of the parallelogram whose sides are the vectors $\mathbf{i} + 3\mathbf{j} - \mathbf{k}$, $2\mathbf{i} - \mathbf{j} + 2\mathbf{k}$ and determine a unit vector normal to its plane. Find also the equation of the plane.

24 What is the unit vector perpendicular to each of the vectors $2\mathbf{i} - \mathbf{j} + \mathbf{k}$ and $3\mathbf{i} + 4\mathbf{j} - \mathbf{k}$? Find two unit vectors in the plane of these vectors, one perpendicular to each of the given vectors.

25 Find the moment about a point $\mathbf{i} + 2\mathbf{j} - 2\mathbf{k}$ of a force represented by $3\mathbf{i} + 4\mathbf{k}$ acting through the point whose position vector is $\mathbf{i} - \mathbf{j} + 3\mathbf{k}$.

26 The vector $\mathbf{c} = l\mathbf{i} + m\mathbf{j} + n\mathbf{k}$ is perpendicular to the vectors $\mathbf{a} = \mathbf{i} + 2\mathbf{k}$ and $\mathbf{b} = -\mathbf{i} + 2\mathbf{j} + 3\mathbf{k}$. Show that $l:m:n = 4:5:-2$, (i) by solving the equation $\mathbf{c} . \mathbf{a} = 0 = \mathbf{c} . \mathbf{b}$, and (ii) by evaluating $\mathbf{a} \times \mathbf{b}$.

27 Forces $\mathbf{i} - \mathbf{j} + 3\mathbf{k}$, $-3\mathbf{i} + 3\mathbf{j} - 2\mathbf{k}$, and $2\mathbf{i} - 2\mathbf{j} - \mathbf{k}$ act at the points $(1, 1, 1)$, $(2, 3, 0)$, and $(-3, 4, 1)$ respectively. Show that they are equivalent to a couple of moment $\sqrt{138}$ about an axis whose direction ratios are $[-4:1:11]$.

28 If $\mathbf{a} \times \mathbf{b} = \mathbf{n}$ and $\mathbf{n} . \mathbf{c} = \alpha$ where $\mathbf{a} = a_1\mathbf{i} + a_3\mathbf{k}$, $\mathbf{b} = b_1\mathbf{i} + b_2\mathbf{j}$, and $\mathbf{c} = c_2\mathbf{j} + c_3\mathbf{k}$ show that $\alpha = a_1b_2c_3 + a_3b_1c_2$.

29 Show that the line through the point $(0, 4, 4)$ in the direction of the vector $[-1, 2, 1]$ intersects the line through the point $(1, 4, -1)$ in the direction of the vector $[0, 1, -2]$ and find the point of intersection.

30 Find the shortest distance between the lines
$$(x + 3)/2 = (y - 6)/3 = (z - 3)/(-2)$$
and
$$x/2 = (y - 6)/2 = z/(-1).$$

31 Find the triple scalar product of the three vectors $\mathbf{a} = 2\mathbf{i} - \mathbf{j} + \mathbf{k}$, $\mathbf{b} = \mathbf{i} - \mathbf{k}$, and $\mathbf{c} = 3\mathbf{i} + 2\mathbf{j}$.

32 Find values of β if $2\mathbf{i} + \beta\mathbf{j} + \mathbf{k}$, $\beta\mathbf{i} + \mathbf{j} + \mathbf{k}$, and $\beta\mathbf{i} + 3\mathbf{j} - 3\mathbf{k}$, are coplanar vectors.

33 Find the coordinates of the point of intersection of the line $\mathbf{r} = \mathbf{a} + t\mathbf{b}$ and the plane $\mathbf{r} . \mathbf{n} = 8$ when $\mathbf{a} = 8\mathbf{i} + 7\mathbf{j} + 12\mathbf{k}$, $\mathbf{b} = 3\mathbf{i} + 4\mathbf{j} + 6\mathbf{k}$, and $\mathbf{n} = 4\mathbf{i} + 3\mathbf{k}$.

34 Show that the system of equations
$$3x - 2y + z = 1$$
$$2x + 4y - 2z = 3$$
$$7x + 6y - 3z = 7$$
is consistent but indeterminate and illustrate the result geometrically.

35 Show that the system of equations

$$2x + 3y - 7z = 2$$
$$x - y \quad\;\; = 4$$
$$3x + 2y - 7z = 1$$

is inconsistent and illustrate geometrically.

36 Prove that $(\mathbf{a} \times \mathbf{b}) . (\mathbf{a} \times \mathbf{c}) = (\mathbf{b} . \mathbf{c})a^2 - (\mathbf{a} . \mathbf{b})(\mathbf{a} . \mathbf{c})$.

37 ABC is a triangle with a right angle at C. If $CB = a$, $CA = b$, and $AB = c$ use the scalar product of vectors to prove Pythagoras' theorem.

38 P_1, P_2, and P_3 are the points whose position vectors are $\mathbf{r}_1$, $\mathbf{r}_2$, and $\mathbf{r}_3$ respectively. Show that the equation of the plane through the points is given by

$$(\mathbf{r} - \mathbf{r}_1) . (\mathbf{r}_2 - \mathbf{r}_1) \times (\mathbf{r}_3 - \mathbf{r}_1) = 0$$

and express the result in the form of a determinant. Find the equation of the plane through the points $(2, 1, 4)$, $(2, 0, 1)$, and $(-1, 3, 0)$.

39 Given $\mathbf{r} = \cos t\mathbf{i} - \sin 2t\mathbf{j} + t^2\mathbf{k}$ find $d\mathbf{r}/dt$, $d^2\mathbf{r}/dt^2$ and $|d^2\mathbf{r}/dt^2|$.

40 The position vector of a point at time t is

$$\mathbf{r} = 2t\mathbf{i} + \tfrac{1}{6} t^3\mathbf{j} - 2\mathbf{k}.$$

Find the velocity and acceleration and show that

$$4v^2 = 16 + f^4$$

where v and f are the magnitudes of the velocity and acceleration.

41 If $\mathbf{r} = \mathbf{a} \sin \omega t + \mathbf{b} \cos \omega t$ where $\mathbf{a}$ and $\mathbf{b}$ are constant vectors, show that $d^2\mathbf{r}/dt^2 = -\omega^2\mathbf{r}$.

42 Show that $\mathbf{r} = 2\mathbf{i} \cos t + 2\mathbf{j} \sin t$ lies on a circle for all t. Show that the equation of the tangent at the point t_1 is

$$\mathbf{r} = 2\mathbf{i} \cos t_1 + 2\mathbf{j} \sin t_1 + \alpha(2\mathbf{j} \cos t_1 - 2\mathbf{i} \sin t_1)$$

where α is a parameter.

43 By differentiating $\mathbf{a} . \mathbf{a}$ show that $\mathbf{a} . (d\mathbf{a}/dt) = a(da/dt)$. If $\mathbf{a}$ is of constant length deduce that $d\mathbf{a}/dt$ and $\mathbf{a}$ are at right angles.

44 A particle moves on a circle $r = a$ with a variable angular velocity ω. Find the components of acceleration.

45 Verify the expressions $\dot{r}$ and $r\dot{\theta}$ for the velocity components in polar coordinates by differentiating and resolving the relationships $x = r \cos \theta$, $y = r \sin \theta$. Also verify the acceleration components

$$\ddot{r} - r\dot{\theta}^2, \qquad \frac{1}{r}\frac{d}{dt}(r^2\dot{\theta}).$$

46 If $\mathbf{a} = 2t\mathbf{i} - t^2\mathbf{j} + 2\mathbf{k}$, $\mathbf{b} = 3\mathbf{i} + t\mathbf{j} - 2t\mathbf{k}$, and $\mathbf{c} = t^2\mathbf{k}$ find $d(\mathbf{a} . \mathbf{b})/dt$, $d(\mathbf{a} \times \mathbf{b})/dt$, and $d(\mathbf{a} . \mathbf{b} \times \mathbf{c})/dt$, (i) by differentiating the products, and (ii) by first evaluating the products.

47 A body moves on the curve $r = ae^{\theta \cot \alpha}$ where a and α are constants (the equiangular spiral) in such a way that $d\theta/dt$ is constant and equal to ω. Find the polar components of the velocity and show that the resultant velocity makes a constant angle with the radius vector. Find also the polar components of the acceleration.

48 A particle is moving such that its acceleration at time t is $2\mathbf{i} + t\mathbf{k}$. If it initially passes through the point $(1, 2, 3)$ with a velocity $2\mathbf{j} - \mathbf{k}$, find the parametric equations of its path.

49 Obtain the eigenvalues and normalised eigenvectors of the matrices

$$\text{(i)} \begin{pmatrix} 1 & -8 \\ 2 & 11 \end{pmatrix} \qquad \text{(ii)} \begin{pmatrix} 1 & 1 & 1 \\ 2 & 2 & 3 \\ 3 & 3 & 2 \end{pmatrix}$$

50 Given that the matrix

$$\begin{pmatrix} 11 & -6 & 2 \\ -6 & 10 & -4 \\ 2 & -4 & 6 \end{pmatrix}$$

has an eigenvalue 18 and corresponding eigenvector

$$\begin{pmatrix} 2/3 \\ -2/3 \\ 1/3 \end{pmatrix},$$

find the other eigenvalues and eigenvectors.

51 Find the eigenvalues and three mutually orthogonal eigenvectors of the symmetric matrix

$$\begin{pmatrix} 0 & 1 & 1 \\ 1 & 0 & 1 \\ 1 & 1 & 0 \end{pmatrix}.$$

52 Find the angle between the normals to the surfaces $x^2 + y^2 + z^2 = 14$ and $z = x^2 y^2 - 2$ at the point $(1, -2, 3)$.

53 P is a fixed point $2\mathbf{i} - \mathbf{j} + \mathbf{k}$ and Q is the general point $x\mathbf{i} + y\mathbf{j} + z\mathbf{k}$. The function $\phi(x, y, z)$ is the square of the distance PQ. Evaluate grad ϕ, and find the directional derivative of ϕ in the direction $\mathbf{i} + 2\mathbf{j} + \mathbf{k}$ when Q has position vector $3\mathbf{i} + \mathbf{j} - 2\mathbf{k}$.

54 A particle is constrained to move along the helix

$$\mathbf{r}(s) = \cos \omega s \, \mathbf{i} + \sin \omega s \, \mathbf{j} + \alpha s \, \mathbf{k}$$

where $\omega^2 + \alpha^2 = 1$. It is acted upon by a force field $\mathbf{F} = \nabla \phi$ where $\phi = \log r$ and $r = |\mathbf{r}|$. Show that the work done by the force on the particle during an incremental displacement dr is

$$\mathbf{F} \cdot \mathbf{dr} = \frac{\alpha^2 s}{1 + \alpha^2 s^2} \, ds$$

and hence when the particle moves from $s = 0$ to $s = 2\pi/\omega$ the total work done is $\log \sqrt{(1 + 4\pi^2\alpha^2/\omega^2)}$.

10
Complex Numbers and Complex Variables

10.0 INTRODUCTION

From a historical point of view the theory of complex numbers arose from the study of the solution of algebraic equations particularly of the quadratic equation $ax^2 + bx + c = 0$. The equation $x^2 - 4x - 5 = 0$ or $(x - 2)^2 = 9$ has two unequal roots given by $x - 2 = \pm 3$; the equation $x^2 - 6x + 9 = 0$ or $(x - 3)^2 = 0$ has two equal roots $x = 3$ but the equation $x^2 - 4x + 13 = 0$ or $(x - 2)^2 = -9$ has roots which involve the finding of a number whose square is -9. Such a number can be represented by the introduction of a symbol i (sometimes j is used) such that $i^2 = -1$ or $i = \sqrt{-1}$ and the two roots of $x^2 - 4x + 13 = 0$ can then be formally written $x - 2 = \pm i3$. The invented number $i3$ was called an *imaginary number*, the other numbers such as 3, -3, or $\sqrt{3}$ with which familiarity was already established being called *real numbers*. The numbers of the form $2 \pm i3$ were called *complex numbers*. The main reason for the introduction of such numbers was for uniformity so that it could be said that any quadratic equation always had two roots, real and unequal, real and equal, or complex.

The term imaginary is an unfortunate legacy from the past and implies that the numbers have some mystical character. From a mathematical point of view the imaginary numbers or the more general complex numbers are considered well-behaved providing they obey a consistent set of algebraic rules, while from an engineering viewpoint the new numbers are of immediate interest and acceptable providing that they can be successfully used in an engineering context. For the latter it is more advantageous to ignore the historical aspect of complex numbers and to consider the symbol i as an operator. Note that i here has no connexion with the unit vector $\mathbf{i}$ (of $\mathbf{i}$, $\mathbf{j}$, $\mathbf{k}$) of Chapter 9.

10.1 SYMBOL i AS AN OPERATOR

Consider any vector $\overline{OP}$ lying in a plane and of length a units, i.e. $\mathbf{a} = \overline{OP}$. The symbol (-1) may be thought of as an operation which when applied to $\overline{OP}$ has the effect of turning $\overline{OP}$ through two right angles in the anticlockwise direction, this being the mathematically positive direction, and leads to another vector $\overline{OQ}$ such that

$$\overline{OQ} = (-1)\overline{OP} = (-1)\mathbf{a} = -\mathbf{a}. \tag{1}$$

A 'mid-way' stage in this operation would be to turn the vector $\overline{OP}$ through one right angle in the anticlockwise direction. Suppose that the symbol i is given to this operation then $i\overline{OP}$ will be represented, in Fig. 10.1, by the vector $\overline{OR}$ such that

$$\overline{OR} = i\overline{OP} = i\mathbf{a}. \tag{2}$$

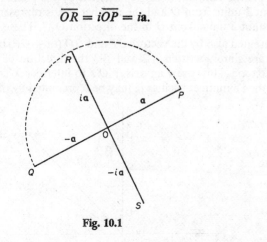

Fig. 10.1

A further application of the operator i to the vector $\overline{OR}$ will then turn it through a right angle to the position of $\overline{OQ}$ so that

$$-\mathbf{a} = \overline{OQ} = i(i\overline{OP}) = i^2\overline{OP} = i^2\mathbf{a}. \tag{3}$$

Equation (3) shows the effect of a double application of the operator i. That is ii or i^2 turns the vector through two right angles and has the same effect as the operation -1 of equation (1). In an algebraic sense (3) may be interpreted as showing the equivalence of i^2 and -1 and in this sense i could be said to be $\sqrt{-1}$. As mentioned in section 10.0, the statement $i = \sqrt{-1}$ may be used as a starting point for the historical development of a complex number theory. The operational approach is simply to consider i as an operation that turns a vector in a plane through one right angle and i^2 (or -1) as the operator that turns the same vector through two right angles.

A third application of the operator i will turn $\overline{OQ}$ into $\overline{OS}$, i.e.

$$\overline{OS} = i(i^2\overline{OP}) = i^3\overline{OP} = -i\overline{OP} \tag{4}$$

and
$$i^3 = ii^2 = -i. \tag{5}$$

The operator i^3 applied to $\overline{OP}$ has the same effect as the operator $(-i)$ applied to $\overline{OP}$. The first turns $\overline{OP}$ through three right angles in an anti-clockwise direction, the second turns $\overline{OP}$ through one right angle in a clockwise direction and in either case the vector $\overline{OS}$ results.

10.1.1 The Argand diagram

In Fig. 10.2 let $X'OX$ produced be an axis on which all real numbers from $-\infty$ to $+\infty$ can be represented, O being the origin zero. For example the number $+3$ is represented by the point P on the line OX distant 3 units from O and the number -3 is represented by the point Q distant 3 units from O in the direction OX'. These numbers may be represented also by the vectors $\overline{OP}$ and $\overline{OQ}$ (or $-\overline{OP}$) but by convention they are simply written $+3$ and -3 the direction of the vector being understood. Now draw an axis $Y'OY$ to intersect $X'OX$ at right angles at O then a number such as $i2$ may be represented by the point R distant

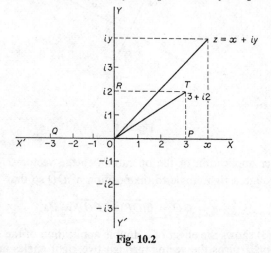

Fig. 10.2

2 units from O in the direction OY and $-i2$ by the point S distant 2 units from O in the direction OY'. Again, the number $i2$ may be represented by the vector $\overline{OR}$ being the vector $+2$ turned through one right angle in the anticlockwise direction, i.e. $\overline{OR} = i(+2) = i2$.

Consider the addition of the two numbers 3 and $i2$. The sum is formally written $3 + i2$ and may be represented by the point T or by the vector $\overline{OT}$, the sum of the vectors $\overline{OP}$ and $\overline{OR}$ (see section 9.3.3). Thus in Fig. 10.2 the complex number $3 + i2$ is represented either by the point T or by the vector $\overline{OT}$.

The axis $X'OX$ containing all the real numbers from $-\infty$ to $+\infty$ is usually called the *axis of real numbers* or the *real axis* and the axis $Y'OY$ containing all the imaginary numbers from $-i\infty$ to $+i\infty$ is called the *axis of imaginary numbers* or the *imaginary axis*. It is usual to let any arbitrary real number be denoted by x and to let any arbitrary imaginary number be denoted by iy (where y is a real number). The sum of these numbers is then denoted by z, i.e.

$$z = x + iy.$$

The two numbers x and y are called the *real part* and the *imaginary part* of z respectively and written *Re* z and *Im* z.

The geometrical interpretation that any complex number z may be represented in the Cartesian plane by a point whose coordinates are (x, y) where the x-axis contains all the real numbers and the y-axis contains all the imaginary numbers is called the *Argand diagram*. The Cartesian plane is often called the z-plane.

Note 1 There is rarely any confusion between z the complex number and z the spatial coordinate of (x, y, z) because the algebra of complex numbers is applied to problems essentially of a two-dimensional nature.

Note 2 Although y is called the imaginary part of z it is a number as real as x. Some writers call iy the imaginary part of z.

Note 3 Too much emphasis should not be placed on the representation of a complex number as a vector because of the operator i in the representation. For instance if $\mathbf{e}_1$ and $\mathbf{e}_2$ are two orthogonal unit base vectors in a Cartesian coordinate system then by equation (33) of section 9.6.1, $\mathbf{e}_1 \cdot \mathbf{e}_1 = 1$ and $\mathbf{e}_2 \cdot \mathbf{e}_2 = 1$ but if $\mathbf{n}$ is a unit vector along OX in the Argand diagram so that $i\mathbf{n}$ is the unit vector along OY then $\mathbf{n} \cdot \mathbf{n} = 1$ whereas $(i\mathbf{n}) \cdot (i\mathbf{n}) = -1$, i.e. a different product is obtained depending on the direction taken.

Note 4 The real numbers are a particular case of the complex numbers.

Note 5 Numerical complex numbers are frequently written in the form $3 + 2i$ instead of $3 + i2$.

10.2 THE ALGEBRA OF COMPLEX NUMBERS

The rules of common algebra may be applied to complex numbers providing that any product i^2 is replaced by -1.

10.2.1 Equality of complex numbers

Two complex numbers are equal

$$x_1 + iy_1 = x_2 + iy_2$$

if and only if $x_1 = x_2$ and $y_1 = y_2$, as is immediately obvious by plotting the two points on the Argand diagram. The zero complex number is $0 + i0$ represented by the point O in the Argand diagram. A complex number is zero if and only if its real and its imaginary parts are zero.

10.2.2 Addition and subtraction of complex numbers

The sum of the complex numbers z_1 and z_2 is given by

$$(x_1 + iy_1) + (x_2 + iy_2) = (x_1 + x_2) + i(y_1 + y_2)$$

and is identical with vector addition.

The difference of two complex numbers z_1 and z_2 is the sum of z_1 and $-z_2$, i.e.

$$z_1 - z_2 = z_1 + (-z_2)$$
$$= (x_1 + iy_1) - (x_2 + iy_2)$$
$$= (x_1 - x_2) + i(y_1 - y_2).$$

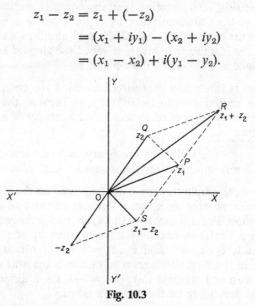

Fig. 10.3

In Fig. 10.3 the points P and Q represent z_1 and z_2 respectively, R is the sum $z_1 + z_2$ and S is the difference $z_1 - z_2$. It is sometimes an advantage to represent $z_1 - z_2$ in vector form by $\overline{OS}$ or its equivalent vector $\overline{QP}$. (Compare with section 9.3.3.)

10.2.3 Multiplication of complex numbers

Given two complex numbers z_1 and z_2 the product is another complex number z_3 such that

$$z_3 = z_1 z_2$$
$$= (x_1 + iy_1)(x_2 + iy_2)$$
$$= (x_1 x_2 - y_1 y_2) + i(x_1 y_2 + x_2 y_1). \tag{6}$$

The result (6) shows that multiplication is commutative, i.e.

$$z_1 z_2 = z_2 z_1.$$

The product of two complex numbers is real providing that

$$x_1 : y_1 = -x_2 : y_2.$$

In particular the numbers $x_1 + iy_1$ and $x_1 - iy_1$ whose product is $x_1^2 + y_1^2$ are called *conjugate* complex numbers. If $z = x + iy$ is any general complex number the conjugate is denoted by $\bar{z}$, so that $\bar{z} = x - iy$. In the Argand diagram $\bar{z}$ is the reflection of z in the real axis. Note that

$$\overline{z_1 \pm z_2} = \bar{z}_1 \pm \bar{z}_2$$

$$\overline{z_1 z_2} = \bar{z}_1 \bar{z}_2.$$

Example (i) $(3 + 2i) + (5 - 7i) = 8 - 5i$

$$(3 + 2i) - (5 - 7i) = -2 + 9i$$

$$6(1 - 5i) = 6 - 30i$$

$$3(2 - 3i) - (7 - 2i) + (1 + 7i) = (6 - 9i) - (6 - 9i)$$

$$= 0 + i0.$$

Example (ii) $(4 + 2i)(3 - i) = (12 + 2) + i(6 - 4)$

$$= 14 + 2i.$$

Example (iii) $i(1 + 2i) = i + 2i^2$

$$= -2 + i.$$

Note that if $(1 + 2i)$ in example (iii) is represented by the vector $\overrightarrow{OP}$ then $i(1 + 2i)$ is represented by the vector $\overrightarrow{OQ}$ obtained by turning $\overrightarrow{OP}$ through one right angle in the positive sense.

10.2.4 Division of complex numbers

The value of $(x_1 + iy_1)/(x_2 + iy_2)$ is defined as a third value $X + iY$ such that

$$x_1 + iy_1 = (x_2 + iy_2)(X + iY)$$

$$= (x_2 X - y_2 Y) + i(x_2 Y + Xy_2).$$

From the equality of complex numbers it follows that

$$x_2 X - y_2 Y = x_1$$

and $x_2 Y + Xy_2 = y_1$

and hence $X + iY = \dfrac{x_1 x_2 + y_1 y_2}{x_2^2 + y_2^2} + \dfrac{i(x_2 y_1 - x_1 y_2)}{x_2^2 + y_2^2}.$

This result can be obtained by an alternative but equivalent process as follows. The numerator and denominator of the quotient

$$(x_1 + iy_1)/(x_2 + iy_2)$$

are each multiplied by the complex conjugate of the denominator so as to make an equivalent quotient with a real denominator.

Thus
$$X + iY = \frac{(x_1 + iy_1)(x_2 - iy_2)}{(x_2 + iy_2)(x_2 - iy_2)}.$$
$$= \frac{(x_1 x_2 + y_1 y_2) + i(x_2 y_1 - x_1 y_2)}{x_2^2 + y_2^2}. \tag{7}$$

The result (7) need not be remembered but each division required should be evaluated using this alternative process.

Note that (7) reduces to x_1/x_2 if $y_1 = 0 = y_2$.

Example (i)
$$\frac{(3 + 2i)}{(4 - 3i)} = \frac{(3 + 2i)(4 + 3i)}{(4 - 3i)(4 + 3i)}$$
$$= \frac{(12 - 6) + i(9 + 8)}{(16 + 9)}$$
$$= (6 + 17i)/25.$$

Example (ii)
$$\frac{(1 + i)(2 - i)}{(1 - i)(3 + i)} = \frac{3 + i}{4 - 2i}$$
$$= \frac{(3 + i)(4 + 2i)}{16 + 4}$$
$$= (1 + i)/2.$$

10.3 POLAR FORM OF A COMPLEX NUMBER

Any complex number $z = x + iy$ may be represented by the point P on the Argand diagram as in Fig. 10.4, i.e. by the point whose Cartesian coordinates are (x, y).

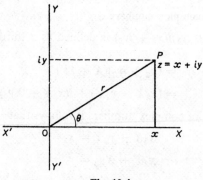

Fig. 10.4

It is frequently an advantage to represent the point in terms of its polar coordinates (r, θ). Now,
$$z = x + iy = r \cos \theta + ir \sin \theta$$
$$= r (\cos \theta + i \sin \theta). \tag{8}$$

In abbreviated form this is often written r cis θ or $r\underline{/\theta}$, so that
$$r(\cos \theta + i \sin \theta) \equiv r \text{ cis } \theta \equiv r\underline{/\theta}.$$

10.3.1 Modulus and argument

The polar coordinate r is a positive quantity being the magnitude of the vector $\overline{OP}$. It is called the *modulus* of z and is written mod z or $|z|$, i.e.

$$\text{mod } z = |z| = r = \sqrt{(x^2 + y^2)}. \tag{9}$$

The polar coordinate θ, giving the direction of the vector $\overline{OP}$, is called the *argument* or *amplitude* of z and written arg z, amp z, or am z. The value of θ is found from

$$\tan \theta = y/x \tag{10}$$

the quadrant in which θ lies being determined from the equations

$$x = r \cos \theta, \quad y = r \sin \theta, \quad -180° < \theta \leqslant 180°.$$

It is sometimes convenient simply to plot the complex point on the Argand diagram and hence to determine the quadrant in which θ must lie. The value of θ within the above range is often called the *principal value* of arg z. Note that $-1 = 1/\underline{180°}$ and that if $z = r/\underline{\theta}$ then $\bar{z} = r/\underline{-\theta}$.

Example (i) Express $2 + 3i$ in the polar form.

Solution $\qquad\qquad r = \sqrt{(4 + 9)} = \sqrt{13}$

and $\qquad\qquad\quad \tan \theta = 3/2$

$$\theta = 56° \ 19'.$$

In abbreviated form $2 + 3i \equiv \sqrt{13}/\underline{56° \ 19'} \equiv \sqrt{13} \text{ cis } (56° \ 19')$.

Example (ii) Express $-2 - 3i$ in polar form.

Solution Here, $\quad r = \sqrt{13}$

and $\qquad\qquad \tan \theta = -3/-2$

$$\theta = -(180° - 56° \ 19') = -123° \ 41'.$$

10.3.2 Multiplication and division in polar form

Let

$$z_1 = r_1(\cos \theta_1 + i \sin \theta_1)$$

and $\qquad\qquad z_2 = r_2(\cos \theta_2 + i \sin \theta_2)$

then the product $z_1 z_2$ is given by

$$\begin{aligned}
z_1 z_2 &= r_1 r_2 \, (\cos \theta_1 + i \sin \theta_1)(\cos \theta_2 + i \sin \theta_2) \\
&= r_1 r_2 [(\cos \theta_1 \cos \theta_2 - \sin \theta_1 \sin \theta_2) \\
&\quad + i(\sin \theta_1 \cos \theta_2 + \cos \theta_1 \sin \theta_2)] \\
&= r_1 r_2 \, [\cos (\theta_1 + \theta_2) + i \sin (\theta_1 + \theta_2)]
\end{aligned} \tag{11}$$

or in abbreviated form

$$(r_1\underline{/\theta_1})(r_2\underline{/\theta_2}) \equiv r_1r_2\underline{/(\theta_1 + \theta_2)} \equiv r_1r_2 \text{ cis } (\theta_1 + \theta_2) \tag{12}$$

Thus to multiply two complex numbers in polar form multiply the moduli and *add* the arguments. Similarly,

$$\begin{aligned}
\frac{z_1}{z_2} &= \frac{r_1(\cos\theta_1 + i\sin\theta_1)}{r_2(\cos\theta_2 + i\sin\theta_2)} \\
&= \frac{r_1}{r_2}\frac{(\cos\theta_1 + i\sin\theta_1)(\cos\theta_2 - i\sin\theta_2)}{\cos^2\theta_2 + \sin^2\theta_2} \\
&= \frac{r_1}{r_2}\frac{\cos(\theta_1 - \theta_2) + i\sin(\theta_1 - \theta_2)}{1}
\end{aligned}$$

or, in abbreviated form,

$$\frac{r_1\underline{/\theta_1}}{r_2\underline{/\theta_2}} \equiv \frac{r_1}{r_2}\underline{/\theta_1 - \theta_2}. \tag{13}$$

Thus, to divide two complex numbers in polar form, divide the moduli and *subtract* the arguments.

The argument of a product, or quotient, of two complex numbers should always be given within the range of values $-180°$ to $+180°$. Note the following results

$$|z_1z_2| = |z_1||z_2|; \quad |z_1/z_2| = |z_1|/|z_2|; \quad z\bar{z} = |z|^2 = |\bar{z}|^2 = r^2.$$

Example (i) In polar form $(-1 + i)(1 + i\sqrt{3})$ may be written $(\sqrt{2}\underline{/135°})(2\underline{/60°})$ or $2\sqrt{2}\underline{/195°}$. Using the principal value for the argument this is $2\sqrt{2}\underline{/-165°}$.

Example (ii) Express $(-i)/(2 + 2i)(-3 + \sqrt{3}i)$ in polar form by two different methods.

Solution

(a)
$$\begin{aligned}
\frac{-i}{(2 + 2i)(-3 + \sqrt{3}i)} &= \frac{-i}{(-6 - 2\sqrt{3}) - i(6 - 2\sqrt{3})} \\
&= \frac{i}{\sqrt{6}[(\sqrt{6} + \sqrt{2}) + i(\sqrt{6} - \sqrt{2})]} \\
&= \frac{(\sqrt{6} - \sqrt{2}) + i(\sqrt{6} + \sqrt{2})}{16\sqrt{6}} \\
&\equiv \frac{\sqrt{16}\underline{/\tan^{-1}(\sqrt{6} + \sqrt{2})/(\sqrt{6} - \sqrt{2})}}{16\sqrt{6}} \\
&\equiv \frac{1}{4\sqrt{6}}\underline{/\tan^{-1}(2 + \sqrt{3})} \\
&\equiv \frac{1}{4\sqrt{6}}\underline{/75°}.
\end{aligned}$$

(b) $\dfrac{-i}{(2 + 2i)(-3 + \sqrt{3}i)} \equiv \dfrac{1\underline{/-\pi/2}}{(2\sqrt{2}\underline{/\pi/4})(2\sqrt{3}\underline{/5\pi/6})}$

$$\equiv \dfrac{1}{4\sqrt{6}} \underline{/-(\pi/2 + \pi/4 + 5\pi/6)}$$

$$\equiv \dfrac{1}{4\sqrt{6}} \underline{/-19\pi/12}$$

$$\equiv \dfrac{1}{4\sqrt{6}} \underline{/+5\pi/12} \quad \text{or} \quad \dfrac{1}{4\sqrt{6}} \underline{/75°}.$$

10.3.3 Representation of multiplication and division on the Argand diagram

Given a complex number z_1 represented by the point P or the vector $\overline{OP}$ in the Argand diagram, an examination of equation (12) shows that the effect of multiplying by z_2 is to rotate the vector $\overline{OP}$ through an angle arg z_2 in the positive (anticlockwise) direction and to change the magnitude of $\overline{OP}$ in the ratio $r_1 r_2 : r_1$, i.e. $r_2 : 1$ or $|z_2| : 1$. Similarly from (13) the effect of dividing z_1 by z_2 is to rotate the vector $\overline{OP}$ through an angle arg z_2 in the negative (clockwise) direction and to change the magnitude of $\overline{OP}$ in the ratio $1 : |z_2|$.

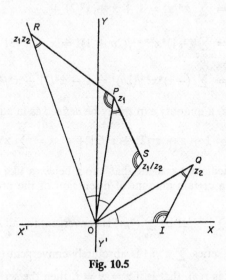

Fig. 10.5

In Fig. 10.5, P represents z_1, Q represents z_2, and I represents the point $1 + i0$. Then R will represent the point $z_1 z_2$ and S will represent the point z_1/z_2 where R and S are such that the triangles OQI, ORP, and OPS are similar so that

$$\frac{OR}{OP} = \frac{OQ}{1} \quad \text{or} \quad OR = |z_1||z_2|,$$

$$\frac{OS}{OP} = \frac{1}{OQ} \quad \text{or} \quad OS = |z_1|/|z_2|,$$

and $\underline{/ROP} = \underline{/POS} = \underline{/QOI} = \arg z_2.$

10.3.4 Inequalities

Complex numbers cannot be ordered in the same way as real numbers so that a statement such as $z_1 > z_2$ or $x_1 + iy_1 > x_2 + iy_2$ is meaningless. However, numbers can be partially ordered by the size of their moduli, e.g. $|3 + 2i| > |1 - i|$. From Fig. 10.3, it is evident that

$$|z_1 + z_2| \leqslant |z_1| + |z_2|$$
$$|z_1 - z_2| \leqslant |z_1| + |z_2|$$
$$|z_1 - z_2| \geqslant ||z_1| - |z_2||.$$

These results should be compared with those of example (ii) of section 9.3.3.

10.4 COMPLEX EXPONENTIAL QUANTITY

In Chapters 4 and 5 it was found that for all real x

$$\exp x = e^x = \sum_{n=0}^{\infty} x^n/n! = 1 + x + x^2/2! + \dots \tag{14}$$

$$\sin x = \sum_{n=0}^{\infty} (-1)^n x^{2n+1}/(2n+1)! = x - x^3/3! + x^5/5! + \dots \tag{15}$$

and $\cos x = \sum_{n=0}^{\infty} (-1)^n x^{2n}/(2n)! = 1 - x^2/2! + x^4/4! + \dots \tag{16}$

For complex z, a quantity $\exp z$ may be *defined* as in equation (14), i.e.

$$\exp z = 1 + z + z^2/2! + z^3/3! + \dots = \sum_{n=0}^{\infty} z^n/n!. \tag{17}$$

It is not immediately evident that $\exp z$ behaves like an exponential quantity and a certain amount of discussion of the properties of (17) is required.

The series $\sum_{n=0}^{\infty} |z|^n/n!$ or $\sum_{n=0}^{\infty} r^n/n!$ is convergent for all r, i.e. for all $|z|$, and hence the series $\sum_{n=0}^{\infty} z^n/n!$ is absolutely convergent (see Chapter 3). Note that if z is real, that is if $z = x = r$, then the series converges to e^x, i.e.

$$\exp z = \exp x = e^x.$$

If z is replaced by iy then (17) gives

$$\exp (iy) = 1 + (iy) + (iy)^2/2! + (iy)^3/3! \ldots$$
$$= (1 - y^2/2! + y^4/4! - \ldots) + i(y - y^3/3! + y^5/5! - \ldots)$$
$$= \cos y + i \sin y \tag{18}$$

using (15) and (16).

To establish the result

$$\exp (z_1 + z_2) = \exp (z_1) \exp (z_2) \tag{19}$$

requires a theorem concerning the multiplication of absolutely convergent series the proof of which cannot be given here. The theorem states that if $S_1 = \sum_{n=0}^{\infty} a_n$ and $S_2 = \sum_{n=0}^{\infty} b_n$ are two absolutely convergent series, then

$$S_1 S_2 = \left(\sum_{n=0}^{\infty} a_n \right) \left(\sum_{n=0}^{\infty} b_n \right)$$
$$= \sum c_n$$

where $c_n = \sum_{t=0}^{\infty} a_t b_{n-t}$.

To form the result (19), let $S_1 = \exp z_1$ and $S_2 = \exp z_2$ so that

$$a_n = z_1^n/n!$$

and
$$b_n = z_2^n/n!,$$

then
$$c_n = \sum_{t=0}^{\infty} [z_1^t/t!][z_2^{n-t}/(n-t)!]$$

$$= (1/n!) \sum_{t=0}^{\infty} \binom{n}{t} z_1^t z_2^{n-t}$$

$$= (z_1 + z_2)^n/n!$$

the last line following from the binomial theorem. Hence by the quoted theorem

$$\left(\sum_{0}^{\infty} z_1^n/n! \right) \left(\sum_{0}^{\infty} z_2^n/n! \right) = \sum_{0}^{\infty} (z_1 + z_2)^n/n!$$

and the result (19) follows at once.

Using (19)

$$\exp z = \exp (x + iy)$$
$$= \exp x \exp (iy)$$
$$= e^x (\cos y + i \sin y) \tag{20}$$

and $\exp z$ is a complex number of modulus e^x and argument y. The complex number $\exp z$ of (17), the sum of the series $\sum_{0}^{\infty} z^n/n!$, has the properties of an exponential quantity and will be designated e^z.

Thus (19) may be written

$$e^{z_1 + z_2} = e^{z_1} e^{z_2} \tag{21}$$

and (20) becomes

$$\begin{aligned}
e^z = \exp z &= e^x (\cos y + i \sin y) \\
&= e^x \exp (iy) \\
&= e^x e^{iy}
\end{aligned} \tag{22}$$

using (18) in the form

$$e^{iy} = \cos y + i \sin y. \tag{23}$$

The equation (19) may be extended to read

$$e^{z_1 + z_2 + z_3 + \cdots + z_n} = e^{z_1} e^{z_2} e^{z_3} \cdots e^{z_n}. \tag{24}$$

The polar form of the complex number may now be written

$$z = x + iy = re^{i\theta} = r(\cos \theta + i \sin \theta). \tag{25}$$

The exponential form of the complex number is called *Euler's form* and θ is now expressed in radian measure.

Example $i = e^{i\pi/2}$; $-1 = e^{i\pi}$; $1 = e^{i0} = e^{2i\pi} = e^{2ki\pi}$, k an integer;
$$\sqrt{3} - i = 2e^{-i\pi/6}; \quad -1 + i = \sqrt{2}e^{3i\pi/4};$$
$$2 + i = \sqrt{5}e^{i\theta}, \quad \theta = \tan^{-1}(\tfrac{1}{2}) = 0.46.$$

Unless stated otherwise the argument θ is always expressed in principal value form, i.e. $-\pi < \theta \leqslant +\pi$. It should be noted, however, that given an argument θ, then

$$e^{i\theta} = e^{(i\theta + 2ik\pi)}, \quad k \text{ an integer.} \tag{26}$$

10.5 DE MOIVRE'S THEOREM

This theorem states that for all n

$$(\cos \theta + i \sin \theta)^n = \cos n\theta + i \sin n\theta. \tag{27}$$

There are three cases to consider:

(i) n a positive integer,
(ii) n a negative integer, and
(iii) n rational but not an integer.

Case (i)
From (25) of section 10.4,

$$\cos \theta + i \sin \theta = e^{i\theta}$$

and so
$$\begin{aligned}
(\cos \theta + i \sin \theta)^n &= (e^{i\theta})^n \\
&= (e^{i\theta})(e^{i\theta})(e^{i\theta}) \ldots (e^{i\theta}) \\
&= e^{in\theta}
\end{aligned}$$

using (24) of section 10.4. Hence,

$$(\cos \theta + i \sin \theta)^n = \cos n\theta + i \sin n\theta. \tag{28}$$

Case (ii)

Let $n = -m$ where m is a positive integer and write

$$(\cos \theta + i \sin \theta)^n = (e^{i\theta})^n$$
$$= (e^{i\theta})^{-m}$$
$$= 1/(e^{i\theta})^m$$
$$= 1/e^{im\theta}$$

using equation (24) of section 10.4.
From equation (21) of section 10.4 with $z_2 = -z_1$

$$1 = e^0 = e^{z_1} e^{-z_1}$$

or
$$e^{-z_1} = 1/(e^{z_1}). \tag{29}$$

Hence,
$$(\cos \theta + i \sin \theta)^n = 1/(e^{im\theta})$$
$$= e^{-im\theta}$$
$$= e^{in\theta}$$
$$= \cos n\theta + i \sin n\theta. \tag{30}$$

Case (iii)

Let $n = p/q$ where p and q are integers without a common factor (other than unity), then

$$(\cos \theta + i \sin \theta)^n = (e^{i\theta})^n$$
$$= (e^{i\theta})^{p/q}$$
$$= (e^{ip\theta})^{1/q}, \text{ since } p \text{ is an integer}$$
$$= [(e^{ip\theta/q})^q]^{1/q}, \text{ since } q \text{ is an integer}$$
$$= (e^{ip\theta/q})$$
$$= e^{in\theta}$$
$$= (\cos n\theta + i \sin n\theta). \tag{31}$$

Thus, for all n

$$(e^{i\theta})^n = e^{in\theta} \tag{32}$$

which is Euler's form of de Moivres theorem. Note that if n is a rational fraction p/q then $e^{ip\theta/q}$ is only one of the possible values of $(e^{i\theta})^{p/q}$. In general terms, from (26) of section 10.4

$$e^{ip\theta/q} = (e^{i\theta + 2ik\pi})^{p/q} \tag{33}$$

Example (i) $(1 + i)^{27} = (\sqrt{2} \, e^{i\pi/4})^{27}$
$$= 2^{27/2} \, e^{27i\pi/4}$$
$$= 2^{27/2} \, e^{6i\pi + 3i\pi/4}$$
$$= 2^{27/2} (\cos 3\pi/4 + i \sin 3\pi/4)$$
$$= 2^{13} (-1 + i).$$

Example (ii) Find $\cos 4\theta$ and $\sin 4\theta$ in terms of powers of $\cos \theta$ and $\sin \theta$ respectively.

Solution Since $(\cos \theta + i \sin \theta)^4 = \cos 4\theta + i \sin 4\theta$ then

$$\cos 4\theta + i \sin 4\theta = \cos^4 \theta + 4i \cos^3 \theta \sin \theta$$
$$- 6 \cos^2 \theta \sin^2 \theta - 4i \cos \theta \sin^3 \theta + \sin^4 \theta$$

after using the binomial expansion.

Equating real and imaginary parts (for equality of complex numbers),

$$\cos 4\theta = \cos^4 \theta - 6 \cos^2 \theta \sin^2 \theta + \sin^4 \theta$$
$$= 1 - 8 \cos^2 \theta + 8 \cos^4 \theta$$

and $\qquad \sin 4\theta = 4 \cos^3 \theta \sin \theta - 4 \cos \theta \sin^3 \theta$
$$= 4 \sin \theta \cos \theta (\cos^2 \theta - \sin^2 \theta).$$

10.5.1 The *n*th roots of unity

From equation (26) of section 10.4, with $\theta = 0$, $1 = e^{2ik\pi}$ where k is an integer $k = 0, 1, 2, 3, \ldots$ Suppose the *n*th roots are denoted by $z_0, z_1, z_2, \ldots$ then

$$z_k = \sqrt[n]{1} = (e^{2ik\pi})^{1/n}$$
$$= e^{2ik\pi/n} \qquad \text{from equation (32),}$$
$$= \cos (2k\pi/n) + i \sin (2k\pi/n) \qquad (34)$$

The expression (34) has n distinct values with argument in the range $(-\pi, +\pi)$, for any n successive values of the integer k. For $k = 0, 1, 2, \ldots, (n - 1)$ the successive roots are

$$z_0 = e^0 = 1; \quad z_1 = e^{2i\pi/n}; \quad z_2 = e^{4i\pi/n}; \ldots;$$
$$z_{n-1} = e^{2(n-1)i\pi/n} = e^{-2i\pi/n}.$$

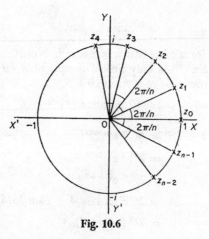

Fig. 10.6

The other values of k give values of z_k which simply repeat those already found, for example $k = n$ and $k = n + 1$ give $z_n = e^{2in\pi/n} = e^{2i\pi}$ $= 1 = z_0$ and $z_{n+1} = e^{2(n+1)i\pi/n} = e^{2i\pi}e^{2i\pi/n} = e^{2i\pi/n} = z_1$. Thus it will be seen that there are n values for the nth roots of unity, no more and no less. Put another way, it follows that the equation $z^n = 1$ has n distinct roots when n is a positive integer. If n is even, two of these roots are the real roots ± 1 and the rest are all complex; if n is odd then one of these roots is real, i.e. $+1$, and the rest are complex.

The roots given by (34) may be conveniently illustrated by plotting on the Argand diagram. The first root is the point $1 + i0$ and the others are distributed on the circumference of the unit circle, (i.e. the circle of radius 1, centre $z = 0$) at equal angular intervals of $2\pi/n$ as shown in Fig. 10.6.

Example (i) Find the fourth roots of unity.

Solution Here,

$$z_k = \sqrt[4]{1} = 1e^{2ik\pi/4} = 1e^{ik\pi/2}$$

and the roots may be represented in abbreviated form by

$$1\underline{/0}; \quad 1\underline{/\pi/2}; \quad 1\underline{/\pi}; \quad 1\underline{/3\pi/2}$$

or $\qquad 1 + i0; \quad 0 + i; \quad -1 + i0; \quad \text{and} \quad 0 - i.$

Example (ii) Find the fifth roots of unity.

Solution The roots are given by

$$z_k = 1e^{2ik\pi/5}, \quad k = 0, 1, 2, 3, 4$$

and may be written

$$1\underline{/0} = 1; \quad 1\underline{/2\pi/5} = 1\underline{/72°}; \quad 1\underline{/144°}; \quad 1\underline{/216°}; \quad 1\underline{/288°}.$$

These roots are plotted on the Argand diagram in Fig. 10.7.

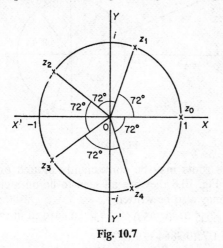

Fig. 10.7

10.5.2 The nth roots of a complex number

If a complex number is expressed in the Euler form then the roots may be found by the method of section 10.5.1. Suppose $a = re^{i\theta}$ or $re^{i(\theta + 2k\pi)}$ where $r = |a|$ and $\theta = \arg a$ and k is an integer, then

$$\sqrt[n]{a} = a^{1/n}$$
$$= (r^{1/n})e^{i(\theta + 2k\pi)/n}.$$

The roots may then be represented in the form $(\sqrt[n]{r})\underline{/(\theta + 2k\pi)/n}$. In the Argand diagram the roots all lie on the circle of radius $\sqrt[n]{r}$, the first root (for $k = 0$) has the argument (θ/n) and the other $(n - 1)$ roots are distributed symmetrically around this circle at angular intervals of $(2\pi/n)$.

Example Find the cube roots of $-5 + 2i$.

Solution Write

$$-5 + 2i \equiv (29)^{\frac{1}{2}} \operatorname{cis} \theta$$

where

$$\tan \theta = -\tfrac{2}{5}$$

and

$$\theta = 180° - 21° 48'$$
$$= 158° 12'.$$

The three cube roots are then

$$z_k \equiv 29^{\frac{1}{6}} \operatorname{cis} (158° 12' + k360°)/3, \qquad k = 0, 1, 2$$
$$\equiv 29^{\frac{1}{6}} \operatorname{cis} (52° 44' + k120°)$$
$$\equiv 1\cdot753\underline{/52° 44' + k120°}$$

i.e. $1\cdot75\underline{/52° 44'}$; $\quad 1\cdot75\underline{/172° 44'}$; $\quad 1\cdot75\underline{/292° 44'}$.

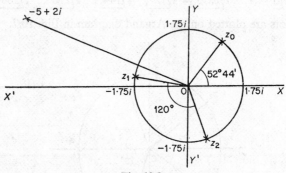

Fig. 10.8

In this form the roots may be conveniently plotted on the Argand diagram and in Fig. 10.8 they are all seen to lie on a circle of radius $1\cdot75$. The roots may also be written

$$1\cdot75(0\cdot6055 + i0\cdot7959); \quad 1\cdot75(-0\cdot9919 + i0\cdot1265)$$

and

$$1\cdot75(0\cdot3864 - i0\cdot9223).$$

10.5.3 The roots of an algebraic equation

The roots found in section 10.5.2 are the roots of the equation $z^n = a$. It is seen that there are n roots. It is a fundamental result in complex algebra that any equation of the form

$$f(z) \equiv z^n + b_1 z^{n-1} + b_2 z^{n-2} + \ldots + b_n = 0 \tag{35}$$

where n is a positive integer, has exactly n roots, any multiple root being counted the requisite number of times. The equation $z^n = a$ is simply a particular case of (35).

Suppose that $z = \alpha + i\beta$ is a root of (35) then

$$f(z) \equiv f(\alpha + i\beta) \equiv (\alpha + i\beta)^n + b_1(\alpha + i\beta)^{n-1} + \ldots + b_n = 0$$

and, providing that the b_r are real, evaluation of the binomial expressions leads to

$$\phi(\alpha, \beta) + i\psi(\alpha, \beta) = 0 \tag{36}$$

where ϕ and ψ are known functions.

Equation (36) is satisfied by

$$\phi(\alpha, \beta) = 0 = \psi(\alpha, \beta)$$

and hence it follows that

$$\phi(\alpha, \beta) - i\psi(\alpha, \beta) = 0. \tag{37}$$

Equation (37) shows that $\alpha - i\beta$ is also a root of

$$f(z) = 0$$

and that complex roots of an algebraic equation with real coefficients occur in conjugate pairs. This result can be usefully employed when solving such equations.

Example (i) The roots of the quadratic equation $ax^2 + bx + c = 0$ with $4ac - b^2 < 0$ are given by

$$x = \alpha + i\beta \quad \text{and} \quad x = \alpha - i\beta$$

where $\alpha = -b/2a$ and $\beta = \sqrt{(4ac - b^2)}/2a$.

Example (ii) Solve the equation

$$z^4 - 4z^2 + 8z + 35 = 0$$

given that $2 + i\sqrt{3}$ is one root.

Solution Two of the roots are $2 + i\sqrt{3}$ and $2 - i\sqrt{3}$ and hence $(z - 2 + i\sqrt{3})(z - 2 - i\sqrt{3})$ is a factor of $z^4 - 4z^2 + 8z + 35$, i,e, $z^2 - 4z + 7$ is a factor.

By inspection,

$$(z^2 - 4z + 7)(z^2 + 4z + 5) = 0$$

and the other two roots are given by

$$z = -2 \pm \sqrt{(4 - 5)} = -2 \pm i.$$

It is important to note that the proviso that the coefficients b_r are real is necessary for the derivation of (37). When the b_r are complex the roots of (35), while still n in number, do not occur in complex conjugate pairs. For example, the roots of

$$z^2 + 2iz + 3 = 0$$

are given by $z = i$ and $z = -3i$.

10.6 TRIGONOMETRIC AND HYPERBOLIC FUNCTIONS

From equation (23) of section 10.4

$$e^{i\theta} = \cos \theta + i \sin \theta$$

and
$$e^{-i\theta} = \cos \theta - i \sin \theta.$$

Adding and subtracting leads to

$$\cos \theta = \frac{e^{i\theta} + e^{-i\theta}}{2} \tag{38}$$

and
$$\sin \theta = \frac{e^{i\theta} - e^{-i\theta}}{2i}. \tag{39}$$

Recalling the definition of the hyperbolic functions of section 4.5.1

$$\cosh \phi = \frac{e^{\phi} + e^{-\phi}}{2} \tag{40}$$

and
$$\sinh \phi = \frac{e^{\phi} - e^{-\phi}}{2} \tag{41}$$

and replacing ϕ by $i\theta$ gives two important relationships,

$$\left.\begin{array}{l} \cosh i\theta = \dfrac{e^{i\theta} + e^{-i\theta}}{2} = \cos \theta \\[2mm] \sinh i\theta = \dfrac{e^{i\theta} - e^{-i\theta}}{2} = i \sin \theta \end{array}\right\} \tag{42}$$

Similarly, replacing θ in equations (38) and (39) by $i\theta$ gives two other relationships

$$\left.\begin{array}{l} \cos i\theta = \dfrac{e^{-\theta} + e^{\theta}}{2} = \cosh \theta \\[2mm] \sin i\theta = \dfrac{e^{-\theta} - e^{\theta}}{2i} = i \sinh \theta \end{array}\right\} \tag{43}$$

To each trigonometric identity in θ there corresponds an equivalent hyperbolic identity obtained by replacing θ by $i\theta$ and using equations (43).

Example (i) $\cos(A + B) = \cos A \cos B - \sin A \sin B$

and so $\cos(iA + iB) = \cos iA \cos iB - \sin iA \sin iB.$

Using equations (43),

$$\cosh(A + B) = \cosh A \cosh B - i^2 \sinh A \sinh B$$
$$= \cosh A \cosh B + \sinh A \sinh B.$$

Example (ii) Since

$$\sin 2\theta = 2 \sin \theta \cos \theta$$

then, using equations (43),

$$i \sinh 2\theta = 2i \sinh \theta \cosh \theta$$
or $$\sinh 2\theta = 2 \sinh \theta \cosh \theta.$$

10.7 FUNCTIONS OF A COMPLEX VARIABLE

The complex quantity z or $x + iy$ has, so far, been considered fixed, e.g. as $3 + 2i$, and represented on the Argand diagram by a fixed point. If, however, x and y are permitted to take on different values then z will take up different positions in the z-plane and is then called a *complex variable*.

For example, if $z = x + iy$ and $|z| = 1$ then z is represented by any point on the circle of unit radius, centre the origin, i.e. on the unit circle on which x and y vary according to the rule $x^2 + y^2 = 1$.

Instead of z being constrained to lie on some particular curve of the z-plane it is often required that z takes a position interior (or exterior) to a curve, i.e. a position within some region of the z-plane. For example, if $z = x + iy$ and $|z| < 1$ then the complex variable z is represented by any point interior to the unit circle $x^2 + y^2 = 1$ of the Argand diagram.

Any expression that contains the variable z, such as z^2, $1/z$, e^z and takes a single value for any given value of z is called a *function of the complex variable z*. This function is usually abbreviated to $f(z)$ or $w(z)$, i.e.

$$w = f(z) = f(x + iy). \tag{44}$$

Example (i) If $w = z^2$ then

$$w = (x + iy)^2$$
$$= x^2 - y^2 + 2ixy.$$

Example (ii) If $w = 1/z$ then

$$w = 1/(x + iy)$$
$$= (x - iy)/(x^2 + y^2).$$

These two examples show that a function of a complex variable is itself a complex variable, i.e.

$$w = u + iv = f(x + iy) \tag{45}$$

and $u = u(x, y)$, $v = v(x, y)$.

10.7.1 Mapping function

If z is represented by a point on the Argand diagram then w will be represented by another point of the same Argand diagram. For example if $z = 1 + i$ then the result above shows that $w = z^2$ is represented by the point $0 + 2i$. However, it is usual to indicate the point z on one Argand diagram, i.e. the z-plane, and to indicate the corresponding point w on a second Argand diagram, i.e. the w-plane with axes of u and v. To each point of the z-plane there corresponds one point on the w-plane. As z covers all points on some curve L_1 or within some region R_1 of the z-plane, the points w cover corresponding points on some curve L_2 or some region R_2 of the w-plane. The functional relationship $w = f(z)$ is said to *map* or to *transform* the curve L_1, or the region R_1, into the curve L_2, or the region R_2, and is called a *mapping* or *transformation function*.

Example (i) Consider the mapping function $w = z^2$ and the imaginary axis $x = 0$, $0 < y < \infty$.

Solution The curve L_1, the line $x = 0$, $0 < y < \infty$, is mapped into the curve L_2 given by

$$u + iv = (x^2 - y^2) + 2ixy, \quad \text{with} \quad x = 0, 0 < y < \infty$$

or

$$u = -y^2; \quad v = 0; \quad 0 < y < +\infty.$$

Thus, as z moves outwards from the origin along the positive imaginary axis, w moves outwards along the negative real axis.

Example (ii) Examine the region into which the first quadrant of the z-plane is mapped by the function $w = z^2$.

Solution Let $z = re^{i\theta}$ and $w = \rho e^{i\phi}$,

then

$$\rho e^{i\phi} = r^2 e^{2i\theta}$$

i.e.

$$\left. \begin{array}{l} \rho = r^2 \\ \phi = 2\theta \end{array} \right\}.$$

Figure 10.9 shows the region of the mapping, the first quadrant of the z-plane is specified by $0 < \theta < \pi/2$, $0 < r < \infty$ and the corresponding region of the w-plane by $0 < \phi < \pi$, $0 < \rho < \infty$. The region of the w-plane is the upper half of the w-plane.

Example (iii) Find the region into which the semi-infinite strip $0 \leqslant y \leqslant \pi$, $-\infty < x \leqslant 0$ of the z-plane is mapped by the mapping function $w = e^z$.

Solution In this case

$$w = \rho e^{i\phi} = e^z = e^{x+iy}$$

i.e.
$$\left.\begin{array}{l} \rho = e^x \\ \phi = y \end{array}\right\}.$$

In Fig. 10.10 the semi-infinite strip is represented by the letters A, B, C, D.

Along AB; $\qquad y = 0,\quad -\infty < x \leqslant 0$

hence $\qquad\qquad \phi = 0,\quad 0 < \rho \leqslant 1,$

which is represented by $A_1 B_1$ in the w-plane.

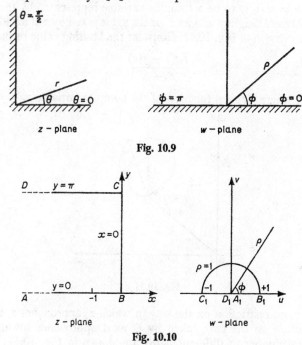

Fig. 10.9

Fig. 10.10

Along BC; $\qquad x = 0,\quad 0 \leqslant y \leqslant \pi$

hence $\qquad\qquad \rho = 1,\quad 0 \leqslant \phi \leqslant \pi,$

which is represented by the arc $B_1 C_1$.

Along CD; $\qquad y = \pi,\quad 0 \geqslant x > -\infty$

hence $\qquad\qquad \phi = \pi,\quad 1 \geqslant \rho > 0$

which is represented by $C_1 D_1$.

At an interior point of the semi-infinite strip, $x < 0$ and hence $\rho = e^x < 1$.

Thus the interior of the semi-infinite strip $0 \leqslant y \leqslant \pi$, $-\infty < x \leqslant 0$ is mapped into the region $0 \leqslant \phi \leqslant \pi$, $0 < \rho \leqslant 1$, i.e. into the interior of the upper half of the unit circle of the w-plane. It may be noted that the boundary $ABCD$ is mapped into the boundary $A_1B_1C_1D_1$ and that the points A and D representing points at infinity in the z-plane map into coincident points at the origin of the w-plane.

The subject of mapping has many applications in engineering wherever a two-dimensional flow problem exists, notably in the fields of electricity, electrostatics, fluid dynamics, and elasticity.

10.8 DIFFERENTIATION OF A COMPLEX FUNCTION

Suppose $z = x + iy$ to be a complex variable represented by the point P of the Argand diagram and let z' be the value taken by z at an adjacent point Q as shown in Fig. 10.11. Consider the limiting value of the ratio

$$\frac{f(z') - f(z)}{z' - z} \tag{46}$$

as $z' \to z$, where $f(z)$ is a function of the complex variable z.

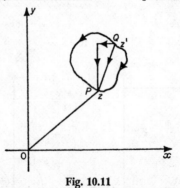

Fig. 10.11

There is no restriction on the way in which z' approaches z, that is, no restriction on the path taken by Q as it approaches towards the point P (a number of different paths are shown in Fig. 10.11) and it would be expected that the limit of (46) (assuming it to exist) would be different for different paths. In general this is true but it is a remarkable fact that there exists a class of functions for which the limit of (46) is unique, i.e. is the same for all paths taken from Q to P. Such functions are called *analytic functions* and it is not altogether surprising that the theory of analytic functions has been extensively studied and has many applications in both theoretical and practical engineering.

10.8.1 The derivative of $f(z)$

When the limit of (46) exists and is unique it is denoted by $f'(z)$ or df/dz and is called the derivative of $f(z)$ with respect to z.

Thus,

$$\frac{df}{dz} = f'(z) = \lim_{\Delta z \to 0} \frac{f(z + \Delta z) - f(z)}{\Delta z} \tag{47}$$

where $\Delta z = z' - z$.

Certain conditions must be satisfied by $f(z)$ in order that the limit in equation (47) is unique. Let $f(z) = u(x, y) + iv(x, y)$, then equation (47) may be written

$$\frac{df}{dz} = f'(z) = \lim_{\substack{\Delta x \to 0 \\ \Delta y \to 0}} \left[\frac{u(x + \Delta x, y + \Delta y) - u(x, y)}{\Delta x + i\Delta y} \right]$$
$$+ \lim_{\substack{\Delta x \to 0 \\ \Delta y \to 0}} i \left[\frac{v(x + \Delta x, y + \Delta y) - v(x, y)}{\Delta x + i\Delta y} \right]. \tag{48}$$

It is possible to examine the limiting process as $z' \to z$ in a general manner but it is sufficient to consider two particular cases. For the first let $\Delta y = 0$ (i.e. let $z' - z$ be real) and for the second let $\Delta x = 0$ (i.e. let $z' - z$ be imaginary). When $\Delta y = 0$, equation (48) gives

$$f'(z) = \lim_{\Delta x \to 0} \left[\frac{u(x + \Delta x, y) - u(x, y)}{\Delta x} \right]$$
$$+ i \left[\lim_{\Delta x \to 0} \frac{v(x + \Delta x, y) - v(x, y)}{\Delta x} \right]$$
$$= \frac{\partial u}{\partial x} + i \frac{\partial v}{\partial x}, \tag{49}$$

and when $\Delta x = 0$, equation (48) gives

$$f'(z) = \lim_{\Delta y \to 0} \left[\frac{u(x, y + \Delta y) - u(x, y)}{i\Delta y} \right]$$
$$+ \lim_{\Delta y \to 0} \left[\frac{v(x, y + \Delta y) - v(x, y)}{\Delta y} \right]$$
$$= \frac{1}{i} \frac{\partial u}{\partial y} + \frac{\partial v}{\partial y}$$

i.e.

$$f'(z) = \frac{\partial v}{\partial y} - i \frac{\partial u}{\partial y} \tag{50}$$

it being assumed that the partial derivatives exist. The condition that $f'(z)$ has the same value in equations (49) and (50) leads to

$$\frac{\partial u}{\partial x} = \frac{\partial v}{\partial y} \quad \text{and} \quad \frac{\partial u}{\partial y} = -\frac{\partial v}{\partial x}. \tag{51}$$

showing that u and v, the real and imaginary parts of $f(z)$, are connected by two partial differential equations. These equations are called the *Cauchy-Riemann equations*.

Because the equations (51) have been obtained by considering two particular limiting processes it is only to be expected that these are *necessary* conditions, i.e. *if* a function is analytic *then* the conditions (51) are satisfied. However, it may be stated, without proof, that providing the four partial derivatives exist throughout a region and are continuous at all points of the region then the conditions (51) are also *sufficient*, i.e. *if* the continuous derivatives satisfy (51) *then* the function $f(z)$ is analytic.

Example (i) If $f(z) = z^2$ then $u = x^2 - y^2$, $v = 2xy$ and

$$\frac{\partial u}{\partial x} = 2x, \quad \frac{\partial v}{\partial y} = 2x$$

$$\frac{\partial u}{\partial y} = -2y, \quad \frac{\partial v}{\partial x} = 2y.$$

The derivatives exist and are continuous everywhere, the Cauchy-Riemann conditions are satisfied everywhere and hence $f(z) = z^2$ is analytic everywhere.

Example (ii) If $f(z) = 1/z$ then $u = x/(x^2 + y^2)$, $v = -y/(x^2 + y^2)$ and

$$\frac{\partial u}{\partial x} = \frac{y^2 - x^2}{(x^2 + y^2)^2}, \quad \frac{\partial v}{\partial y} = \frac{y^2 - x^2}{(x^2 + y^2)^2}$$

$$\frac{\partial u}{\partial y} = \frac{-2xy}{(x^2 + y^2)^2}, \quad \frac{\partial v}{\partial x} = \frac{+2xy}{(x^2 + y^2)^2}.$$

The derivatives exist and are continuous everywhere except at the point $x = 0 = y$. The Cauchy-Riemann conditions are satisfied everywhere except at this point. The function $1/z$ is therefore analytic everywhere except at the point $z = 0$.

Example (iii) If $f(z) = \bar{z} = x - iy$ then $u = x$, $v = -y$ and

$$\frac{\partial u}{\partial x} = 1, \quad \frac{\partial v}{\partial y} = -1$$

$$\frac{\partial u}{\partial y} = 0, \quad \frac{\partial v}{\partial x} = 0.$$

The derivatives exist and are continuous everywhere but the Cauchy-Riemann conditions are not true anywhere. The function $\bar{z}$ is not analytic.

Example (iv) If $f(z) = |z|^2 = x^2 + y^2$ then $u = x^2 + y^2$, $v = 0$ and

$$\frac{\partial u}{\partial x} = 2x, \quad \frac{\partial v}{\partial y} = 0$$

$$\frac{\partial u}{\partial y} = 2y, \quad \frac{\partial v}{\partial x} = 0.$$

The derivatives exist and are continuous everywhere but the Cauchy-Riemann conditions are satisfied only at the origin $x = 0 = y$. The function $|z|^2$ is analytic only at the point $z = 0$.

10.8.2 Harmonic functions

If a function $f(z)$ is analytic within a region it is an important fact that $f(z)$ then has derivatives *of all orders*. The proof of this statement is beyond the scope of this book but, in particular, the second order derivatives exist and hence from (51),

$$\frac{\partial^2 u}{\partial x^2} = \frac{\partial^2 v}{\partial x \, \partial y} = \frac{\partial^2 v}{\partial y \, \partial x} = -\frac{\partial^2 u}{\partial y^2}$$

i.e.
$$\frac{\partial^2 u}{\partial x^2} + \frac{\partial^2 u}{\partial y^2} = 0 \tag{52}$$

and similarly

$$\frac{\partial^2 v}{\partial x^2} + \frac{\partial^2 v}{\partial y^2} = 0. \tag{53}$$

The functions u and v and hence $u + iv$ are solutions of the second order partial differential equation

$$\frac{\partial^2 V}{\partial x^2} + \frac{\partial^2 V}{\partial y^2} = 0. \tag{54}$$

The equation (54) is the two-dimensional form of a famous equation called *Laplace's equation* and any solution is called an *harmonic function*. When $u + iv$ is analytic the two functions u and v are called *harmonic conjugate functions* (note that this is a different use of the word conjugate to that used in section 10.2.3).

Many problems in engineering can be reduced to the mathematical problem of finding a function that satisfies the two-dimensional Laplace equation within some region and satisfies certain conditions on the boundary of the region. In view of equations (52) and (53) it may be anticipated that the solution of such problems can be facilitated by the use of complex variable analysis.

Using the notation of Section 9.12

$$\nabla = \frac{\partial}{\partial x}\mathbf{i} + \frac{\partial}{\partial y}\mathbf{j}$$

and
$$\nabla \cdot \nabla V = \left(\frac{\partial}{\partial x}\mathbf{i} + \frac{\partial}{\partial y}\mathbf{j}\right) \cdot \left(\frac{\partial V}{\partial x}\mathbf{i} + \frac{\partial V}{\partial y}\mathbf{j}\right)$$

$$= \frac{\partial^2 V}{\partial x^2} + \frac{\partial^2 V}{\partial y^2}$$

Following the usual vector notation $\nabla \cdot \nabla$ is written ∇^2 and hence Laplace's equation has the alternative operational form

$$\nabla^2 V = 0. \tag{55}$$

Example (i) Show that the function $u = \cosh 2x \cos 2y$ is harmonic. Use the Cauchy-Riemann equations to deduce the harmonic conjugate function v that makes $w(x, y) = u(x, y) + iv(x, y)$ an analytic function and express w as a function of the complex variable $z = x + iy$.

Solution Differentiating,

$$\frac{\partial^2 u}{\partial x^2} = 4 \cosh 2x \cos 2y; \frac{\partial^2 u}{\partial y^2} = -4 \cosh 2x \cos 2y$$

and hence $\nabla^2 u = \dfrac{\partial^2 u}{\partial x^2} + \dfrac{\partial^2 u}{\partial y^2} = 0$ and u is an harmonic function. Now, from the first equation (51)

$$\frac{\partial v}{\partial y} = \frac{\partial u}{\partial x} = 2 \sinh 2x \cos 2y$$

and $\qquad\qquad v = \sinh 2x \sin 2y + g(x)$

where $g(x)$ is an arbitrary function of x only. The second equation (51), $\dfrac{\partial v}{\partial x} = -\dfrac{\partial u}{\partial y}$ leads to

$$2 \cosh 2x \sin 2y + g'(x) = 2 \cosh 2x \sin 2y$$

and it follows that

$$g'(x) = 0$$

i.e. $\qquad\qquad g(x) = C = \text{constant.}$

Thus $\qquad\qquad v = \sinh 2x \sin 2y + C$

where C is an arbitrary constant.
The analytic function $w = u + iv$ is given by

$$w = \cosh 2x \cos 2y + i \sinh 2x \sin 2y + iC$$
$$= \cosh (2x + i2y) + iC$$

i.e. $\qquad\qquad w = \cosh (2z) + iC$

Example (ii) Show that $u = y^3 - 2y - 3x^2y$ and $v = x^3 + 2x - 3xy^2 + C$ are harmonic conjugate functions.
    Express $w = u + iv$ in the form $f(z)$ where $z = x + iy$ and find C if $f(0) = 1$.

Solution Differentiating,

$$\partial^2 u/\partial x^2 = -6y; \partial^2 u/\partial y^2 = 6y$$

and $\qquad\qquad \partial^2 v/\partial x^2 = 6x; \partial^2 v/\partial y^2 = -6x$

and hence $\nabla^2 u = 0$, $\nabla^2 v = 0$, and u, v, are harmonic functions.

Also $\qquad\qquad \partial u/\partial x = -6xy = \partial v/\partial y$

and
$$\partial u/\partial y = 3y^2 - 2 - 3x^2 = -(-3y^2 + 2 + 3x^2) = -\partial v/\partial x$$

and u, v, are harmonic conjugate functions.

Further, $w = u + iv$
$$= y^3 - 2y - 3x^2y + ix^3 + 2ix - 3ixy^2 + iC$$
$$= f(x + iy).$$

Setting $y = 0$ in the last result gives
$$f(x) = ix^3 + 2ix + iC$$
and so $f(z) = iz^3 + 2iz + iC$.

With $f(0) = 1$, then $iC = 1$, i.e. $C = -i$, and finally
$$w = f(z) = i(z^3 + 2z) + 1.$$

EXERCISE 10

1 Express the following complex numbers in the form $a + ib$, where a and b are real:

  (i) $i(c + id)$,   (ii) $(1 + i)(2 - 3i)$,
  (iii) $(1 + 2i)/(1 - i)$,   (iv) $(1 + i)^3/(1 - i)$,
  (v) $1/(1 + 2i)^3$,   (vi) $(1 + i)(3 + 2i)/(2 - i)$.

2 Given that $z = x + iy$, find the value, in terms of x and y, and in the form $a + ib$, of (i) z^4; (ii) $(1 + z)/(1 - z)$; (iii) $z + 1/2z$.

3 In the relationship $(R + ipL)(S - i/pC) = P/Q$ all the quantities are real except i. Show that $p = \sqrt{(R/LSC)}$ and find R in terms of C, L, P, Q, and S.

4 Express in the form $re^{i\theta}$, giving the principal value of θ:

  (i) $+3$;   (ii) -3;   (iii) $1 + i$;   (iv) $i - 1$;   (v) $\sqrt{3}/2 - i/2$;
  (vi) $-8(i\sqrt{3} - 1)$;   (vii) $(3 + i)/(2i - 1)$;   (viii) $(1 + i)^9$;
  (ix) $(1 + e^{i\pi/3})(1 + e^{3i\pi/2})$.

5 Simplify $2/(\cos \pi/4 + i \sin \pi/4)$; $(\sqrt{3} - i)^{15}$.

6 Solve the equation $(x^2 + 1)^2 = 4(2x - 1)$ given that $2i - 1$ is one root.

7 Find all the solutions of $z^6 + 1 = 0$ and indicate their positions on the Argand diagram.

8 Find the cube roots of $8(i\sqrt{3} - 1)$.

9 Solve the equation $z^5 + 1 = 0$ and hence deduce that
$$z^5 + 1 = (z + 1)(z^2 - 2z \cos \pi/5 + 1)(z^2 - 2z \cos 3\pi/5 + 1).$$

10 Use de Moivre's theorem to show that
$$\cos 5\theta = 16 \cos^5 \theta - 20 \cos^3 \theta + 5 \cos \theta$$
$$\sin 5\theta = 16 \sin^5 \theta - 20 \sin^3 \theta + 5 \sin \theta.$$

11 Writing $z = \cos\theta + i\sin\theta$ show that $2\cos n\theta = z^n + z^{-n}$ and $2i\sin n\theta = z^n - z^{-n}$ where n is any integer. By expanding $(z - 1/z)^4(z + 1/z)^3$ deduce that

$$2^6 \sin^4\theta \cos^3\theta = \cos 7\theta - \cos 5\theta - 3\cos 3\theta + 3\cos\theta.$$

12 Derive the following identities from the equivalent trigonometrical identities

$$\cosh^2\phi - \sinh^2\phi = 1,$$
$$\cosh^2\phi + \sinh^2\phi = \cosh 2\phi$$
$$= 2\cosh^2\phi - 1$$
$$= 1 + 2\sinh^2\phi,$$
$$\tanh 2\phi = 2\tanh\phi/(1 + \tanh^2\phi).$$

13 Given that

$$\cos(x + iy) = \cos\theta + i\sin\theta$$

where x, y, and θ are real prove that

$$\cos 2x + \cosh 2y = 2.$$

14 Show that the solutions of $\cos z = -3$ are the solutions of the simultaneous equations

$$\cos x \cosh y = -3$$
$$\sin x \sinh y = 0.$$

Hence show that the solutions are

$$z = (2r + 1)\pi \pm i\cosh^{-1} 3.$$

15 Describe geometrically the path or region of the z-plane if

(i) $Re\, z = 2$ (ii) $Im\,(z + iz) = 0$ (iii) $0 < \arg z < \pi/3$
(iv) $|z| < 1$ (v) $|z - 3| = 2$ (vi) $|z + 3i| > 4$
(vii) $z\bar{z} = 2$ (viii) $|z/(z - 1)| = \sqrt{2}$
(ix) $1 \leqslant |z| \leqslant 2,\ \pi/4 < \arg z \leqslant \pi/2$.

16 Test whether the following functions are analytic and find $f'(z)$ where possible, stating any restriction on z.

(i) $f(z) = x^2 + y^2 - 2ixy$ (ii) $f(z) = 2x - 3y + i(3x + 2y)$
(iii) $f(z) = \sin(1/z)$ (iv) $f(z) = xe^z$.

17 Write the analytic function $f(z)$ in the form $u(r, \theta) + iv(r, \theta)$ where $z = re^{i\theta}$, and from the Cauchy-Riemann equations in Cartesian form deduce the equations in polar form

$$\frac{\partial u}{\partial r} = \frac{1}{r}\frac{\partial v}{\partial\theta} \quad \text{and} \quad \frac{1}{r}\frac{\partial u}{\partial\theta} = -\frac{\partial v}{\partial r}.$$

18 Determine value for a, b, c, and d if

$$(x^2 + axy + by^2) + i(cx^2 + dxy + y^2)$$

is analytic everywhere.

19 Investigate the paths into which the real and imaginary axes of the z-plane are mapped under the transformation $w = (z + 1)^2$. What region in the z-plane is transformed into the semicircle $u^2 + v^2 \leqslant 1$, $v > 0$ in the w-plane?

20 A, B, and C are the points 1, 2, $(2 + i)$ respectively, in the z-plane. Find the paths in the w-plane which correspond to the straight lines AB, BC, and AC when $w = 1 + 2/z$. Sketch on the Argand diagram. (Hint. For AC, evaluate $u + v$ and $(u - 1)^2 + v^2$.)

21 When $w = e^z$ show that the straight line $x = ky$, $y > 0$ in the z-plane is transformed into an equiangular spiral in the w-plane.

22 Show that $u = y^3 - 3x^2y$ is a harmonic function and use the Cauchy-Riemann equations to find the harmonic conjugate function v where $w = u + iv$ is analytic. Express w as a function of z.

23 Use the Cauchy-Riemann equations to show that if $w = u + iv = f(z)$ is analytic then each member of the family of curves $u(x, y) = $ constant cuts each member of the family of curves $v(x, y) = $ constant orthogonally. Illustrate by sketching the families of curves given by the transformation

(i) $w = z^2$; (ii) $w = 1/z$; (iii) $z = \sin w$.

24 Show that $u = x + 3x^2y - y^3$ is an harmonic function. Find the harmonic conjugate function v such that $w = u + iv$ is analytic. Express w in the form $f(z)$ where $z = x + iy$.

11
Ordinary Differential Equations

11.0 INTRODUCTION

An ordinary differential equation is a relation connecting one independent variable say x, one dependent variable say y, and differential coefficients of y with respect to x. Many relationships in engineering are most easily expressed in the form of a differential equation. For example $dq/dt = -kq$ is the differential equation formed from the problem of a leaky condenser discharging at a rate proportional to the charge q.

The *order* of a differential equation is the order of its highest derivative. The *degree* of a differential equation is the power of its highest order derivative. For example

$$\left(\frac{d^2y}{dx^2}\right)^2 = \left(\frac{dy}{dx}\right)^3 + ye^z$$

is of order 2 and degree 2.

The *solution* of a differential equation is obtained by integration. The equation is solved when y is expressed in terms of x by a relationship of the form $y = f(x)$ or $\phi(x, y) = 0$. A differential equation of order n will in general have a solution involving n arbitrary constants. This solution is called the *general* solution. If boundary conditions are prescribed, these arbitrary constants can be specified.

The differential equation

$$\frac{dy}{dx} = f(x), \quad \text{or} \quad Dy = f(x) \tag{1}$$

has been solved, in Chapter 6, by integration, to give

$$y = \int f(x)\, dx + C. \tag{2}$$

Similarly the equation

$$\frac{d^n y}{dx^n} = f(x) \tag{3}$$

may be solved by integrating n times giving

$$y = C_0 + C_1 x + C_2 x^2 + \ldots + C_{n-1} x^{n-1} + F_n(x) \tag{4}$$

where $C_0, C_1, C_2, \ldots, C_{n-1}$ are arbitrary constants and $F_n(x)$ is the result of integrating $f(x)$ n times.

11.1 FIRST ORDER EQUATIONS

11.1.1 One variable absent

(i) If the equation involves dy/dx and x only, i.e. if $\dfrac{dy}{dx} = f(x)$, then integrating both sides

$$y = \int f(x)\, dx + C.$$

Example Solve $x^3 \dfrac{dy}{dx} = 2x^2 + 3$.

Solution Divide by x^3, $x \neq 0$

$$\frac{dy}{dx} = \frac{2}{x} + \frac{3}{x^3}$$

then

$$y = 2 \log |x| - \frac{3}{2x^2} + C,$$

where C is an arbitrary constant.

(ii) If the equation involves dy/dx and y only, i.e. $dy/dx = g(y)$, $g(y) \neq 0$, then

$$\frac{dx}{dy} = \frac{1}{g(y)},$$

and integrating both sides

$$x = \int \frac{dy}{g(y)} + C.$$

Example Solve $\dfrac{dy}{dx} = ay$, a given non-zero constant. \hfill (5)

Solution Rearranging and integrating,

$$x = \int \frac{dy}{ay} + C$$

$$= \frac{1}{a} \log |y| + C$$

or

$$\log |y| = ax - aC$$

and this is one form of solution of (5), where C is an arbitrary constant. Now

$$y = e^{ax - aC}$$

$$= e^{ax}\, e^{-aC}$$

or

$$y = A e^{ax}, \hfill (6)$$

where $A = e^{-aC}$ is also an arbitrary constant.

When a logarithm appears in an integral it is often more convenient to write the arbitrary constant in logarithmic form. Thus

$$ax = \log|y| + \log B$$

then

$$e^{ax} = By$$

or

$$y = Ae^{ax}, \quad A = 1/B.$$

The solution $y = Ae^{ax}$ is well known as the solution of equation (5), since e^x is defined in Chapter 4 as the particular solution of $dy/dx = y$ such that when $x = 0$, $y = 1$.

11.1.2 Variables separable

Equations of this type are of the form

$$g(y)\frac{dy}{dx} = f(x) \tag{7}$$

or

$$g(y)\,dy = f(x)\,dx,$$

i.e. the variables x and y may be separated. Then the complete solution is given by

$$\int g(y)\,dy = \int f(x)\,dx + C. \tag{8}$$

Example (i) Solve $\tan x(dy/dx) = \cot y$.

Solution Re-write the equation as $\tan y\,dy = \cot x\,dx$, then

$$\int \frac{\sin y}{\cos y}\,dy = \int \frac{\cos x}{\sin x}\,dx + C,$$

or

$$-\log|\cos y| = \log|\sin x| - \log c,$$

or

$$\cos y \sin x = c,$$

where the arbitrary constant is taken in the most convenient form, $C = -\log c$.

Example (ii) A mass m falls vertically from rest from a distance h above the ground. Determine the velocity with which it hits the ground if the air resistance is proportional to the square of the velocity.

Solution Let the distance x be measured as positive downwards.

$$\text{Velocity } v = dx/dt.$$

$$\text{Acceleration} = \frac{dv}{dt} = \frac{dv}{dx}\frac{dx}{dt} = v\frac{dv}{dx}. \tag{9}$$

From Newton's Second Law of motion for constant mass,

$$mg - mkv^2 = mv\frac{dv}{dx}$$

where the constant of proportionality is mk.

Hence
$$\int \frac{v}{g - kv^2} \, dv = \int dx + C$$

or
$$-\frac{1}{2k} \log |g - kv^2| = x + C.$$

The initial condition $x = 0$, $v = 0$ determine $C = (-\log g)/2k$ so that
$$x = \frac{1}{2k} \log \left| \frac{g}{g - kv^2} \right|.$$

Hence
$$e^{2kx} = g/(g - kv^2),$$

and
$$v^2 = \frac{g}{k} (1 - e^{-2kx})$$

gives the velocity at distance x from the start.

In particular, when $x = h$,
$$v^2 = \frac{g}{k} (1 - e^{-2kh}).$$

Note that when $h \to \infty$, $v^2 \to g/k$. That is, no matter from what height the mass falls, the velocity will never exceed $\sqrt{(g/k)}$.

11.1.3 Homogeneous equations
These are equations of the form
$$\frac{dy}{dx} = f\left(\frac{y}{x}\right) \tag{10}$$

and are reduced to the form of equation (7) in which the variables are separable, by the substitution $u = y/x$. If $y = ux$, then
$$\frac{dy}{dx} = u + x \frac{du}{dx},$$

and substituting for y and dy/dx in equation (10), leads to the equation
$$u + x \frac{du}{dx} = f(u).$$

In this equation the variables separate.

Example (i) Solve the differential equation
$$x \frac{dy}{dx} = y + y \log \left(\frac{y}{x}\right).$$

Solution Write $y = ux$, $dy/dx = u + x \, du/dx$, then the equation becomes
$$x^2 \frac{du}{dx} = ux \log u.$$

If $x \neq 0$, divide by x and separate the variables so that the equation now is

$$\frac{du}{u \log u} = \frac{dx}{x}.$$

Integrating, $\log \log |u| = \log |x| + \log c$, or $\log u = cx$. That is, $u = e^{cx}$ and the solution of the equation is

$$y = xe^{cx},$$

where c is an arbitrary constant.

Example (ii) Solve the differential equation

$$\frac{dy}{dx} = \frac{9x + y}{x + y}$$

Solution Writing $y = ux$, $dy/dx = u + x\, du/dx$ gives

$$u + x\frac{du}{dx} = \frac{9 + u}{1 + u},$$

and hence

$$x\frac{du}{dx} = \frac{9 - u^2}{1 + u}.$$

Separating the variables,

$$\int \frac{dx}{x} = \int \frac{(1 + u)\, du}{(3 - u)(3 + u)} = \int \left\{ \frac{\frac{2}{3}}{3 - u} - \frac{\frac{1}{3}}{3 + u} \right\} du.$$

Integrating,

$$\log x = -\tfrac{2}{3} \log (3 - u) - \tfrac{1}{3} \log (3 + u) + \tfrac{1}{3} \log c,$$

where c is an arbitrary constant. Hence

$$x^3(3 - u)^2(3 + u) = c,$$

or

$$x^3(3 - y/x)^2(3 + y/x) = c,$$

or

$$(3x - y)^2(3x + y) = c.$$

Example (iii) Solve the differential equation

$$\frac{dy}{dx} = \frac{9x + y + 2}{x + y + 1}.$$

Solution This equation can be made homogeneous by changing the variables to X and Y where $X = x + y + 1$ and $Y = 9x + y + 2$. Then $dX = dx + dy$, $dY = 9dx + dy$ and

$$\frac{dy}{dx} = \left(9 - \frac{dY}{dX}\right) \Big/ \left(\frac{dY}{dX} - 1\right)$$

and the equation becomes

$$\frac{dY}{dX} = \frac{9X + Y}{X + Y}$$

which is solved in example (ii). Hence its solution is

$$(3X - Y)^2(3X + Y) = c,$$

and the required solution is

$$(6x - 2y - 1)^2(12x + 4y + 5) = c,$$

where c is an arbitrary constant.

11.1.4 Exact equations

An ordinary differential equation which can be written in the form

$$M(x, y)\, dx + N(x, y)\, dy = dF(x, y), \tag{11}$$

such that the left side is an exact differential, is an *exact* equation. That is, there exists a function $G(x, y)$ such that

$$\frac{\partial G}{\partial x} = M, \qquad \frac{\partial G}{\partial y} = N. \tag{12}$$

Hence (11) may be written

$$\frac{\partial G}{\partial x}\, dx + \frac{\partial G}{\partial y}\, dy = dF(x, y),$$

or

$$dG = dF.$$

Integrating,

$$G(x, y) = F(x, y) + C.$$

Differentiating equations (12) partially with respect to y and x respectively, leads to

$$\frac{\partial^2 G}{\partial y \partial x} = \frac{\partial M}{\partial y} = \frac{\partial^2 G}{\partial x \partial y} = \frac{\partial N}{\partial x},$$

and a necessary condition for exactness in (11) is

$$\frac{\partial M}{\partial y} = \frac{\partial N}{\partial x}. \tag{13}$$

That is (11) cannot be exact unless equation (13) is true. It can be proved that the equation is always exact when equation (13) is true, although the proof is beyond the scope of this book, and equation (13) is both a necessary and sufficient condition that (11) is exact.

Example (i) Solve $y^2 dx + 2xy\,dy = e^x\,dx$

Solution Re-write the equation as

$$d(xy^2) = d(e^x),$$

and integrating

$$xy^2 = e^x + C$$

where C is an arbitrary constant.

It can be proved that the left side of (11) can always be made exact by multiplication by a suitable *integrating factor* $\rho(x, y)$, which has the property that

$$\frac{\partial}{\partial y}\left[\rho(x, y)M(x, y)\right] = \frac{\partial}{\partial x}\left[\rho(x, y)N(x, y)\right].$$

It is not easy to determine ρ from this equation. However, certain combinations of differentials may be recognised.

Example (ii) Solve $x\,dy - y\,dx = xy^2 dx$.

Solution Divide by y^2 and transpose to give

$$\frac{y\,dx - x\,dy}{y^2} = -x\,dx,$$

i.e. the integrating factor is $1/y^2$ so that

$$d\!\left(\frac{x}{y}\right) + x\,dx = 0.$$

Integration gives

$$\frac{x}{y} + \frac{x^2}{2} = C$$

where C is an arbitrary constant.

11.1.5 First order linear equations

The differential equation is of the type

$$\frac{dy}{dx} + P(x)y = Q(x) \tag{14}$$

First, let $P(x) = a$, a a given constant, so that

$$\frac{dy}{dx} + ay = Q(x) \tag{15}$$

The left side is not an exact differential but can be made so by multiplying by an integrating factor. If $Q(x) = 0$, equation (15) has the solution $y = Ae^{-ax}$, or $ye^{ax} = A$. i.e. $d(ye^{ax}) = 0$. This suggests that the left side of equation (15) can be made an exact differential by multiplication by e^{ax}.

Multiplying both sides of equation (15) by e^{az} ($\neq 0$) gives

$$e^{az}\frac{dy}{dx} + ae^{az}y = e^{az}Q(x)$$

or $$\frac{d}{dx}(ye^{az}) = e^{az}Q(x).$$

Integrating,

$$ye^{az} = \int e^{az}Q(x)\,dx + C$$

or $$y = e^{-az}\int e^{az}Q(x)\,dx + Ce^{-az} \tag{16}$$

where C is an arbitrary constant.

Example Solve

$$R\frac{dq}{dt} + \frac{q}{C} = V$$

with $q = 0$ at $t = 0$, R, C, and V are given constants. Note that C is not an arbitrary constant.

Solution The equation may be written

$$\frac{dq}{dt} + \frac{q}{RC} = \frac{V}{R}.$$

Multiply by $e^{t/RC}$ to give

$$\frac{d}{dt}(qe^{t/RC}) = \frac{V}{R}e^{t/RC}$$

Integrating, $$qe^{t/RC} = VCe^{t/RC} + k$$

where k is an arbitrary constant.

Hence $$q = VC + ke^{-t/RC}.$$

The initial condition $q = 0$, $t = 0$ give the value of $k = -VC$ and

$$q = VC(1 - e^{-t/RC}).$$

Secondly, consider the more general equation (14). This can be made exact by multiplication by an integrating factor $z(x)$ to be determined. For let $P(x) = dz/dx$, or $z = \int P(x)\,dx$, then equation (14) becomes

$$\frac{dy}{dx} + \frac{dz}{dx}y = Q(x).$$

Multiplication by e^z gives

$$\frac{d}{dx}(e^z y) = e^z Q(x),$$

from which on integration,

$$e^z y = \int e^z Q(x)\,dx + C \qquad (17)$$

where C is an arbitrary constant.

Since
$$z = \int P(x)\,dx$$

is known, y is given in terms of x. Hence, multiplying (14) by the integrating factor $\exp(\int P\,dx)$ makes the left side of (14) exactly integrable.

Example (i) Solve the differential equation

$$x \log x \frac{dy}{dx} + y = 2 \log x.$$

Solution Divide by $x \log x$ to bring the equation to the standard form of equation (14)

$$\frac{dy}{dx} + \frac{y}{x \log x} = \frac{2}{x}. \qquad (18)$$

Here, $P(x) = 1/x \log x$, $\int P\,dx = \log |\log x|$ so that $\exp(\int P\,dx) = \log x$. Multiply equation (18) by the integrating factor $\log x$ and find

$$\log x \frac{dy}{dx} + \frac{y}{x} = \frac{2}{x} \log x,$$

or
$$\frac{d}{dx}(y \log x) = \frac{2}{x} \log x.$$

Integrating, $y \log x = (\log x)^2 + C$

and $y = \log x + C/\log x$

where C is an arbitrary constant.

Example (ii) Solve $1 + y \frac{dx}{dy} = e^{-z} \sec^2 x \frac{dx}{dy}.$

Solution Multiply by dy/dx to obtain the standard form of equation (14),

$$\frac{dy}{dx} + y = e^{-z} \sec^2 x.$$

The integrating factor is e^z, and hence

$$\frac{d}{dx}(ye^z) = \sec^2 x.$$

It follows that $\qquad ye^z = \tan x + C,$

or $\qquad\qquad\qquad y = e^{-z}\tan x + Ce^{-z}$

where C is an arbitrary constant.

11.2 SOME SPECIAL TYPES OF SECOND ORDER EQUATIONS

Some types of second order differential equations can be reduced to first order by a suitable change of variable.

11.2.1 Equations with the dependent variable absent

This type of equation involves d^2y/dx^2, dy/dx and x only. It is reduced to a first order equation by the substitutions

$$\frac{dy}{dx} = p, \qquad \frac{d^2y}{dx^2} = \frac{dp}{dx}, \tag{19}$$

so that the original equation becomes one involving dp/dx, p, and x, which is of the first order in p. This may be solved for p as a function of x and, by further integration, y obtained as a function of x.

Example Solve the differential equation

$$x\frac{d^2y}{dx^2} + \frac{dy}{dx} = 0. \tag{20}$$

Solution Using (19), equation (20) becomes

$$x\frac{dp}{dx} + p = 0,$$

so that

$$\int \frac{dp}{p} + \int \frac{dx}{x} = 0.$$

Integrating leads to

$$\log|p| + \log|x| = \log A,$$

or $\qquad\qquad\qquad p = A/x$

where A is an arbitrary constant.

Hence $\qquad\qquad\qquad \dfrac{dy}{dx} = \dfrac{A}{x},$

and integrating once more,

$$y = B + A\log|x|.$$

This is the solution of equation (20) where B is a second arbitrary constant.

11.2.2 Equations with the independent variable absent

This type of equation involves d^2y/dx^2, dy/dx, and y only. The substitutions used are

$$\frac{dy}{dx} = p, \qquad \frac{d^2y}{dx^2} = \frac{dp}{dx} = \frac{dp}{dy}\frac{dy}{dx} = p\frac{dp}{dy}. \qquad (21)$$

The differential equation now involves p, dp/dy, and y and is of the first order in p. Its solution gives p in terms of y and a further integration leads to y in terms of x.

Example (i) Solve the differential equation

$$y(d^2y/dx^2) + (dy/dx)^2 = 0 \qquad (22)$$

Solution The substitutions of (21) used in (22) give

$$y(pdp/dy) + p^2 = 0 \qquad (23)$$

or $\qquad\qquad dp/p + dy/y = 0.$

Integrating,

$$p = A/y,$$

where A is an arbitrary constant. Then

$$p = dy/dx = A/y$$

and $\qquad\qquad y^2 = 2Ax + B$

where B is another arbitrary constant.

An elementary application of this method is given by the solution of the equation of motion of a particle moving with simple harmonic motion as follows in the next example.

Example (ii) A particle of mass m moves in a straight line under a force which is always directed towards a fixed point in the line and which is proportional to the distance of the particle from the fixed point. Examine the motion.

Solution Let the fixed point be the origin O and the fixed straight line be the x-axis. Let the force per unit mass at distance x from O be n^2x where n is some constant and let t be the time, then the equation of motion of the particle is

$$m\frac{d^2x}{dt^2} = -mn^2x,$$

or $\qquad\qquad \dfrac{d^2x}{dt^2} = -n^2x. \qquad (24)$

Using substitutions corresponding to those in (21), i.e.

$$\frac{dx}{dt} = p, \qquad \frac{d^2x}{dt^2} = p\frac{dp}{dx},$$

equation (24) becomes

$$p\frac{dp}{dx} + n^2x = 0.$$

Integrating, $\frac{1}{2}p^2 + \frac{1}{2}n^2x^2 = \frac{1}{2}a^2n^2$, where $\frac{1}{2}a^2n^2$ is chosen to be the arbitrary constant. Then $p^2 = n^2(a^2 - x^2)$ and

$$p = \frac{dx}{dt} = \pm n\sqrt{(a^2 - x^2)}.$$

Integrating once more,

$$\sin^{-1}(x/a) = \pm(nt + \varepsilon),$$

or
$$x = \pm a\sin(nt + \varepsilon).$$

Since a is arbitrary there is no need for the double $\pm$ sign and the general solution of equation (24) is

$$x = a\sin(nt + \varepsilon).$$

Thus the motion is simple harmonic with period $2\pi/n$. The maximum displacement on either side of O is a, which is the amplitude of the motion. The angle ε is the phase angle.

11.3 LINEAR DIFFERENTIAL EQUATIONS WITH CONSTANT COEFFICIENTS

These equations have the general form

$$a_n\frac{d^ny}{dx^n} + a_{n-1}\frac{d^{n-1}y}{dx^{n-1}} + a_{n-2}\frac{d^{n-2}y}{dx^{n-2}} + \ldots + a_0y = V(x), \quad (25)$$

where $a_0, a_1, a_2, \ldots, a_n$ are given real constants and V is a given function of x.

11.3.1 The D-notation

Let D be the operator d/dx so that $Dy = dy/dx$ (see section 2.1.3.) then

$$\frac{d^2y}{dx^2} = \frac{d}{dx}\left(\frac{dy}{dx}\right)$$

is written D^2y, and D^m means the operator d^m/dx^m where m is a positive integer. Then, if m and n are positive integers,

$$D^{m+n} = D^{n+m} = D^m[D^n] = D^n[D^m].$$

The operator D is an example of a *linear* operator which is defined as follows. An operator ϕ is a linear operator if

(i) $\phi(y_1 + y_2) = \phi(y_1) + \phi(y_2)$ and   (26)

(ii) $\phi(ky) = k\phi(y)$, k constant.   (27)

It is simple to show that $\phi = D$ satisfies these equations. Also, suppose $y = y_1$ and $y = y_2$ both satisfy the equation

$$\phi(y) = 0,   (28)$$

then $y = Ay_1 + By_2$ also satisfies equation (28), where A and B are arbitrary constants. For, substituting $y = Ay_1 + By_2$ in equation (28),

$$\phi(y) = \phi(Ay_1 + By_2) = \phi(Ay_1) + \phi(By_2) \text{ from (26)}$$
$$= A\phi(y_1) + B\phi(y_2) \text{ from (27)}$$
$$= 0,$$

since both y_1 and y_2 satisfy equation (28).

Similarly it may be shown that if $y_1, y_2, y_3, \ldots, y_n$ are all solutions of equation (28) then so also is

$$y = A_1 y_1 + A_2 y_2 + A_3 y_3 + \ldots + A_n y_n,$$

where $A_1, A_2, A_3, \ldots, A_n$ are arbitrary constants.

An important property of linear operators is that they may be factorised. For example,

$$\frac{d^2 y}{dx^2} - \frac{dy}{dx} - 6y = \frac{d}{dx}\left(\frac{dy}{dx} + 2y\right) - 3\left(\frac{dy}{dx} + 2y\right),$$

which may be written as

$$\left(\frac{d}{dx} - 3\right)\left(\frac{dy}{dx} + 2y\right).$$

Writing the operator d/dx as D leads to

$$(D^2 - D - 6)y = (D - 3)(D + 2)y,$$

or, the operator D satisfies the equation

$$D^2 - D - 6 = (D - 3)(D + 2).$$

Any polynomial $P(D)$ in the linear operator D is itself a linear operator, for if

$$P(D) = a_n D^n + a_{n-1} D^{n-1} + \ldots + a_1 D + a_0   (29)$$

where $a_0, a_1, a_2, \ldots, a_n$ are constants, then $P(D)$ satisfies equations (26) and (27) since each term on the right of (29) satisfies these equations.

11.3.2 The solution of differential equations with constant coefficients

Equation (25) may now be written

$$(a_n D^n + a_{n-1} D^{n-1} + \ldots + a_1 D + a_0)y = V(x)\Big\}$$
or $$\phi(y) = V(x)\Big\}   (30)$$

where the linear operator ϕ is $P(D)$ as defined in equation (29).

The technique for solving equation (30) is as follows:

(i) let y_c be the complete solution containing n arbitrary constants of the *reduced* or *homogeneous* equation

$$\phi(y) = 0;$$

(ii) let y_p be any particular integral of equation (30) so that $\phi(y_p) = V(x)$. Then the complete solution of equation (30) is

$$y = y_c + y_p \tag{31}$$

since
$$\phi(y) = \phi(y_c + y_p)$$
$$= \phi(y_c) + \phi(y_p)$$
$$= 0 + V(x).$$

The rest of this chapter is devoted to finding y_c, called the *complementary function*, and y_p called the *particular integral*, for different forms of the function $V(x)$. The complementary function contains the arbitrary constants of integration. The particular integral is *any* function, no matter how simple nor how found, which satisfies the differential equation. For example it is easy to see that the function $y_p = x$ is a particular integral of the differential equation.

$$\frac{d^2y}{dx^2} + y = x. \tag{32}$$

The complementary function is $A \sin (x + B)$ from the example (i) in section 11.2.2, and contains the two arbitrary constants. Hence the *complete primitive* of equation (32) is

$$y = A \sin (x + B) + x. \tag{33}$$

Expanding $\sin (x + B)$,

$$y = A \sin x \cos B + A \cos x \sin B + x;$$

and since A, B are arbitrary constants, so are $A \cos B$ and $A \sin B$ and the solution (33) may be written

$$y = C \sin x + E \cos x + x \tag{34}$$

where C, E are arbitrary constants.

11.4 THE COMPLEMENTARY FUNCTION

Consider the solution of the second order differential equation

$$(D^2 + aD + b)y = 0 \tag{35}$$

with a and b given real constants. This may be written

$$(D - \lambda)(D - \mu)y = 0, \tag{36}$$

where λ and μ may be real and distinct, real and equal, or complex in which case they are conjugate, (see section 10.5.3). To solve equation (36), let

$$(D - \mu)y = u(x) \tag{37}$$

where u is a function to be found. From equations (36) and (37), $(D - \lambda)u = 0$ and from equation (6) of section 11.1.1, $u = Ae^{\lambda x}$, where A is an arbitrary constant. Substituting this value of u in equation (37),

$$(D - \mu)y = Ae^{\lambda x} \tag{38}$$

and this is a first-order equation in y solved as in section 11.1.5. Multiplying equation (38) by the integrating factor $e^{-\mu x}$ gives

$$D(ye^{-\mu x}) = Ae^{(\lambda - \mu)x},$$

and integrating gives

$$ye^{-\mu x} = \int Ae^{(\lambda - \mu)x}\, dx + B$$

or

$$y = e^{\mu x} \int Ae^{(\lambda - \mu)x}\, dx + Be^{\mu x} \tag{39}$$

where A, B are arbitrary constants. This is the general solution of (36). Three cases arise, depending on the nature of the roots λ, μ of the equation

$$D^2 + aD + b = 0$$

Case (i) λ, μ real and distinct.

Equation (39) becomes

$$y = e^{\mu x} \frac{A}{\lambda - \mu} e^{(\lambda - \mu)x} + Be^{\mu x}$$

or

$$y = Ce^{\lambda x} + Be^{\mu x} \tag{40}$$

with B and C arbitrary constants.

Case (ii) $\lambda = \mu$, roots real and equal.

Equation (39) becomes

$$y = e^{\lambda x} \int A\, dx + Be^{\lambda x}$$

or

$$y = e^{\lambda x}(Ax + B). \tag{41}$$

Case (iii) λ, μ complex.

Let $\lambda = \alpha + i\beta$, $\mu = \alpha - i\beta$. From equation (40), $y = Ce^{\lambda x} + Be^{\mu x}$, since the roots are distinct, or

$$y = e^{\alpha x}(E \cos \beta x + F \sin \beta x) \tag{42}$$

with E, F real arbitrary constants.

Example (i) Solve the equation

$$\frac{d^2y}{dx^2} - 5\frac{dy}{dx} + 6y = 0.$$

Solution Write as $(D^2 - 5D + 6)y = 0$

or $(D - 2)(D - 3)y = 0,$

and from equation (40) the solution is

$$y = Ae^{2x} + Be^{3x}.$$

where A and B are arbitrary constants.

Example (ii) Solve $(D^2 + 2D + 1)y = 0.$

Solution The given equation may be written

$$(D + 1)^2 y = 0,$$

and from (41) the solution is

$$y = (Ax + B)e^{-x}$$

where A and B are arbitrary constants.

Example (iii) Solve $(D^2 - 2D + 5)y = 0.$

Solution The quadratic in D will not factorise to give real factors. Since

$$D^2 - 2D + 5 = (D - 1)^2 + 4,$$

the values of λ and μ are $1 \pm 2i$. From equation (42) the solution is

$$y = e^x(A \cos 2x + B \sin 2x),$$

where A and B are arbitrary constants.

The method of solution of the nth order differential equation $P(D)y = 0$, where $P(D)$ is a polynomial in D of degree n, is an extension of the method for the second order equation.

Example (iv) Solve $(D - 2)(D - 3)^2(D^2 + 4)y = 0.$

Solution The roots of $P(D) = 0$ are 2, 3, 3, $\pm 2i$, and

$$y = Ae^{2x} + (Bx + C)e^{3x} + E \cos 2x + F \sin 2x$$

where A, B, C, E, and F are arbitrary constants of integration of the fifth order differential equation.

Example (v) A particle of mass m moving in a straight line is acted on by a restoring force which is proportional to its displacement from a fixed point in the line and also by a resistance which is proportional to the velocity. Write down the equation of motion of the particle and solve it.

Solution As in example (ii) of section 11.2.2, the equation of motion of the particle may be written in the form

$$m \frac{d^2x}{dt^2} = -mn^2x - 2mk \frac{dx}{dt},$$

where the last term represents the resistance proportional to the velocity and k is constant. Let D represent the operator d/dt, then the equation of motion may be written

$$(D^2 + 2kD + n^2)x = 0. \tag{43}$$

The roots of the quadratic $D^2 + 2kD + n^2 = 0$ are

$$-k \pm \sqrt{(k^2 - n^2)}. \tag{44}$$

Three possibilities must now be considered:

(i) If $k > n$, $\sqrt{(k^2 - n^2)}$ is real, and the roots are real, distinct, and both negative since $k > \sqrt{(k^2 - n^2)}$. Let them be $-\lambda$ and $-\mu$. The solution of equation (43) is then

$$x = Ae^{-\lambda t} + Be^{-\mu t} \tag{45}$$

where A and B are arbitrary constants.

(ii) if $k = n$ the roots (44) are each equal to $-k$ and the solution of equation (43) is

$$x = e^{-kt}(At + B) \tag{46}$$

where A and B are arbitrary constants.

(iii) if $k < n$ the roots (44) are complex and may be written $-k \pm i\beta$ where $\beta = \sqrt{(n^2 - k^2)}$ is real. The general solution of equation (43) in this case is

$$x = ae^{-kt} \sin(\beta t + \alpha) \tag{47}$$

where a and α are arbitrary constants.

In all three cases (45), (46), and (47) the value of x approaches zero as $t \to \infty$ because of the negative exponential factor in each. When the value of k, called the *damping factor*, is equal to n, the rate of decay of x is greatest. In this case the motion is said to be *critically damped*. When $k > n$ the motion is heavily damped or overdamped. When $k < n$ the motion is oscillatory but the factor e^{-kt} damps the amplitude of the vibration which dies away as $t \to \infty$, and the period is increased.

Certain measuring instruments are constructed so that the motion of the pointer is critically damped in order that the pointer shall return to its equilibrium position as quickly as possible. Sketches of three possible distance-time graphs are shown in Fig. 11.1.

Damped oscillations are important also in electricity where the differential equation for a simple oscillatory circuit may be shown to be of the form of equation (43). For let q be the charge on the plate of a condenser of capacity C which is connected through an inductance L and resistance R, as in Fig. 11.2, and let there be no external e.m.f. so that the algebraic sum of the voltage drops is zero.

Let j be the current in the circuit then the two equations

$$L\frac{dj}{dt} + Rj + \frac{q}{C} = 0 \qquad \text{and} \qquad j = dq/dt$$

give the second order equation

$$\frac{d^2q}{dt^2} + \frac{R}{L}\frac{dq}{dt} + \frac{q}{LC} = 0 \tag{48}$$

This equation may be compared with equation (43) with q instead of x and where $2k = R/L$, $n^2 = 1/LC$, so that

(i) if $R/2L \geqslant 1/\sqrt{(LC)}$ the charge is not oscillatory and its behaviour is sketched in Fig. 11.1 (a) and (b).

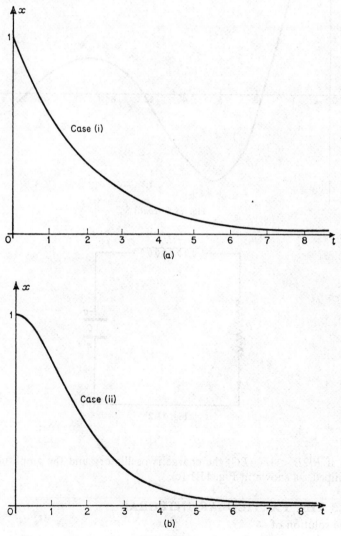

(a)

(b)

Fig. 11.1

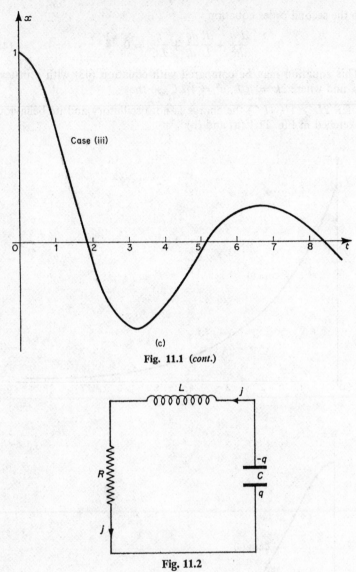

Fig. 11.1 (cont.)

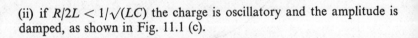

Fig. 11.2

(ii) if $R/2L < 1/\sqrt{(LC)}$ the charge is oscillatory and the amplitude is damped, as shown in Fig. 11.1 (c).

11.5 THE PARTICULAR INTEGRAL
The solution of

$$(D^2 + aD + b)y = V(x) \qquad (49)$$

is $y = y_c + y_p$ where y_c is found as in Section 11.4 and y_p is any particular integral which satisfies equation (49) and contains no arbitrary constants. The form of the particular integral depends on the form of $V(x)$. In this section the particular integral corresponding to $V(x)$ as a polynomial in x, an exponential function or as a circular function will be investigated.

11.5.1 $V(x)$ a polynomial in x

There are several methods for finding particular integrals, one of which consists in assuming a particular integral of the same form as $V(x)$, and a second defines an inverse operator D^{-1}. A third method, adopted here, consists in finding a particular integral of a set of equations. This method is chosen as being the simplest to apply and the simplest to justify. It is explained by examples.

Example·(i) Find a particular integral of the equation

$$(D^2 + 2D - 3)y = 9x^3. \tag{50}$$

Solution Differentiate both sides of equation (50) repeatedly until the right side is zero as follows:

$$(D^2 + 2D - 3)Dy = 27x^2 \tag{51}$$
$$(D^2 + 2D - 3)D^2y = 54x \tag{52}$$
$$(D^2 + 2D - 3)D^3y = 54 \tag{53}$$
$$(D^2 + 2D - 3)D^4y = 0. \tag{54}$$

From equation (54) a particular integral of the set of equations is given by $D^4y = 0$, from which it follows that

$$D^{n+4}y = 0, \quad n \geqslant 0. \tag{55}$$

With this particular integral, substituting from equation (55) in equation (53), from equation (55) and (53) in equation (52) and so on, the following results are obtained:

from equation (53), $D^5y + 2D^4y - 3D^3y = 54$,

i.e. $$D^3y = -18; \tag{56}$$

similarly, using equations (55) and (56) in equation (52) leads to

$$D^2y = -18x - 12; \tag{57}$$

and using equations (56) and (57) in equation (51) gives

$$Dy = -9x^2 - 12x - 14. \tag{58}$$

Finally, using equations (57) and (58) in equation (50) gives

$$-18x - 12 - 18x^2 - 24x - 28 - 3y = 9x^3,$$

or $$y_p = -3x^3 - 6x^2 - 14x - 40/3.$$

The complete solution of equation (50) is $y_c + y_p$ where

$$y_c = Ae^{-3x} + Be^x$$

and A, B are arbitrary constants.

Example (ii) Find a particular integral for the equation

$$(D^2 + D)y = x^2. \tag{59}$$

Solution Differentiate equation (59) repeatedly after having re-written it in the form

$$(D + 1)Dy = x^2, \tag{60}$$

to give

$$(D + 1)D^2y = 2x, \tag{61}$$

$$(D + 1)D^3y = 2, \tag{62}$$

$$(D + 1)D^4y = 0. \tag{63}$$

A particular integral is $D^{n+4}y = 0, n \geqslant 0$.

Substituting

in equation (62) gives $D^3y = 2$,

in equation (61) gives $2 + D^2y = 2x$ or $D^2y = 2x - 2$,

in equation (60) gives $2x - 2 + Dy = x^2$, or $Dy = x^2 - 2x + 2$,

and this is a particular integral of equation (59), from which, by integration,

$$y_p = \tfrac{1}{3}x^3 - x^2 + 2x.$$

There is no need to add an arbitrary constant of integration as this would be absorbed in the arbitrary constant A of the complementary function which is $y_c = A + Be^{-x}$.

Example (iii) Solve the differential equation

$$\frac{d^3y}{dx^3} + \frac{d^2y}{dx^2} = 2 + 4x^3. \tag{64}$$

Solution This is a third order equation and the methods of the previous sections apply. The equation is

$$(D + 1)D^2y = 2 + 4x^3 \tag{65}$$

from which the complementary function is

$$y_c = A + Bx + Ce^{-x}$$

where A, B, and C are arbitrary constants. To find the particular integral, differentiate equation (65) four times and obtain

$$(D + 1)D^3y = 12x^2$$

$$(D + 1)D^4y = 24x$$

$$(D + 1)D^5y = 24$$

$$(D + 1)D^6y = 0.$$

Hence a particular integral is $D^{n+6}y = 0$, $n \geqslant 0$.

Substituting back, $D^5 y = 24$,

$$D^4 y = 24x - 24,$$
$$D^3 y = 12x^2 - 24x + 24,$$
$$D^2 y = 2 + 4x^3 - 12x^2 + 24x - 24.$$

Integrating twice,

$$Dy = x^4 - 4x^3 + 12x^2 - 22x,$$
$$y_p = \tfrac{1}{5}x^5 - x^4 + 4x^3 - 11x^2.$$

The complete primitive of equation (64) is therefore

$$y = y_c + y_p = Ce^{-x} + A + Bx - 11x^2 + 4x^3 - x^4 + \tfrac{1}{5}x^5.$$

11.5.2 $V(x)$ of exponential form

Consider the second order differential equation

$$(D^2 + aD + b)y = e^{\alpha x} \tag{66}$$

where a, b, and α are given constants.
Since $D(e^{\alpha x}) = \alpha e^{\alpha x}$ and $D^2(e^{\alpha x}) = \alpha^2 e^{\alpha x}$,

$$(D^2 + aD + b)e^{\alpha x} = (\alpha^2 + a\alpha + b)e^{\alpha x}. \tag{67}$$

Hence

$$(D^2 + aD + b)\left\{\frac{e^{\alpha x}}{\alpha^2 + a\alpha + b}\right\} = e^{\alpha x}$$

provided $\alpha^2 + a\alpha + b \neq 0$, and a particular integral of equation (66) is $e^{\alpha x}/(\alpha^2 + a\alpha + b)$ since this function satisfies equation (66).

More generally if the differential equation is of order n, such that $P(D)$ is a polynomial of degree n, the equation

$$P(D)y = e^{\alpha x} \tag{68}$$

has a particular integral

$$y_p = \frac{e^{\alpha x}}{P(\alpha)} \tag{69}$$

provided $P(\alpha) \neq 0$.

Example Solve $(D^2 - 3D + 2)y = 20 \cosh 3x$. \hfill (70)

Solution Since $20 \cosh 3x = 10e^{3x} + 10e^{-3x}$, there are two particular integrals corresponding to $10e^{3x}$ and 10^{-3x} respectively.

Now $P(D) = D^2 - 3D + 2$, so that when $\alpha = 3$, $P(3) = 2$, and when $\alpha = -3$, $P(-3) = 20$. Hence using equation (69),

$$y_p = 10e^{3x}/2 + 10e^{-3x}/20.$$

The complementary function derives from

$$(D - 1)(D - 2)y = 0,$$

and is $y_c = Ae^x + Be^{2x}$, so that the complete solution of equation (70) is

$$y = Ae^x + Be^{2x} + 5e^{3x} + \tfrac{1}{2}e^{-3x}$$

where A and B are arbitrary constants.

Suppose now that $P(\alpha) = 0$. This means that the term $e^{\alpha x}$ on the right of equation (68) occurs in the complementary function. It is necessary under this condition to use the theorem that

$$P(D)(e^{kx}y) = e^{kx}P(D + k)y \tag{71}$$

where k is a constant. This is proved by induction as follows. Since $P(D)$ is a polynomial in D it is only necessary to prove

$$D^n(e^{kx}y) = e^{kx}(D + k)^n y. \tag{72}$$

Assume that, for some fixed arbitrary value of n, equation (72) is true, then

$$\begin{aligned}
D^{n+1}(e^{kx}y) &= DD^n(e^{kx}y) = D[e^{kx}(D + k)^n y] \\
&= e^{kx}D(D + k)^n y + ke^{kx}(D + k)^n y \\
&= e^{kx}(D + k)^n(D + k)y \\
&= e^{kx}(D + k)^{n+1}y.
\end{aligned}$$

That is, if equation (72) is true for some fixed arbitrary value of n then it is true for $n + 1$. When $n = 1$ equation (72) is true, since

$$D(e^{kx}y) = e^{kx}Dy + ke^{kx}y = e^{kx}(D + k)y.$$

Hence equation (72) is true for all positive integers $n \geqslant 1$, and thus equation (71) is true for all polynomials $P(D)$.

Let the differential equation be

$$P(D)y = e^{\alpha x} \tag{73}$$

where α is a given constant and $P(\alpha) = 0$. Write $y = e^{\alpha x}u$, where u is a function of x to be found, and equation (73) becomes

$$P(D)(e^{\alpha x}u) = e^{\alpha x}.$$

Using the result (71) with $k = \alpha$ leads to

$$e^{\alpha x}P(D + \alpha)u = e^{\alpha x}.$$

Dividing by $e^{\alpha x}$ ($\neq 0$) gives

$$P(D + \alpha)u = 1,$$

and u is found using the method of section 11.5.1.

Example Solve $(D - 2)(D + 1)y = e^{2x}$. $\tag{74}$

Solution The complementary function is

$$y_c = Ae^{2x} + Be^{-x},$$

where A and B are arbitrary constants. To find the particular integral, since $P(2) = 0$, write $y_p = e^{2x}u$ in equation (74), then

$$(D - 2)(D + 1)(e^{2x}u) = e^{2x}$$

becomes $\qquad e^{2x}(D - 2 + 2)(D + 1 + 2)u = e^{2x},$

or $\qquad\qquad (D + 3)Du = 1.$

Differentiating, a particular integral is $D^2u = 0$ and all higher derivatives are zero, and $u = \frac{1}{3}x$. Hence $y_p = \frac{1}{3}xe^{2x}$ and the complete solution of equation (74) is

$$y = (A + \tfrac{1}{3}x)e^{2x} + Be^{-x}.$$

11.5.3 $V(x)$ is $\cos \omega x$ or $\sin \omega x$

Since $e^{i\omega x} = \cos \omega x + i \sin \omega x$, it is necessary to solve $P(D)y = e^{i\omega x}$ using the methods of section 11.5.2, choosing the real part of the solution if $V(x) = \cos \omega x$ and the imaginary part if $V(x) = \sin \omega x$. For if the equation is $P(D)y = R_1(x) + iR_2(x)$, the solution is $y = y_1 + iy_2$ where $P(D)y_1 = R_1(x)$ and $P(D)y_2 = R_2(x)$, since

$$P(D)\{y_1 + iy_2\} = P(D)y_1 + iP(D)y_2 = R_1(x) + iR_2(x).$$

In particular, if $R_2(x)$ is $\sin \omega x$, the solution is $Im(y)$.

Example (i) Solve $(D + 1)^2(D + 2)y = \cos 3x$.

Solution The complementary function is

$$y_c = (A + Bx)e^{-x} + Ce^{-2x}$$

where A, B and C are arbitrary constants.

For the particular integral, consider

$$(D + 1)^2(D + 2)y = e^{3ix}.$$

Now $P(D) = (D + 1)^2(D + 2)$ so that

$$P(3i) = (3i + 1)^2(3i + 2)$$
$$= -2(17 + 6i).$$

Hence $\qquad y_p = Re\left[\dfrac{1}{-2(17 + 6i)}\, e^{3ix}\right]$

$$= Re\left[-\tfrac{1}{2}\,\dfrac{17 - 6i}{17^2 + 6^2}\,(\cos 3x + i \sin 3x)\right]$$

$$= -\tfrac{1}{2}\left(\dfrac{17}{325}\cos 3x + \dfrac{6}{325}\sin 3x\right).$$

This may be written $-\frac{1}{2}\cos (3x - \theta)$ where $\cos \theta = 17/325$ and $\sin \theta = 6/325$. The complete solution is therefore

$$y = (A + Bx)e^{-x} + Ce^{-2x} - \tfrac{1}{2}\cos (3x + \theta)$$

where θ is defined above.

Example (ii) Solve $(D^2 - 2D + 2)y = e^x \sin x$.   (75)

Solution The complementary function is obtained from the roots of $D^2 - 2D + 2 = 0$, or $(D - 1)^2 + 1 = 0$, and from case (iii) of section 11.4, with $\lambda = 1 + i$, $\mu = 1 - i$,

$$y_c = e^x(A \cos x + B \sin x).$$

To find the particular integral consider

$$(D^2 - 2D + 2)y = e^x e^{ix} = e^{(1 + i)x}.   (76)$$

With the notation of section 11.5.2, $\alpha = 1 + i$, $P(D) = (D - 1)^2 + 1$, $P(\alpha) = i^2 + 1 = 0$.

Hence writing $y = e^{(1 + i)x}u(x)$ in equation (76) leads to

$$(D + 2i)Du = 1,   (77)$$

and differentiating this once leads to

$$(D + 2i)D^2u = 0,$$

so that a particular integral of equation (77) is $D^{n+2}u = 0$, $n \geqslant 0$. Substituting back in equation (77) gives $Du = 1/2i$, and integration shows that

$$u = x/2i = -\tfrac{1}{2}ix$$

Thus $$y_p = Im[-\tfrac{1}{2}ixe^xe^{ix}]$$

$$= Im[-\tfrac{1}{2}ixe^x (\cos x + i \sin x)]$$

$$= -\tfrac{1}{2}xe^x \cos x.$$

The complete solution of section (75) is therefore

$$y = e^x (A \cos x + B \sin x - \tfrac{1}{2} x \cos x)$$

where A and B are arbitrary constants.

11.5.4 $V(x)$ is $x^n e^{\alpha x}$

In this case it is necessary to write $y = e^{\alpha x}u(x)$ and solve as in sections 11.5.2 or 11.5.3.

Example Solve $d^2y/dx^2 + n^2y = x \cos nx$.   (78)

Solution Re-write equation (78) as

$$(D^2 + n^2)y = xe^{inx}.   (79)$$

The complementary function is

$$y_c = A \cos nx + B \sin nx,$$

where A and B are arbitrary constants. To find the particular integral write $y = e^{inx}u$ in equation (79) and

$$(D^2 + n^2)(e^{inx}u) = xe^{inx}$$

becomes $[(D + in)^2 + n^2]u = x$

or $(D + 2in)Du = x.$

Hence $(D + 2in)D^2u = 1$

and $(D + 2in)D^3u = 0,$

and a particular integral is $D^{n+3}u = 0$, $n \geqslant 0$. Substituting back and integrating leads to the value of u as

$$u = -\frac{ix^2}{4n} + \frac{x}{4n^2}.$$

Then $y_p = Re\left[\left(-\frac{ix^2}{4n} + \frac{x}{4n^2}\right)(\cos nx + i \sin nx)\right]$

$$= \frac{x}{4n^2}\cos nx + \frac{x^2}{4n}\sin nx.$$

The complete primitive of equation (78) is, therefore

$$y = A\cos nx + B\sin nx + \frac{x}{4n^2}\cos nx + \frac{x^2}{4n}\sin nx,$$

where A and B are arbitrary constants.

11.6 APPLICATIONS

Many vibrational problems in engineering such as occur in oscillating electrical networks, the bending of beams, torsional problems, the extension of springs, give rise to second order differential equations of the types solved in section 11.5. For example, if the electrical network of Fig. 11.2 has an applied voltage $V(t)$, equation (48) is replaced by

$$\frac{d^2q}{dt^2} + \frac{R}{L}\frac{dq}{dt} + \frac{q}{LC} = \frac{V(t)}{L}.$$

When $V(t)$ is a polynomial in t, an exponential, or a circular function, this equation is solved by one of the methods of section 11.5. When $V(t)$ is not of these forms further methods, not discussed here, may be used to find the solution.

11.7 ISOCLINES. GRAPHICAL INTEGRATION OF $dy/dx = f(x, y)$

Suppose the solution of the differential equation

$$dy/dx = f(x, y) \tag{80}$$

is obtainable in the form

$$y = \phi(x, c) \tag{81}$$

where c represents the constant of integration. The graph of $y = \phi(x, c)$ for any given value of c is called an *integral curve*. Different values of c give the members of the family of integral curves.

Example (i) Sketch the family of integral curves of the differential equation $dy/dx = 3x^2 - 2x$.

Solution Here

$$y = x^3 - x^2 + c$$

and the integral curves are cubic curves of the form

$$y - c = x^2(x - 1)$$

and are shown in Fig. 11.3.

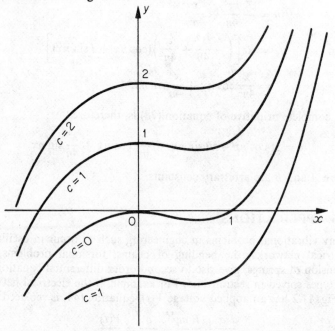

Fig. 11.3

Let such a family as (81) be represented by the curves as shown in Fig. 11.4.

The Fig. 11.4 also shows other curves, such as AA', BB', and CC' which are the loci of those points, on the integral curves, that have a common slope dy/dx or $\phi'(x, c)$. Such curves are called *isoclines* and Fig. 11.4 indicates isoclines AA', and BB' for which $\phi'(x, c)$ is zero and the isocline CC' for which $\phi'(x, c)$ is negative unity. The isoclines are curves whose equations are given by

$$\frac{dy}{dx} = m$$

i.e.

$$f(x, y) = m. \tag{82}$$

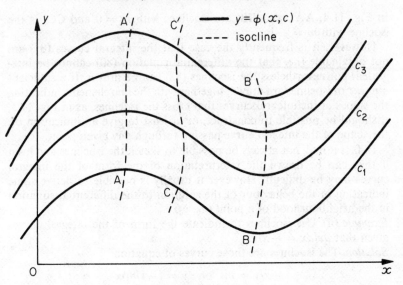

Fig. 11.4

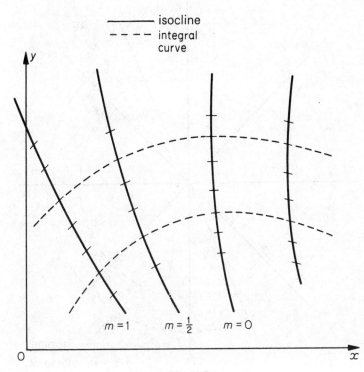

Fig. 11.5

In Fig. 11.4, AA', and BB' are isoclines with $m = 0$ and CC' is the isocline with $m = -1$.

However, it is frequently the case that the integral curves (81) are not obtainable (i.e. that the differential equation (80) cannot be integrated) but nevertheless the isoclines (82) can be drawn. If a sufficient number of isoclines are drawn, together with the line elements indicating the slopes of the integral curves that cross the isoclines, as in Fig. 11.5, it should be possible to construct, or at least to give an indication of, the shape of the integral curve passing through any given point.

In fact it may not always be possible to sketch the isoclines and even if they can be drawn the interpretation of the form of the integral curves may be difficult. However it usually is possible to derive some indication of the behaviour of the solution to the differential equation in the neighbourhood of a point (x_0, y_0).

Example (ii) Use isoclines to indicate the form of the integral curves given that $dy/dx = -x/y$.

Solution The isoclines are those curves of equation

$$-x/y = m \quad \text{or} \quad y = (-1/m)x$$

where m is the slope of the integral curves that cut the isoclines.

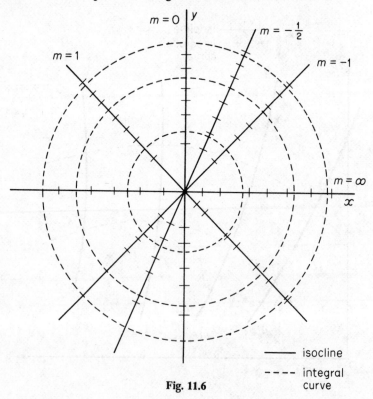

Fig. 11.6

— isocline
---- integral curve

The isoclines are straight lines of slope $(-1/m)$ passing through the origin. Since $(m)(-1/m) = -1$, the integral curves must cut the isoclines at right angles. The Fig. 11.6 shows some of the isoclines and the associated slopes on the integral curves.

In this case the integral curves are obviously a family of circles of equations $x^2 + y^2 = c^2$.

In Example (ii) the differential equation can be integrated by the method of section 11.1.2 to give

$$\int y \, dy = -\int x \, dx$$

i.e.
$$\tfrac{1}{2}y^2 = -\tfrac{1}{2}x^2 + \text{constant}$$

which is of the form $x^2 + y^2 = c^2$.

Example (iii) Draw the isoclines for the differential equation $dy/dx = x + y^2$. Construct the portion of the integral curves, for $x > 0$, passing through (a) the point (0, 0), (b) the point (1, 1).
Solution The equations of the isoclines are given by

$$x + y^2 = m$$

where m is the slope of the integral curves where they cross the isoclines.

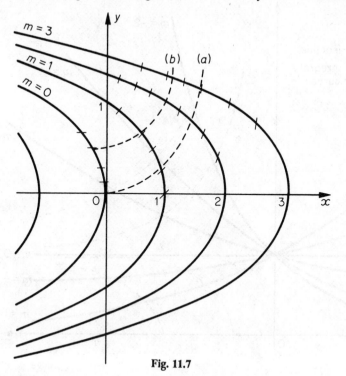

Fig. 11.7

The equation

$$y^2 = m - x$$

is a parabola whose axis is the x-axis and whose vertex is $(m, 0)$. The family of isoclines is shown in Fig. 11.7. The Fig. 11.7 also shows the line elements, i.e. the portions of the integral curves where they cross the isoclines. The integral curve through $(0, 0)$ has the shape of curve (a) and the integral curve through $(1, 1)$ has the shape of curve (b). Clearly, if isoclines are drawn at closer intervals it should be possible to get a better approximation to the shapes of the integral curves about the two points chosen. The Fig. 11.7 also indicates that for $x > 1$ the solution y increases rapidly with increasing x.

Example (iv) Sketch the isoclines and solution curves of

$$x \frac{dy}{dx} + y = \frac{1}{y}, x, y > 0$$

Solution The isoclines are given by the equation

$$mx + y = 1/y$$

i.e.

$$y(y + mx) = 1$$

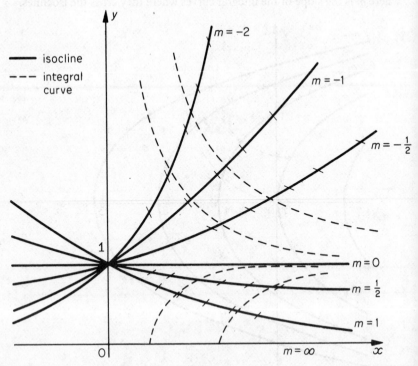

Fig. 11.8

Comparison with $YX = 1$ shows that $y(y + mx) = 1$ is an hyperbola with asymptotes $y = 0$, and $y = -mx$. Re-writing in the form $1 - y^2 = mxy$ shows that the branch of the hyperbola through (0, 1) is the one required. The isoclines and associated integral curves are shown in Fig. 11.8.

EXERCISE 11

1 Solve the following differential equations:

(i) $5\dfrac{dy}{dx} - 3y = 0$;

(ii) $\dfrac{dy}{dx} = e^{x-y}$;

(iii) $x^2(y^2 + 1) + y\sqrt{(x^3 + 1)}\dfrac{dy}{dx} = 0$;

(iv) $\sqrt{(2xy)}\dfrac{dy}{dx} = 1$;

(v) $\sin x \dfrac{dx}{dy} + \cosh^2 y = 0$;

(vi) $y\sqrt{(2x^2 + 3)}\dfrac{dy}{dx} + x\sqrt{(4 - y^2)} = 0$.

2 Solve the differential equations

(i) $\dfrac{dy}{dx} - (1 + e^y) \sin x = 0$;

(ii) $y\dfrac{dy}{dx} - x^2 = x^2 y$;

(iii) $e^y\left(\dfrac{dy}{dx} + 1\right) = 1$, given that $y = 1$ when $x = 0$;

(iv) $(1 - x)y\dfrac{dy}{dx} = 4 - y^2$, and show that the particular curve of the system which passes through the origin has equation $y^2 = 4x(2 - x)$.

3 Solve the differential equations

(i) $(x + 1)\dfrac{dy}{dx} + 2y = 4(x + 1)^2$, $y = 0$ when $x = 1$;

(ii) $\dfrac{dy}{dx} + 2y \tan x = 3$;   (iii) $x \log x \dfrac{dy}{dx} + y = x \log x$;

(iv) $\cot x + y\dfrac{dx}{dy} = \dfrac{dx}{dy}$;   (v) $\cos x \dfrac{dy}{dx} + y \sin x = \cos 2x$.

4 (i) If $dy/dx + 2y \tan x = \sin x$ and $y = 0$ when $x = 0$, show that $y = -2$ when $x = \pi$.

(ii) Solve the differential equation

$$\frac{dy}{dx} + y \cot x = x^2.$$

(iii) Solve the differential equation

$$x \frac{dy}{dx} + y = x^3.$$

5 Use the substitution $y = ux$ to solve

(i) $x(x + y) \dfrac{dy}{dx} = y(x - y);$ (ii) $3x^2 \dfrac{dy}{dx} = x^2 + 2y^2;$

(iii) $x^2 \dfrac{dy}{dx} = y(2x + y);$ (iv) $x^2 + y^2 + 2xy \dfrac{dy}{dx} = 0;$

(v) $(x^2 - xy) \dfrac{dy}{dx} + y^2 = xy.$

6 Show that the following differential equations are exact and find their solutions;

(i) $(x + y)\,dx + (x + y^2)\,dy = 0;$

(ii) $(2xe^y + e^z)\,dx + (x^2 + 1)e^y\,dy = 0;$

(iii) $(2xy + y^2)\,dx + (x^2 + 2xy - y)\,dy = 0.$

7 Solve the differential equations

(i) $y \dfrac{d^2y}{dx^2} + \left(\dfrac{dy}{dx}\right)^2 + 1 = 0$

given that when $x = 0$, $y = 1$ and $dy/dx = 0$;

(ii) $(1 - x^2) \dfrac{d^2y}{dx^2} - x \dfrac{dy}{dx} = 2;$

(iii) $\dfrac{d^2y}{dx^2} + k^2 \left(\dfrac{dy}{dx}\right)^2 = 1,$

where k is a given constant.

8 A particle moves in a straight line, its distance x from a fixed point of the line at time t is given by the equation

$$\frac{d^2x}{dt^2} + k\left(\frac{dx}{dt}\right)^2 = \mu x,$$

where k and μ are positive constants. If initially the particle is at rest at the point $x_0 = 1/(2k)$, obtain an expression for the velocity, and show that the particle moves with constant acceleration $\mu/(2k)$ in the direction of increasing x.

9 Obtain the complete primitive for each of the following:

(i) $(D^2 - 3D - 10)y = 0$;   (ii) $(D^2 + 3D)y = 0$;
(iii) $(D^2 + 4D + 4)y = 0$;   (iv) $(D^2 - 4D + 13)y = 0$;
(v) $(D^4 + 4m^4)y = 0$, where m is constant.

10 Find the most general solution of the following equations such that
(a) $y = 0$ when $x = 0$; (b) $y = 1$, $dy/dx = 0$ when $x = 0$.

(i) $\dfrac{d^2y}{dx^2} + 4\dfrac{dy}{dx} + 3y = 0$;   (ii) $\dfrac{d^2y}{dx^2} - 2\dfrac{dy}{dx} + y = 0$.

11 Determine particular integrals for the equations

(i) $(D^2 + 4D + 3)y = x^2$;   (ii) $(D^2 + D)y = 3x + 7$.

12 Obtain the complete primitives of the following equations

(i) $(D^2 + 2D + 5)y = 4 + 2x$; (ii) $(D^2 - 4D + 4)y = x^2$;
(iii) $(5D^2 - 4D - 1)y = 2x^2$;   (iv) $(D^4 + 3D^2)y = x^2$.

13 Determine particular integrals for the following equations:

(i) $(D^2 + 4D + 3)y = e^x$;   (ii) $(D^2 + 4D + 3)y = 3e^{-x}$;
(iii) $(D^2 - 2D + 1)y = e^x$;   (iv) $(D^2 + 5D + 6)y = xe^{-x}$;
(v) $(D^2 + 4)y = x^2e^x$;   (vi) $(3D^2 - 10D + 3)y = 2e^{3x}$.

14 Solve the following differential equations:

(i) $\dfrac{d^2y}{dx^2} + 2\dfrac{dy}{dx} + 5y = (4 + 2x)e^x$;

(ii) $\dfrac{d^2y}{dx^2} + 2\dfrac{dy}{dx} = x^2e^{2x}$;

(iii) $\dfrac{d^3y}{dx^3} - 2\dfrac{d^2y}{dx^2} + \dfrac{dy}{dx} = xe^x$.

15 Determine particular integrals for the differential equations:

(i) $(D^2 + 3D)y = 5\cos x$;   (ii) $(D^2 + 9)y = 2\cos 3x$;
(iii) $(D^2 - 2D + 1)y = e^x \sin x$; (iv) $(D^2 - 2D + 2)y = e^x \sin x$;
(v) $(D^4 + 2D^2 + 1)y = \sin x$.

16 Solve the following differential equations

(i) $(D^2 + 9)y = 4\cos 2x$;   (ii) $(D + 4)y = \cos x$;
(iii) $(D^2 + 5D + 6)y = \cos x + \sin 2x$.

17 (i) Solve $\dfrac{d^2x}{dt^2} - 4\dfrac{dx}{dt} + 4x = 0$, given that $x = 0$, $dx/dt = 2$ when $t = 0$.

(ii) Solve $\dfrac{d^2x}{dt^2} + x = 3\sin 2t$, given that $x = dx/dt = 0$ when $t = 0$.

18 (i) Solve the differential equation

$$\frac{d^2y}{dx^2} + 6\frac{dy}{dx} + 9y = e^{-3x},$$

given that when $x = 0$, $y = 0$ and $dy/dx = -1$.

(ii) Find the most general solution of the differential equation $(D^2 - 3D - 4)^2y = 0$, such that $dy/dx = 0$ when $x = 0$ and the ratio y/x tends to zero as x tends to zero.

19 Solve the equations

(i) $(D - 1)^2y = xe^x$, given that $y = 1$, $Dy = \frac{1}{2}$ when $x = 0$;

(ii) $\frac{d^2y}{dx^2} + 4\frac{dy}{dx} + 3y = 8e^x$, given that when $x = 0$, $y = 0$ and $dy/dx = 4$.

20 Solve the differential equations

(i) $\frac{d^2y}{dx^2} + 3\frac{dy}{dx} + 2y = 4e^{-x}$, given that y and dy/dx are zero when x is zero;

(ii) $\frac{d^2y}{dx^2} + 2\frac{dy}{dx} + 10y = \sin 3x$, given that $y = 0$, $dy/dx = 1$ when $x = 0$.

21 Solve the differential equations

(i) $(D^2 - 2D + 1)y = x^3$;   (ii) $(D^2 - 10D + 29)y = e^x \cos 3x$;
(iii) $(9D^2 + 6D + 1)y = 0$;   (iv) $(D^2 + 2D + 1)y = x + 1$;
(v) $(D^2 + 3D + 2)y = e^x$;   (vi) $(D^2 + 2D + 5)^2y = 0$.

22 Integrate the differential equations

(i) $\frac{d^2y}{dx^2} + 3\frac{dy}{dx} - 4y = 2\cos 2x$;

(ii) $\frac{d^2y}{dx^2} = 2\frac{dy}{dx} + 3y$;

(iii) $\frac{d^2y}{dx^2} + 4\frac{dy}{dx} = 2\sin 2x$;

(iv) $5\frac{d^2y}{dx^2} - 3\frac{dy}{dx} - 2y = x^2e^x$;

(v) $\frac{d^2y}{dx^2} - 4\frac{dy}{dx} + 5y = x(e^x \sin x + x)$.

23 Sketch the isoclines of the differential equation

$$dy/dx = (x - y)/x$$

for $x > 0$. Use these to sketch the solution curves for $x > 0$.

24 Use the method of section 11.1.3 to show that the general solution of the differential equation $dy/dx = 1 + (y/x)$ is $y = x \log Ax$ where A is a constant. Sketch the isoclines of the differential equation, show the solution curve passing through $(1, 0)$, and compare this with the actual solution curve through that point.

25 Sketch the isoclines of the differential equation

$$dy/dx = (x^2 + y^2)/x$$

and use them to sketch the solution curve near to the point $(2, 0)$.

12
Analytical Properties of Algebraic Equations

12.0 INTRODUCTION

This chapter is concerned with some of the simpler analytical properties of algebraic equations particularly those of polynomial type. These properties are of use in subsequent chapters.

12.1 THE REMAINDER THEOREM

This states that if $f(x)$ is a function possessing a Taylor series at the point $x = a$ then the remainder on dividing by $(x - a)$ is $f(a)$.

Proof

The Taylor series for $f(x)$ is

$$f(x) = f(a) + (x - a)f'(a) + (x - a)^2 f''(a)/2! + \ldots$$

$$= f(a) + (x - a)\phi(x)$$

where

$$\phi(x) = f'(a) + (x - a)f''(a)/2! + \ldots$$

Hence
$$f(x)/(x - a) = \phi(x) + f(a)/(x - a)$$

in which $\phi(x)$ is called the *quotient* and $f(a)$ is called the *remainder*.

In particular if $f(x)$ is a polynomial of degree n then $\phi(x)$ is a polynomial of degree $n - 1$.

Example (i) If $x^4 - 7x^3 + x - 2$ is divided by $(x + 2)$ the remainder is $(-2)^4 - 7(-2)^3 + (-2) - 2 = 68$.

Example (ii) If $x \sin 2x$ is divided by $(x - \pi/4)$ the remainder is $(\pi/4)(\sin \pi/2) = \pi/4$.

12.1.1 The factor theorem

This states that if $(x - a)$ is a factor of $f(x)$ then $f(a)$ is zero and is an immediate consequence of the remainder theorem.

Example (i) Let $f(x)$ be $x^3 + 2x^2 - 5x - 6$, then by trial

$$f(2) = 8 + 8 - 10 - 6 = 0$$

and $(x - 2)$ is a factor of $f(x)$. By inspection

$$f(x) = (x - 2)(x^2 + 4x + 3)$$
$$= (x - 2)(x + 1)(x + 3).$$

Example (ii) Let $f(x)$ be $1 - \sin x$ then by trial $f(\pi/2) = 0$ and $(x - \pi/2)$ is a factor of $1 - \sin x$. But

$$f(x) = 1 - \sin x$$
$$= f(\pi/2) + (x - \pi/2)f'(\pi/2) + (x - \pi/2)^2 f''(\pi/2)/2! + \ldots$$
$$= 0 + 0 + (x - \pi/2)^2/2! + \ldots$$

which shows that $(x - \pi/2)$ is a double factor of $1 - \sin x$.

12.2 SOLUTION OF EQUATIONS. MULTIPLE ROOTS

It follows immediately from Taylor series and section 12.1 that the equation $f(x) = 0$ has a root at $x = a$ if $f(a) = 0$. This is a simple root if $f'(a) \neq 0$; a double root if $f'(a) = 0$, $f''(a) \neq 0$; a triple root if $f'(a) = f''(a) = 0$, $f'''(a) \neq 0$; and so on. The condition that $f(x) = 0$ has n equal roots $x = a$ is that $f(a) = f'(a) = f''(a) = \ldots = f^{(n-1)}(a) = 0$, $f^{(n)}(a) \neq 0$.

If $f(x)$ is a polynomial of order n then the equation $f(x) = 0$ can have at most n roots $x = x_1$, $x = x_2$, $x = x_3$, $\ldots$, $x = x_n$ because successive division of $f(x)$ by the factors $(x - x_1)$, $(x - x_2)$, $\ldots$ $(x - x_n)$ gives a final quotient which is constant. On the other hand a general equation $f(x) = 0$ where $f(x)$ has a Taylor expansion can have an infinity of roots. For example, the equation $\sin x = 0$ has an infinity of roots given by $x = r\pi$, where r is an integer.

12.3 DETACHED COEFFICIENTS. SYNTHETIC DIVISION

It is often necessary to divide a polynomial by a linear expression $(x - \alpha)$ in order to examine the quotient and remainder. There are two ways in which the familiar long division process may be displayed. The first is by using detached coefficients in the long division process itself and the other is a condensed version of this display called *synthetic division*. To illustrate the methods consider the division of $3x^4 - 7x^3 + 4x^2 - 3x - 5$ by $(x - 2)$.

12.3.1 Division using detached coefficients

The reader should have no difficulty in following the long division process employed.

$$
\begin{array}{r}
1 \quad -2)\ 3 \quad -7 \quad +4 \quad -3 \quad -5 \quad (3 \quad -1 \quad +2 \quad +1 \\
3 \quad -6 \\
\hline
-1 \quad +4 \\
-1 \quad +2 \\
\hline
+2 \quad -3 \\
+2 \quad -4 \\
\hline
+1 \quad -5 \\
+1 \quad -2 \\
\hline
-3
\end{array}
$$

The quotient is $3x^3 - x^2 + 2x + 1$ and the remainder is -3. This process starts by dividing $3x^4$ by x to give the first coefficient 3 of the quotient, this coefficient is then multiplied by the coefficient -2 of the divisor and *subtracted* from the next coefficient -7 of the dividend to give the term -1 of the third row. This process is repeated until no further division is possible.

12.3.2 Synthetic division

The process of section 12.3.1 may be conveniently contracted by omitting the irrelevant detail and changing the sign of the coefficient -2 of the divisor so as to make *additions* instead of subtractions. The process is displayed below:

$$
\begin{array}{r|rrrr}
 & 3 & -7 & +4 & -3 & -5 \\
+2 & & +6 & -2 & +4 & +2 \\
\hline
 & 3 & -1 & +2 & +1 & \boxed{-3}
\end{array}
$$

The procedure is as follows:

  (i) write the dividend in detached coefficient form, any absent terms being shown by a zero coefficient,
 (ii) show the multiplier $+2$ (associated with the divisor, $x - 2$),
(iii) carry the first coefficient of the dividend, i.e. 3, into the last row,
 (iv) multiply the first coefficient of the last row by $+2$ and place the product $+6$ in the second place of the second row,
  (v) *add* the two terms of the second column to give -1 as the second term of the last row,
 (vi) multiply the second coefficient of the last row by the multiplier $+2$ and *add* to the third coefficient of the dividend,

(vii) continue the process,
(viii) the last term of the last row is the remainder and the preceding terms form the coefficients of the quotient.

Example Use synthetic division to obtain the quotient and remainder when $3x^5 + 9x^4 - x^2 + 11x$ is divided by $(x + 3)$.

Solution

$$
\begin{array}{r|rrrrrr}
 & 3 & +9 & +0 & -1 & +11 & +0 \\
-3 & & -9 & 0 & 0 & +3 & -42 \\
\hline
 & 3 & 0 & 0 & -1 & +14 & \boxed{-42} \\
\end{array}
$$

The quotient is $3x^4 - x + 14$ and the remainder is -42. Note the zero coefficients in the dividend corresponding to the coefficients of x^3 and constant term.

12.3.3 Division of a polynomial by $(ax - b)$

Let $f(x)$ be a polynomial of degree n then

$$\frac{f(x)}{ax - b} = \frac{1}{a}\left[\frac{f(x)}{x - b/a}\right]$$

$$= \frac{1}{a}\left[\phi(x) + \frac{f(b/a)}{x - b/a}\right]$$

$$= \frac{1}{a}\phi(x) + \frac{f(b/a)}{ax - b}$$

where $\phi(x)$ is a polynomial of degree $n - 1$.
The process is then seen to be as follows:

 (i) divide $f(x)$ by the expression $(x - b/a)$,
(ii) the coefficients of the quotient obtained are a times too large but the remainder term is correct.

Example (i) Divide $21x^4 - 7x^3 + 3x^2 - 10x - 1$ by $(3x - 1)$.

Solution

$$
\begin{array}{r|rrrrr}
 & 21 & -7 & +3 & -10 & -1 \\
\frac{1}{3} & & +7 & 0 & +1 & -3 \\
\hline
 & 21 & 0 & +3 & -9 & \boxed{-4} \\
\end{array}
$$

The quotient is $7x^3 + x - 3$ and the remainder is -4.

Example (ii) Divide $16x^4 + 24x^3 - 3x - 2$ by $(2x + 3)$.

Solution

$$
\begin{array}{r|rrrrr}
 & 16 & +24 & +0 & -3 & -2 \\
-\frac{3}{2} & & -24 & 0 & 0 & +\frac{9}{2} \\
\hline
 & 16 & 0 & 0 & -3 & \boxed{+\frac{5}{2}} \\
\end{array}
$$

The quotient is $8x^3 - \frac{3}{2}$ and the remainder is $+\frac{5}{2}$.

12.3.4 Repeated division

Repeated division of a polynomial by a linear expression may be easily displayed by the synthetic division process.

Example Divide $6x^5 + 80x^2 - 1$ by $(x + 2)^3$.

Solution

$$
\begin{array}{r|rrrrrr}
 & 6 & 0 & 0 & +80 & 0 & -1 \\
-2 & & -12 & +24 & -48 & -64 & +128 \\
\hline
 & 6 & -12 & +24 & +32 & -64 & +127 \\
-2 & & -12 & +48 & -144 & +224 & \\
\hline
 & 6 & -24 & +72 & -112 & +160 & \\
-2 & & -12 & +72 & -288 & & \\
\hline
 & 6 & -36 & +144 & -400 & &
\end{array}
$$

Hence

$$\frac{6x^5 + 80x^2 - 1}{(x + 2)^3} = 6x^2 - 36x + 144 - \frac{400}{(x + 2)} + \frac{160}{(x + 2)^2}$$
$$+ \frac{127}{(x + 2)^3}.$$

It may be noted that these terms are the partial fractions relevant to $(x + 2)^3$.

Continued division may be used to express the given polynomial as a polynomial in the linear divisor. For example, continuing the example above

$$
\begin{array}{r|rrr}
 & 6 & -36 & +144 \\
-2 & & -12 & +96 \\
\hline
 & 6 & -48 & +240 \\
-2 & & -12 & \\
\hline
 & 6 & -60 &
\end{array}
$$

from which it follows that

$$6x^5 + 80x^2 - 1 = 6(x + 2)^5 - 60(x + 2)^4 + 240(x + 2)^3$$
$$-400(x + 2)^2 + 160(x + 2) + 127.$$

From the Taylor series it is evident that the coefficients $+127$, $+160$, -400, $+240$, -60, and $+6$ are the successive terms $f(-2)$, $f'(-2)$, $f''(-2)$, $f'''(-2)$, $f^{(4)}(-2)$ and $f^{(5)}(-2)$ respectively.

12.3.5 Division of a polynomial by a quadratic expression

Synthetic division may be extended to include the division of a polynomial of degree greater than two by a quadratic expression $ax^2 +
bx + c$. Consider the long division of $3x^4 + 5x^3 - 15x^2 + 10x - 1$
by $x^2 + 2x - 3$. In detached coefficient form this is

$$
\begin{array}{r}
1 + 2 - 3)3 \quad +5 \quad -15 \quad +10 \quad -1(3 \quad -1 \quad -4 \\
3 \quad +6 \quad -9 \quad \\
\hline
-1 \quad -6 \quad +10 \\
-1 \quad -2 \quad +3 \\
\hline
-4 \quad +7 \quad -1 \\
-4 \quad -8 \quad +12 \\
\hline
+15 \quad -13
\end{array}
$$

and the quotient is $3x^2 - x - 4$ the remainder being $(15x - 13)$.
As in section 12.3.2 this process may be contracted by omitting un-
necessary repeated coefficients and changing the signs of the coefficients $+2$ and -3 in the divisor so that any subtraction is replaced
by addition. The display is similar to that of synthetic division by a
linear expression but an additional row of coefficients is required. The
first row of coefficients is obtained from the dividend. The second row
is obtained from the products of the multiplier -2 with the coefficients
of the quotient and the third row consists of the products of the multiplier $+3$ and the coefficients of the quotient. The last row consists of
the coefficients of the quotient and remainder. The condensed array
is shown below

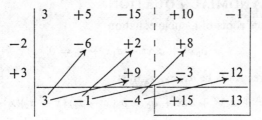

and the procedure is as follows:

(i) carry the first coefficient of the dividend into the first place of the
last row,

(ii) multiply this coefficient by the multiplier -2 and enter the product
in the second place of the second row. Multiply the same coefficient
by the second multiplier $+3$ and enter the product in the third
place of the third row,

(iii) *add* the elements of the second column and enter in the second place of the last row,
(iv) repeat the process of (ii) and (iii) entering the products into the third and fourth places of the second and third rows respectively,
 (v) repeat the process until the last coefficient of the third row is found. Add the coefficients of the last two columns,
(vi) the coefficients of the linear remainder are the last two coefficients of the last row. The coefficients of the quotient are the remaining ones of the last row.

The procedure may easily be adapted to include a divisor with a coefficient of x^2 other than unity. As in section 12.3.3, the scheme is written for a divisor $x^2 + (b/a)x + (c/a)$, the remainder will then have the correct coefficients but coefficients of the quotient will be a times too large.

Example Divide $3x^3 + 2x^2 - x + 5$ by $3x^2 - x + 2$.

Solution The scheme becomes

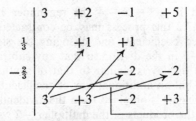

and the quotient is $(x + 1)$, the remainder $(-2x + 3)$.

12.4 RELATIONSHIPS BETWEEN THE ROOTS OF A POLYNOMIAL EQUATION

Consider the roots of a cubic equation

$$a_3x^3 + a_2x^2 + a_1x + a_0 = 0.$$

Let the three roots be α_1, α_2, α_3 then

$$a_3x^3 + a_2x^2 + a_1x + a_0 \equiv a_3(x - \alpha_1)(x - \alpha_2)(x - \alpha_3)$$

and comparison of like coefficients leads to the following relationships

$$\left.\begin{aligned} \alpha_1 + \alpha_2 + \alpha_3 &= -a_2/a_3 \\ \alpha_2\alpha_3 + \alpha_3\alpha_1 + \alpha_1\alpha_2 &= a_1/a_3 \\ \alpha_1\alpha_2\alpha_3 &= -a_0/a_3 \end{aligned}\right\} \tag{1}$$

The left sides of equations (1) are the sums of the products of the roots taken one at a time, two at a time, and three at a time respectively.

Example (i) If α_1, α_2, α_3 are the roots of $2x^3 + 3x^2 - 17x + 12 = 0$, then

$$\alpha_1 + \alpha_2 + \alpha_3 = -3/2$$

$$\alpha_2\alpha_3 + \alpha_3\alpha_1 + \alpha_1\alpha_2 = -17/2$$

$$\alpha_1\alpha_2\alpha_3 = -6.$$

Extension of this result to the n roots of an nth degree polynomial equation

$$a_n x^n + a_{n-1}x^{n-1} + a_{n-2}x^{n-2} + \ldots + a_1 x + a_0 = 0$$

leads to the relationships

$$\left.\begin{aligned}
\Sigma\alpha_i &= -a_{n-1}/a_n \\
\Sigma\alpha_i\alpha_j &= +a_{n-2}/a_n \quad i \neq j \\
\Sigma\alpha_i\alpha_j\alpha_k &= -a_{n-3}/a_n \quad i \neq j \neq k \\
&\qquad\cdot\qquad\cdot \\
&\qquad\cdot\qquad\cdot \\
&\qquad\cdot\qquad\cdot \\
\alpha_1\alpha_2\alpha_3 \ldots \alpha_n &= (-1)^n a_0/a_n
\end{aligned}\right\} \tag{2}$$

in which the left sides are the sums of the products of the roots taken one at a time, two at a time, three at a time, up to n at a time.

It should be noted that the relationships do not in themselves solve the original equations. For example, elimination of α_2 and α_3 in equations (1) simply leads to

$$a_3\alpha_1{}^3 + a_2\alpha_1{}^2 + a_1\alpha_1 + a_0 = 0.$$

However, if some additional information about the roots is known, then these relationships may help to solve the equation.

Example (ii) Find the roots of $8x^3 + 18x^2 - 17x + 3 = 0$ given that one root is double one of the others.

Solution Suppose the roots are α, 2α, and β then,

$$3\alpha + \beta = -18/8 \quad \text{(i)}$$

$$2\alpha^2 + 3\alpha\beta = -17/8 \quad \text{(ii)}$$

$$2\alpha^2\beta = -3/8 \quad \text{(iii).}$$

Eliminating β between (i) and (ii) gives

$$56\alpha^2 + 54\alpha - 17 = 0$$

which leads to

$$\alpha = 1/4 \text{ or } \alpha = -17/14.$$

From (i) $\beta = -3$ or $\beta = 39/28$ and of the two possibilities $\alpha = \tfrac{1}{4}$, $\beta = -3$ satisfies (iii) but $\alpha = -17/14$, $\beta = 39/28$ does not.

12.4.1 Diminishing the roots of an equation

The polynomial equation

$$a_n x^n + a_{n-1} x^{n-1} + \ldots + a_0 = 0$$

and the polynomial equation

$$a_n (x - k)^n + b_{n-1}(x - k)^{n-1} + \ldots + b_0 = 0$$

where the b_i are obtained, as in section 12.3.4, by dividing the first equation n times by $(x - k)$, where k is a constant, have the same roots $\alpha_1, \alpha_2, \alpha_3, \ldots, \alpha_n$. The equation

$$a_n y^n + b_{n-1} y^{n-1} + b_{n-2} y^{n-2} + \ldots + b_0 = 0$$

therefore has roots $(\alpha_1 - k), (\alpha_2 - k), \ldots, (\alpha_n - k)$. This process of forming a new equation whose roots are each k less than the original is called diminishing the roots by k. Similarly the roots can be increased by k.

Example Diminish the roots of $x^3 - 6x^2 + 9x - 4 = 0$ by 2.

Solution By section 12.3.4

	1	−6	+9	−4
+2		+2	−8	+2
	1	−4	+1	−2
+2		+2	−4	
	1	−2	−3	
+2		+2		
+2	1	0		

and hence the roots of $y^3 - 3y - 2 = 0$ are each 2 less than those of the given equation.

12.5 LOCATION OF THE ROOTS OF A POLYNOMIAL EQUATION WITH REAL COEFFICIENTS

The process of root location is usually a trial and error one but there are a number of properties that may be used as aids to location and classification of roots.

(a) If $x_2 > x_1$ and $f(x_2), f(x_1)$ are of opposite signs then there is at least one root of $f(x) = 0$ in the range $x_1 < x < x_2$. This is obvious geometrically.

Example (i) If $f(x) = x^3 + 2x + 1$, then, since $f(-1) < 0$ and $f(0) > 0$, there is at least one root between $x = -1$ and $x = 0$.

(b) If α and β are successive real zeros of a polynomial $f(x)$ then $f'(x)$ has an odd number of real zeros in the interval $\alpha < x < \beta$.

This property follows from Rolle's theorem of section 5.1.1 because $f(\alpha) = 0 = f(\beta)$, $f(x)$ is continuous and $f'(x)$ exists. Stated differently the real roots of $f(x) = 0$ are separated by the real roots of $f'(x) = 0$.

Example (ii) If $f(x) = x^3 - 7x + 2$, then $f'(x) = 3x^2 - 7$ and the latter has two roots $x = \pm\sqrt{(7/3)}$. Since $\underset{x \to -\infty}{\text{Lim}} f(x) < 0$, $f(-\sqrt{(7/3)}) > 0$, $f(+\sqrt{(7/3)}) < 0$ and $\underset{x \to +\infty}{\text{Lim}} f(x) > 0$ there are three changes of sign for $f(x)$ and it follows from (a) that $f(x) = 0$ has three real roots. One of these roots is in the range $-\infty < x < -\sqrt{(7/3)}$, one in the range $-\sqrt{(7/3)} < x < +\sqrt{(7/3)}$, and one in the range $\sqrt{(7/3)} < x < +\infty$.

It may be stated that if $f'(x) = 0$ has $(n - 1)$ unequal roots α_1, α_2, $\alpha_3, \ldots, \alpha_{n-1}$ in ascending order and if the signs of the terms of the series $f(-\infty), f(\alpha_1), f(\alpha_2), f(\alpha_3), \ldots, f(\alpha_{n-1}), f(+\infty)$ alternate then $f(x) = 0$ has n unequal roots.

(c) Since complex roots occur in conjugate pairs (see section 10.5.3) then a polynomial of even degree has an even number (including zero) of real roots and a polynomial of odd degree has an odd number of real roots (i.e. at least one real root).

(d) The Descartes' rule of signs states that if $f(x)$ is a polynomial written in descending powers of x then

(i) the number of positive real roots of $f(x) = 0$ is not larger than the number of *variations* of $f(x)$ and,

(ii) the number of negative real roots is not larger than the number of variations of $f(-x)$.

In applying the rule, $f(x)$ is said to have a variation if two successive terms have opposite signs.

Example (iii) Show that $x^5 - 11x^3 + 2x^2 + 12 = 0$ has two complex roots, two positive real roots, and one negative real root.

Solution Since $f(-\infty) < 0$, $f(0) = 12 > 0$, $f(2) = -36 < 0$, and $f(+\infty) > 0$ then, by (a), there must be at least one negative real root and at least two positive real roots. Descartes' rule of signs (d) states that $f(x) = 0$ has not more than two positive real roots and not more than one negative real root because $f(x)$ has two variations and $f(-x)$ has one variation.

There are therefore exactly two positive real roots and one negative real root and hence two complex roots.

(e) A rough graph is helpful in the location of the roots of an equation $f(x) = 0$. For example, a sketch of the curve $y = x^3 - 3x - 1$ from $x = -2$ to $x = +2$ will show that there is one root between -2 and -1, one root between -1 and 0, and one root between 1 and 2.

Alternatively a sketch of the two graphs of $y = x^3$ and $y = 1 + 3x$ will show the three intersection points.

12.5.1 Procedure after location of a root

If a root $x = \alpha$ of a polynomial equation is known or can be exactly located then the polynomial may be divided by $(x - \alpha)$ to give a quotient and a zero remainder (the evaluation of the latter serves as a check on the working). An attempt may then be made to locate the zeros of the quotient and hence the further roots of the original equation.

It is likely that the roots of a polynomial cannot be found exactly and in this case an approximate value must be obtained. Methods of approximate root location will be discussed in the next chapter.

EXERCISE 12

1 Divide $3x^6 - x^5 + 4x^3 - 1$ by (i) $x - 2$, and (ii) $3x - 1$.

2 Factorise the function $4x^4 - 6x^3 + 4x^2 - 3x + 1$.

3 Express
$$(3x^7 - 36x^6 + 162x^5 - 324x^4 + 244x^3 - 13x^2 + 51x - 58)/(x - 3)^4$$
as the sum of a quotient and partial fractions.

4 Find the quotient and remainder when $4x^6 - 7x^4 + 24x^3 - 18x^2 + 13$ is divided (i) by $x^2 - 2x + 3$ and (ii) by $2x^2 + 3x - 1$.

5 Using the relationships between the roots find the equation whose roots are the squares of those of $3x^3 + 2x - 1 = 0$.

6 Solve the equation $12x^3 - 20x^2 + x + 3 = 0$ given that the sum of two of the roots is 2.

7 Show that $6y^3 + 19y^2 + 11y - 11 = 0$ is the equation obtained by diminishing by three the roots of $6x^3 - 35x^2 + 59x - 35 = 0$. Show that there is a root $y = \frac{1}{2}$ and hence solve completely the equation in x.

8 Show that $f(x) \equiv 6x^4 - 7x^3 - x - 6 = 0$ has one negative real root, one positive real root, and two complex roots.

9 An approximate root of $x^3 + x^2 - 7x + 1 = 0$ is $x = 2 \cdot 10$ correct to two decimal places. Divide the polynomial by $(x - 2 \cdot 10)$ and find approximations to the other two roots.

10 Show that $x^3 - 3x - 1 = 0$ has three real roots, one between $x = -2$ and $x = -1$, one between $x = -1$ and $x = 0$, and one between $x = 1$ and $x = 2$. Show that the substitution $x = r \cos \theta$ and the identity $\cos 3\theta = 4 \cos^3 \theta - 3 \cos \theta$ transforms the given equation to $\cos 3\theta = \frac{1}{2}$ provided that $r = 2$. Solve this equation for θ and hence find all the roots of the given equation.

11 Show that the method of Question 10 applied to the equation $x^3 - 3x - 4 = 0$ leads to roots $x = 2 \cos \theta$ where θ is a solution of $\cos 3\theta = 2$. Solve for θ using the method of Question 14, Exercise 10 and hence find all the roots of the given equation.

13
Numerical Methods I

13.0 INTRODUCTION

The roots of an equation $f(x) = 0$, or the solution of a system of equations, may be found by either graphical or numerical processes. The former may be used if accuracy is not important, there being a limit to the accuracy with which a graph can be drawn and read. The numerical processes start with an approximation to the solution, often obtained from a rough graph, and then use a systematic scheme of numerical computation to improve the value. Such a process stops when the value is found to the required degree of accuracy. Any numerical process should, as far as possible, include some form of checking procedure so as to eliminate the propagation of errors. Among the numerical methods there is the *iterative method*. This starts with an approximation to the required value and by a given form of calculation produces a better approximation, which is then improved by the *same* form of calculation, the process terminating when two successive iterations give identical value to the degree of accuracy demanded (e.g. to the same number of decimal places). This method is the basis of most high speed computer programs, the computer performing a large number of iterations in a short space of time. It is possible for an iterative process to lead away from the required root, that is the process could diverge instead of converge. Under certain conditions the process may diverge from the required root but converge to some unwanted root. Usually it is necessary to ensure that certain criteria are satisfied by the equation or system of equations before commencing an iterative process. If these criteria are not satisfied then it may be necessary to employ an alternative process.

13.1 GRAPHICAL METHODS

When accuracy is not important, a root of the equation $f(x) = 0$ may be found by plotting the graph of the function $y = f(x)$ over some range of values $a \leqslant x \leqslant b$ for which $f(a)$ and $f(b)$ are of opposite signs. The root is then found by reading the abscissa of the crossing point with the x-axis. This is a process that needs no elaboration.

13.1.1 Intersection of two graphs

Instead of drawing the graph of $y = f(x)$ and finding the intersections with the x-axis it is often an advantage to write $f(x)$ in the form

$\phi(x) - \psi(x)$ and then draw the graphs of the two functions $y = \phi(x)$ and $y = \psi(x)$. The abscissae of the intersection points give the required roots of $f(x) = 0$.

Example The roots of the equation $1 + x^2 - 3 \tan x = 0$ are obtained from the intersections of $y = 1 + x^2$ and $y = 3 \tan x$. These curves are easily drawn to any reasonable accuracy and the abscissae of the intersections points obtained. The curves are indicated in Fig. 13.1 and from this rough sketch the smallest root is found to be approximately 0·4.

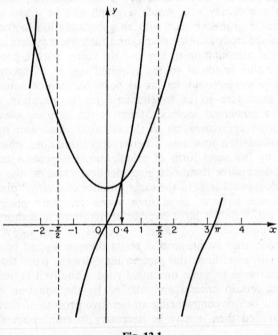

Fig. 13.1

13.1.2 Repeated plotting on a larger scale

An approximation $x = x_0$ to a root of $f(x) = 0$ may be improved by drawing the part of the graph $y = f(x)$, or of the parts of the graphs $y = \phi(x)$ and $y = \psi(x)$, near to $x = x_0$ to a larger scale and locating the root with additional accuracy. Instead of drawing the curves accurately it is usual to replace them by straight lines drawn through the end points of the interval about $x = x_0$. This introduces an error into the new estimate of the root so that the process is repeated until a sufficiently accurate result is obtained.

Example Continuing the example of section 13.1.1 the following table may be constructed

x	$\phi(x) = 1 + x^2$	$\psi(x) = 3 \tan x$
0·3	1·09	0·93
0·4	1·16	1·27

from which it is seen that the curves of $y = \phi(x)$ and $y = \psi(x)$ intersect between $x = 0·3$ and $x = 0·4$. Replacing the curves by straight lines through the end points of each curve as in Fig. 13.2 gives the improved value of the root as 0·36.

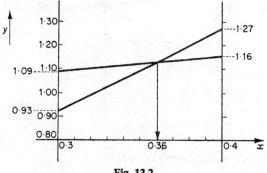

Fig. 13.2

An estimate of the error involved is found by calculating $f(0·36)$. This is 0·0004, and the result 0·36 may be accurate enough to stop the process at this stage. Improved accuracy may be obtained by drawing the straight lines through the end points corresponding to $x = 0·360$ and $x = 0·361$. This leads to the estimate $x = 0·3601$.

At this stage it may be claimed that the result $x = 0·360$ is correct to three decimal places.

13.2 NUMERICAL METHODS OF SOLVING $f(x) = 0$

The graphical solution of section 13.1.2 can be represented numerically as an iterative process known as the *rule of false position* and this is discussed in section 13.2.1. In addition two other methods of iteration are considered. One is discussed in section 13.2.2 and the other, a simple but also a very powerful process, known as Newton's method, is described in section 13.2.3.

13.2.1 Rule of false position or method of linear interpolation

Suppose a root of the equation $f(x) \equiv \phi(x) - \psi(x) = 0$ has been located between the values x_0 and $x_0 + a$ (the smaller the value of a

the better), then reference to Fig. 13.3 shows that a better approximation to the root will be given by $x = x_1 = x_0 + h$ where, using similar triangles,

$$\frac{h}{a - h} = \left| \frac{\psi(x_0) - \phi(x_0)}{\phi(x_0 + a) - \psi(x_0 + a)} \right| = \frac{A}{B} \text{ (say)}$$

or

$$\frac{h}{a} = \frac{A}{A + B}.$$

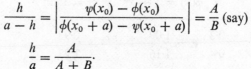

Fig. 13.3

The second approximation to the root is

$$x_1 = x_0 + \left(\frac{A}{A + B} \right) a \tag{1}$$

It may be noted that if $\psi(x) = 0$ then $\phi(x) = f(x)$ and A, B are simply $|f(x_0)|$ and $|f(x_0 + a)|$ respectively.

Example (i) Calculate the root of $1 + x^2 - 3 \tan x = 0$ that lies in the range $0 < x < 1$.

Solution Let $\phi(x) = 1 + x^2$ and $\psi(x) = 3 \tan x$, then it is easily seen that there is a root between 0 and 1. Suppose the first approximation to be $x_0 = 0$, then the following table may be constructed and the next approximation found by an application of equation (1):

x	$\phi(x)$	$\psi(x)$	$\phi - \psi$
0	1	0	1
1	2	4·67	−2·67

With $a = 1$, $A = 1$, $B = 2.67$, then $h = (1/3.67) \times 1$

$$= 0.27$$

$$\simeq 0.3.$$

The second approximation is then 0·3. Proceeding as before:

x	$\phi(x)$	$\psi(x)$	$\phi - \psi$
0·3	1·090	0·928	+0·162
0·4	1·160	1·268	−0·108

and here, $a = 0·1$, $A = 0·162$, $B = 0·108$, from which
$$h = (162/270) \times 0·1 = 0·06.$$
The third approximation is 0·36, then

x	$\phi(x)$	$\psi(x)$	$\phi - \psi$
0·360	1·129 60	1·129 21	+0·000 39
0·361	1·130 32	1·132 64	−0·002 32

and here, $a = 0·001$, $A = 0·000\ 39$, $B = 0·002\ 32$, and hence
$$h = (39/271) \times 0·001 = 0·000\ 1.$$
The fourth approximation is 0·360 1.

Note 1 The difference $\phi - \psi$ not only provides the values of A and B but the change in sign of the difference shows that there is a root between the two values of x used in the calculation.

Note 2 With this method it is necessary to retain more decimal places in each successive calculation, five figure tables were used in the last approximation.

Example (ii) Find the real root of
$$f(x) \equiv x^3 - 6x^2 + 12x - 24 = 0$$
to two decimal places.

Solution Since $f(4) = -8$ and $f(5) = +11$, there is a root between $x = 4$ and $x = 5$. Diminish the roots by 4 using the results of section 12.4.1:

then $F(z) \equiv z^3 + 6z^2 + 12z - 8 = 0$ has a root between 0 and 1.

Now successively apply (1) to $F(z)$ starting from the value $z_0 = 0$. Then

z	$F(z) \equiv z^3 + 6z^2 + 12z - 8$
0	-8
1	$+11$

and here, $a = 1$, $A = 8$, $B = 11$, hence

$$h = (8/19) \times 1 = 0 \cdot 4.$$

The second approximation is $z_1 = 0 \cdot 4$, then

z	$F(z)$
0·4	$-2 \cdot 176$
0·6	$+1 \cdot 576$

and here, $a = 0 \cdot 2$, $A = 2 \cdot 176$, $B = 1 \cdot 576$, hence

$$h = (2176/3752) \times 0 \cdot 2 = 0 \cdot 12.$$

The third approximation is $z_2 = 0 \cdot 52$, then

z	$F(z)$
0·52	$+0 \cdot 003\ 0$
0·51	$-0 \cdot 186\ 7$

and here $a = 0 \cdot 01$, $A = 0 \cdot 186\ 7$, $B = 0 \cdot 003\ 0$, and hence

$$h = (1867/1897) \times 0 \cdot 01 = 0 \cdot 009\ 8.$$

(Note that the sign of $F(z)$ when $z = 0 \cdot 52$ shows that the third approximation has overshot the root.)

The fourth approximation is $z_3 = 0 \cdot 519(8)$. At this stage it may be claimed that the root is $z = 0 \cdot 52$ correct to two decimal places. The root of $f(x) = 0$ is then $4 \cdot 52$.

In practice this method is seldom used after the second stage. A value having been found the approximation is improved by using a more powerful method such as the Newton–Raphson method of section 13.2.3.

13.2.2 Another iterative method

There is a useful iterative method available when $f(x) = 0$ can be written in the form $x = \phi(x)$. The iterative procedure is given by

$$x_{n+1} = \phi(x_n)$$

where x_n is the nth iterate and the procedure ends when the required accuracy has been obtained in two successive iterations.

The figures 13.4(a) and (b) illustrate two convergent iterative processes $x_0, x_1, x_2, x_3 \ldots$, one converging directly to the root and the other oscillating about the root value. The Fig. 13.4(c) illustrates a process that diverges from the root value. It is obviously possible with this process to start with an approximation very close to the required value and then to calculate values that are further and further away from the true value. It is a necessary and sufficient condition for convergence that $|\phi'(x)| < 1$ in the neighbourhood of the root. This criterion is satisfied for the curves of Fig. 13.4(a) and (b) but not for Fig. 13.4(c).

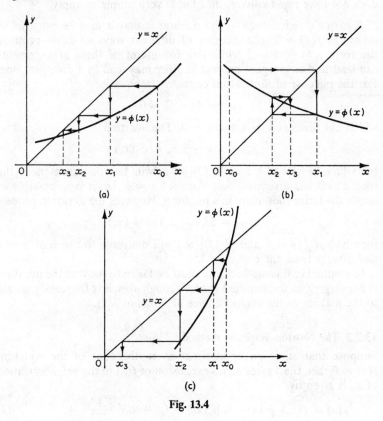

Fig. 13.4

Example Find the root of the equation

$$x^3 - 2x^2 + 3x - 1 = 0$$

near to 0·5.

Solution Write

$$x_{n+1} = \tfrac{1}{3}(1 + 2x_n^2 - x_n^3)$$

and start from $x_0 = 0$. The iteration proceeds as follows:

$$x_0 = 0 \qquad\qquad x_5 = 0.430$$
$$x_1 = 0.3 \qquad\qquad x_6 = 0.430\ 1$$
$$x_2 = 0.38 \qquad\qquad x_7 = 0.430\ 14$$
$$x_3 = 0.41 \qquad\qquad x_8 = 0.430\ 16$$
$$x_4 = 0.42 \qquad\qquad x_9 = 0.430\ 160$$

The root is therefore $0.430\ 2$ correct to 4 decimal places. This procedure does not have rapid convergence but is very simple to apply.

A major disadvantage of the method is that it may be possible to rearrange $f(x) = 0$ in a number of different ways so as to produce the form $x = \phi(x)$ and while one (or more) of these arrangements can lead to a convergent process another may lead to a divergent one. For the purpose of illustration consider the equation

$$x^2 - 4x + 3 = 0$$

which has roots at $x = 1$ and $x = 3$. The iteration

$$x_{n+1} = 4 - 3/x_n \equiv \phi_1(x_n)$$

for which $|\phi_1'(1)| > 1$ and $|\phi_1'(3)| < 1$, will be convergent near the root $x = 3$ but divergent near the root $x = 1$, i.e. it is impossible to locate the latter root using this iteration. However the iteration process

$$x_{n+1} = 3/(4 - x_n) \equiv \phi_2(x_n)$$

for which $|\phi_2'(1)| < 1$ and $|\phi_2'(3)| > 1$ will converge to the root $x = 1$ and diverge from the root $x = 3$.

In employing this method care must be taken to see that the iteration is converging to the required root. A rough sketch is frequently an aid to the making of the correct choice of function $\phi(x)$.

13.2.3 The Newton–Raphson method

Suppose that x_0 is an approximation to the root of the equation $f(x) = 0$ then the Taylor series expansion of $f(x)$ in the neighbourhood of x_0 is given by

$$f(x) = f(x_0) + (x - x_0)f'(x_0) + (x - x_0)^2\frac{f''(x_0)}{2!} + \ldots \qquad (2)$$

If x_1 is a value of x nearer to the exact root, then $(x_1 - x_0)$ will be small, $f(x_1)$ is approximately zero, and providing that $f''(x_0)$ is not too large the Taylor series may be replaced by

$$0 = f(x_0) + (x_1 - x_0)f'(x_0)$$

from which
$$x_1 = x_0 - \frac{f(x_0)}{f'(x_0)}. \qquad (3)$$

This is the Newton–Raphson rule and repeating for successive approximations $x_1, x_2, x_3, \ldots$ leads to the iterative relationship

$$x_{n+1} = x_n - \frac{f(x_n)}{f'(x_n)} \tag{4}$$

where x_n is the nth iterate. The geometrical interpretation is shown in Fig. 13.5.

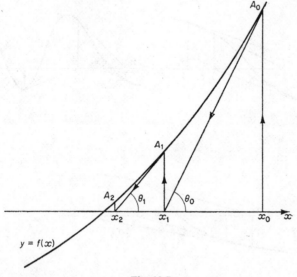

Fig. 13.5

The ordinate at A_0 is $f(x_0)$ and, since $\tan \theta_0 = f'(x_0)$, the tangent at A_0 intersects the x-axis in the point x_1. The process illustrated in Fig. 13.5 is seen to be convergent.

As a working rule the number of reliable decimal places in the quotient $f(x_n)/f'(x_n)$ is usually double the number of zeros between the decimal point and the first significant figure (i.e. the number of correct decimal places is approximately doubled at each iteration). The final answer is obtained when two successive iterates agree to the requisite number of decimal places.

It is evident from Fig. 13.5 that the convergence will be rapid providing that $f'(x)$ is large near to the root but becomes slower as $f'(x)$ approaches zero. In the latter case it becomes necessary to compute $f(x_n)$ and $f'(x_n)$ to high accuracy to obtain an accurate value of x_{n+1}.

Under certain conditions the convergence may be too slow for hand computation or may even diverge. The difficulty is frequently overcome by plotting a rough curve to a larger scale near to the root or roots and examining the geometrical situation. Different cases of convergence are illustrated in Fig. 13.6.

In Fig. 13.6(a) the process is convergent, the iterations being successively on either side of the root. In Fig. 13.6(b) the process converges to the root at R and neither of the roots at P or Q are located, in this case a starting approximation much nearer to P (or Q) will lead to a convergent process to P (or Q). In Fig. 13.6(c) the process is divergent starting from the approximation x_0. As in the previous case the root can be located by a convergent process starting from an approximation nearer to the root itself.

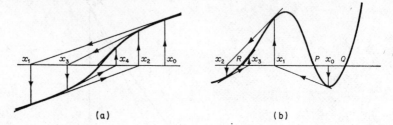

(a)   (b)

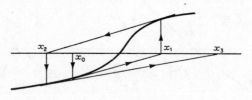

(c)

Fig. 13.6

Example (i) Find the smallest root of $1 + x^2 - 3 \tan x = 0$.

Solution Write

$$f(x) = 1 + x^2 - 3 \tan x$$

then

$$f'(x) = 2x - 3 \sec^2 x.$$

Take the first approximation to be $x_0 = 0 \cdot 3$ (being the second approximation obtained by the rule of false position) then

$$f(0 \cdot 3) = \quad 0 \cdot 16$$

$$f'(0 \cdot 3) = -2 \cdot 58,$$

and

$$x_1 = 0 \cdot 3 - (0 \cdot 16 / -2 \cdot 58)$$

$$= 0 \cdot 3 + 0 \cdot 06$$

$$= 0 \cdot 36.$$

With $x_1 = 0.36$,
$$x_2 = 0.36 - f(0.36)/f'(0.36)$$
$$= 0.36 - 0.000\ 4/(-2.705\ 1)$$
$$= 0.36 + 0.000\ 147\ 87$$
i.e. $x_2 = 0.360\ 148 \simeq 0.360\ 15$.

Using five figure accuracy, $f(0.360\ 15)$ is $0.000\ 08$. The result is probably correct to the fifth decimal place but a further iteration would be necessary to confirm this.

Example (ii) Find the root of
$$x^3 - 6x^2 + 12x - 24 = 0$$
near to $x = 4$.

Solution This is example (ii) of section 13.2.1 in which the rule of false position was used to show that $x = 4.52$ correct to two decimal places after three iterations.

Let $f(x) = x^3 - 6x^2 + 12x - 24$
and $f'(x) = 3x^2 - 12x + 12$
$$= 3(x - 2)^2$$
and start with $x_0 = 4$, then
$$x_1 = 4.0 - (-8/12) = 4.6$$
$$x_2 = 4.6 - (1.576/20.28) = 4.6 - 0.077\ 8 = 4.52 \text{ (say)}$$
$$x_3 = 4.52 - (0.003\ 10/19.051\ 3)$$
$$= 4.52 - 0.000\ 162\ 7 = 4.519\ 84$$
$$x_4 = 4.519\ 84 - (-0.000\ 02/19.048\ 79)$$
$$= 4.519\ 84 + 0.000\ 001\ 049\ 9$$
$$= 4.519\ 841\ 049\ 9$$

After four iterations the root is $4.519\ 84$ correct to five decimal places.

A useful application of the Newton–Raphson rule is to the calculation of square roots. If $x = \sqrt{a}$ then x is a root of $f(x) \equiv x^2 - a = 0$ and the rule then states that if x_n is the nth iterate
$$x_{n+1} = x_n - (f(x_n)/f'(x_n))$$
$$= x_n - (x_n^2 - a)/2x_n$$
i.e. $x_{n+1} = \frac{1}{2}(x_n + a/x_n)$.   (5)

Thus, if x_n is an approximation to the square root of a number a then the mean of x_n and the quotient a/x_n is a better approximation. Each iteration is correct usually to twice as many decimal places as the preceding iteration.

Example (iii) Find $\sqrt{5}$.

Solution Take the first approximation as $x_0 = 2$, then

$$x_1 = \tfrac{1}{2}(2 + 2 \cdot 5) = 2 \cdot 25$$
$$x_2 = \tfrac{1}{2}(2 \cdot 25 + 2 \cdot 222\ 2) = 2 \cdot 236\ 1$$
$$x_3 = \tfrac{1}{2}(2 \cdot 236\ 1 + 2 \cdot 236\ 035\ 96) = 2 \cdot 236\ 067\ 98$$

The result $2 \cdot 236\ 1$ is correct to four decimal places.

Note 1 A major advantage of the Newton–Raphson method is that any error made is not disastrous providing that the process, starting from the incorrect value, remains convergent.

Note 2 If $f'(x)$ is large the rate of convergence is large and the denominator term $f'(x)$ may be conveniently replaced by a suitable constant. For instance, in example (ii) above, $f'(x)$ could be replaced by 20, the successive iterates being $x_0 = 4$, $x_1 = 4 \cdot 4$, $x_2 = 4 \cdot 509$, $x_3 = 4 \cdot 519\ 1$.

Note 3 A measure of the accuracy of the iterate x_n is given by the value of $f(x_n)$. This value is called the *residual* and usually denoted by R.

Note 4 If the coefficients of the equation $f(x) = 0$ are known only to a certain accuracy then the approximations to the roots of the equation cannot be calculated with certainty to a greater accuracy. For instance if the coefficients are known to be correct only to two decimal places then there is no advantage in calculating the roots to many more decimal places.

Note 5 If (a) a small change in the value of the coefficients of $f(x)$ leads to a large change in the value of the roots, or

(b) small values of the residual R can be obtained even when the values of x_n are widely different from the true value of the root, then the equation $f(x) = 0$ is said to be *ill-conditioned*. Equations that are ill-conditioned are difficult to deal with by any method of numerical analysis. In the Newton–Raphson method it frequently means that $f(x)$ and $f'(x)$ are both small and that the calculations in each iteration must be carried out with the retention of many more decimal places than can be claimed with accuracy in the result of the iteration.

13.3 NUMERICAL SOLUTION OF LINEAR ALGEBRAIC SIMULTANEOUS EQUATIONS

The elimination method of Gauss was used in section 8.6.1 to solve a set of linear simultaneous algebraic equations. The numbers given there were simple integers and the calculations were performed using only mental arithmetic. The matrix scheme can be used even when the elements are decimal but it is more convenient to draw up a computational scheme and to introduce a system of checking. This scheme will be discussed in section 13.3.1. As in section 13.2 it is possible to form an iterative process that will converge to the solution of the set of equations and two such methods are discussed in sections 13.3.2 and 13.3.3.

13.3.1 Gaussian elimination method

Given a set of three equations in three unknowns x, y, and z:

$$\left. \begin{array}{l} 4 \cdot 12x - 9 \cdot 68y + 2 \cdot 01z = 4 \cdot 93 \\ 1 \cdot 88x - 4 \cdot 62y + 5 \cdot 50z = 3 \cdot 11 \\ 1 \cdot 10x - 0 \cdot 96y + 2 \cdot 72z = 4 \cdot 02 \end{array} \right\} \qquad (6)$$

the basic process is first to eliminate x and y and form an equation in z only, second to use this to form an equation in y only and finally to form an equation in x only. In terms of matrix algebra this is equivalent to transforming the coefficient matrix into an equivalent triangular matrix.

A simple check is introduced, and assuming no numerical mistakes are made (or left uncorrected) the errors in the final answer are due solely to rounding off errors in the multiplication and division used. Finally an estimate of the error may be formed and a correction applied. The computational scheme proceeds as follows:

Equation	x	y	z	b	σ	$R^{(1)}$	$R^{(2)}$	Operation
(6.1)	4·12	−9·68	2·01	4·93	1·38	0·006	0·000 1	
(6.2)	1·88	−4·62	5·50	3·11	5·87	−0·001	0·000 3	
(6.3)	1·10	−0·96	2·72	4·02	6·88	0·000	0·000 4	
(6.4)	1·000	−2·350	0·488	1·197	0·335	0·001 5		(1) ÷ 4·12
(6.5)	−1·880	4·418	−0·917	−2·250	−0·630	−0·002 8		(4) × −1·88
(6.6)	0·000	−0·202	4·583	0·860	5·240	−0·003 8		(5) + (2)
(6.7)	−1·100	2·585	−0·537	−1·317	−0·369	−0·001 7		(4) × −1·10
(6.8)	0·000	1·625	2·183	2·703	6·511	−0·001 7		(3) + (7)
(6.9)		1·000	1·343	1·663	4·007	−0·001 0		(8) ÷ 1·625
(6.10)		0·202	0·271	0·336	0·809	−0·000 2		(9) × 0·202
(6.11)		0·000	4·854	1·196	6·049	−0·004 0		(10) + (6)
(6.12)		1·000		0·246	1·246	−0·000 8		(11) ÷ 4·854
(6.13)		1·000		1·333	2·334	+0·000 1		(9) − (12) × 1·343
(6.14)	1·000			4·210	5·212	+0·002 1		(4) + (13) × 2·350 − (12) × 0·488

The first estimate of the solution is given by the b column of (6.12), (6·13), and (6.14). The columns marked R are columns of residuals but are not computed until after the first estimate is found. The columns x, y, z, and b together form the coefficients of the augmented matrix of the set of equations. The column marked σ is the check column and the first three entries in this column are simply the row sums of the entries in the previous four columns. The elimination process proceeds by performing the operations as indicated in the last column, the aim being to eliminate the successive variables until the final coefficient matrix is in a triangular form. As an aid to computation it is useful to reduce to unity the coefficient of the variable being eliminated. At each stage of the elimination the entry in the σ column is computed by performing the operations upon the previous entries of that column. This entry is then checked against the sum of the other entries of the row. For example, in forming row (6.9) the entry σ_9 is given by $\sigma_8 \div 1.625$, i.e. $6.511/1.625$ or 4.007. This entry is then checked against the sum of the other entries in that row, namely 4.006 which indicates that there is probably no major numerical error.

The first estimate of the solution is

$$x_1 = 4.210, \; y_1 = 1.333, \text{ and } z_1 = 0.246$$

with the third decimal place in doubt. The column of residuals $R^{(1)}$ is now computed as follows. The first entry is obtained by inserting the estimates x_1, y_1, and z_1 into the original equation (6.1) written in the form

$$4.12x_1 - 9.68y_1 + 2.01z_1 - 4.93.$$

The second and third entries in the $R^{(1)}$ column are found in like manner from the original equations (6.2) and (6.3). The first three entries are 0.006, -0.001, and 0.000. Note that if the estimates had been exact then the residuals would all have been zero. The remaining entries in the $R^{(1)}$ column are computed by performing the appropriate operation on the previous entries. Finally, the entries b of (6.12), (6.13), and (6.14) are adjusted by *subtracting* the corresponding residual value. The second estimate of the solution is then

$$x_2 = 4.210 - 0.002\,1 = 4.207\,9$$

$$y_2 = 1.333 - 0.000\,1 = 1.332\,9$$

$$z_2 = 0.246 + 0.000\,8 = 0.246\,8$$

The first three entries in the second column of residual values may now be found by inserting these new estimates into the original equations. The new residuals are $0.000\,1$, $0.000\,3$, and $0.000\,4$ respectively. The rest of the column may then be computed and the third estimates found.

The above method of solution solves the set of equations in one computational process and, providing the equations are not ill-conditioned and sufficient accuracy is maintained in all multiplications and divisions, the first estimate obtained should be very close to the correct answer. In contrast to this process it is possible to formulate an iterative process starting from a rough estimate of the solution and then gradually converging to the correct result by successive calculations of the same kind. There are two basic methods in common use, both modifications of a method originally due to Gauss.

13.3.2 The Gauss–Jacobi method

Suppose that a set of equations in three unknowns is given in the form

$$\left.\begin{array}{l} a_{11}x + a_{12}y + a_{13}z = b_1 \\ a_{21}x + a_{22}y + a_{23}z = b_2 \\ a_{31}x + a_{32}y + a_{33}z = b_3 \end{array}\right\} \tag{7}$$

and that the equations are already in the form, or by suitable adjustment of the augmented matrix may be put into the form, in which the diagonal terms of the coefficient matrix are large compared with the other terms. Then the equations may be written

$$\left.\begin{array}{l} x^{(r+1)} = (b_1 - a_{12}y^{(r)} - a_{13}z^{(r)})/a_{11} \\ y^{(r+1)} = (b_2 - a_{21}x^{(r)} - a_{23}z^{(r)})/a_{22} \\ z^{(r+1)} = (b_3 - a_{31}x^{(r)} - a_{32}y^{(r)})/a_{33} \end{array}\right\} \tag{8}$$

where $x^{(r)}$, $y^{(r)}$, and $z^{(r)}$ are the rth iterates. The procedure starts by assuming an estimate $x^{(0)}$, $y^{(0)}$, and $z^{(0)}$, which, in the absence of any better estimate may be taken as 0, 0, and 0. The second estimates may then be found from equations (8). The results are set out in table form and it is usually soon evident whether the process is convergent or not. It is possible for the process to be slowly convergent (as for example when the diagonal coefficients are not much larger than the other coefficients of the matrix) or even to diverge.

For purposes of illustration of the method, consider the following example.

Example Solve the set of equations

$$\left.\begin{array}{l} 8x - y + z - 18 = 0 \\ 2x + 5y - 2z - 3 = 0 \\ x + y - 3z + 6 = 0 \end{array}\right\}. \tag{9}$$

Solution The iterative scheme is applied to the equations

$$\left.\begin{array}{l} x = (y - z + 18)/8 \\ y = (-2x + 2z + 3)/5 \\ z = (y + x + 6)/3 \end{array}\right\}. \tag{10}$$

Then,

$$\left.\begin{array}{l} x^{(r+1)} = 0 \cdot 125y^{(r)} - 0 \cdot 125z^{(r)} + 2 \cdot 250 \\ y^{(r+1)} = -0 \cdot 400x^{(r)} + 0 \cdot 400z^{(r)} + 0 \cdot 600 \\ z^{(r+1)} = 0 \cdot 333y^{(r)} + 0 \cdot 333x^{(r)} + 2 \cdot 000 \end{array}\right\} \tag{11}$$

and the iteration starts with $x^{(0)} = y^{(0)} = z^{(0)} = 0$. Successive iterations are performed and the results are given in the table below

Iteration	0	1	2	3	4
x	0	2·250	2·075	1·943 9	2·011 4
y	0	0·600	0·500	0·949 6	0·965 7
z	0	2·000	2·949	2·858 2	2·964 4

$$\tag{12}$$

The process is clearly convergent, the actual values being $x = 2$, $y = 1$, and $z = 3$.

13.3.3 The Gauss–Seidel method

This method is a modification of the Gauss–Jacobi method. The set of equations (7) are written in the form

$$\left.\begin{aligned}
x^{(r+1)} &= (b_1 - a_{12}y^{(r)} - a_{13}z^{(r)})/a_{11} \\
y^{(r+1)} &= (b_2 - a_{21}x^{(r+1)} - a_{23}z^{(r)})/a_{22} \\
z^{(r+1)} &= (b_3 - a_{31}x^{(r+1)} - a_{32}y^{(r+1)})/a_{33}
\end{aligned}\right\} \tag{13}$$

which takes the place of equations (8) of section 13.3.2.

The procedure starts as before with an initial estimate $x^{(0)}$, $y^{(0)}$, and $z^{(0)}$ and then $x^{(1)}$ is found from the first of the equations (13), $y^{(1)}$ from the second and $z^{(1)}$ from the third. It will be noted that this calculation for the second estimate differs from the previous method in that the current values of the variables at each stage of the iteration are used in proceeding to the next stage of the iteration. The Gauss–Jacobi iteration proceeds by whole steps whereas the Gauss–Seidel method proceeds by single steps. The advantage of the latter is that convergence is more rapid but there is the disadvantage that the calculations are not so simple to perform.

Example Starting from the set of equations (10) with the first estimates $x^{(0)} = y^{(0)} = z^{(0)} = 0$, solve the equations (9).

Solution The following table may be formed using equations (13),

Iteration	0	1	2	3	4	5
x	0	2·250	1·881 3	1·997 24	1·997 86	1·999 83
y	0	−0·300	0·907 5	0·972 94	0·996 88	0·999 37
z	0	2·650	2·929 6	2·990 06	2·998 25	2·999 73

In five iterations the answer is close to the exact result of $x = 2$, $y = 1$, and $z = 3$. The result of the third iteration is better than that of the fourth iteration using the Gauss–Jacobi method.

Both the iterative methods are suitable for use with a high speed computer.

13.4 NUMERICAL SOLUTION OF DIFFERENTIAL EQUATIONS

The numerical solution of a differential equation of the nth order is usually obtained by solving an equivalent system of n first order differential equations. Consider for example, a general third order differential equation expressed in functional form

$$\frac{d^3y}{dx^3} = f\left(x, y, \frac{dy}{dx}, \frac{d^2y}{dx^2}\right) \tag{14}$$

Using the substitutions $dy/dx = z$; $dz/dx = w$ means that the equation (14) may be replaced by the system of equations

$$\left.\begin{aligned}
\frac{dy}{dx} &= z(x, y) \\[1ex]
\frac{dz}{dx} &= w(x, y, z) \\[1ex]
\frac{dw}{dx} &= f(x, y, z, w)
\end{aligned}\right\} \tag{15}$$

The solution of (14) may then be considered as equivalent to solving the three simultaneous equations (15). It is necessary then to discuss the numerical solution of a general first order differential equation of the form

$$\frac{dy}{dx} = f(x, y)$$

13.4.1 Euler's method of solution

Consider the equation

$$\frac{dy}{dx} = f(x, y) \tag{16}$$

together with its boundary conditions

$$y = y_0; \; x = x_0 \tag{17}$$

Suppose the solution of (16) is represented by a function

$$y = \phi(x), \; x = x_0, y = y_0,$$

then the solution would be represented graphically by the integral curve passing through the point (x_0, y_0). Such a curve is illustrated in Fig. 13.7. If the function $\phi(x)$ can be found analytically then the

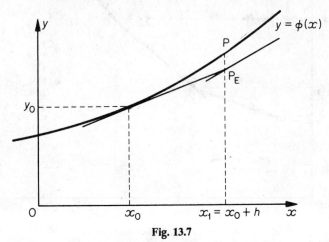

Fig. 13.7

equation (16) is solved and there is no need for a numerical solution. In practice however the function $\phi(x)$ may not be available so that the Fig. 13.7 is simply a theoretical solution curve. If $x_1(=x_0 + h)$ is the abscissa of a point P on the solution curve near to (x_0, y_0) then the actual ordinate of P is $\phi(x_1)$ or $\phi(x_0 + h)$.

The Euler method of solution is a simple computation that gives an approximation to the actual value $\phi(x_0 + h)$. Analytically the approximation is to take the first two terms in the Taylor expansion of $\phi(x_0 + h)$ (see section 5.2), i.e.

$$\phi(x_0 + h) \simeq \phi(x_0) + h\phi'(x_0) = y_0 + h\left(\frac{dy}{dx}\right)_{x_0, y_0}$$

This value is then taken as the value y_1 corresponding to x_1, i.e.

$$y_1 = y_0 + hf(x_0, y_0) \tag{18}$$

Graphically the process is to draw the tangent line to the integral curve at the point (x_0, y_0), as in Fig. 13.7, and to construct the point P_E where the tangent line intersects the ordinate at $x_1 = x_0 + h$. For this reason the Euler method is sometimes called the *tangent method* of approximation. The whole process may now be repeated starting from the value (x_1, y_1) to compute the next approximation given by (x_2, y_2)

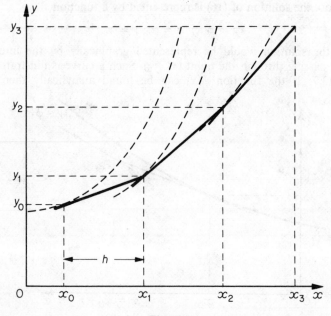

Fig. 13.8

where $y_2 = y_1 + hf(x_1, y_1), x_1 = x_0 + h, x_2 = x_0 + 2h$ (19)

The successive points (x_0, y_0), (x_1, y_1), (x_2, y_2) are then approximations to the actual points on the solution curve through (x_0, y_0) as in Fig. 13.8.

If y_n is the approximation to the solution of (16) at the point $x_n = x_0 + nh$ starting from the point (x_0, y_0), then the next point is (x_{n+1}, y_{n+1}) such that

$$y_{n+1} = y_n + hf(x_n, y_n) \qquad (20)$$

Example (i) Find an approximation for $y(0\cdot3)$ if

$$dy/dx = 2x/y$$

and $y(0) = 2$.

Solution

Here

$$f(x, y) = 2x/y \text{ and } x_0 = 0, y_0 = 2$$

Choosing $\qquad h = 0\cdot1,$

then $\qquad x_1 = 0\cdot1$

and $\qquad \begin{aligned} y_1 &= y_0 + hf(x_0, y_0) \\ &= 2 + (0\cdot1)(0) \\ &= 2; \end{aligned}$

then $\qquad x_2 = 0\cdot2$

and $\qquad \begin{aligned} y_2 &= y_1 + hf(x_1, y_1) \\ &= 2 + (0\cdot1)(0\cdot2/2) \\ &= 2\cdot01; \end{aligned}$

then $\qquad x_3 = 0\cdot3$

and $\qquad \begin{aligned} y_3 &= y_2 + hf(x_2, y_2) \\ &= 2\cdot01 + (0\cdot1)(0\cdot4/2\cdot01) \\ &= 2\cdot0299. \end{aligned}$

Hence $y(0\cdot3)$ is approximately $2\cdot03$.

The differential equation of Example (i) can be solved analytically by separation of variables (see 11.1.2) and the solution is $y^2/2 = x^2 + A$. With $y(0) = 2$, then $A = 2$ and hence the particular solution is $y^2 = 2(x^2 + 2)$. The actual value of $y(0\cdot3)$ is thus $\sqrt{4\cdot18}$ or $2\cdot045$ to three decimal places. The solution of Example (i) may be written in table form as follows

n	0	1	2	3
x_n	0	0·1	0·2	0·3
y_n	2	2·0	2·01	2·03
h	0·1	0·1	0·1	
$f(x_n, y_n)$	0	0·1	0·199	
y_{n+1}	2·0	2·01	2·0299	

Example (ii) Find an approximation to $y(0.6)$ if

$$\frac{dy}{dx} = x + y^2$$

and $y(0) = 0$. (Use an interval $h = 0.2$).

Solution Setting the solution out in tabular form and working to three decimal places

n	0	1	2	3
x_n	0	0·2	0·4	0·6
y_n	0	0·0	0·04	0·120
h	0·2	0·2	0·2	
$f(x_n, y_n) = x_n + y^2_n$	0	0·2	0·402	
$y_{n+1} = y_n + hf(x_n, y_n)$	0	0·04	0·120	

The result is $y(0.6) \simeq 0.120$.

There is no simple analytical solution to the differential equation of Example (ii), but this is the differential equation of Example (iii) of section 11.7 and Fig. 11.7 indicates the solution using isoclines. When $x = 0.6$ on the relevant solution curve (*a*) which passes through (0, 0), y is approximately 0·2. On this curve, it is seen that y increases rapidly with x and it is to be expected that Euler's method would seriously underestimate the true value of $y(0.6)$.

13.4.2 The Improved Euler method

The Euler method of 13.4.1 is not very accurate, and the Fig. 13.8 illustrates how the successive approximations (x_1, y_1), (x_2, y_2), (x_3, y_3) can deviate from the correct positions on the integral curve through (x_0, y_0). An improvement to the basic Euler method is the following:

(i) Use the basic Euler method as in Fig. 13.9 to estimate the value of y at x_1 $(=x_0 + h)$. Call this point $(x_1, y_1{}^P)$.

(ii) Use this latter result to estimate the value of the slope of the integral curve at the point $(x_1, y_1{}^P)$. Call this y_1'.

(iii) Compute the average of the slopes of the integral curve at (x_0, y_0), i.e. y_0', and the integral curve at $(x_1, y_1{}^P)$, i.e. y_1'.

(iv) Use the basic Euler method at the point (x_0, y_0) with the *average* slope value, i.e. $\frac{1}{2}(y_0' + y_1')$, to estimate (x_1, y_1).

(v) Repeat the whole process to estimate the next point (x_2, y_2).

In the Fig. 13.9 the ordinate length y_1 is given by

$$y_1 = y_0 + \tfrac{1}{2}h[f(x_0, y_0) + f(x_1, y_1{}^P)]$$

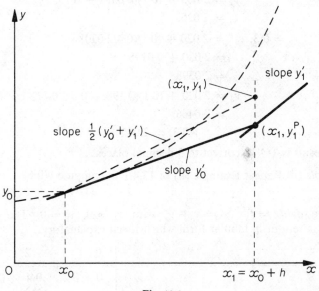

Fig. 13.9

The formula corresponding to (20) is

$$y_{n+1} = y_n + \tfrac{1}{2}h[f(x_n, y_n) + f(x_{n+1}, y^P{}_{n+1})] \tag{21}$$

where
$$y^P{}_{n+1} = y_n + hf(x_n, y_n) \tag{22}$$

and
$$x_n = x_0 + nh$$

The superfix P, as in $y_1{}^P$, is used to indicate the *predicted value* of the ordinate y_1. This value is then *corrected* by using the average of the slopes. The technique is a simple case of the *predictor-corrector* method of finding an approximation.

Example (i) Repeat Example (i) of 13.4.1 to find $y(0\cdot3)$.

Solution

Here $dy/dx = f(x, y) = 2x/y$ and $x_0 = 0$, $y_0 = 2$. With $h = 0.1$,

then $\qquad\qquad x_1 = 0.1,\ y_1^P = 2 + (0.1)(0) = 2$

and $\qquad f(x_1, y_1^P) = 2(0.1)/2 = 0.1$.

Hence $\qquad\qquad y_1 = 2 + (0.1)(0 + 0.1)/2$
$$= 2.005.$$

Then $\qquad\qquad x_2 = 0.2,\ y_2^P = 2.005 + (0.1)(0.2/2.005)$
$$= 2.015$$

and $\qquad f(x_2, y_2^P) = (0.4)/2.015 = 0.1985$.

Hence $\qquad\qquad y_2 = 2.005 + (0.1)(0.0997 + 0.1985)/2$
$$= 2.020.$$

Then $\quad x_3 = 0.3,\ y_3^P = 2.020 + (0.1)(0.4/2.020)$
$$= 2.020 + 0.0198$$
$$= 2.0398,$$

and $\qquad\qquad y_3 = 2.020 + (0.1)(0.198 + 0.6/2.04)/2$
$$= 2.0446.$$

Hence, $\qquad\qquad y(0.3) \simeq 2.045$

This result is $\sqrt{4.18}$ correct to 3 decimal places.

Example (ii) Repeat Example (ii) of 13.4.1 to compute $y(0.6)$.

Solution

Here $dy/dx = f(x, y) = x + y^2$ with $x_0 = 0$, $y_0 = 0$. The calculation is set out in tabular form which is self-explanatory.

n	0	1	2	3
x_n	0	0.2	0.4	0.6
y_n	0	0.02	0.080	0.183
h	0.2	0.2	0.2	
$f(x_n, y_n) = x_n + y_n^2$	0.0	0.2	0.406	
x_{n+1}	0.2	0.4	0.6	
$y_{n+1}^P = y_n + hf(x_n, y_n)$	0.0	0.06	0.161	
$f(x_{n+1}, y_{n+1}^P) = x_{n+1} + (y_{n+1}^P)^2$	0.2	0.404	0.626	
$y_{n+1} = y_n + \tfrac{1}{2}h[f(x_n, y_n) + f(x_{n+1}, y_{n+1}^P)]$	0.02	0.080	0.183	

The improved Euler method gives $y(0\cdot6) = 0\cdot183$ as the solution of $dy/dx = x + y^2$, $y(0) = 0$, a result which should be a better estimate than the value $0\cdot120$ of 13.4.1.

Example (iii) Use the modified Euler method with $h = 0\cdot1$ and work to two decimal places, to estimate $y(\frac{1}{2})$ given that

$$\frac{dy}{dx} = 1 + y, \quad y(0) = 0.$$

Solve the equation analytically and hence obtain an estimate to e.

Solution

Here $f(x, y) = 1 + y$ and the result is tabulated below

n	0	1	2	3	4	5
x_n	0	0·1	0·2	0·3	0·4	0·5
y_n	0	0·10	0·22	0·35	0·49	0·65
h	0·1	0·1	0·1	0·1	0·1	
$f(x_n, y_n)$	1·0	1·10	1·22	1·35	1·49	
x_{n+1}	0·1	0·2	0·3	0·4	0·5	
y_{n+1}^P	0·1	0·21	0·34	0·48	0·64	
$f(x_{n+1}, y_{n+1}^P)$	1·1	1·21	1·34	1·48	1·64	
y_{n+1}	0·10	0·22	0·35	0·49	0·65	

Euler's improved method gives the estimate

$$y(\tfrac{1}{2}) = 0\cdot65.$$

Analytically,
$$\frac{dy}{dx} = 1 + y$$

or
$$\int \frac{dy}{1 + y} = \int dx$$

and hence $\log(1 + y) = x + \text{constant}$.

The solution is $1 + y = Ae^x$, where A is a constant, and since $y(0) = 0$ it follows that $A = 1$, i.e. $y = e^x - 1$. Using the calculated estimate

$$e^{\frac{1}{2}} = 1 + y(\tfrac{1}{2})$$
$$\simeq 1 + 0\cdot65$$
$$= 1\cdot65$$

and hence, $e \simeq (1 \cdot 65)^2 \simeq 2 \cdot 72$.

This is the actual value of e, correct to two decimal places.

13.5 ERROR AND PROPAGATION OF ERROR

It has been assumed in the preceding sections that the supplied data are exact. For example the coefficients of equation (6) of 13.3.1 were assumed to be precisely those numbers. In any practical problem it is likely that such coefficients have been obtained as the result of an experiment, for example as the readings on a meter. The data are not exact but are approximations to some probably unknown values. Most engineering measurements however precise they may be are subject to some form of approximation and it is necessary to know how the use of the approximation, instead of the exact value, affects subsequent calculations performed with the data.

Suppose the exact (but possibly unknown) value of some quantity is N and the approximation to this value is n, then the difference between n and N is called the *absolute error* in n, i.e. if this error is denoted by ε then
$$\varepsilon = n - N \tag{23}$$

13.5.1 Errors

Errors may be introduced in many different ways;

   (i) Human error is obvious and is particularly common,
  (ii) Round-off error is due to the need to express a number in decimal form and with a convenient number of decimal places. It is often introduced into a calculation when an exact fraction is written in decimal form. For example, the numbers $0 \cdot 3$, $0 \cdot 33$, $0 \cdot 333$ are successive approximations to the exact number $1/3$,
 (iii) Errors may be introduced due to the nature of the method of computation used,
  (iv) There may be errors inherent in the data being used, due perhaps to experimental measurement.

The errors due to (i) may be reduced by checking and cross-checking and those due to (iv) may need a full statistical analysis. The chief concern here is with errors due to (ii).

13.5.2 Round-off errors

The number $130/63$ can be expressed as a non-terminating decimal $2 \cdot 063492063 \ldots$. In order to use such a number in a computation it will usually be written as $2 \cdot 06$, $2 \cdot 0635$, $2 \cdot 063492$ with successively more accurate approximations. The approximations are the result of *rounding-off* to 2, 4, or 6 decimal places. Sometimes the term *significant figures* may be used instead. The three approximations above would then have 3, 5, and 7 significant figures respectively.

Example (i) The numbers $12 \cdot 32$, $0 \cdot 0134$, $125 \cdot 0$ are said to have 4, 3, and 4, significant figures respectively.

It is useful to represent numbers in the form $M \times 10^m$, where $1 \leqslant M < 10$ and m is an integer, to indicate the number of significant figures. Thus in the above example the numbers are respectively $1\cdot232 \times 10^1$; $1\cdot34 \times 10^{-2}$; and $1\cdot250 \times 10^2$. All the numbers retained in M are significant. Note that the zeros of the second number are not significant whereas the zero of the third number is significant. In the former case the zeros merely locate the decimal point but in the latter case the zero serves to distinguish the number from any other number such as $125\cdot4$ or $125\cdot9$.

Example (ii) Express $438\cdot968$ correct to 2 and 1 decimal places. Also express the number correct to 4, 3, 2, and 1 significant figures.

Solution

The number is $438\cdot97$ to 2 decimal places

and $439\cdot0$ to 1 decimal place.

The number is $4\cdot390 \times 10^2$ to 4 significant figures,

and $4\cdot40 \times 10^2$ to 3 significant figures,

and $4\cdot4 \times 10^2$ to 2 significant figures,

and 4×10^2 to 1 significant figure.

The round-off rule to n significant figures is to retain n significant digits and discard the remainder. If the discarded number is

   (i) less than half a unit in the nth place leave the nth digit unchanged,
  (ii) greater than half a unit in the nth place increase the nth digit by 1,
 (iii) exactly half a unit in the nth place, add either 0 or 1 to the nth digit to make it an even number.

Example (iii) The number $13\cdot1535$ is $13\cdot154$ to 5 significant figures; $13\cdot15$ to 4 significant figures; and $13\cdot2$ to three significant figures.
Example (iv) Express $14\cdot2545$ to 5, 4, and 3 significant figures.
Solution Rounded-off to 5 significant figures the number is $14\cdot254$, the discarded remainder being exactly $\frac{1}{2}$ unit in the 5th place.

Rounded-off to 4 significant figures the number is $14\cdot25$, the discarded remainder $0\cdot0045$ being less than $\frac{1}{2}$ unit in the 4th place.

To 3 significant figures the number is $14\cdot3$, the discarded number $0\cdot0545$ being greater than $\frac{1}{2}$ unit in the 3rd place.

13.5.3 Absolute, relative, and percentage error

The absolute error has already been defined in equation (23) as $\varepsilon = n - N$. If only the magnitude of the error is important then

$$|\varepsilon| = |n - N| \tag{24}$$

is called the *absolute error modulus*.

The absolute error is not necessarily a good indicator of the error.

For example, two measurements may be one metre and one kilometre, each with an absolute error of one centimetre but it is obvious that the former error is more significant than the latter. A better indicator in this case is the *relative error r* given by

$$r = \varepsilon/N \tag{25}$$

In the above illustration the relative errors are

$$r = 1/100 = 0.01 \text{ (for } N = 1 \text{ metre)}$$

and $$r = 1/100000 = 0.00001 \text{ (for } N = 1 \text{ kilometre)}$$

In practice the number N may not be known but only its approximation n, in this case the equation (25) is written

$$
\begin{aligned}
r &= \varepsilon/N \\
&= \varepsilon/(n - \varepsilon) = (\varepsilon/n)(1 - \varepsilon/n)^{-1} \\
&= (\varepsilon/n)(1 + \varepsilon/n + \varepsilon^2/n^2 + \ldots)
\end{aligned}
$$

Thus a good indicator of the relative error is

$$r \simeq \varepsilon/n \tag{26}$$

The percentage error is simply $100r$. The percentage error for an absolute error of 1 centimetre in one metre is 1% and for one centimetre in one kilometre is 0.001%.

13.5.4 Accuracy

This is not simply the absence of error. For example a clock may be said to be accurate if it does not gain or lose more than, say, 3 seconds in 12 hours. In any 12-hour interval the absolute error ε of the clock is then such that $-3 \leqslant \varepsilon \leqslant +3$, i.e. ε lies within a band of values $(-3, +3)$. This band is called the *error interval*. If the clock was correct at midnight then the accurate time T hours, when the clock-time is $t(<12$ hours) would be in the range

$$T = t \pm 3/3600 \text{ or } t - 3/3600 \leqslant T \leqslant t + 3/3600.$$

Example Suppose the number 2.67 is the result of a rounding-off to two decimal places. The absolute error bound is 0.005, i.e. the correct number N lies within the band of values

$$(2.67 - 0.005) \quad \text{to} \quad (2.67 + 0.005)$$

i.e. $$2.665 < N < 2.675$$

13.5.5 Propagation of errors

The basic algebraical operations are those of addition and subtraction, multiplication and division and these lead to propagation of errors. Errors may also be propagated by the use of other functional relationships, indices, powers, roots, etc.

(a) *Addition and subtraction*

Suppose N_1 and N_2 are exact, then N_3 where

$$N_3 = N_1 \pm N_2$$

is also exact. If n_1 and n_2 are approximations to N_1 and N_2 with absolute errors $\varepsilon_1 = n_1 - N_1$ and $\varepsilon_2 = n_2 - N_2$ then n_3 is the approximation to N_3 with an error ε_3 given by

$$\begin{aligned}
\varepsilon_3 &= n_3 - N_3 \\
&= (n_1 \pm n_2) - (N_1 \pm N_2) \\
&= (n_1 - N_1) \pm (n_2 - N_2)
\end{aligned}$$

i.e. $\qquad\qquad\qquad \varepsilon_3 = \varepsilon_1 \pm \varepsilon_2$ \hfill (28)

Thus, in an addition (subtraction) the absolute errors are added (subtracted).

In many cases it is sufficient to use the absolute error moduli $|\varepsilon_1|$, $|\varepsilon_2|$, and $|\varepsilon_3|$ and the result

$$|\varepsilon_3| \leqslant |\varepsilon_1| + |\varepsilon_2|. \qquad\qquad (29)$$

The result (29) simply states that the worst possible error will occur when the errors are additive. For example, in a subtraction the errors are additive when they are of opposite signs.

Example (i) The two numbers 59·82 and 0·56 have been obtained by a rounding-off process. Estimate the error in the addition

$$59{\cdot}82 + 0{\cdot}56 = 60{\cdot}38$$

Solution The maximum error modulus in 59·82 is $|\varepsilon_1| = 0{\cdot}005$ and the maximum error modulus in 0·56 is $|\varepsilon_2| = 0{\cdot}005$ also. If the error modulus in the addition is $|\varepsilon_3|$, then

$$|\varepsilon_3| \leqslant 0{\cdot}005 + 0{\cdot}005 = 0{\cdot}01.$$

The result of the addition is then $60{\cdot}38 \pm 0{\cdot}01$, i.e. the second decimal place is in doubt. A more meaningful answer would be 60·4 correct to 3 significant figures. The relative error is $0{\cdot}01/60{\cdot}38$, i.e. 0·0002.

Example (ii) The numbers 70·03 and 69·25 have been rounded-off to 2 decimal places. Express the difference of the two numbers to a meaningful number of significant figures.

Solution The difference 0·78 has an error ε_3 and

$$|\varepsilon_3| \leqslant 0{\cdot}005 + 0{\cdot}005 = 0{\cdot}01$$

The last digit is again in doubt and the difference lies in the band of values 0·77 to 0·79. Consequently the result is 0·8 correct to 1 significant figure. Notice that the original numbers are of comparable size and each has 4 significant figures but the difference has only 1 significant figure. There can be a large loss of significant figures in a subtraction and this can be a source of serious error in an involved calculation.

(b) *Multiplication and division*

Suppose that $n_3 = n_1n_2$ is an approximation to the exact result $N_3 = N_1N_2$ and that

$$\varepsilon_1 = n_1 - N_1$$

$$\varepsilon_2 = n_2 - N_2$$

$$\varepsilon_3 = n_3 - N_3$$

It follows that

$$
\begin{aligned}
\varepsilon_3 &= n_3 - N_3 \\
&= n_1n_2 - N_1N_2 \\
&= n_1n_2 - (n_1 - \varepsilon_1)(n_2 - \varepsilon_2) \\
&= n_1n_2 - (n_1n_2 - \varepsilon_1n_2 - \varepsilon_2n_1 + \varepsilon_1\varepsilon_2) \\
&= \varepsilon_1n_2 + \varepsilon_2n_1 - \varepsilon_1\varepsilon_2
\end{aligned}
$$

If, as is frequently the case, $\varepsilon_1\varepsilon_2$ is small compared to the other term then

$$\varepsilon_3 \simeq \varepsilon_1n_2 + \varepsilon_2n_1 \tag{30}$$

is a measure of the absolute error involved. Dividing both sides of (30) by $n_3(= n_1n_2)$ then an estimate of the relative error involved in the multiplication is given by

$$\frac{\varepsilon_3}{n_3} \simeq \frac{\varepsilon_1n_2}{n_1n_2} + \frac{\varepsilon_2n_1}{n_1n_2}$$

i.e.

$$r_3 \simeq r_1 + r_2 \tag{31}$$

Similarly, in a division, suppose that $n_3 = n_1/n_2$ is an approximation to $N_3 = N_1/N_2$ then

$$
\begin{aligned}
\varepsilon_3 &= n_3 - N_3 \\
&= n_1/n_2 - N_1/N_2 \\
&= n_1/n_2 - (n_1 - \varepsilon_1)/(n_2 - \varepsilon_2) \\
&= n_1/n_2 - (n_1/n_2)[(1 - \varepsilon_1/n_1)(1 - \varepsilon_2/n_2)^{-1}] \\
&= n_1/n_2 - (n_1/n_2)[1 + \varepsilon_2/n_2 - \varepsilon_1/n_1 - \varepsilon_1\varepsilon_2/n_1n_2 + \ldots] \\
&= (n_2/n_2)[\varepsilon_1/n_1 - \varepsilon_2/n_2] + \ldots
\end{aligned}
$$

An estimate of the relative error in a division is given by

$$r_3 \simeq r_1 - r_2 \tag{32}$$

Thus in a multiplication (division) an estimate of the error is obtained by adding (subtracting) the relative errors.

In many numerical calculations it may not be known whether the error is positive or negative and the only meaningful error measure is the absolute error modulus $|\varepsilon|$ or the *relative error modulus* $|r|$. The two estimates (31) and (32) are frequently replaced by their cruder, but simpler, forms

$$|r_3| \leqslant |r_1| + |r_2|. \tag{33}$$

Note that $|r_1 + r_2| \leqslant |r_1| + |r_2|$ and $|r_1 - r_2| \leqslant |r_1| + |r_2|$ (see also Section 10.3.4).

Example (iii) The numbers 1·3 and 2·4 are the result of a rounding-off process. Give the product 1·3 × 2·4 to a meaningful number of significant figures.

Solution If the numbers are assumed accurate then 1·3 × 2·4 = 3·12. Due to the rounding-off process each number has an absolute error modulus of 0·05. The relative error moduli are

$$|r_1| = \frac{0·05}{1·3} = 0·038 \quad \text{and} \quad |r_2| = \frac{0·05}{2·4} = 0·021$$

respectively. From (33) an estimate of the relative error modulus of the product is

$$|r_3| \leqslant |r_1| + |r_2| = 0·059$$

The absolute error modulus is then estimated as

$$|\varepsilon_3| \leqslant 0·059 \times 3·12 = 0·1841.$$

The product $N = 1·3 \times 2·4$ has an estimated value given by

$$3·12 - 0·18 \leqslant N \leqslant 3·12 + 0·18$$

i.e. $2·94 \leqslant N \leqslant 3·30$

The result of the multiplication to a meaningful number of significant figures is thus $N = 3$.

Instead of working in terms of the relative error (with the simple rule of addition (subtraction) in the case of multiplication (division) it is often convenient to work with the absolute error.

Example (iv) Estimate 3·2/0·411 if the numbers are the result of rounding-off.

Solution The quotient is N and

$$N = \frac{3·2 \pm 0·05}{0·411 \pm 0·0005}$$

$$= \frac{(3·2 \pm 0·05)(1 \pm 0·0012)^{-1}}{0·411}$$

$$= (3·2 \pm 0·05)(1 \mp 0·0012 \ldots)/0·411$$

or $N = [3·2 \pm (0·05 + 3·2 \times 0·0012)]/0·411$ (for maximum possible error).

Hence, $N \simeq 7·786 \pm 0·131$ and N has a value in the range 7·66 to 7·92. Thus it is possible to claim that N has the value 8 correct to 1 significant figure.

(c) *Functional relationships*

Suppose that $n_2 = f(n_1)$ is an approximation to the exact result $N_2 = f(N_1)$ where $f(x)$ is some functional relationship of x. If n_1 has an absolute error ε_1, and n_2 an absolute error ε_2 then

$$\varepsilon_2 = n_2 - N_2 = f(n_1) - f(N_1)$$
$$= f(n_1) - f(n_1 - \varepsilon_1)$$
$$= f(n_1) - [f(n_1) - \varepsilon_1 f'(n_1) + \frac{1}{2!}\varepsilon_1^2 f''(n_1) \ldots .]$$

assuming the Taylor expansion of section 5.2. Thus

$$\varepsilon_2 \simeq \varepsilon_1 f'(n_1) \qquad (34)$$

Example (v) If $1\cdot 377$ is the result of a rounding-off to 4 significant figures, estimate the absolute error modulus of $\log 1\cdot 377$ and hence give the result to an appropiate number of figures.
Solution From (34)

$$|\varepsilon_1| = 0\cdot 0005$$

and $$f'(n_1) = 1/1\cdot 377.$$

The estimate for $|\varepsilon_2|$ is then

$$|\varepsilon_2| \simeq (0\cdot 0005)(1/1\cdot 377)$$
$$\simeq 0\cdot 0004.$$

Using 4 figure tables

$$\log 1\cdot 377 \simeq 0\cdot 3200 \pm 0\cdot 0004.$$

The result is $0\cdot 320$ correct to 3 significant figures.

13·6 ERRORS IN ITERATIVE PROCESSES

An iterative process for calculating a root of $f(x) = 0$ is often written in the form

$$x_{n+1} = F(x_n). \qquad (35)$$

For example, in 13.2.2 $F(x_n)$ is $\phi(x_n)$ where

$$x - \phi(x) \equiv f(x)$$

while, in 13.2.3, $F(x_n)$ is the expression given by (4), i.e.

$$F(x_n) = x_n - f(x_n)/f'(x_n).$$

Suppose the absolute error in the iterate x_n is ε_n then, if X is the correct value, (35) leads to

$$X + \varepsilon_{n+1} = F(X + \varepsilon_n)$$
$$= F(X) + \varepsilon_n F'(X) + \frac{1}{2!}\varepsilon_n^2 F''(X) + \ldots$$

which, with $X = F(X)$ reduces to

$$\varepsilon_{n+1} \simeq \varepsilon_n F'(X) + \tfrac{1}{2}\varepsilon_n^2 F''(X).$$

For the iterative method of 13.2.2, $F(X) \equiv \phi(X)$

i.e. $$F'(X) \equiv \phi'(X)$$

and $$\varepsilon_{n+1} \simeq \varepsilon_n \phi'(X) + \tfrac{1}{2}\varepsilon_n^2 \phi''(X)$$

i.e. $$\varepsilon_{n+1} \simeq \varepsilon_n \phi'(X). \tag{36}$$

From (36) it may be noted that if $|\phi'(X)| < 1$ then

$$|\varepsilon_{n+1}| < |\varepsilon_n|$$

and the error decreases as the process proceeds i.e. the process is convergent.

For the iterative method of 13.2.3

$$F(X) = X - f(X)/f'(X)$$

and hence

$$
\begin{aligned}
F'(X) &= 1 - f'(X)/f'(X) + f(X)f''(X)/\{f'(X)\}^2 \\
&= f(X)f''(X)/\{f'(X)\}^2 \\
&= 0
\end{aligned}
$$

since $f(X) = 0$.

It follows that

$$
\begin{aligned}
F''(X) &= f'(X)f''(X)/\{f'(X)\}^2 \\
&= f''(X)/f'(X)
\end{aligned}
$$

and so $$\varepsilon_{n+1} \simeq A\varepsilon_n^2 \tag{37}$$

where $$A = f''(X)/2f'(X).$$

From (37) it may be noted that the error in the $(n + 1)$th iterate is dependent on the square of the error in the nth iterate. This is one reason why the Newton-Raphson method of 13.2.3. is usually a rapidly convergent process. However the form of A also shows that if $f''(X)/f'(X)$ is too large (i.e. if $f'(X)$ is too small and/or $f''(X)$ is too large) then the process could diverge. The equation (37) shows that if $A\varepsilon_n > 1$, i.e. $\varepsilon_{n+1} > \varepsilon_n$ then this divergent process is possible. For example if $A = 15$ and $\varepsilon_0 = 0.1$ then (37) gives $\varepsilon_1 \simeq 0.15$, $\varepsilon_2 \simeq 0.33$, and $\varepsilon_3 \simeq 1.5$ and the process diverges. Such a divergent process is illustrated in Fig. 13.6(c) of section 13.2.3. It may be possible to prevent this divergency by starting from a better approximation. For example, if $A = 15$ but $\varepsilon_0 = 0.05$ then $\varepsilon_1 \simeq 0.04$, $\varepsilon_2 \simeq 0.02$, and $\varepsilon_3 \simeq 0.006$ and the process converges.

EXERCISE 13

1 Show that the equation $x + 2 \sin x = 1$ has a root between 0 and 1 and use the rule of false position to find it to two decimal places.

2 Show graphically that $2 \cos 2x = e^z$ has only one positive root but an infinity of negative roots. Use the rule of false position to find, to two places of decimals, the two roots of numerically least value.

3 Find the root of Question 1, correct to four places of decimals, by the Newton–Raphson method.

4 Show that $4x^3 + 3x^2 - 6x - 1 = e^{-z}$ has two negative roots and one positive root, and use the Newton–Raphson method to find the negative root between -1 and 0 to three decimal places.

5 Evaluate $\sqrt{7}$ to four significant figures by the Newton–Raphson method.

6 Show that if x_r is an approximation to $\sqrt[3]{a}$ then a better approximation is given by
$$x_{r+1} = \tfrac{1}{3}[2x_r + a/x_r^2].$$
Evaluate $\sqrt[3]{10}$ correct to four significant figures.

7 Use the iteration procedure, $x_{r+1} = \phi(x_r)$, to find the positive root of the equation $6x = \log(x + 2)$ correct to four decimal places.

8 Find, to four decimal places, the smallest positive root of $x^3 - 3x + 1 = 0$ using (i) the iteration procedure of section 13.2.2, and (ii) the Newton–Raphson method. Find also the negative root of the equation.

9 Use Gaussian elimination to solve the following equations
$$1 \cdot 23x + 0 \cdot 86y + 0 \cdot 39z = 3 \cdot 50$$
$$0 \cdot 06x - 2 \cdot 01y - 0 \cdot 26z = 2 \cdot 62$$
$$0 \cdot 39x - 0 \cdot 24y + 1 \cdot 53z = -1 \cdot 98$$

Use a check column and calculate the first estimates to three places of decimals. Calculate the residuals and form the second estimates.

10 Solve the equations of Question 9 using (i) the Gauss–Jacobi method, and (ii) the Gauss–Seidel method.

11 Obtain x_1, x_2, x_3, and x_4 correct to three places of decimals by the Gauss–Seidel method, given
$$20x_1 - x_2 + x_3 - x_4 = 8$$
$$x_1 - 40x_2 + 4x_3 + x_4 = 4$$
$$-x_1 + x_2 + 10x_3 = 6$$
$$x_2 - x_3 + 8x_4 = -5.$$

12 Use Euler's method with step length $0 \cdot 2$ to solve

$$dy/dx + y^2 = 0, \quad y(0) = 1$$

for values of x up to $x = 1$ (work to two decimal places). Integrate the equation and compare the approximate value of $y(1)$ with the actual value.

13 Solve numerically the differential equation

$$dy/dx = xy + y^2$$

with initial value $y(0) = 1$ using the improved Euler method. Use a step size of $0 \cdot 1$ for x up to $0 \cdot 4$. Work to three decimal places and give the answer to two decimal places.

14 Solve numerically the differential equation

$$dy/dx = x - y^2$$

with initial value $y(0) = 1$, using the improved Euler method with step size $0 \cdot 2$ for x up to $1 \cdot 0$ and calculate to three decimal places. From a graph of the results find approximately the minimum value of y and the value of x at which y reaches this minimum.

15 If all the numbers are rounded-off find the maximum possible error of the following calculations and give the answer to an appropriate number of decimal places

$$(2 \cdot 00 \times 4 \cdot 00) + 3 \cdot 62 - 9 \cdot 82$$

16 If $x = 0 \cdot 56$ is the result of rounding-off give the value of e^{-3x} to an appropriate number of decimal places (tables give $e^{-1 \cdot 68}$ as $0 \cdot 1864$).

14
Numerical Methods II

14.1 INTERPOLATION

The reader is probably familiar with the idea of interpolation as applied to a table of values of a variable. For example, suppose that the function $f(x)$ is tabulated at intervals of $0{\cdot}1$ from $x = 1{\cdot}0$ to $x = 2{\cdot}0$ and that it is required to find $f(1{\cdot}05)$. The function $f(x)$ may be a known function, but complicated, or it may be an unknown function the tabulation having being obtained as the result of an experiment. Let $f(1{\cdot}0) = 2{\cdot}718\,28$ and $f(1{\cdot}1) = 3{\cdot}004\,17$ then an approximation to $f(1{\cdot}05)$ can be obtained by assuming that the graph of $f(x)$ in the range $1{\cdot}0 \leqslant x \leqslant 1{\cdot}1$ is a straight line so that $f(1{\cdot}05) = 2{\cdot}718\,28 + \frac{1}{2}(3{\cdot}004\,17 - 2{\cdot}718\,28) = 2{\cdot}861\,22$.

The general process of finding a value of the function at other than the tabulated values is called *interpolation*, the simple process above is called *linear interpolation* and is unlikely, in the example given, to be accurate beyond the second decimal place.

A better approximation to the required value is obtained by using all of the known tabulated values instead of just the two values on either side of the interpolated point. This can be done by replacing the function $f(x)$ by a simpler function $p(x)$ that takes the values of $f(x)$ at all of the tabulated points and from which the value of $p(1{\cdot}05)$ can be found by a simple computation. There are many types of replacement functions $p(x)$, the simplest of which is a series of powers of x, i.e. a polynomial function. Other types of replacement functions that are in common use are series of trigonometric functions, or series of exponential functions, or series of more complicated (but well known) mathematical functions.

It is in this wider sense, that is in the finding of the best type of replacement function, that the word interpolation is often used nowadays. The polynomial functions can lead to accurate interpolation formulae and these are the only functions to be discussed here.

14.2 DIFFERENCE TABLE

Suppose $y_0, y_1, y_2, \ldots, y_n$ is a set of values of a (known or unknown) function $f(x)$ corresponding to a set of values $x_0, x_1, x_2, \ldots, x_n$ written in ascending order of magnitude so that $x_0 < x_1 < \ldots < x_{n-1} < x_n$. The variable x is called the *argument* and the variable values y_r are called the *entries*, the corresponding values being conveniently displayed in table form. It is required to estimate the value of the entry that would be obtained for a value of the argument in the range $x_0 < x < x_n$.

Any reliable interpolation formula used to find this estimate will employ all the given information, that is it will be based on the values of the $n + 1$ entries.

Instead of using the values directly it is found convenient to use relationships dependent on these values. The relationship $y_{r+1} - y_r$ is called the *first difference* of y_r, and is denoted by Δy_r, i.e.

$$\Delta y_r = y_{r+1} - y_r. \tag{1}$$

The relationship $\Delta y_{r+1} - \Delta y_r$, being the difference of the first difference of y_r, is called the *second difference* of y_r and is denoted by $\Delta(\Delta y_r)$ or $\Delta^2 y_r$, i.e.

$$\Delta^2 y_r = \Delta y_{r+1} - \Delta y_r. \tag{2}$$

For example,

$$\begin{aligned}
\Delta^2 y_0 &= \Delta y_1 - \Delta y_0 \\
&= (y_2 - y_1) - (y_1 - y_0) \\
&= y_2 - 2y_1 + y_0,
\end{aligned}$$

and similarly, $\Delta^2 y_1 = y_3 - 2y_2 + y_1.$

In general terms

$$\Delta^2 y_r = y_{r+2} - 2y_{r+1} + y_r, \qquad r = 0, 1, 2, \ldots, n - 2. \tag{3}$$

In a similar manner the *third differences* may be formed so that

$$\begin{aligned}
\Delta^3 y_r &= \Delta^2 y_{r+1} - \Delta^2 y_r \\
&= (y_{r+3} - 2y_{r+2} + y_{r+1}) - (y_{r+2} - 2y_{r+1} + y_r) \\
&= y_{r+3} - 3y_{r+2} + 3y_{r+1} - y_r; \; r = 0, 1, 2, \ldots, n - 3. \tag{4}
\end{aligned}$$

Clearly, successive differences may be calculated, each difference being dependent on a selection of the given entries $y_0, y_1, y_2, \ldots, y_n$.

x	y	Δy	$\Delta^2 y$	$\Delta^3 y$	$\Delta^4 y$	$\Delta^5 y$
x_0	y_0					
		Δy_0				
x_1	y_1		$\Delta^2 y_0$			
		Δy_1		$\Delta^3 y_0$		
x_2	y_2		$\Delta^2 y_1$		$\Delta^4 y_0$	
		Δy_2		$\Delta^3 y_1$		$\Delta^5 y_0$
x_3	y_3		$\Delta^2 y_2$		$\Delta^4 y_1$	
		Δy_3		$\Delta^3 y_2$		
x_4	y_4		$\Delta^2 y_3$			
		Δy_4				
x_5	y_5					

Table 14.1

The polynomial interpolation formula, dependent on the $n + 1$ entries, may then be expressed in terms of these differences. For convenience the differences are set out in tabulated form, entries in successive difference columns being placed in the table mid-way between the two entries used in forming that difference as in Table 14.1.

Example Form the difference table for the function

$$y = x^2 - 2$$

for $x = 0$ to $x = 6$ at intervals of 1 unit.

Solution The values of y are tabulated and successive differences found as below:

x	y	Δy	$\Delta^2 y$	$\Delta^3 y$
0	−2			
		1		
1	−1		2	
		3		0
2	2		2	
		5		0
3	7		2	
		7		0
4	14		2	
		9		0
5	23		2	
		11		
6	34			

In this example all the differences after the second are zero. It is no coincidence that the second differences of a second degree polynomial are the same. For, consider a polynomial of degree s, i.e.

$$y = a_s x^s + a_{s-1} x^{s-1} + \ldots + a_1 x + a_0$$

and suppose a difference table is formed for equal intervals h of the argument x, then $x_{r+1} = x_r + h$ and

$$\begin{aligned}
\Delta y_r &= y_{r+1} - y_r \\
&= [a_s(x_r + h)^s + \ldots] - [a_s x_r^s + \ldots] \\
&= a_s(x_r^s + s x_r^{s-1} h + \ldots - x_r^s) + \ldots \\
&= a_s s h x_r^{s-1} + \text{terms in lower powers of } x_r,
\end{aligned}$$

and

$$\begin{aligned}
\Delta^2 y_r &= \Delta y_{r+1} - \Delta y_r \\
&= a_s s h [(x_r + h)^{s-1} - x_r^{s-1}] + \ldots \\
&= a_s s(s - 1) h^2 x_r^{s-2} + \text{terms in lower powers of } x_r.
\end{aligned}$$

Each successive difference is a polynomial of degree one less than its predecessor. Finally,

$$\Delta^s y_r = a_s(s!)h^s$$
$$= \text{constant.} \qquad (5)$$

The equation (5) shows that the sth difference of a polynomial of degree s is constant and it follows that all subsequent differences are zero. The converse may be used to deduce a polynomial relationship as in the following example (see also example 2 of Exercise 14).

Example In the following difference table

x	y	Δy	$\Delta^2 y$	$\Delta^3 y$
0	4			
		19		
1	23		18	
		37		6
2	60		24	
		61		6
3	121		30	
		91		6
4	212		36	
		127		6
5	339		42	
6	508	169		

the third difference column is constant and there is an exact polynomial relationship between x and y of degree three in x. The observant reader will see that $y = (x + 2)^3 - 4$.

It is possible for the values of the argument x to be given at unequal intervals and in this case the above tables are modified by employing *divided differences*. Only equal intervals of the argument are considered in this book.

14.2.1 Errors in a difference table

The first step in any interpolation procedure is the formation of a difference table from the given data and since this involves a large number of additions, subtractions, and entering of results it is a source of numerical error. A simple check can be obtained by adding the entries of any column of differences and comparing the result with the difference between the last and first entries of the previous column.

For example, adding the entries of a second difference column

$$\Delta^2 y_0 + \Delta^2 y_1 + \Delta^2 y_2 + \ldots + \Delta^2 y_5$$
$$= (\Delta y_1 - \Delta y_0) + (\Delta y_2 - \Delta y_1) + \ldots + (\Delta y_6 - \Delta y_5)$$
$$= \Delta y_6 - \Delta y_0.$$

An error is frequently made in the entry column y and care must be taken when transferring the data from its given state (e.g. as the result of an experiment) to a difference table. Suppose that the entry

y_3 of Table 14.1 is subject to an error ε, then Table 14.2 shows how this error is propagated throughout the table.

The error in (say) the third difference column appears in four terms with error coefficients given by the binomial coefficients of $(1 - \varepsilon)^3$.

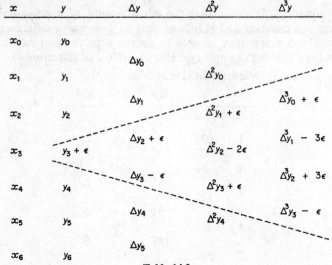

x	y	Δy	$\overset{2}{\Delta} y$	$\overset{3}{\Delta} y$
x_0	y_0			
		Δy_0		
x_1	y_1		$\Delta^2 y_0$	
		Δy_1		$\Delta^3 y_0 + \epsilon$
x_2	y_2		$\Delta^2 y_1 + \epsilon$	
		$\Delta y_2 + \epsilon$		$\Delta^3 y_1 - 3\epsilon$
x_3	$y_3 + \epsilon$		$\Delta^2 y_2 - 2\epsilon$	
		$\Delta y_3 - \epsilon$		$\Delta^3 y_2 + 3\epsilon$
x_4	y_4		$\Delta^2 y_3 + \epsilon$	
		Δy_4		$\Delta^3 y_3 - \epsilon$
x_5	y_5		$\Delta^2 y_4$	
		Δy_5		
x_6	y_6			

Table 14.2

Example Find the errors in the following table.

x	y	Δy	$\Delta^2 y$	$\Delta^3 y$	Error in $\Delta^3 y$
0	34·8				
		38·6			
1	73·4		21·6		
		60·2		4·8	
2	133·6		26·4		
		86·6		4·8	
3	220·2		31·2		
		117·8		4·3	$= 4\cdot8 - 0\cdot5$
4	338·0		35·5		
		153·3		6·3	$= 4\cdot8 + 1\cdot5$
5	491·3	— — — — — —	— 41·8 —	— — — — —	
		195·1		3·3	$= 4\cdot8 - 1\cdot5$
6	686·4		45·1		
		240·2		5·3	$= 4\cdot8 + 0\cdot5$
7	926·6		50·4		
		290·6		3·8	
8	1 217·2		55·2		
		345·8			
9	1 563·0				
		(1 528·2)	(307·2)	(32·6)	

Solution Adding each difference column and checking against the difference of the first and last entries of the preceding column reveals an error in the last column, i.e. $55 \cdot 2 - 21 \cdot 6 = 33 \cdot 6$. Recalculation of $\Delta^3 y$ shows that the last entry should be $4 \cdot 8$ and not $3 \cdot 8$. The appearance of the $\Delta^3 y$ column suggests now that all the entries should be $4 \cdot 8$. The values $4 \cdot 3, 6 \cdot 3, 3 \cdot 3$, and $5 \cdot 3$ are $4 \cdot 8 - 0 \cdot 5, 4 \cdot 8 + 3(0 \cdot 5), 4 \cdot 8 - 3(0 \cdot 5)$, and $4 \cdot 8 + 0 \cdot 5$ respectively. Comparison with Table 14.2 shows that there is an error $\varepsilon = -0 \cdot 5$ in the entry $491 \cdot 3$. The corrected value is $491 \cdot 8$. It is left to the reader to complete the corrected table.

14.3 THE NEWTON-GREGORY FORMULAE OF INTERPOLATION

Let $y_{-n}, y_{-n+1}, \ldots, y_{-1}, y_0, y_1, \ldots, y_{n-1}, y_n$ be values of the entries corresponding to equally spaced values $x_{-n}, x_{-n+1}, \ldots, x_{-1}, x_0, x_1, \ldots, x_{n-1}, x_n$ in the argument. The unique polynomial function $y = p(x)$ satisfying any $n + 1$ successive values is of degree n and could have the form

$$p(x) = A_0 + A_1 x + A_2 x^2 + \ldots + A_n x^n \tag{6}$$

the $n + 1$ unknowns A_i being found from the $n + 1$ linear equations obtained by substituting the corresponding x and y values in function (6).

14.3.1 The forward interpolation formula

Suppose the $n + 1$ successive values are (x_i, y_i), $i = 0, 1, 2, \ldots, n$ then it is more convenient to write function (6) in the form

$$p(x) = a_0 + (x - x_0)a_1 + (x - x_0)(x - x_1)a_2 + \ldots$$
$$\ldots + (x - x_0)(x - x_1) \ldots (x - x_{n-1})a_n \tag{7}$$

and to use the $n + 1$ values to calculate the a_i. Let the equal intervals between successive arguments be h, then

$$x_r - x_s = (r - s)h. \tag{8}$$

The unknowns are then determined from the $n + 1$ equations

$$y_r = a_0 + rha_1 + r(r - 1)h^2 a_2 + r(r - 1)(r - 2)h^3 a_3 + \ldots$$
$$+ r(r - 1)(r - 2) \ldots (r - n + 1)h^n a_n$$
$$= a_0 + \sum_{t=1}^{n} r(r - 1)(r - 2) \ldots (r - t + 1)h^t a_t \tag{9}$$

with $r = 0, 1, 2, \ldots, n$.
Forming successive differences leads to

$$\Delta y_r = y_{r+1} - y_r$$
$$= ha_1 + \sum_{t=2}^{n} [t][r(r - 1)(r - 2) \ldots (r - t + 2)]h^t a_t \tag{10}$$

$$\Delta^2 y_r = \Delta y_{r+1} - \Delta y_r$$

$$= 2h^2 a_2$$

$$+ \sum_{t=3}^{n} [t(t-1)][r(r-1)(r-2)\ldots(r-t+3)]h^t a_t \quad (11)$$

$$\Delta^3 y_r = \Delta^2 y_{r+1} - \Delta^2 y_r$$

$$= 3 \cdot 2h^3 a_3$$

$$+ \sum_{t=4}^{n} [t(t-1)(t-2)][r(r-1)(r-2)$$

$$\ldots(r-t+4)]h^t a_t \quad (12)$$

$$\vdots \qquad \vdots$$

$$\Delta^{n-1} y_r = (n-1)!h^{n-1} a_{n-1} + [n(n-1)(n-2)\ldots 2][r]h^n a_n \quad (13)$$

$$\Delta^n y_r = n!h^n a_n \quad (14)$$

From equations (9) to (14) it follows that

$$a_0 = y_0; \quad a_1 = \Delta y_0/h; \quad a_2 = \Delta^2 y_0/2!h^2; \quad \ldots;$$

$$a_n = \Delta^n y_0/n!h^n \quad (15)$$

and hence in equation (7)

$$p(x) = y_0 + \frac{(x - x_0)}{h} \Delta y_0 + \frac{(x - x_0)(x - x_1)}{h^2} \frac{\Delta^2 y_0}{2!} + \cdots$$

$$+ \frac{(x - x_0)(x - x_1)\ldots(x - x_{n-1})}{h^n} \frac{\Delta^n y_0}{n!} \quad (16)$$

in which x is any value in the range $x_0 \leqslant x \leqslant x_n$.

If the argument is written

$$x = x_0 + uh \quad (17)$$

where $0 \leqslant u \leqslant n$, then, with the use of (8), equation (16) may be expressed in the form

$$y = f(x) = f(x_0 + uh) \simeq p(x_0 + uh)$$

$$= y_0 + u\Delta y_0 + u(u-1) \frac{\Delta^2 y_0}{2!} + \cdots$$

$$+ u(u-1)(u-2)\ldots(u-n+1) \frac{\Delta^n y_0}{n!} \quad (18)$$

where $p(x)$ is the polynomial approximating the actual (unknown) function $f(x)$. An examination of Table 14.1 and equation (18) shows that the 'path' of this interpolation proceeds through the table from y_0 along the diagonal of the leading differences of y_0, i.e. the formula uses those values *forward* (in the table) from y_0. The formula (18) is frequently called the *forward Newton-Gregory interpolation formula* and is used mainly for interpolation near the beginning of the table where u is small. The coefficients of equation (18) are binomial coefficients and the formula may be written

$$y \simeq y_0 + \sum_{r=1}^{n} \binom{u}{r} \Delta^r y_0 \tag{19}$$

where $\binom{u}{r}$ is $\dfrac{u(u-1)(u-2) \ldots (u-r+1)}{r!}$.

Example (i) In the following table use the result (18) to find (a) $f(2.4)$, (b) $f(8.7)$.

x	2	4	6	8	10
$f(x)$	9·68	10·96	12·32	13·76	15·28

Solution Form a difference table and note that all differences beyond the second are zero.

x	$y = f(x)$	Δy	$\Delta^2 y$
2	9·68		
		1·28	
4	10·96		0·08
		1·36	
6	12·32		0·08
		1·44	
8	13·76		0·08
		1·52	
10	15·28		

(a) From (17) with $x = 2.4$, $x_0 = 2$, $h = 2$, then $u = 0.2$ and equation (18) gives

$$f(2.4) \simeq 9.68 + 0.2 \times 1.28 + (0.2)(-0.8)\frac{(0.08)}{2} + 0 + 0 \ldots$$

$$= 9.929\ 6$$

(b) From (17) with $x = 8.7$, $x_0 = 2$, then $u = 3.35$ and equation (18) gives

$$f(8.7) \simeq 9.68 + (3.35)(1.28) + (3.35)(2.35)\frac{(0.08)}{2}$$

$$= 14.282\ 9.$$

Example (ii) In the following table of e^x use the formula (18) to calculate (a) $e^{0.12}$, and (b) $e^{2.00}$.

x	0·1	0·6	1·1	1·6	2·1
e^x	1·105 2	1·822 1	3·004 2	4·953 0	8·166 2

Solution Form a difference table and note that in this case there is no difference column that is constant (this is to be expected since e^x cannot be represented by a polynomial function).

x	$y = e^z$	Δy	$\Delta^2 y$	$\Delta^3 y$	$\Delta^4 y$
0·1	1·105 2				
		0·716 9			
0·6	1·822 1		0·465 2		
		1·182 1		0·301 5	
1·1	3·004 2		0·766 7		0·196 2
		1·948 8		0·497 7	
1·6	4·953 0		1·264 4		
		3·213 2			
2·1	8·166 2				

(a) with $x = 0·12$, $x_0 = 0·10$, $h = 0·5$, then $u = 0·04$ and (18) gives

$$e^{0·12} \simeq 1·105\,2 + (0·04)(0·716\,9) + (0·04)(-0·96)\frac{(0·465\,2)}{2}$$

$$+ (0·04)(-0·96)(-1·96)\frac{(0·301\,5)}{6}$$

$$+ (0·04)(-0·96)(-1·96)(-2·96)\frac{(0·196\,2)}{24}$$

$$= 1·126\,9.$$

The correct value to 5 decimal places is known to be 1·127 50.
(b) with $x = 2·00$, $x_0 = 0·10$, $h = 0·5$, then $u = 3·8$ and equation (18) gives

$$e^{2·00} \simeq 1·105\,2 + 2·724\,22 + 2·474\,86 + 0·962\,39 + 0·125\,25$$

$$= 7·391\,9 \text{ (to four decimal places).}$$

The correct value to four decimal places is known to be 7·389 1.
In example (i) the Newton-Gregory formula (18) for $p(x)$ is identical with the function $f(x)$, which is a quadratic function, and the results for $f(2·4)$ and $f(8·7)$ will therefore be correct to the number of decimal places retained. In example (ii) the function e^z is replaced by a fourth degree polynomial which takes the value of e^z at the five given entries. Because the successive differences decrease, the higher differences are relatively small so that any term in the formula has a smaller contribution than its predecessor. For the range of values given (i.e. $x = 0·1$ to $x = 2·1$) the quartic polynomial is a reasonable approximation to the actual function e^z. The error involved in the estimation for $e^{0·12}$ is about 0·05 per cent and in the estimation for $e^{2·00}$ is about 0·04 per cent.

Example (iii) Use the formula (18) to estimate the value of $f(2·5)$ in the following table of $f(x)$.

x	2	3	4	5	6
$f(x)$	1	2	6	24	120

Solution Form the difference table from the given data

x	$f(x)$	Δy	$\Delta^2 y$	$\Delta^3 y$	$\Delta^4 y$
2	1				
		1			
3	2		3		
		4		11	
4	6		14		53
		18		64	
5	24		78		
		96			
6	120				

With $x = 2\cdot5$, $x_0 = 2$, $h = 1$, then $u = 0\cdot5$ and formula (18) gives

$$f(2\cdot5) \simeq 1 + 0\cdot5 - 0\cdot375 + 0\cdot687\,5 - 2\cdot070\,3$$
$$= -0\cdot257\,8.$$

The function $f(x)$ in this example is well-known and is called the *Gamma function* or the *generalized factorial function,* $f(x) = (x - 1)!$. The value of $f(2\cdot5)$ is known to be $1\cdot329\,3$. The estimated value is obviously incorrect, the reason being that the successive differences are increasing rapidly and the series obtained from formula (18) is a divergent one. The values of $f(x)$ are increasing so rapidly that it is impossible for a fourth degree polynomial to approximate to the function even though it is exact at the given five values. In other words, the information provided (i.e. the given data) is too meagre to describe the behaviour of the function.

Although the exponential function of example (ii) is also a rapidly increasing function the data are given with an interval of x small enough to be able to interpolate with reasonable accuracy within the range of the table. If here, the interval of the argument x had been too large (say $h = 1$) the interpolation formula would not have provided an accurate estimation except possibly at values extremely close to the given entries. The difference table below shows how rapidly the differences are increasing for $h = 1$.

x	e^z	Δy	$\Delta^2 y$	$\Delta^3 y$	$\Delta^4 y$
0	1				
		1·718 3			
1	2·718 3		2·952 5		
		4·670 8		5·073 1	
2	7·389 1		8·025 6		8·717 6
		12·696 4		13·790 7	
3	20·085 5		21·816 3		
		34·512 7			
4	54·598 2				

Another form of the interpolation formula may be used when interpolation is required at a value near the end of the table.

14.3.2 The Newton–Gregory formula for backward interpolation

Suppose that (x_0, y_0) is at the end of the table then it is required to find a polynomial that satisfies the $(n + 1)$ values (x_{-i}, y_{-i}), $i = 0, 1, 2, 3,$ ..., n. Proceeding as in section 14.3.1 it is convenient to write function (6) in the form

$$p(x) = b_0 + (x - x_0)b_1 + (x - x_0)(x - x_{-1})b_2 + \ldots$$
$$+ (x - x_0)(x - x_{-1}) \ldots (x - x_{-(n-1)})b_n \quad (20)$$

with

$$x_{-r} - x_{-s} = -(r - s)h, \quad (21)$$

from which it follows that

$$y_{-r} = b_0 + \sum_{t=1}^{n} (-1)^t r(r - 1)(r - 2) \ldots (r - t + 1)h^t b_t. \quad (22)$$

Taking successive differences of equation (22) and using the notation of Table 14.3 leads to

$$b_0 = y_0; \; hb_1 = \Delta y_{-1}; \; 2!h^2 b_2 = \Delta^2 y_{-2}; \; \ldots; \; n!h^n b_n = \Delta^n y_{-n}. \quad (23)$$

x	y	Δy	$\Delta^2 y$	$\Delta^3 y$	$\Delta^4 y$
x_{-n}	y_{-n}				
		Δy_{-n}			
x_{-n+1}	y_{-n+1}		$\Delta^2 y_{-n}$		
- - - - - - -	- - - - - - -	- - - - - - -	- - - - - - -	- - - - - - -	- - - - - - -
- - - - - - -	- - - - - - -	- - - - - - -	- - - - - - -	- - - - - - -	- - - - - - -
- - - - - - -	- - - - - - -	- - - - - - -	- - - - - - -	- - - - - - -	- - - - - - -
x_{-2}	y_{-2}		$\Delta^2 y_{-3}$		$\Delta^4 y_{-4}$
		Δy_{-2}		$\Delta^3 y_{-3}$	
x_{-1}	y_{-1}		$\Delta^2 y_{-2}$		
		Δy_{-1}			
x_0	y_0				

Table 14.3

Substituting into equation (20) and using

$$x = x_0 + uh \quad (24)$$

gives the *backward interpolation formula* in the form

$$y = f(x) = f(x_0 + uh) \simeq p(x_0 + uh)$$
$$= y_0 + u\Delta y_{-1} + u(u + 1)\frac{\Delta^2 y_{-2}}{2!} + u(u + 1)(u + 2)\frac{\Delta^3 y_{-3}}{3!}$$
$$+ \ldots + u(u + 1)(u + 2) \ldots (u + n - 1)\frac{\Delta^n y_{-n}}{n!} \quad (25)$$

which may be written

$$y \simeq y_0 + \sum_{r=1}^{n} \binom{u + r - 1}{r} \Delta^r y_{-r} \qquad (26)$$

where $$\binom{u}{r} = \frac{u(u - 1)(u - 2) \ldots (u - r + 1)}{r!}.$$

Table 14.4 shows the paths of the forward and backward interpolation formulae through the difference table for $n = 5$.

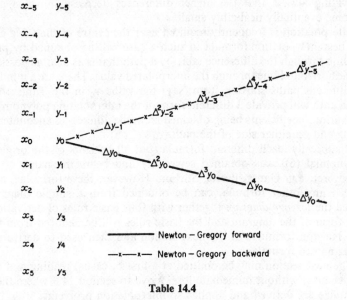

Table 14.4

Example Apply the backward interpolation formula (26) to find $e^{2 \cdot 00}$ in example (ii), section 14.3.1.

Solution Using the difference table of that example and substituting into equation (26) with $x = 2 \cdot 00$, $x_0 = 2 \cdot 10$, $h = 0 \cdot 5$, so that $u = -0 \cdot 2$, leads to

$$e^{2 \cdot 00} \simeq 8 \cdot 166\ 2 - (0 \cdot 2)(3 \cdot 213\ 2)$$
$$- (0 \cdot 2)(0 \cdot 8)\frac{(1 \cdot 264\ 4)}{2}$$
$$- (0 \cdot 2)(0 \cdot 8)(1 \cdot 8)\frac{(0 \cdot 497\ 7)}{6}$$
$$- (0 \cdot 2)(0 \cdot 8)(1 \cdot 8)(2 \cdot 8)\frac{(0 \cdot 196\ 2)}{24}$$
$$= 7 \cdot 392\ 0.$$

14.3.3 Interpolation paths

It has been observed (see Table 14.4) that the two Newton-Gregory formulae use the information of the difference table obtained by proceeding through the table along different paths. The forward formula is useful when interpolating near the beginning of the table while the backward formula is useful when interpolating near the end of the table. Both formulae may be used with a reasonable expectation of accuracy providing that the interval h in the given data is not too large and providing always that the higher differences decrease in value and become eventually negligibly small.

Interpolation is frequently required near the centre of the table and the best interpolation formula in such a case will be obtained by proceeding through the difference table by a path that is as near as possible to the horizontal line through the interpolated value. There are a number of different paths starting from (say) the value y_0 in the Table 14.4. Each path will provide a different form of the interpolating polynomial, the various coefficients being obtained from the differences encountered along and on either side of the path.

Historically, each different formula (but all variants of the original polynomial (6)) was obtained separately and bears the name of its discoverer, e.g. Gauss, Bessel, Stirling. However, these formulae, and indeed any other formula, can be obtained from a simple diagram called the *lozenge diagram* together with four basic rules of derivation. The form of the diagram and the basic rules will be deduced from the two Newton formulae in the next section and then used to derive the other named formulae.

The next section may be omitted at a first or casual reading and the reader may, without immediate loss, move to section 14.5 where these formulae are derived and applied to interpolation problems.

14.4 THE LOZENGE DIAGRAM

The Table 14.5 shows the lozenge diagram which is, in essence, the difference table about the point y_0 together with the binomial coefficients of the Newton-Gregory formulae displayed according to a definite pattern. The coefficient $\begin{pmatrix} u - s \\ r \end{pmatrix}$ is placed immediately above the difference $\Delta^r y_s$. It may be noted that $\begin{pmatrix} u - s \\ 0 \end{pmatrix} = 1$ for all s.

The two Newton formulae

$$y = f(x_0 + uh) \simeq p(x_0 + uh)$$

$$= y_0 + \begin{pmatrix} u \\ 1 \end{pmatrix} \Delta y_0 + \begin{pmatrix} u \\ 2 \end{pmatrix} \Delta^2 y_0 + \begin{pmatrix} u \\ 3 \end{pmatrix} \Delta^3 y_0 + \ldots + \begin{pmatrix} u \\ r \end{pmatrix} \Delta^r y_0 + \ldots$$

$$(27)$$

and

$$y = f(x_0 + uh) \simeq p(x_0 + uh)$$

$$= y_0 + \binom{u}{1} \Delta y_{-1} + \binom{u+1}{2} \Delta^2 y_{-2} + \binom{u+2}{3} \Delta^3 y_{-3} + \dots$$

$$+ \binom{u+r-1}{r} \Delta^r y_{-r} \dots \quad (28)$$

may be obtained from Table 14.5 by proceeding along the appropriate diagonal paths and applying the following two rules:

Rule 1 If, on arrival at a difference entry, the slope of the step was negative then form the product of that difference and the binomial coefficient immediately *above* it; but, if the slope was positive then form the product of the difference and the binomial coefficient immediately *below* it.

Rule 2 Add the successive products formed on proceeding through the table from left to right.

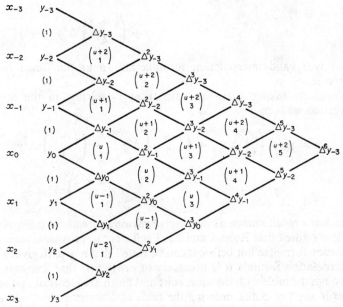

Table 14.5

Example (i) Starting from y_0 and proceeding via the diagonal path of negative slope then

$$y = y_0 + \binom{u}{1} \Delta y_0 + \binom{u}{2} \Delta^2 y_0 + \binom{u}{3} \Delta^3 y_0 + \dots$$

which is equation (27).

Example (ii) Starting from y_0 and proceeding via the diagonal of positive slope

$$y = y_0 + \binom{u}{1} \Delta y_{-1} + \binom{u+1}{2} \Delta^2 y_{-2} + \binom{u+2}{3} \Delta^3 y_{-3} + \dots$$

which is equation (28).

Example (iii) The backward Newton-Gregory formula starting at y_1 and using Rules 1 and 2 is

$$y = y_1 + \binom{u-1}{1} \Delta y_0 + \binom{u}{2} \Delta^2 y_{-1} + \binom{u+1}{3} \Delta^3 y_{-2} + \dots .$$

Suppose now that a zig-zag path passing through y_0, Δy_0, $\Delta^2 y_{-1}$, $\Delta^3 y_{-1}$, $\Delta^4 y_{-2}$, is considered, then the two rules would give the formula

$$y = y_0 + \binom{u}{1} \Delta y_0 + \binom{u}{2} \Delta^2 y_{-1} + \binom{u+1}{3} \Delta^3 y_{-1}$$
$$+ \binom{u+1}{4} \Delta^4 y_{-2} + \dots \quad (29)$$

This is a valid interpolation formula known as the *Gauss forward formula*.

Similarly taking a zig-zag path from y_0 starting in the opposite direction leads to

$$y = y_0 + \binom{u}{1} \Delta y_{-1} + \binom{u+1}{2} \Delta^2 y_{-1} + \binom{u+1}{3} \Delta^3 y_{-2}$$
$$+ \binom{u+2}{4} \Delta^4 y_{-2} \dots \quad (30)$$

which is a result known as the *Gauss backward formula* of interpolation.

It is evident that Rules 1 and 2 are sufficient to provide certain interpolation formulae but before it can be stated that any path gives a valid interpolation formula it is necessary to prove first that any two paths through the table with the same start and finish are identical and second that if any two paths, entering the table at different points, end at the same point then the derived formulae are identical. For example, in Table 14.5 a path that starts at y_0 and ends at $\Delta^6 y_{-3}$ will give a formula that is a sixth degree polynomial satisfying the seven values (x_{-3}, y_{-3}), . . ., (x_3, y_3) and the second requirement would show that the same polynomial should be found by any other path starting at (say) y_2 and ending at $\Delta^6 y_{-3}$. To prove the first requirement it is necessary only to consider the paths taken round one complete lozenge. Consider the paths 1 and 2 in Fig. 14.1.

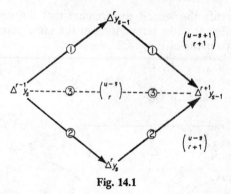

Fig. 14.1

The contribution from path 1 is

$$\binom{u-s}{r}\Delta^r y_{s-1} + \binom{u-s+1}{r+1}\Delta^{r+1} y_{s-1} \tag{31}$$

and the contribution from path 2 is

$$\binom{u-s}{r}\Delta^r y_s + \binom{u-s}{r+1}\Delta^{r+1} y_{s-1} \tag{32}$$

That the contributions (31) and (32) are identical is easily shown by substituting

$$\Delta^r y_s = \Delta^r y_{s-1} + \Delta^{r+1} y_{s-1}$$

in (32) and using the fact that

$$\binom{u-s+1}{r+1} = \binom{u-s}{r} + \binom{u-s}{r+1}.$$

Since the contributions (31) and (32) are identical, then each is equivalent to

$$\frac{1}{2}\left[\Delta^r y_s + \Delta^r y_{s-1}\right]\binom{u-s}{r} + \frac{1}{2}\left[\binom{u-s+1}{r+1} + \binom{u-s}{r+1}\right]\Delta^{r+1} y_{s-1}. \tag{33}$$

Examination of (33) shows that the step from $\Delta^{r-1} y_s$ to $\Delta^{r+1} y_{s-1}$ can be taken by the horizontal path 3 of Fig. 14.1 providing the following two rules are obeyed:

Rule 3 If in a horizontal step from left to right, the path passes through a factor then add the product of that factor and the average of the differences immediately above and below.

Rule 4 If in a horizontal step from left to right, the path passes through a column entry then add the product of that entry and the average of the factors immediately above and below.

Finally to justify the second requirement that the same formula is obtained by different paths terminating at the same point in the table, consider the two paths of Fig. 14.2.

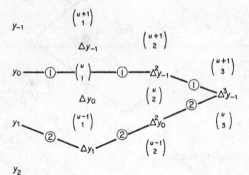

Fig. 14.2

The contributions are

Path 1: $y_0 + \frac{1}{2}[\Delta y_{-1} + \Delta y_0] \binom{u}{1} + \frac{1}{2}\left[\binom{u+1}{2} + \binom{u}{2}\right]\Delta^2 y_{-1}$

$$+ \binom{u+1}{3}\Delta^3 y_{-1} \qquad (34)$$

Path 2: $y_1 + \binom{u-1}{1}\Delta y_1 + \binom{u-1}{2}\Delta^2 y_0 + \binom{u}{3}\Delta^3 y_{-1} \qquad (35)$

In (34), put

$$\Delta^2 y_{-1} = \Delta^2 y_0 - \Delta^3 y_{-1},$$
$$\Delta y_{-1} = \Delta y_0 - \Delta^2 y_{-1}$$
$$= \Delta y_0 - \Delta^2 y_0 + \Delta^3 y_{-1},$$

and in (35), put

$$y_1 = y_0 + \Delta y_0,$$
$$\Delta y_1 = \Delta y_0 + \Delta^2 y_0,$$

then the contributions are

Path 1: $y_0 + \binom{u}{1}\Delta y_0 + \frac{1}{2}\Delta^2 y_0\left[\binom{u+1}{2} + \binom{u}{2} - \binom{u}{1}\right]$

$$+ \frac{1}{2}\Delta^3 y_{-1}\left[\binom{u}{1} - \binom{u+1}{2} + 2\binom{u+1}{3} - \binom{u}{2}\right] \qquad (36)$$

Path 2: $y_0 + \Delta y_0\left[1 + \binom{u-1}{1}\right] + \Delta^2 y_0\left[\binom{u-1}{1} + \binom{u-1}{2}\right]$

$$+ \binom{u}{3}\Delta^3 y_{-1} \qquad (37)$$

The coefficient of $\Delta^3 y_{-1}$ in (36) is

$$\frac{1}{2}\left[u - \frac{(u+1)u}{2} + \frac{2(u+1)(u)(u-1)}{6} - \frac{u(u-1)}{2}\right]$$

$$= \frac{1}{2}\left[\frac{u(u-1)(u-2)}{3}\right]$$

$$= \binom{u}{3}$$

which is the coefficient in (37). It is left to the reader to show that the other coefficients of (36) and (37) are identical.

14.5 THE INTERPOLATION FORMULAE

Using the lozenge diagram of Table 14.5 and the four basic rules it is now a simple matter to derive any interpolation formula by choosing a path through the table from left to right. Note that a step from right to left can be included if the product is subtracted instead of added. Of the many formulae that can be obtained, four have already been mentioned, the two Newton-Gregory and the two Gauss formulae (see (27), (28), (29), (30), of section 14.4).

It has been noted that the Newton-Gregory formulae are useful for interpolation near the ends of the table. The Gaussian formulae are useful near the centre of the table but are superseded by two other formulae which proceed through the table via horizontal paths.

14.5.1 The Stirling formula

This is a formula obtained from the path starting at y_0 (see Table 14.5) and proceeding through the table on a horizontal path. Using rules 2, 3, and 4,

$$y = y_0 + \frac{1}{2}\binom{u}{1}[\Delta y_{-1} + \Delta y_0] + \frac{1}{2}\left[\binom{u+1}{2} + \binom{u}{2}\right]\Delta^2 y_{-1}$$

$$+ \frac{1}{2}\binom{u+1}{3}[\Delta^3 y_{-1} + \Delta^3 y_{-2}]$$

$$+ \frac{1}{2}\left[\binom{u+2}{4} + \binom{u+1}{4}\right]\Delta^4 y_{-2} + \ldots$$

$$= y_0 + \frac{u(\Delta y_{-1} + \Delta y_0)}{2} + \frac{u^2}{2}\Delta^2 y_{-1} + \frac{u(u^2-1)}{3!}\frac{(\Delta^3 y_{-2} + \Delta^3 y_{-1})}{2}$$

$$+ \frac{u^2(u^2-1)}{4!}\Delta^4 y_{-2} + \ldots . \tag{38}$$

Note that this is just the average of the two Gauss formulae (29) and (30) of section 14.4.

14.5.2 The Bessel formula

This interpolation formula is obtained by taking a path that starts midway between the entries y_0 and y_1 and proceeds on a horizontal path through the table.

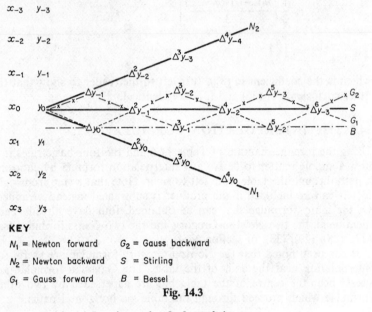

KEY

N_1 = Newton forward G_2 = Gauss backward

N_2 = Newton backward S = Stirling

G_1 = Gauss forward B = Bessel

Fig. 14.3

From Table 14.5, using rules 2, 3, and 4,

$$y = \tfrac{1}{2}(y_0 + y_1) + \frac{1}{2}\left[\binom{u}{1} + \binom{u-1}{1}\right]\Delta y_0$$

$$+ \frac{1}{2}\binom{u}{2}[\Delta^2 y_{-1} + \Delta^2 y_0] + \frac{1}{2}\left[\binom{u+1}{3} + \binom{u}{3}\right]\Delta^3 y_{-1}$$

$$+ \frac{1}{2}\binom{u+1}{4}[\Delta^4 y_{-2} + \Delta^4 y_{-1}] + \ldots$$

$$= \frac{y_0 + y_1}{2} + (u - \tfrac{1}{2})\Delta y_0 + \frac{u(u-1)}{2}\frac{(\Delta^2 y_{-1} + \Delta^2 y_0)}{2}$$

$$+ \frac{u(u-\tfrac{1}{2})(u-1)}{3!}\Delta^3 y_{-1} + \ldots \tag{39}$$

A particularly simple form of (39) may be used to compute the value of y midway between the given values y_0 and y_1. With $u = \tfrac{1}{2}$, the coefficients of all the odd differences are zero and the formula becomes

$$y = \frac{y_0 + y_1}{2} - \frac{1}{8}\frac{\Delta^2 y_{-1} + \Delta^2 y_0}{2} + \frac{3}{128}\frac{\Delta^4 y_{-2} + \Delta^4 y_{-1}}{2} + \ldots$$

$$+ (-1)^n \frac{[1 \cdot 3 \cdot 5 \cdot \ldots \cdot (2n-1)]^2}{2^{2n}(2n)!}\frac{\Delta^{2n} y_{-n} + \Delta^{2n} y_{-n+1}}{2} + \ldots \tag{40}$$

The result (40) is called the formula for *interpolating to halves*.

The paths taken through the table to derive the named formulae of interpolations are shown together in Fig. 14.3.

Example (i) Use Stirling's formula to compute $\tan(89° 26')$ given the following table of values of $\tan x$.

x	$89° 21'$	$89° 23'$	$89° 25'$	$89° 27'$	$89° 29'$
$\tan x$	88·14	92·91	98·22	104·17	110·90

Solution Form the difference table and apply formula (38) with $x = 89° 26'$, $x_0 = 89° 25'$, $h = 2'$, and $u = 0·5$.

x	y	Δy	$\Delta^2 y$	$\Delta^3 y$	$\Delta^4 y$
$89° 21'$	88·14				
		4·77			
$89° 23'$	92·91		0·54		
		5·31		0·10	
$89° 25'$	98·22		0·64		0·04
		5·95		0·14	
$89° 27'$	104·17		0·78		
		6·73			
$89° 29'$	110·90				

Retaining all decimal places,

$$\tan(89° 26') \simeq 98·22 + 0·5(5·63) + \tfrac{1}{2}(0·25)(0·64)$$
$$+ \tfrac{1}{6}(0·5)(-0·75)(0·12) + \tfrac{1}{24}(0·25)(-0·75)(0·04)$$
$$= 98·220 + 2·815 + 0·080 - 0·007\,5 - 0·000\,312\,5$$
$$\simeq 101·107.$$

Example (ii) Use Bessel's formula to estimate $\sqrt[3]{46·24}$ given the following table of $y = \sqrt[3]{x}$.

x	41	45	49	53
$y = \sqrt[3]{x}$	3·448 2	3·556 9	3·659 3	3·756 3

Solution Form the difference table and use formula (39) with $x = 46·24$, $x_0 = 45$, $h = 4·0$, and $u = 0·31$.

x	y	Δy	$\Delta^2 y$	$\Delta^3 y$
41	3·448 2			
		0·108 7		
45	3·556 9		−0·006 3	
		0·102 4		0·000 9
49	3·659 3		−0·005 4	
		0·097 0		
53	3·756 3			

Then,

$$\sqrt[3]{46\cdot24} \simeq 3\cdot608\ 1 + (-0\cdot19)(0\cdot102\ 4)$$
$$+ \tfrac{1}{2}(0\cdot31)(-0\cdot69)(-0\cdot005\ 85)$$
$$+ \tfrac{1}{6}(0\cdot31)(-0\cdot19)(-0\cdot69)(0\cdot000\ 9)$$
$$= 3\cdot608\ 1 - 0\cdot019\ 456 + 0\cdot000\ 626 + 0\cdot000\ 006$$
$$\simeq 3\cdot589\ 3.$$

Example (iii) Form a cubic polynomial that may be used to replace $\sin x$ in the range $0\cdot90 \leqslant x \leqslant 1\cdot20$ given the four values

x	0·90	1·00	1·10	1·20
$\sin x$	0·783 3	0·841 5	0·891 2	0·932 0

Solution Form the difference table,

x	y	Δy	$\Delta^2 y$	$\Delta^3 y$
0·9	0·783 3			
		0·058 2		
1·0	0·841 5		−0·008 5	
		0·049 7		−0·000 4
1·1	0·891 2 .		−0·008 9	
		0·040 8		
1·2	0·932 0			

and use Bessel's formula (39) with $x_0 = 1\cdot0$, $h = 0\cdot1$, then,

$$\sin (1 + 0\cdot1u) \simeq 0\cdot866\ 35 + (u - \tfrac{1}{2})(0\cdot049\ 7)$$
$$+ \tfrac{1}{2}u(u - 1)(-0\cdot008\ 7)$$
$$+ \tfrac{1}{6}u(u - \tfrac{1}{2})(u - 1)(-0\cdot000\ 4)$$
$$\simeq 0\cdot841\ 50 + 0\cdot054\ 02u - 0\cdot004\ 25u^2 - 0\cdot000\ 07u^3.$$

Putting $u = 10v$ then,

$$y = 0\cdot841\ 50 + 0\cdot540\ 2v - 0\cdot425v^2 - 0\cdot07v^3$$

may be used as the cubic expression to represent $\sin (1 + v)$ in the range $-0\cdot1 \leqslant v \leqslant 0\cdot2$. Finally, replacement of v by $x-1$ gives the cubic polynomial in x that represents $\sin x$.

14.5.3 The accuracy of the formulae

Each formula derived for the function $f(x_0 + uh)$ is an approximating polynomial. If the actual function $f(x)$ is a polynomial in x then, providing the difference table can be taken far enough (i.e. that enough data are given and used) all the formulae are equally accurate and will give an exact result (assuming the original data to be exact).

If the function $f(x)$ is not a polynomial then the difference table should first be examined to see that the successive differences Δy_i, $\Delta^2 y_i$,

$\Delta^3 y_t$, . . . decrease, which means that successive terms in the polynomial approximation also decrease. Some indication of the accuracy to be expected is often given by an examination of the difference columns. If the differences decrease rapidly and the rth difference column is approximately constant then it is reasonable to expect that the formulae will give results as accurate as the given data. If the differences slowly decrease or there is no column of (approximate) constants then the reader should be aware that a problem of accuracy does exist. When the differences increase it is highly probable that the interpolation is impossible, the formula used being part of a divergent series. Interpolation in the latter case might be possible if the differences increase so slowly that the binomial coefficients in the formulae are able to control the size of the terms and ensure a convergent series. The difficulty of knowing whether a result is approximately correct or not is usually removed by an examination of the remainder term of the interpolating series used. The discussion involved in this examination is not considered here.

The choice of interpolating polynomial depends on the position in the table of the interpolated value. It is usual to keep u as small as possible but to use a path through the table that gives a polynomial formula using all the available information. The following statements read in conjunction with Fig. 14.3 may be taken as general rules:

Rule 1 If x_0 is at the start of the table, i.e. y is known for argument values $x_0, x_1, x_2, \ldots$, and interpolation is required near the beginning of the table then the best interpolation formula is the Newton-Gregory forward formula.

Rule 2 If x_0 is at the end of the table, i.e. y is known for argument values $x_0, x_{-1}, x_{-2}, \ldots$, and interpolation is required near the end of the table then the best result is obtained from the Newton-Gregory backward formula of interpolation.

Rule 3 If x_0 is in the centre of the table, i.e. y is known for argument values on either side of x_0 and interpolation is required near the centre of the table then the best formula to use is either that of Stirling or of Bessel.

If interpolation is required at either side of x_0 for the range $-\frac{1}{4} \leqslant u \leqslant +\frac{1}{4}$, then the Stirling formula should be used.

If interpolation is required near the middle of the interval (x_0, x_1) for the range $\frac{1}{4} \leqslant u \leqslant \frac{3}{4}$, then the formula of Bessel should be used.

Example Form the difference table from the given data for $y = \sin x^\circ$ and estimate y when (i) $x = 12$, (ii) $x = 31$, (iii) $x = 37$, (iv) $x = 45$.

x	10	20	30	40	50	60
y	0·173 65	0·342 02	0·500 00	0·642 79	0·766 04	0·866 03

Solution In the following difference table the decimal point is understood, the fifth difference for example being 0·000 18.

x	y	Δy	$\Delta^2 y$	$\Delta^3 y$	$\Delta^4 y$	$\Delta^5 y$
10	17365					
		16837				
20	34202		−1039			
		15798		−480		
30	50000		−1519		45	
		14279		−435		18
40	64279		−1954		63	
		12325		−372		
50	76604		−2326			
		9999				
60	86603					

(i) For $x = 12$, take $x_0 = 10$ then, since $h = 10$, $u = 0·2$ and the forward Newton-Gregory formula may be used. From equation (27) of section 14.4

$$y(12) \simeq 0·173\ 65 + (0·2)(0·168\ 37) + \frac{(0·2)(-0·8)}{2}(-0·010\ 39)$$

$$+ \frac{(0·2)(-0·8)(-1·8)}{6}(-0·004\ 80)$$

$$+ \frac{(0·2)(-0·8)(-1·8)(-2·8)}{24}(0·000\ 45)$$

$$+ \frac{(0·2)(-0·8)(-1·8)(-2·8)(-3·8)}{120}(0·000\ 18)$$

$$= 0·173\ 65 + 0·033\ 674 + 0·000\ 831\ 2 - 0·000\ 230\ 4$$
$$- 0·000\ 001\ 5 + 0·000\ 004\ 5$$
$$= 0·207\ 93$$

The result correct to five decimal places is known to be 0·207 91. The estimate is in error by less than 0·01 per cent.

(ii) For $x = 31$, take $x_0 = 30$, $h = 10$, then $u = 0·1$ and the Stirling formula (38) of section 14.5.1 gives

$$y(31) \simeq 0·5 + (0·1)(0·150\ 385) + \frac{(0·01)}{2}(-0·015\ 19)$$

$$+ \frac{(0·1)(-0·99)}{6}(0·004\ 575) + \frac{(0·01)(-0·99)}{24}(0·000\ 45)$$

$$= 0·500\ 00 + 0·015\ 038\ 5 - 0·000\ 075\ 95$$
$$- 0·000\ 075\ 49 - 0·000\ 000\ 19$$
$$= 0·514\ 89.$$

The correct result is 0·515 04 and the estimate is in error by less than 0·03 per cent.

(iii) For $x = 37$, $x_0 = 30$, $h = 10$, then $u = 0·7$ and Bessel's formula (39) gives

$$y(37) \simeq 0·571\ 395 + (0·2)(0·142\ 79) + \frac{(0·7)(-0·3)}{2}(-0·017\ 365)$$

$$+ \frac{(0·7)(0·2)(-0·3)}{6}(-0·004\ 35)$$

$$+ \frac{(0·7)(-0·51)(-1·3)}{24}(0·000\ 54)$$

$$+ \frac{(0·7)(0·2)(-0·51)(-1·3)}{120}(0·000\ 18)$$

$$= 0·571\ 395 + 0·028\ 558 + 0·001\ 823\ 325$$

$$+ 0·000\ 030\ 45 + 0·000\ 010\ 442 + 0·000\ 002$$

$$= 0·601\ 82.$$

The correct result, to five decimal places, is 0·601 81 and there is an error less than 0·002 per cent in this estimate.

Using the Lozenge diagram the reader should verify the derivation of the sixth term in the formula.

(iv) For $x = 45$, $x_0 = 40$, $h = 10$, then $u = 0·5$ and using Bessel's formula (40) gives

$$y(45) \simeq 0·704\ 415 - \tfrac{1}{8}(-0·021\ 40) + \tfrac{3}{128}(0·000\ 72)$$

$$= 0·704\ 415 + 0·002\ 675 + 0·000\ 017$$

$$= 0·707\ 11.$$

In deriving the average coefficient 0·000 72 of the third term of $y(45)$ it has been assumed that $\Delta^5 y_r$ is constant at 0·000 18, i.e. the coefficient 0·000 72 is the average of 0·000 63 and 0·000 81. The estimated value is correct to five decimal places.

Finally it is possible to give an indication of the superiority that the central difference formulae of Stirling and Bessel have over the Newton-Gregory formulae by a comparison of the coefficients of the terms. Consider the case of $u = 0·3$ in the formulae (27) of section 14.4, (38) of section 14.5.1, and (39) of section 14.5.2. The three formulae based on the value $x = x_0$ are as follows:

Newton-Gregory:

$$y(x_0 + 0·3h) = y_0 + 0·3\Delta y_0 - 0·105\Delta^2 y_0 + 0·059\ 5\Delta^3 y_0$$

$$- 0·040\ 16\Delta^4 y_0 + \ldots$$

Stirling:

$$y(x_0 + 0{\cdot}3h) = y_0 + 0{\cdot}3\frac{(\Delta y_{-1} + \Delta y_0)}{2} + 0{\cdot}045\Delta^2 y_{-1}$$

$$- 0{\cdot}045\ 5\frac{(\Delta^3 y_{-2} + \Delta^3 y_{-1})}{2} - 0{\cdot}003\ 412\ 5\Delta^4 y_{-2}$$

$$+ \ldots$$

Bessel:

$$y(x_0 + 0{\cdot}3h) = \frac{y_0 + y_1}{2} - (0{\cdot}2)\Delta y_0 - 0{\cdot}105\frac{(\Delta^2 y_{-1} + \Delta^2 y_0)}{2}$$

$$+ 0{\cdot}007\Delta^3 y_{-1} + 0{\cdot}019\ 3\frac{(\Delta^4 y_{-2} + \Delta^4 y_{-1})}{2} + \ldots$$

It can be observed that after the first few terms the coefficients in the Stirling formula decrease more rapidly than those of the Bessel formula which in turn decrease more rapidly than those of the Newton-Gregory formula. The contribution from the higher order terms of the Stirling formula is much less than that of the Newton-Gregory formula. If the differences are only decreasing slowly or if there are round-off errors in the initial data (which are magnified in the higher difference columns, see section 14.2.1) it is preferable, whenever possible, to use a central difference formula such as that of Stirling or Bessel.

14.6 INVERSE INTERPOLATION

Inverse interpolation applied to a set of tabulated values (x_i, y_i), $i = -n, \ldots, -1, 0, +1, \ldots, +n$ requires an estimation of x for some given y in the range $y_{-n} < y < y_{+n}$. In essence the process reduces to solving the interpolation formula $y = p(x)$ for x, given the value of y. The formulae of section 14.5 have all been expressed in the form $y = p(x_0 + uh)$ and it usually easier to solve for u and then to determine x. The process is simply that of determining the root of a polynomial equation, a process discussed in some detail in Chapter 13.

The interpolation formula used depends on the position in the table of the value required. Only one method of attack is discussed here namely the iterative method of section 13.2.2. The equation $p(x_0 + uh) =$ constant is written in the form

$$u = \phi(u)$$

and the iterative process

$$u_{r+1} = \phi(u_r)$$

is used.

Example Find the value of x for which $y = 10$ in the following table

x	2·8	2·9	3·0	3·1	3·2
y	8·252 7	9·114 6	10·067 7	11·121 5	12·286 6

Solution Form the difference table

x	y	Δy	$\Delta^2 y$	$\Delta^3 y$	$\Delta^4 y$
2·8	8·252 7				
		0·861 9			
2·9	9·114 6		0·091 2		
		0·953 1		0·009 5	
3·0	10·067 7		0·100 7		0·001 1
		1·053 8		0·010 6	
3·1	11·121 5		0·111 3		
		1·165 1			
3·2	12·286 6				

Because the value $y = 10$ is near the centre value the best interpolation formula to use is that of Stirling. With $x_0 = 3$, $h = 0·1$, and $x = 3·0 + 0·1u$ then equation (38) leads to the approximation

$$10 = 10·067 \ 7 + u(1·003 \ 45) + u^2(0·050 \ 35)$$

$$+ \frac{u(u^2 - 1)}{6}(0·010 \ 05) + \frac{u^2(u^2 - 1)}{24}(0·001 \ 1),$$

which may be rewritten in the form

$$-u = \frac{0·067 \ 7}{1·003 \ 45} + u^2 \frac{(0·050 \ 35)}{1·003 \ 45} + u(u^2 - 1)\frac{(0·001 \ 675)}{1·003 \ 45}$$

$$+ u^2(u^2 - 1)\frac{(0·000 \ 046)}{1·003 \ 45}$$

$$= \phi(u).$$

For the purposes of calculation, put $u = -v$ then

$$v = 0·067 \ 467 + v^2(0·050 \ 177) + v(1 - v^2)(0·001 \ 669)$$
$$- v^2(1 - v^2)(0·000 \ 046)$$

and the iteration proceeds as follows

$v_0 = 0$	$v_3 = 0·067 \ 810 \ 6$
$v_1 = 0·067 \ 467$	$v_4 = 0·067 \ 810 \ 6$
$v_2 = 0·067 \ 807$	

The value of u may therefore be taken as $-0·067 \ 81$ from which $x = x_0 + uh = 2·993 \ 2$ to four decimal places. The table is, in fact, one for $y = \cosh x$ and the result correct to four decimal places is known to be $2·993 \ 3$. The error in the calculated value is $0·003$ per cent.

14.7 NUMERICAL DIFFERENTIATION

This is the process of calculation of the derivative dy/dx from given data (x_i, y_i), when the actual relationship between x and y is unknown. The procedure is to replace the true relationship $y = f(x)$ by the best approximating polynomial $y = p(x)$ and then to differentiate the latter.

The interpolation polynomial to be used depends as usual on the position in the table of the value at which the derivative is required.

The interpolation formulae obtained in the previous sections have all been expressed in terms of the variable u rather than x, in the form

$$y(x) = f(x_0 + uh) \simeq p(x) \qquad (41)$$

and differentiation gives

$$\frac{dy}{dx} = \frac{dy}{du}\frac{du}{dx} = \frac{1}{h}\frac{dy}{du} \simeq \frac{1}{h}\frac{dp}{du}. \qquad (42)$$

Further derivatives may be obtained, for example

$$\frac{d^2y}{dx^2} \simeq \frac{1}{h^2}\frac{d^2p}{du^2}. \qquad (43)$$

Example From the data given, compute (i) dy/dx when $x = 3\cdot21$, (ii) d^2y/dx^2 when $x = 3\cdot8$.

x	3·0	3·5	4·0	4·5
y	1·484 30	1·550 23	1·607 46	1·658 01

Solution Construct the difference table

x	y	Δy	$\Delta^2 y$	$\Delta^3 y$
3·0	1·484 30			
		0·065 93		
3·5	1·550 23		−0·008 70	
		0·057 23		0·002 02
4·0	1·607 46		−0·006 68	
		0·050 55		
4·5	1·658 01			

(i) To find dy/dx at $x = 3\cdot21$ near the start of the table use the Newton-Gregory formula (27) based on $x_0 = 3$, $h = 0\cdot5$, and $u = 0\cdot42$ then,

$$y \simeq 1\cdot484\ 30 + u(0\cdot065\ 93) + \frac{u(u-1)}{2}(-0\cdot008\ 70)$$

$$+ \frac{u(u-1)(u-2)}{6}(0\cdot002\ 02)$$

and with (42),

$$0\cdot5\ dy/dx \simeq [0\cdot065\ 93 - (2u-1)(0\cdot004\ 35)$$
$$+ (3u^2 - 6u + 2)(0\cdot000\ 336)]_{u=0\cdot42}$$
$$= 0\cdot065\ 93 + 0\cdot000\ 696 + 0\cdot000\ 003$$
$$= 0\cdot066\ 629.$$

Thus $dy/dx \simeq 0.133\,2$, when $x = 3.21$.

(ii) To find d^2y/dx^2 at $x = 3.8$ near the centre of the table use the Bessel formula (39) based on $x_0 = 3.5$, $h = 0.5$, and $u = 0.6$, then

$$y \simeq 1.578\,845 + (u - \tfrac{1}{2})(0.057\,23) + \tfrac{1}{2}u(u - 1)(-0.007\,69)$$

$$+ \tfrac{1}{6}(u)(u - \tfrac{1}{2})(u - 1)(0.002\,02)$$

and

$$0.5\,dy/dx \simeq 0.057\,23 + (u - \tfrac{1}{2})(-0.007\,69)$$

$$+ \tfrac{1}{6}(3u^2 - 3u + \tfrac{1}{2})(0.002\,02)$$

and so,

$$(0.5)^2 \frac{d^2y}{dx^2} \simeq [-0.007\,69 + (u - \tfrac{1}{2})(0.002\,02)]_{u=0.6}$$

or

$$\left[\frac{d^2y}{dx^2}\right]_{x=3.8} \simeq (-0.007\,488)/0.25$$

$$= -0.029\,9(52).$$

The given data are those of the function

$$y = \log_{10}(10x + 0.5)$$

from which

$$\left[\frac{dy}{dx}\right]_{x=3.21} = 0.133\,22 \quad \text{and} \quad \left[\frac{d^2y}{dx^2}\right]_{x=3.8} = -0.029\,30$$

and the estimated value in (i) is in error by approximately 0.03 per cent while that in (ii) is in error by over 2 per cent.

The process of successive differentiation has the effect of exaggerating the errors involved in the original approximations, i.e. if the data are accurate to a specified number of decimal places then the derivative obtained by the above process is usually accurate to a lesser number of decimal places.

14.7.1 Maxima and minima

To find the approximate turning value of a function it is necessary to form the approximate derivative, equate to zero and solve the resultant equation for u (and hence for x).

Example Find the turning point, given the following data.

x	2	3	4	5	6
y	31.187 5	12.027 5	2.865 3	3.705 2	14.544 0

Solution Form the difference table in the usual way. A rough sketch shows that there is a minimum in the region $x = 4.5$.

x	y	Δy	$\Delta^2 y$	$\Delta^3 y$	$\Delta^4 y$
2	31·187 5				
		−19·160 0			
3	12·027 5		9·997 8		
		−9·162 2		0·004 3	
4	2·865 3		10·002 1		−0·007 5
		0·839 9		−0·003 2	
5	3·705 2		9·998 9		
		10·838 8			
6	14·544 0				

Using Bessel's formula (39) of section 14.5.2 with $x_0 = 4.0$, $h = 1$, then

$$y \simeq 3.285\ 25 + (u - \tfrac{1}{2})(0.839\ 9) + \frac{(u)(u-1)}{2}(10.000\ 5)$$
$$+ \frac{u(u - \tfrac{1}{2})(u - 1)}{6}(-0.003\ 2)$$

and $h\dfrac{dy}{dx} \simeq 0.839\ 9 + (u - \tfrac{1}{2})(10.000\ 5) + (3u^2 - 3u + \tfrac{1}{2})(-0.000\ 53)$.

An approximation to the value of u corresponding to the minimum point is given by solving the approximate equation

$$0.001\ 6u^2 - 10.002\ 1u + 4.160\ 6 = 0.$$

Writing this in the form

$$u = 0.416\ 0 + 0.000\ 16u^2$$

the approximate root is 0.416 from which it follows that $x \simeq 4.416$. It is left to the reader to show that the corresponding minimum value is given by $y = 2$.

14.8 NUMERICAL INTEGRATION

This is the process of forming a definite integral of y from data given by a set of tabulated values of x and y. The process is frequently called *integration by quadrature*. The data may be experimental, in which case the actual relationship $y = f(x)$ is unknown but the integral $\int f(x)\,dx$ is required over some range of argument within the range of given data.

Frequently however it is necessary to compute the integral $\displaystyle\int_a^b f(x)\,dx$ for a known function $f(x)$ for which the methods of integration discussed in Chapter 6 do not apply. In this case the values of $f(x)$ are tabulated at intervals of the argument x within the range (a, b) so as to form the required data.

An interpolating polynomial is now used to replace the function $f(x)$, and the approximation obtained by term by term integration. Instead of using one polynomial to replace $f(x)$ in the whole range (a, b) the usual procedure is to write

$$\int_a^b f(x)\,dx = \int_a^{a_1} f(x)\,dx + \int_{a_1}^{a_2} f(x)\,dx + \ldots + \int_{a_r}^b f(x)\,dx$$

where $a < a_1 < a_2 < \ldots < a_r < b$ and then to replace $f(x)$ of each sub-range by interpolation polynomials of the same kind. The error in the approximation depends on how closely the interpolation polynomials approximate the function $f(x)$. Since integration is a summation process it is reasonable to suppose that errors in the original data, some of which may be positive and some negative, tend to be smoothed out.

There are many different kinds of quadrature formulae each based on integration of a particular kind of interpolating polynomial. The simplest to use and often the most effective are those based on the forward Newton-Gregory polynomial.

14.8.1 Quadrature formulae based on the Newton–Gregory polynomial

Consider a function y given at n equal intervals of the argument x from x_0 to $x_0 + nh$ then, since $x = x_0 + uh$, $dx = h\,du$, the integration of the forward Newton-Gregory formula (27) leads to

$$\int_{x_0}^{x_0 + nh} y\,dx = h \int_{u=0}^{n} \left[y_0 + \binom{u}{1}\Delta y_0 + \binom{u}{2}\Delta^2 y_0 + \binom{u}{3}\Delta^3 y_0 + \ldots \right.$$

$$\left. \binom{u}{r}\Delta^r y_0 + \ldots \right] du \tag{44}$$

The general term of equation (44) involves the integration with respect to u of $\binom{u}{r}$, i.e.

$$\frac{1}{r!} \int_0^n u(u-1)(u-2)\ldots(u-r+1)\,du.$$

It is possible to integrate such an expression for arbitrary r but for the simple quadrature formulae based on equation (44) there is no need for such generality.

Suppose it is required to evaluate the integral of $f(x)$ over an interval from $x = a$ to $x = b$. The first step is to subdivide the interval (a, b) into a number of equal parts. If there are N such parts the argument values may be written

$$a = x_0, x_1, x_2, \ldots, x_{N-1}, x_N = b$$

with $x_r - x_{r-1} = h$ and $b - a = Nh$.

The second step is to take the argument (and ordinate) values in groups of p at a time, i.e.

$$(x_0, x_1, \ldots, x_{p-1}); (x_{p-1}, \ldots, x_{2p-1}); \ldots (\quad \ldots x_N)$$

and then to apply the result (44) to each group in turn with $n = p - 1$. A different integration formula will result for each different value of p. The most useful results are those obtained with $p = 2$, $p = 3$, and $p = 7$.

14.8.2 The trapezoidal rule

This is the case with $p = 2$, when the $N + 1$ argument values are grouped in pairs $(x_0, x_1), (x_1, x_2) \ldots (x_{N-1}, x_N)$ and equation (44) is applied with $n = 1$. The polynomial in the integrand of equation (44) is assumed to be of the first degree, all differences after the first being taken as zero. Thus

$$\int_{x_0}^{x_1} y \, dx \simeq h \int_0^1 (y_0 + u\Delta y_0)du = h(y_0 + \tfrac{1}{2}\Delta y_0)$$
$$= h[y_0 + \tfrac{1}{2}(y_1 - y_0)]$$
$$= \tfrac{1}{2}h(y_0 + y_1),$$
$$\int_{x_1}^{x_2} y \, dx \simeq \tfrac{1}{2}h(y_1 + y_2),$$
$$\vdots$$
$$\int_{x_{N-1}}^{x_N} y \, dx \simeq \tfrac{1}{2}h(y_{N-1} + y_N).$$

Hence

$$\int_a^b y \, dx \simeq \int_{x_0}^{x_N} y \, dx = \tfrac{1}{2}h(y_0 + 2y_1 + 2y_2 + \ldots + 2y_{N-1} + y_N).$$
$$(45)$$

This is the *trapezoidal rule*. It is a simple result to apply but is not particularly accurate unless h is made extremely small (and N very large). Geometrically, the process of integration

$$\int_a^b f(x) \, dx$$

can be represented as the area under the curve $y = f(x)$ and the trapezoidal rule simply replaces the curve by a set of chords through the points (x_r, y_r) and (x_{r+1}, y_{r+1}) with $r = 0, 1, \ldots, N - 1$ and then approximates to the actual area under the curve by adding together the trapezium areas as in Fig. 14.4.

14.8.3 Simpson's rule

This is the case with $p = 3$, for which the $N + 1$ argument values are grouped, three at a time, i.e.

$$(x_0, x_1, x_2); (x_2, x_3, x_4); \ldots, (x_{N-2}, x_{N-1}, x_N)$$

It is evident that N, the number of intervals, must be even and the number of ordinates must be odd. The integration (44) is applied to

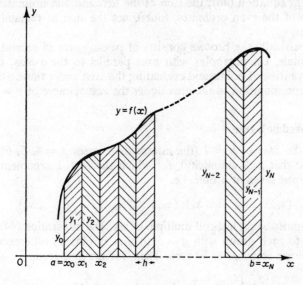

Fig. 14.4

each group with $n = 2$. The integrand is assumed to be of the second degree all differences after the second being omitted. This leads to

$$\int_{x_0}^{x_2} y \, dx \simeq h \int_0^2 [y_0 + u\Delta y_0 + \tfrac{1}{2}u(u - 1)\Delta^2 y_0] \, du$$

$$= h[2y_0 + 2\Delta y_0 + \tfrac{1}{3}\Delta^2 y_0]$$

$$= h[2y_0 + 2(y_1 - y_0) + \tfrac{1}{3}(y_2 - 2y_1 + y_0)]$$

$$= \tfrac{1}{3}h[y_0 + 4y_1 + y_2]$$

and to

$$\int_{x_2}^{x_4} y \, dx \simeq \frac{h}{3}[y_2 + 4y_3 + y_4],$$

$$\vdots \qquad \vdots$$

$$\int_{x_{N-2}}^{x_N} y \, dx \simeq \frac{h}{3}[y_{N-2} + 4y_{N-1} + y_N].$$

Adding the results gives

$$\int_a^b y\,dx \simeq \frac{h}{3}\left[(y_0 + y_N) + 2(y_2 + y_4 + \ldots + y_{N-2})\right.$$
$$\left. + 4(y_1 + y_3 + \ldots + y_{N-1})\right] \qquad (46)$$

This simple and reasonably accurate formula is known as *Simpson's rule*. The reader should note the grouping of the terms inside the square bracket in equation (46): the sum of the first and last ordinates, twice the sum of the even ordinates, four times the sum of remaining odd ordinates.

Geometrically the process consists of passing arcs of second degree polynomials, i.e. parabolas with axes parallel to the y-axis, through successive trios of points and evaluating the area under these parabolas as an approximation to the area under the actual curve of $y = f(x)$.

14.8.4 Weddle's rule

This is the case of $p = 7$ (the intermediate cases $p = 4, 5, 6$ lead to formulae that lack simplicity), for which the $N + 1$ argument values are grouped seven at a time, i.e.

$$(x_0, x_1, x_2, \ldots, x_6);\ (x_6, \ldots, x_{12});\ \ldots (\ldots, x_N).$$

Here N must be an integral multiple of six. The integration (44) is now applied to each group with $n = 6$ so that in (44) all differences after the sixth are assumed to be zero. Then, for the first group

$$\int_{x_0}^{x_6} y\,dx \simeq h \int_{u=0}^{6} \left[y_0 + u\Delta y_0 + \tfrac{1}{2}u(u - 1)\Delta^2 y_0\right.$$
$$\left. + \tfrac{1}{6}u(u - 1)(u - 2)\Delta^3 y_0 + \ldots + \binom{u}{6}\Delta^6 y_0\right]du$$
$$= h[6y_0 + 18\Delta y_0 + 27\Delta^2 y_0 + 24\Delta^3 y_0 + \tfrac{123}{10}\Delta^4 y_0$$
$$+ \tfrac{33}{10}\Delta^5 y_0 + \tfrac{41}{140}\Delta^6 y_0].$$

Before proceeding with the substitution in terms of the ordinates $y_0, y_1, y_2, \ldots, y_6$, the last term is amended to $(3\Delta^6 y_0)/10$ instead of $(41\Delta^6 y_0)/140$. This introduces an error of $(h\Delta^6 y_0)/140$ into the evaluation, which will be insignificant providing that h and $\Delta^6 y_0$ are small. Replacing the differences by their expressions in $y_0, y_1, y_2, \ldots, y_6$ leads, after some reduction, to

$$\int_{x_0}^{x_6} y\,dx \simeq \tfrac{3}{10}h(y_0 + 5y_1 + y_2 + 6y_3 + y_4 + 5y_5 + y_6).$$

Similarly, for the second group of terms

$$\int_{x_6}^{x_{12}} y\,dx \simeq \tfrac{3}{10}h(y_6 + 5y_7 + y_8 + 6y_9 + y_{10} + 5y_{11} + y_{12})$$

Repeating for each successive group of terms and adding leads to

$$\int_a^b y \, dx = \int_{x_0}^{x_N} y \, dx$$

$$= \tfrac{3}{10}h \left[\sum_{i=0}^{N} c_i y_i \right] \tag{47}$$

where

$$c_i = 1, 5, 1, 6, 1, 5; \, 2, 5, 1, 6, 1, 5; \, 2, \ldots; \ldots; \ldots, 5;$$
$$2, 5, 1, 6, 1, 5, 1.$$

This is *Weddle's rule*. The coefficients c_i are most simply remembered in groups, i.e.

first group: 1, 5, 1, 6, 1, 5

all interior groups: 2, 5, 1, 6, 1, 5

last group: 2, 5, 1, 6, 1, 5, 1

The geometrical interpretation is that the portion of the curve $y = f(x)$ through each successive group of seven points is replaced by an arc of a sixth degree polynomial through these seven points. The rule is more accurate but requires more calculation than does Simpson's rule.

14.8.5 Quadrature formulae based on other interpolation polynomials

Further formulae can be produced dependent on the integration of the Stirling, Gauss, and Bessel interpolation polynomials. These results, while providing rapidly converging quadrature formulae, have the disadvantage of difficulty in practical application. Such formulae are not discussed here partly because of this disadvantage and partly because they are superseded by more sophisticated techniques which have application to computer programs.

14.8.6 Application of trapezoidal, Simpson's and Weddle's rule

As stated earlier if, as the result of an experiment, data are available in the form (x_i, y_i), $i = 0, 1, 2, \ldots, N$ then an approximation to the integral of $f(x)$ from x_0 to x_N can be obtained by immediately applying the results of the previous sections. If it is required to evaluate the integral of some known function then the data must be computed before proceeding. In each case the accuracy depends, in geometrical terms, on the accuracy with which the areas under the polynomials approximate to the areas under the actual curve. An estimate of the error involved may be easily found.

In the case of the trapezoidal rule the principal part of the error in each integration is the first term neglected in the integral (44), i.e. the error in

$$\int_{x_0}^{x_1} y \, dx$$

is the first term of

$$h \int_0^1 \left\{ \binom{u}{2} \Delta^2 y_0 + \binom{u}{3} \Delta^3 y_0 + \ldots \right\} du$$

or

$$\frac{-h}{12} \Delta^2 y_0.$$

The total error in the use of the trapezoidal rule is

$$\frac{-h}{12} (\Delta^2 y_0 + \Delta^2 y_1 + \ldots + \Delta^2 y_{N-1}) \tag{48}$$

In using Simpson's rule the actual error in the integral from x_0 to x_2 is

$$h \int_0^2 \left\{ \binom{u}{3} \Delta^3 y_0 + \binom{u}{4} \Delta^4 y_0 + \ldots \right\} du$$

Now,

$$\int_0^2 \binom{u}{3} du = \frac{1}{3!} \int_0^2 u(u-1)(u-2) \, du$$

$$= \frac{1}{3!} [4 - 8 + 4]$$

$$= 0$$

and the principal part of the error is given by

$$h \int_0^2 \binom{u}{4} \Delta^4 y_0 \, du$$

which reduces to

$$-\frac{h}{90} \Delta^4 y_0.$$

The total error involved in the use of Simpson's rule is then

$$-\frac{h}{90} [\Delta^4 y_0 + \Delta^4 y_2 + \Delta^4 y_4 + \ldots + \Delta^4 y_{N-2}]. \tag{49}$$

The fact that the error depends only on the fourth order differences means that Simpson's rule will be exact when $f(x)$ is a polynomial of the first, second or third degree.

In applying Weddle's rule the principal error arises from the omission of the term $-\dfrac{h}{140} \Delta^6 y_0$.

The total error is then

$$-\frac{h}{140} [\Delta^6 y_0 + \Delta^6 y_6 + \ldots]. \tag{50}$$

Weddle's rule will be exact for the integral of $f(x)$ when the latter is a polynomial of up to the fifth degree.

This chapter is concluded by examples of evaluation of an integral by each of the three rules.

Example (i) Estimate the value of $\int_0^{0.6} \sin x \, dx$ using a sub-division of 0·1 in the argument, (i) by the trapezoidal rule, (ii) by Simpson's rule, and (iii) by Weddle's rule. Compare the results with the exact value obtained by integration.

Solution This example is used merely as an illustration of the numerical processes; the correct value of the integral may be obtained using tables of trigonometric functions, i.e.

$$\int_0^{0.6} \sin x \, dx = \left[-\cos x \right]_0^{0.6}$$

$$= 1 - \cos 0.6$$

$$= 1 - 0.825 \, 34$$

$$= 0.174 \, 66 \text{ to five decimal places.}$$

To find estimates by the numerical processes, a table of values is required. This is given below to five decimal places.

x	0	0·1	0·2	0·3	0·4	0·5	0·6
$\sin x$	0·000 00	0·099 83	0·198 67	0·295 52	0·389 42	0·479 43	0·564 64

(i) Evaluating by the trapezoidal rule (45) of section 14.8.2 with $N = 6$, $h = 0.1$, gives an approximate value of

$$\tfrac{1}{2}(0.1)[0.564 \, 64 + 2(1.462 \, 87)] = \tfrac{1}{2}(0.1)(3.490 \, 38)$$

$$= 0.174 \, 52 \text{ to five decimal places.}$$

(ii) Evaluating by Simpson's rule (46) of section 14.8.3 with $N = 6$, $h = 0.1$ gives an approximate value of

$$\tfrac{1}{3}(0.1)[0.564 \, 64 + 2(0.588 \, 09) + 4(0.874 \, 78)] = \tfrac{1}{3}(0.1)(5.239 \, 94)$$
$$= 0.174 \, 665$$

or 0·174 67 to five decimal places.

(iii) Evaluating by Weddle's rule (47) of section 14.8.4 with $N = 6$, $h = 0.1$ gives an approximation of

$$\tfrac{3}{10}(0.1)[0.0 + 0.499 \, 15 + 0.198 \, 67 + 1.773 \, 12$$
$$+ \, 0.389 \, 42 + 2.397 \, 15 + 0.564 \, 64] = \tfrac{3}{10}(0.1)(5.822 \, 15)$$

$$= 0.174 \, 664$$

or 0·174 66 to five decimal places.

The first two results are in error by approximately 0·08 per cent, 0·006 per cent. The evaluation by Weddle's rule is correct to five decimal places.

Example (ii) Evaluate the integral $\int_0^{1\cdot2} \exp(-x^2)\, dx$ using a subdivision of 0·2 in the argument and estimate the error involved.

Solution This integral appears in statistical analysis and cannot be evaluated by using the standard methods of integration. Form the difference table for $y = \exp(-x^2)$ from $x_0 = 0$ to $x_N = 1\cdot2$ at equal intervals of $h = 0\cdot2$.

x	y	Δy	$\Delta^2 y$	$\Delta^3 y$	$\Delta^4 y$	$\Delta^5 y$	$\Delta^6 y$
0	1·000 0						
		−0·039 2					
0·2	0·960 8		−0·069 5				
		−0·108 7		0·023 8			
0·4	0·852 1		−0·045 7		0·005 9		
		−0·154 4		0·029 7		−0·008 6	
0·6	0·697 7		−0·016 0		−0·002 7		+0·001 7
		−0·170 4		0·027 0		−0·006 9	
0·8	0·527 3		+0·011 0		−0·009 6		
		−0·159 4		0·017 4			
1·0	0·367 9		+0·028 4				
		−0·131 0					
1·2	0·236 9						

(i) The trapezoidal rule
From equation (45) of section 14.8.2 with $h = 0\cdot2$, $N = 6$, the approximate value of the integral

$$\int_0^{1\cdot2} \exp(-x^2)\, dx$$

is given by

$$0\cdot1[1\cdot236\ 9 + 2(3\cdot405\ 8)] = 0\cdot804\ 85.$$

(ii) Simpson's rule
From equation (46) of section 14.8.3 with $h = 0\cdot2$ and $N = 6$ the approximation to the integral is

$$\tfrac{1}{3}(0\cdot2)[1\cdot236\ 9 + 2(0\cdot852\ 1 + 0\cdot527\ 3)$$
$$+ 4(0\cdot960\ 8 + 0\cdot697\ 7 + 0\cdot367\ 9)$$
$$= \tfrac{1}{3}(0\cdot2)[12\cdot101\ 3]$$
$$= 0\cdot806\ 7(54).$$

(iii) Weddle's rule
From equation (47) of section 14.8.4 with $h = 0\cdot2$ and $N = 6$ the approximation to the integral is given by

$$0\cdot06(1\cdot000\ 0 + 4\cdot804\ 0 + 0\cdot852\ 1 + 4\cdot186\ 2$$
$$+ 0\cdot527\ 3 + 1\cdot839\ 5 + 0\cdot236\ 9) = 0\cdot806\ 7(60).$$

An estimation of the error involved in each case may be obtained from the equations (48), (49), and (50) and the difference table. From (48), the estimation of the error in the trapezoidal rule is

$$-(0 \cdot 2)(-0 \cdot 091 \ 8)/12 = +0 \cdot 001 \ 5(3)$$

and $0 \cdot 806 \ 4$ should be a better result than $0 \cdot 804 \ 8$.

From (49), the estimation of the error in Simpson's rule is

$$-(0 \cdot 2)(-0 \cdot 006 \ 4)/90 = +0 \cdot 000 \ 0(14)$$

which is zero to four places of decimals. Since round-off errors in the entries are propagated into the higher differences the result estimated by Simpson's rule is probably between $0 \cdot 806 \ 7$ and $0 \cdot 806 \ 8$.

Similarly from (50) the estimated error in Weddle's rule is

$$-(0 \cdot 2)(0 \cdot 001 \ 7)/140 = -0 \cdot 000 \ 0(02)$$

and the result cannot be improved (to four decimal places).

From a practical point of view the most reliable method of improving an estimation is to sub-divide the interval of the argument and recalculate the approximation. An examination of the two results will immediately show how many decimal places may be claimed with accuracy. If in the present example the values of $\exp(-x^2)$ are calculated to four decimal places at the values of $x = 0 \cdot 1, 0 \cdot 3, \ldots, 1 \cdot 1$, inserted in the table and Simpson's rule used with 13 ordinates, the estimated value is found to be $0 \cdot 806 \ 7(36)$. The conclusion would be that, with the data given, the integral has the value $0 \cdot 806 \ 7$ to four decimal places.

This integral can be evaluated by other means and its value is $0 \cdot 806 \ 74$ correct to five decimal places. Any difference between the estimated value using Simpson's rule (or Weddle's rule) and the actual value would be due entirely to round-off errors in the column of entries.

14.9 GENERAL REMARKS

In any numerical work it is worth noting that to avoid introducing further errors the intermediate calculations are performed using as many decimal places as possible, or convenient, but that the result of all the calculations is rounded off to no more than the number of decimal places in the given data. If interpolation is required in a table given to four decimal places, the calculations are performed using at least five decimal places and the result rounded off to four decimal places. It is impossible to claim a result correct to more decimal places than are given in the original information.

The original data frequently have round-off errors present and as in section 14.2.1 these errors are magnified in the higher differences. The result of a calculation may therefore have an error in the final decimal place claimed. Refer also to sections 13.5 and 13.6.

The calculations in this, and the next, chapter have been performed using a desk calculator and, as the reader will find on carrying out these or similar calculations, the mistakes that will occur are due to the operator and not the machine. The commonest mistakes made are to enter the wrong number in the machine, e.g. 8432 instead of 8342, or to fail to account correctly for a decimal point. Such mistakes can only be avoided by checking each entry made and formulating a definite routine for each calculation. For a final check on the work it may be necessary to recalculate completely or, preferably, have someone else repeat the calculations.

Familiarity with a machine is the best aid to correct calculations and this can only come with practice.

EXERCISE 14

1 There is a cubic relationship between x and y given in the following table. Show that there are two errors in the table, locate them and correct the table.

x	0	1	2	3	4	5
y	97·336	132·651	175·616	226·981	287·496	357·911

x	6	7	8	9	10	11
y	438·977	531·441	636·046	753·571	884·736	1030·301

2 Form a difference table from the following data and deduce that $f(x)$ is a quadratic form. Use the table to find the value of $f(0)$. Use the Newton-Gregory forward interpolation formula based on $x = 0$ to find the quadratic function $f(x)$.

x	2	3	4	5	6
$f(x)$	3·00	4·28	5·88	7·80	10·04

3 Use the backward Newton-Gregory interpolation formula to find the cubic function of x which takes the values 2, 11, 32, 71, when x has the values -6, -4, -2, 0 respectively.

4 From the following table use an appropriate interpolation formula to estimate $f(16·4)$ and $f(23·5)$

x	16	18	20	22	24
$f(x)$	261·3	293·7	330·0	372·2	422·3

5 The following is an extract from a table of sinh x. Use the forward Newton-Gregory interpolation formula based on $x = 4·60$ to estimate the value of sinh 4·612. Also evaluate using the Stirling formula based on $x = 4·61$. Compare and comment on the results.

x	4·60	4·61	4·62	4·63
sinh x	49·737	50·237	50·742	51·252

6 Use an appropriate interpolation formula to estimate the values of $f(0·3)$, $f(2·55)$, $f(1·75)$, and $f(1·55)$ given the following data:

x	0	0·5	1·0	1·5	2·0	2·5	3·0
$f(x)$	0	0·480 4	0·865 8	1·131 7	1·301 8	1·407 0	1·471 3

7 Show that the forward and backward Gauss formulae

$$y = y_0 + \binom{u}{1} \Delta y_0 + \binom{u}{2} \Delta^2 y_{-1} + \binom{u+1}{3} \Delta^3 y_{-1} + \ldots$$

$$y = y_1 + \binom{u-1}{1} \Delta y_0 + \binom{u}{2} \Delta^2 y_0 + \binom{u}{3} \Delta^3 y_{-1} + \ldots$$

are identical if they are truncated by neglecting the fourth and higher differences.

Deduce Bessel's formula and use it to find $\cos^{-1} 0.7$ to the nearest minute given that

$$\cos 30° = 0.866\,0, \quad \cos 40° = 0.766\,0,$$
$$\cos 50° = 0.642\,8, \quad \cos 60° = 0.500\,0.$$

8 When x takes the values 12, 22, 32, 42, 52, and 62 the corresponding values of $f(x)$ are 62·146 08, 62·344 11, 62·538 29, 62·728 77, 62·915 69, and 63·099 18. Evaluate approximately $f'(14)$ and $f''(12)$.

9 The function $f(x)$ has a maximum value for $3 < x < 4$. Given the following difference table use Bessel's formula to find the maximum value of $f(x)$.

x	$f(x)$	Δ	Δ^2	Δ^3	Δ^4
3	1·219 710		−0·137 020		−0·024 872
		−0·014 934		0·062 828	
4	1·204 776		−0·074 192		−0·023 366

10 The following is a table of $f(x) = x^2/(1 + x^4)$ from $x = 0$ to $x = 1·2$ at intervals of $0·1$. Find the approximate value of

$$\int_0^{1·2} \frac{x^2}{1 + x^4}\, dx$$

using (i) the trapezoidal rule, (ii) Simpson's rule, and (iii) Weddle's rule. In each case estimate the error involved.

x	$f(x)$	x	$f(x)$
0	0·000 000	0·6	0·318 697
0·1	0·009 999	0·7	0·395 129
0·2	0·039 936	0·8	0·454 030
0·3	0·089 277	0·9	0·489 101
0·4	0·156 006	1·0	0·500 000
0·5	0·235 294	1·1	0·491 051
		1·2	0·468 506

15
Statistics I

15.0 INTRODUCTION

The study of statistical methods divides into two parts, *descriptive* statistics which consists in the collection, presentation, and summarizing of data and statistical *inference* which is concerned with drawing conclusions about the source of data.

15.1 STATISTICAL EXPERIMENTS

In the design of an experiment, two main aims are required. These are (i) that it is possible to judge the validity of the conclusions, and (ii) that the experiment is efficient in that conclusions may be drawn on as few observations as possible.

A statistical experiment must be capable of being repeated, or be such that it can be considered as being repetitive.

Example (i) An unbiased coin tossed 12 times may produce the result

$$H\ T\ T\ T\ H\ T\ H\ T\ T\ H\ T\ T$$

or 8*T*, 4*H* where *T* stands for 'tails' and *H* for 'heads'. The experiment of tossing a coin 12 times can be repeated say 6 times to give results as follows:

T	8	10	11	9	10	9
H	4	2	1	3	2	3

This experiment shows *random* variation from one experiment to the next.

Example (ii) In an agricultural experiment two types *A* and *B* of wheat are grown on ten plots with the following yield in kg/m^2:

A	0·29	0·27	0·28	0·27	0·36	0·30	0·28	0·28	0·33	0·29
B	0·30	0·31	0·28	0·29	0·27	0·30	0·32	0·30	0·27	0·28

This experiment is thus repetitive and shows random variation.

A typical statistical problem occurs in mass production where it is necessary to maintain a given standard. For example, a machine is to produce ball-bearings of 2 mm diameter, and from a set of fifty observations it is to be decided whether or not the standard is being maintained. It is totally impracticable as well as expensive to measure the whole output of the machine. The experiment is repetitive in that many sets of fifty observations may be made. Assuming that the machine is set correctly, the measurements from each set of fifty observations will show random variation.

15.1.1 Definitions

A *population* is any set of objects about which numerical information is required.

An *individual* is any member of the population.

A *sample* is any set of individuals.

An *attribute* is any property which any individual may or may not possess and about which information is required.

Example (i) All the results which would be obtained in an infinite number of tosses of a coin is a population. Twelve such tosses constitute a sample from this population and each result, H or T, is an individual from the population. The attribute considered could be that a head is tossed, and the number of times this happened would be counted. It would be *expected* that in the long run, there would be 50 per cent heads and 50 per cent tails.

Example (ii) The heights of all men in England is a population of heights. The attribute under consideration might be the average height, or the maximum or minimum height. A sample from this population might be the heights of 100 men chosen at random, and the statistical problem to infer properties of the population from the sample.

15.2 PROBABILITY

The whole of statistical methods depend upon probability theory.

15.2.1 Definitions

An *experiment* or *trial* is the establishing of certain conditions which may produce one of several results or cases.

Example (i) toss a coin, (ii) toss a coin twice, (iii) toss two coins, (iv) measure the diameter of a ball-bearing.

An *event* is a possible outcome of an experiment.

A *random variable* is a variable which specifies the result of an experiment. For example in the coin-tossing experiment, the random variable x may be defined as $x = 0$ if a head is thrown, $x = 1$ if a tail is thrown.

Mutually exclusive events: Two events E_1 and E_2 are said to be mutually exclusive if E_2 precludes E_1. More generally, E_1, E_2, E_3, . . ., E_n are mutually exclusive if all pairs (E_i, E_j), $i \neq j$ are mutually exclusive.

For example in the experiment of tossing two coins,

(i) let E_1 be the event H, T, and let E_2 be the event H, H, then E_1 and E_2 are mutually exclusive.

(ii) let E_1 be the event H and T, i.e. HT and TH. Let E_2 be the event H, then E_1 and E_2 are not mutually exclusive.

The event *either E_1 or E_2* is the event that either E_1 or E_2 or both, happens.

The event *both E_1 and E_2* is the event that both E_1 and E_2 happen.

For example, if the experiment is to throw a six-sided die, and E_1 the event that the score is odd, E_2 the event that the score is a multiple of 3. Then the event (either E_1 or E_2) is the event that the score is 1, 3, 5 or 6. The event (both E_1 and E_2) is the event that the score is 3.

Mutually exhaustive events: The events E_1 and E_2 are said to be mutually exhaustive if either E_1 or E_2 must happen. For example, in the experiment of rolling a single die, E_1, the event that the score is even, and E_2 the event that the score is odd, are mutually exhaustive. More generally, $E_1, E_2, E_3, \ldots, E_n$ is a mutually exhaustive set if one of the events $E_1, E_2, E_3, \ldots$ or E_n must happen.

An *elementary event* is an event which cannot be decomposed into simpler events. For example the event E that the score is 5 on rolling a die is an elementary event. The event E that the score is a multiple of 3 can be decomposed into the elementary events E_1, that the score is 3 and E_2, that the score is 6.

15.2.2 Empirical definition of probability

It is not usually possible to say that the outcome of an experiment will be E, but it may be possible to assign a probability to E. Suppose the event E occurs m times in n repetitions of an experiment. Then if as $n \to \infty$ the ratio $m/n \to$ a limit p, this limit is called the *probability* of the event E. This is the empirical or statistical definition of probability.

Example (i) In the experiment of tossing a single coin, let r be the number of heads thrown, and n be the total number of throws. Then the probability of tossing a head with this coin is

Lim [(number of heads)/(total number of throws)]

which may be written in symbols as

$$P\{H\} = \text{Lim } (r/n).$$

Example (ii) In the experiment on wheat yields of section 15.1,

$P\{$wheat yield is between 0·28 and 0·31 kg/m^2

$= \text{Lim [(number of plots with this yield)/(total number of plots)]}.$

Properties

If E is certain,	$m = n$	and	$P\{E\} = 1$	(1)
If E is impossible,	$m = 0$	and	$P\{E\} = 0$	(2)
Since $0 \leqslant m \leqslant n$,			$0 \leqslant p \leqslant 1$.	(3)

15.2.3 Addition theorem

If E_1 and E_2 are mutually exclusive events

$$P \text{ \{either } E_1 \text{ or } E_2\} = P\{E_1\} + P\{E_2\}. \tag{4}$$

Let an experiment be performed n times, and let E_1 be obtained n_1 times and E_2 be obtained n_2 times. Then (either E_1 or E_2) has been obtained $(n_1 + n_2)$ times. Hence

$$
\begin{aligned}
P\{\text{either } E_1 \text{ or } E_2\} &= \text{Lim} \left[(n_1 + n_2)/n \right] \\
&= \text{Lim} \left[n_1/n + n_2/n \right] \\
&= \text{Lim} (n_1/n) + \text{Lim} (n_2/n) \\
&= P\{E_1\} + P\{E_2\}.
\end{aligned}
$$

This result is extended as follows. If $E_1, E_2, E_3, \ldots, E_n$ are n mutually exclusive events,

$$
\begin{aligned}
P\{E_1 \text{ or } E_2 \text{ or } E_3 \text{ or } &\ldots \text{ or } E_n\} \\
&= P\{E_1\} + P\{E_2\} + P\{E_3\} + \ldots + P\{E_n\} \tag{5}
\end{aligned}
$$

Corollary 1 If $E_1, E_2, E_3, \ldots, E_n$ are also mutually exhaustive then

$$P\{E_1\} + P\{E_2\} + P\{E_3\} + \ldots + P\{E_n\} = 1 \tag{6}$$

Proof The event $(E_1 \text{ or } E_2 \text{ or } E_3 \text{ or } \ldots \text{ or } E_n)$ must happen so that

$$P\{E_1 \text{ or } E_2 \text{ or } E_3 \text{ or } \ldots \text{ or } E_n\} = 1$$

and combining this with equation (5) gives equation (6).

Corollary 2 If an experiment must result in one of n mutually exclusive events, i.e. they are mutually exclusive and exhaustive, which are all equally likely, then

$$
\begin{aligned}
P\{E_1\} = P\{E_2\} = P\{E_3\} = \ldots &= P\{E_n\} = 1/(\text{total number of events}) \\
&= 1/n. \tag{7}
\end{aligned}
$$

Proof Since the n events are exhaustive, from equation (6)

$$P\{E_1\} + P\{E_2\} + P\{E_3\} + \ldots + P\{E_n\} = 1,$$

and since they are all equally likely,

$$P\{E_1\} = P\{E_2\} = P\{E_3\} = \ldots = P\{E_n\}.$$

Hence the result (7).

Example If x is the score on rolling a die, then

$$P\{x = 1\} = P\{x = 2\} = P\{x = 3\} = \ldots = P\{x = 6\} = \tfrac{1}{6}.$$

Corollary 3 Let the result of an experiment be one of the finite number of mutually exhaustive and elementary events $E_1, E_2, E_3, \ldots, E_n$, and let all these events be equally likely. Let E be the event defined by the occurrence of m of these elementary events. Then

$$P\{E\} = m/n, \tag{8}$$

since $P\{E\} = P$ {any m of the events $E_1, \ldots, E_n$}

= sum of probabilities of these m events

= m/n.

Thus $$P\{E\} = \frac{\text{number of outcomes favourable to } E}{\text{total number of outcomes}} \tag{9}$$

and this is the result used in calculating probabilities.

Example (i) If a single coin is tossed, the probability of getting a head, i.e. $P\{H\} = \frac{1}{2}$.

Example (ii) In one throw of a die,

P {score is 5} $= \frac{1}{6}$;

P {score is even} $= 3/6 = \frac{1}{2}$.

Example (iii) Find the probability that the score is 8 if two like dice are thrown.

Solution A total score of 8 may be obtained from the pairs (2, 6), (3, 5), (4, 4), (5, 3), (6, 2). The total number of possible events is 36 and there are 5 outcomes favourable to the event that the score is 8.

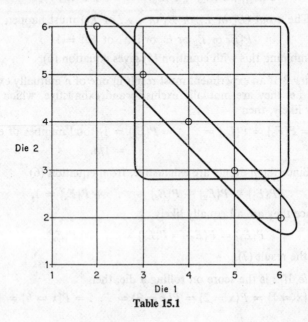

Die 2

Die 1

Table 15.1

Table 15.1 shows the representation of each outcome by the pair (r, s), r being the score on the first die, s the score on the second.

Hence, P {total score is 8} $= 5/36$.

Example (iv) What is the probability of selecting a red ball in one draw from a box containing 5 red and 3 black balls?

Solution The total possible number of outcomes is 8, and of these the number favourable to the event, drawing a red ball, is 5. Hence,

$$P \text{ {ball selected at random is red} } = \tfrac{5}{8}.$$

15.2.4 Use of permutations and combinations

The number of *permutations* of n unlike objects taken r at a time is the number of ordered arrangements of n objects taken r at a time. This is $n(n - 1)(n - 2) \ldots (n - r + 1)$, which is written

$$^nP_r = n!/(n - r)!,$$

where by definition, $n! = n(n - 1)(n - 2) \ldots 3 . 2 . 1$ and $0! = 1$.

Example (i) The number of different arrangements of the letters A, B, C, D is $4! = 24$.

Example (ii) The number of permutations of 2 letters from the 4 letters A, B, C, D is

$$^4P_2 = 4!/2! = 12.$$

The number of *combinations* of n different objects taken r at a time is the number of ways of selecting r objects from the n different objects, without regard to order. The number of arrangements is ^nP_r and each arrangement can be permuted in $r!$ ways. Hence the number of selections is

$$^nP_r/(r!) = n!/(r!)(n - r)!$$

which is written ^nC_r or $\binom{n}{r}$.

The number of ways of dividing n different objects into three groups, the first containing n_1, the second n_2, and the third n_3 objects, where $n_1 + n_2 + n_3 = n$, is $n!/(n_1! \, n_2! \, n_3!)$. This is proved as follows. The first group is selected in $\binom{n}{n_1}$ ways, and once this is fixed there are $(n - n_1)$ objects from which to select n_2, and this can be done in $\binom{n - n_1}{n_2}$ ways. The third group is now fixed. Hence the total number of ways of dividing the n objects is

$$\binom{n}{n_1} \times \binom{n - n_1}{n_2} \times 1 = \frac{n!}{n_1!(n - n_1)!} \frac{(n - n_1)!}{n_2!(n - n_1 - n_2)!}$$

$$= \frac{n!}{n_1! \, n_2! \, n_3!} \quad \text{since } n - n_1 - n_2 = n_3.$$

Extending this argument, the number of ways of dividing n objects into k groups, the first containing n_1, the second n_2 and so on, objects, so that $n_1 + n_2 + n_3 + \ldots + n_k = n$ is

$$\frac{n!}{n_1!\, n_2!\, n_3!\, \ldots \, n_k!}$$

Example (i) A box contains 5 red and 3 blue balls. Find the probability of the event E that two balls selected at random are both red.

Solution The total number of possible outcomes is the number of ways of selecting 2 from 8, which is $\binom{8}{2}$. The number of outcomes favourable to E is the number of ways of selecting 2 red from 5 red balls which is $\binom{5}{2}$. Thus,

$$P\{E\} = \binom{5}{2} \Big/ \binom{8}{2} = 5/14.$$

Example (ii) A box contains 2 red, 4 blue, and 5 white balls. Find the probability of the event E, that 4 balls selected at random contain exactly 2 white balls.

Solution The total number of possible outcomes is the number of ways of selecting 4 from 11, which is $\binom{11}{4}$. The number of outcomes favourable to the event E is the number of ways of selecting exactly 2 white from 5 white and 2 others from the remaining 6 others. This is $\binom{5}{2}\binom{6}{2}$. Hence

$$P\{E\} = \binom{5}{2}\binom{6}{2} \Big/ \binom{11}{4} = 5/11.$$

15.2.5 Use of the complementary event

The events E and $\bar{E}$ are complementary events if $\bar{E}$ occurs whenever E does not. Thus E and $\bar{E}$ are both mutually exclusive and exhaustive. Hence

$$P\{\bar{E}\} = 1 - P\{E\}. \tag{10}$$

If $P\{E\} = p$ and $P\{\bar{E}\} = q$, then

$$p + q = 1 \quad \text{or} \quad q = 1 - p. \tag{11}$$

Example (i) A box contains 2 red, 4 blue, and 5 white balls. Find the probability that 4 balls selected at random contain at least one white ball.

Solution Let E be the event that at least one of the four selected is white, then $\bar{E}$ is the event that none of the four is white. Now

$$P\{\bar{E}\} = \binom{6}{4} \Big/ \binom{11}{4} = 1/22,$$

and hence $P\{E\} = 21/22.$

Example (ii) In an experiment in which a coin is spun 12 times, find the probability that at least 2 heads will be thrown.

Solution Let E be the event of obtaining at least 2 heads. Then the probability of a head (H) in one throw is $\frac{1}{2}$. The number of ways of obtaining exactly 2 heads and 10 tails is $\binom{12}{2}$, and the probability of this is $\binom{12}{2} \left(\frac{1}{2}\right)^2 \left(\frac{1}{2}\right)^{10}$.

There are two methods of obtaining $P\{E\}$.

Method (a) The probability of the event E is the probability of obtaining 2, 3, 4, . . ., 12 heads. Hence

$$P\{E\} = P\{2H \text{ or } 3H \text{ or } . . . \text{ or } 12H\}$$
$$= P\{2H\} + P\{3H\} + . . . + P\{12H\}$$
$$= \binom{12}{2}\left(\frac{1}{2}\right)^2\left(\frac{1}{2}\right)^{10} + \binom{12}{3}\left(\frac{1}{2}\right)^3\left(\frac{1}{2}\right)^9 + . . . + \binom{12}{12}\left(\frac{1}{2}\right)^{12}\left(\frac{1}{2}\right)^0.$$

Method (b) E is the event of obtaining at least 2 heads, so that the complementary event $\bar{E}$ is the event of obtaining no head or one head. Now

$$P\{\bar{E}\} = \left(\frac{1}{2}\right)^{12} + \binom{12}{1}\left(\frac{1}{2}\right)^{11}\left(\frac{1}{2}\right) = 13\left(\frac{1}{2}\right)^{12},$$

and $P\{E\}$ which is $1 - P\{\bar{E}\}$ is given by

$$P\{E\} = 1 - 13\left(\frac{1}{2}\right)^{12}.$$

15.2.6 Conditional probability

The probability that the event E_2 will occur knowing that the event E_1 has already occurred, or is certain to occur, is called the *conditional probability* of E_2, given E_1. This is written $P\{E_2|E_1\}$.

Suppose E_1 and E_2 are outcomes of an experiment which is repeated n times. Let E_1 occur n_1 times, and let the number of times both E_1 and E_2 occur be n_2. Then the probability of E_2 given that E_1 is certain

$$= \text{Lim} \, (n_2/n_1)$$
$$= \text{Lim} \left(\frac{n_2}{n} \Big/ \frac{n_1}{n}\right)$$
$$= \text{Lim} \, (n_2/n)/\text{Lim} \, (n_1/n)$$
$$= P\{\text{both } E_1 \text{ and } E_2\}/P\{E_1\}.$$

Hence the *Multiplication theorem* follows in the form

$$P\{\text{both } E_1 \text{ and } E_2\} = P\{E_1\}P\{E_2|E_1\}. \tag{12}$$

Two events E_1 and E_2 are said to be *independent* if

$$P\{\text{both } E_1 \text{ and } E_2\} = P\{E_1\}P\{E_2\} \tag{13}$$

i.e. if $P\{E_2|E_1\} = P\{E_2\}$.

The definition of independent events is extended to the k events $E_1, E_2, E_3, \ldots, E_k$, which are independent if

$$P\{E_1 \text{ and } E_2 \text{ and } \ldots \text{ and } E_k\} = P\{E_1\}P\{E_2\} \ldots P\{E_k\} \tag{14}$$

Example (i) Two coins are tossed. What is the probability that the first coin shows a tail and the second a head?

Solution Let E_1 be the event tail on the first coin and E_2 be the event head on the second coin. Then since all possible outcomes may be written

$$HH, \quad HT, \quad TH, \quad TT,$$

the required probability

$$P\{\text{both } E_1 \text{ and } E_2\} = \tfrac{1}{4} = \tfrac{1}{2} \times \tfrac{1}{2} = P\{E_1\}P\{E_2\}.$$

Hence E_1 and E_2 are independent events.

Example (ii) Two dice are rolled. What is the probability that the total score is eight if it is known that the dice show at least three points each?

Solution Let E_1 be the event that the dice show at least 3 points each, E_2 the event that the total score is 8, then, using Table 15.1,

$$P\{E_2|E_1\} = P\{\text{both } E_1 \text{ and } E_2\}/P\{E_1\}$$

$$= \frac{3/36}{16/36} = \frac{3}{16}$$

Example (iii) Three dice are rolled. If no two dice show the same face, what is the probability that at least one face shows a six?

Solution Let E_1 be the event that no two faces show the same score, E_2 the event that at least one face shows a six, then $\bar{E}_2$ is the event that no face shows a six. Then

$$P\{E_2|E_1\} = 1 - P\{\bar{E}_2|E_1\}.$$

Since there are 3! ways of arranging sets of three,

$$P\{E_1\} = \binom{6}{3}\left(\frac{1}{6}\right)^3 3! = (6 . 5 . 4)/6^3;$$

or alternatively, $P\{E_1\} = P\{\text{1st is any number and}$
$$\text{2nd is a different number and}$$
$$\text{3rd is different from 1st two}\}$$

$$= \frac{6}{6} \cdot \frac{5}{6} \cdot \frac{4}{6}.$$

Also, $P\{\text{both } \bar{E}_2 \text{ and } E_1\} = \dfrac{5}{6} \cdot \dfrac{4}{6} \cdot \dfrac{3}{6} = \dfrac{5 \cdot 4 \cdot 3}{6^3}.$

Hence, $P\{\bar{E}_2 | E_1\} = P\{\text{both } \bar{E}_2 \text{ and } E_1\} / P\{E_1\}$

$= (5 \cdot 4 \cdot 3)/(6 \cdot 5 \cdot 4) = \tfrac{1}{2}.$

Hence $P\{E_2 | E_1\} = \tfrac{1}{2}.$

15.3 RANDOM VARIABLES

Consider an experiment A which may have several outcomes. It is convenient to introduce a variable x to correspond with each event of A. For example if a coin is spun the random variable may be defined by $x = 0$ if the event is heads, $x = 1$ if the event is tails. If two coins are spun and the number of heads (H) counted, the random variable may be defined as $x = 0$ if no H, $x = 1$ if 1 H (HT or TH), and $x = 2$ if $2H$. In the experiment of wheat yields of section 15.1, x may be defined as the yield in kg/m^2, in the experiment of measuring the diameter of ball-bearings x may be defined as the diameter in mm. The random variable x is defined as a number associated with each event of an experiment, random since its value depends on the outcome of an experiment and hence on chance.

15.3.1 Discrete random variables

A discrete random variable is a random variable that can take only a finite number of distinct values which can be arranged in a definite order.

Let the random variable x take values $x_1, x_2, x_3, \ldots, x_n$ corresponding to all the n events of an experiment.

Let
$$\left. \begin{array}{l} p_1 = P\{x = x_1\} \\ p_2 = P\{x = x_2\} \\ p_3 = P\{x = x_3\} \\ \cdot \qquad \cdot \\ \cdot \qquad \cdot \\ \cdot \qquad \cdot \\ p_n = P\{x = x_n\} \end{array} \right\} \tag{15}$$

Then the set of numbers p_i, $i = 1, 2, 3, \ldots, n$, defines a *frequency function* $f(x)$ which gives the probability that the random variable x will assume any value in its range. A frequency function is often given as a table of values. Thus, if a single die is rolled and x is defined as the score, the table of values is given by Table 15.2

x	1	2	3	4	5	6
$f(x)$	1/6	1/6	1/6	1/6	1/6	1/6

Table 15.2

It is sometimes convenient to graph the frequency function $f(x)$ by a line graph, as in Fig. 15.1.

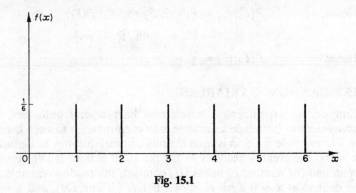

Fig. 15.1

If two dice are rolled, let x be the total score, then Table 15.3 is obtained from Table 15.1.

x	2	3	4	5	6	7	8	9	10	11	12
$f(x)$	1/36	2/36	3/36	4/36	5/36	6/36	5/36	4/36	3/36	2/36	1/36

Table 15.3

and the line graph is shown in Fig. 15.2.

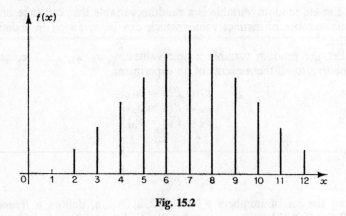

Fig. 15.2

Since $\qquad f(x_i) = P\{x = x_i\} = p_i, \quad i = 1, 2, 3, \ldots, n,$

and there are exactly n events associated with the experiment,

$$\sum_{i=1}^{n} p_i = 1,$$

or $\qquad\qquad \sum_{i} f(x_i) = 1. \qquad\qquad (16)$

The frequency function determines the distribution of any event associated with the experiment. For example

$$P\{x_r \leqslant x \leqslant x_s\} = P\{x = x_r\} + P\{x = x_{r+1}\} + \ldots + P\{x = x_s\}$$

$$= f(x_r) + f(x_{r+1}) + \ldots + f(x_s) \tag{17}$$

An important function associated with the frequency function is the *distribution* function $F(x)$, sometimes called the cumulative distribution function, which is defined by

$$F(x) = \sum_{t \leqslant x} f(t), \tag{18}$$

where the summation is taken over all those values of x less than or equal to a given value. Hence

$$F(x_k) = \sum_{t \leqslant x_k} f(t)$$

$$= f(x_1) + f(x_2) + f(x_3) + \ldots + f(x_k),$$

or $$F(x_k) = p_1 + p_2 + p_3 + \ldots + p_k, \tag{19}$$

i.e. $F(x_k)$ gives the probability that the random variable x is less than or equal to the given value x_k, whereas $f(x_k)$ is the probability that the random variable x is equal to x_k. Hence

$$P\{a \leqslant x \leqslant b\} = F(b) - F(a).$$

The graph of $F(x)$ for the frequency function of Table 15.3 is shown in Fig. 15.3.

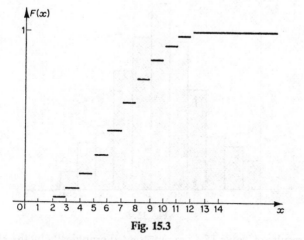

Fig. 15.3

15.3.2 Continuous random variables

If a random variable is capable of taking any value in a given interval or intervals it is called a *continuous* random variable. Thus the random variable x defined as the height of a man from the population of the heights of all men in a given place is a continuous random variable, since it can take any value in an interval say from 1·5 metre to 1·8 metre. Variables which involve measurement such as height, length, temperature, and weight are continuous variables.

As an example consider the set of numbers x representing the diameter in mm of 200 ball-bearings, the diameters being measured to three places of decimals. Values of x are thus regarded as values of a continuous variable whose values have been rounded off to three decimal places, so that a measurement of 0·155 mm means that the actual measurement is between 0·1545 mm and 0·1555 mm. The frequency or number of times the value x_i occurs is denoted by f_i, so that $\sum_i f_i = 200$ and the resulting frequency distribution is shown in table form in Table 15.4.

x	0·115	0·125	0·135	0·145	0·155	0·165	0·175	0·185	0·195	0·205	0·215
f	6	9	11	21	31	39	32	25	12	10	4

Table 15.4

The frequency here denotes the *absolute* frequency with which x occurs, and this table is the type of result usually obtained in an observed frequency distribution. The word frequency more often denotes the *relative* frequency, which is the ratio of absolute frequency to the total number of observations, since relative frequency approaches probability as the number of observations increases indefinitely.

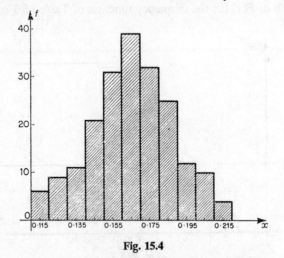

Fig. 15.4

The results of Table 15.4 are represented graphically by the *histogram* of Fig. 15.4 in which areas represent absolute frequencies: for example, the area of the rectangle the centre of whose base is 0·115 represents frequency 6. The total area of the histogram is thus 200. The length of the interval on the x-axis, here given by 0·01, is usually denoted by h and called the *class-interval*.

If the measurement of 200 ball-bearings is considered as 200 runs of an experiment, a further 200 runs would result in a histogram twice the area of that in Fig. 15.4. In order to compare histograms based on different numbers of experiments, it is more convenient to choose the scale on the ordinate-axis to represent relative frequency f_i/n so that the total area is 1. With this choice of unit the histogram approaches a fixed histogram as $n \to \infty$. Also if the measurements x can be made as accurately 'as we please', the class interval h can be chosen as 'small as we please' so that the histogram approaches a smooth curve as $h \to 0$ and $n \to \infty$. If this curve can be represented by $f(x)$, this is the frequency function of the continuous random variable x. For a continuous random variable the frequency function is sometimes called the *probability density function*.

The frequency function for the continuous random variable x is defined as the function $f(x)$ which has the following properties:

(i) $$f(x) \geqslant 0; \tag{20}$$

(compare with $p_i \geqslant 0$ all i).

(ii) $$\int_{-\infty}^{\infty} f(x)\, dx = 1; \tag{21}$$

(compare with $\sum_i p_i = 1$).

(iii) $$\int_{a}^{b} f(x)\, dx = P\{a < x < b\} \tag{22}$$

where a, b are any two values of x; (compare with $P\{x_r \leqslant x \leqslant x_s\}$ $= \sum_{i=r}^{s} f(x_i)$).

The distribution function $F(x)$ is defined by $F(x) = \sum_{t < x} f(t)$, the summation now being the limit of a sum and hence an integral, or

$$F(x) = \int_{-\infty}^{x} f(t)\, dt \tag{23}$$

and $$f(x) = F'(x). \tag{24}$$

The value of statistical methods is in finding mathematical models to fit as closely as possible the observed or empirical data. There are standard frequency functions which are used most often as models, and some of these are demonstrated in the following examples.

Example (i) Consider the possibility of using

$$f(x) = \begin{cases} \lambda e^{-\lambda x} & x \geqslant 0 \\ 0 & x < 0 \end{cases}$$

as a frequency function for x, where λ is a positive constant.

Solution Checking the properties (20) and (21) gives

(a)
$$\lambda e^{-\lambda x} > 0 \quad \text{all } x;$$

(b)
$$\int_{-\infty}^{\infty} f(x)\, dx = \int_{0}^{\infty} \lambda e^{-\lambda x}\, dx$$
$$= \left[-e^{-\lambda x} \right]_{0}^{\infty}$$
$$= 1 \quad \text{for all } \lambda.$$

The distribution function $F(x)$ is given by

$$F(x) = \int_{-\infty}^{x} f(t)\, dt$$
$$= \int_{0}^{x} \lambda e^{-\lambda t}\, dt$$
$$= 1 - e^{-\lambda x}$$
$$\to 1 \quad \text{as} \quad x \to \infty.$$

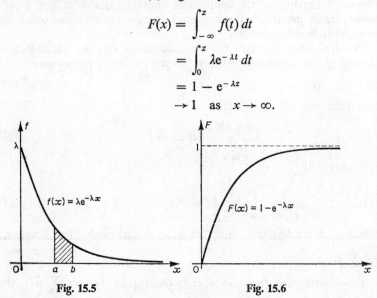

Fig. 15.5   Fig. 15.6

The graphs of $f(x)$ and $F(x)$ are shown in Figs. 15.5 and 15.6, and

$$P\{a < x < b\} = \int_{a}^{b} f(x)\, dx$$
$$= F(b) - F(a)$$
$$= 1 - e^{-\lambda b} - 1 + e^{-\lambda a}$$
$$= e^{-\lambda a} - e^{-\lambda b}.$$

Although the choice of $f(x)$ as a negative exponential appears to be arbitrary, it is found that this function can be used as a good approximation to many observed frequency distributions.

Example (ii) A uniform, or rectangular distribution, is of use when the continuous random variable x is equally likely to take all values between two given values a and b. Consider the possibility of using $f(x) = c$, where c is a constant, as the frequency function for x.

Solution Since, from equation (21), $\int_a^b f(x)\,dx = 1$, then $c = 1/(b-a)$. Hence the frequency function becomes

$$f(x) = \begin{cases} \dfrac{1}{b-a} & \text{for } a < x < b, \\ 0 & \text{for all other values of } x. \end{cases}$$

The distribution function $F(x)$ is given by

$$F(x) = \int_a^x f(t)\,dt = (x-a)/(b-a),$$

and the functions $f(x)$, $F(x)$ are sketched in Figs. 15.7 and 15.8.

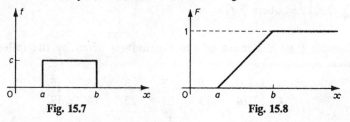

Fig. 15.7   Fig. 15.8

Example (iii) The life x, in hours, of a certain type of electronic valve has the frequency function

$$f(x) = 100/x^2, \quad x > 100;$$
$$f(x) = 0, \quad x < 100.$$

(a) Find the probability that the life of a valve is less than 150 hours.
(b) If the valve is still working after 150 hours, find the probability that it will have to be replaced before 250 hours life.

Solution

(a)
$$P\{x < 150\} = \int_{-\infty}^{150} f(x)\,dx$$
$$= 0 + \int_{100}^{150} \frac{100}{x^2}\,dx$$
$$= \tfrac{1}{3}.$$

(b) Let E_1 be the event $x > 150$ and let E_2 be the event $x < 250$. Then

$$P\{E_2|E_1\} = P\{E_2 \text{ and } E_1\}/P\{E_1\}$$
$$= P\{E_2 \text{ and } E_1\}/[1 - P\{\bar{E_1}\}].$$

Now, from (a), $\qquad\qquad P\{\bar{E_1}\} = \tfrac{1}{3}.$

Also $\qquad P\{E_2 \text{ and } E_1\} = \int_{150}^{250} \frac{100}{x^2}\,dx = \frac{4}{15}.$

Hence $P\{E_2|E_1\} = \tfrac{2}{5}$, and this is the probability that a valve, if it lasts more than 150 hours, will cease to function before 250 hours of its life.

15.4 REDUCTION OF DATA FOR AN OBSERVED FREQUENCY DISTRIBUTION

A mass of raw data needs to be ordered if it is to be easily read and understood. Measures of *location* and *dispersion* are used to describe the distribution in few and simple parameters.

15.4.1 Measures of location

(a) The *mean* or arithmetic average is the most important measure of location. For a discrete random variable, suppose there are f_i values of $x = x_i$, $i = 1, 2, 3, . . ., k$, then the mean m or $\bar{x}$ is defined by $\left(\sum_i f_i x_i\right)/n$, where $\sum_i f_i = n$.

Example Find the mean of the distribution given by the following table:

x	0	1	2	3	4	5
f	14	15	23	22	14	12.

Solution $\qquad\qquad \sum f_i = 100 = n,$

$\bar{x}$ or m

$$= (0 \times 14 + 1 \times 15 + 2 \times 23 + 3 \times 22 + 4 \times 14 + 5 \times 12)/100$$
$$= 2{\cdot}43.$$

For a continuous random variable, as in Table 15.4 of section 15.3.2 it is necessary to assume that approximately all the observations in a given class interval have the value at the mid-point of that interval. If x_i is the mid-point of the ith interval, and f_i is the corresponding class frequency, then as before:

$$\bar{x} \text{ or } m = (\sum f_i x_i)/n. \qquad (25)$$

The mean of the distribution of Table 15.4 is $0{\cdot}165\ 3$.

(b) Suppose there are n observations which can be arranged in order, then the *median* is the value of the middle observation. If $n = 2k + 1$ (odd) the median is the value of the $(k + 1)$th observation. If $n = 2k$ (even) the median is the value of the average of the kth and $(k + 1)$th observations. The median is of use for example, in describing the wage level in a certain industry where a few very high wages affect the mean but not the median. The median wage is such that half the employees receive at least this much.

(c) The *mode* is the observation which has maximum frequency. For example for the set of measurements 4, 4, 5, 5, 5, 6, 7, 7, 8 the mode is 5. A single mode gives a rough and quick measure of location but is not very useful, particularly if the histogram has more than one peak.

15.4.2 Measures of dispersion

These are measures of the spread of the observations.

(a) The *range* is the difference between the largest and smallest measurements in a set of measurements. It is used in quality control.

(b) The *mean deviation* is the average of the deviations of each observation from the mean, all deviations being taken as positive, i.e.

$$\text{mean deviation} = \frac{1}{n} \sum_{i=1}^{n} |x_i - m|$$

or

$$\frac{1}{n} \sum_{i=1}^{k} |x_i - m| f_i.$$

(c) The *variance* and *standard deviation*, which is the square root of the variance, are the most important measures of dispersion. Variance, denoted by s^2, is defined by

$$s^2 = \frac{1}{n} \sum (x_i - m)^2 f_i \tag{26}$$

and the standard deviation s is the square root of this.

(d) The *interquartile range* is defined as $Q_3 - Q_1$, where Q_1 is that value such that $\frac{1}{4}$ of the observations are less than Q_1, Q_2 is the median, Q_3 is such that $\frac{3}{4}$ of the observations are less than Q_3.

15.5 MEASURES OF LOCATION AND DISPERSION FOR A PROBABILITY DISTRIBUTION

15.5.1 The mean

(a) First, consider the mean of a probability distribution for a discrete random variable. Let an experiment be repeated n times, and let there be f_i observations x_i, $i = 1, 2, 3, \ldots, k$. Then the mean of the n observations is m, and using the definition of section 15.4.1,

$$m = \frac{1}{n} \sum_{i=1}^{k} f_i x_i$$

$$= \sum_{i=1}^{k} \left(\frac{f_i}{n} \right) x_i.$$

Now as $n \to \infty$, $f_i/n \to p_i$ where $p_i = P\{x = x_i\}$, and hence $m \to \sum p_i x_i$. Hence the mean of a probability distribution for a discrete random variable is defined as

$$\mu = \sum_i p_i x_i. \tag{27}$$

(b) Second, consider the mean of a probability distribution for a continuous random variable. Let an experiment be repeated n times, and let the observations be grouped into classes of width $h = \Delta x$. Let f_i be the number of observations in the ith interval and x_i be the mid-point of that interval. Then m, the mean of the grouped distribution is given by

$$m = \frac{1}{n} \sum_i f_i x_i$$

$$= \sum_i \left(\frac{f_i}{n}\right) x_i.$$

Now, as $n \to \infty$, the value of f_i/n approaches the probability that x lies in the ith interval, i.e. $f_i/n \to f(x_i) \Delta x$, where $f(x)$ is the frequency function. Hence, as $n \to \infty$, $m \to \sum\{f(x_i) \Delta x\}x_i$. If also $\Delta x \to 0$, then $m \to \int xf(x)\,dx$. Hence the mean for a continuous random variable is defined by

$$\mu = \int xf(x)\,dx, \tag{28}$$

the integration being over the range of values of x for which $f(x)$ is defined.

The mean for any distribution is thus defined by the *first moment* of the distribution. Compare with formula (12) of section 7.2.3.

15.5.2 Dispersion, variance

(a) First consider the variance of a probability distribution for a discrete random variable. For n observations x_i, where x_i occurs f_i times, $i = 1, 2, 3, \ldots, k$,

$$s^2 = \frac{1}{n} \sum_i f_i (x_i - m)^2$$

$$= \sum_i \left(\frac{f_i}{n}\right) (x_i - m)^2.$$

As $n \to \infty$, the value of $(f_i)/n \to p_i = P\{x = x_i\}$. Hence the variance of the probability distribution is defined by

$$\sigma^2 = \sum_i p_i (x_i - \mu)^2. \tag{29}$$

(b) Second, consider the variance of a probability distribution for a continuous random variable. For n observations the variance of the grouped distribution is

$$s^2 = \frac{1}{n} \sum_i f_i (x_i - m)^2$$

$$= \sum_i \left(\frac{f_i}{n}\right) (x_i - m)^2.$$

As $n \to \infty$, the value $(f_i)/n \to f(x_i) \, \Delta x$, and if at the same time $\Delta x \to 0$, the variance of a continuous probability distribution is defined by

$$\sigma^2 = \int (x - \mu)^2 f(x) \, dx. \tag{30}$$

The integration is taken over the range of values of x for which $f(x)$ is defined.

The variance of a distribution is defined by the *second moment* of the distribution. Compare with formula (20) of section 7.2.7.

Example (i) Let x be the score recorded when a die is rolled. Find the mean and variance for the probability distribution.

Solution The variable x is discrete and the following table gives the (rectangular) probability distribution.

x	1	2	3	4	5	6
p	$\frac{1}{6}$	$\frac{1}{6}$	$\frac{1}{6}$	$\frac{1}{6}$	$\frac{1}{6}$	$\frac{1}{6}$

The mean μ and variance σ^2 are given by

$$\mu = \Sigma p_i x_i = \tfrac{1}{6}(1 + 2 + 3 + \ldots + 6) = 3 \cdot 5,$$

$$\sigma^2 = \Sigma p_i (x_i - \mu)^2$$

$$= \tfrac{1}{6}(1 - 3 \cdot 5)^2 + \tfrac{1}{6}(2 - 3 \cdot 5)^2 + \ldots + \tfrac{1}{6}(6 - 3 \cdot 5)^2$$

$$= 35/12.$$

Example (ii) Find the mean and variance for the rectangular distribution given by

$$f(x) = \begin{cases} = \dfrac{1}{2a}, & -a < x < a, \\ = 0, & \text{when } x \text{ is not in the interval } (-a, a). \end{cases}$$

Solution Using formulae (28) and (30), the mean μ and variance σ^2 are given by

$$\mu = \int_{-a}^{a} x f(x) \, dx = \int_{-a}^{a} \tfrac{1}{2}(x/a) \, dx = 0,$$

$$\sigma^2 = \int_{-a}^{a} (x - 0)^2 f(x) \, dx = \tfrac{1}{3} a^2.$$

Example (iii) Find the mean, variance, and median for the negative exponential frequency function $f(x)$, defined by

$$f(x) = \begin{cases} \lambda e^{-\lambda x}, & x \geqslant 0 \\ 0, & x < 0. \end{cases}$$

Solution The mean μ is given by

$$\mu = \int_0^\infty xf(x)\,dx = \int_0^\infty x\lambda e^{-\lambda x}\,dx = 1/\lambda,$$

on integrating by parts.

The variance is σ^2, where

$$
\begin{aligned}
\sigma^2 &= \int_0^\infty (x - 1/\lambda)^2 \lambda e^{-\lambda x}\,dx \\
&= \int_0^\infty x^2 \lambda e^{-\lambda x}\,dx - \frac{2}{\lambda}\int_0^\infty \lambda x e^{-\lambda x}\,dx + \frac{1}{\lambda}\int_0^\infty e^{-\lambda x}\,dx \\
&= \frac{2}{\lambda^2} - \frac{2}{\lambda}\cdot\frac{1}{\lambda} + \frac{1}{\lambda^2} \\
&= \frac{1}{\lambda^2}.
\end{aligned}
$$

The median x_m is given by

$$\int_0^{x_m} \lambda e^{-\lambda x}\,dx = \tfrac{1}{2},$$

since half the population lies each side of the median. Hence

$$\left[-e^{-\lambda x}\right]_0^{x_m} = \tfrac{1}{2},$$

or

$$1 - e^{-\lambda x_m} = \tfrac{1}{2}.$$

Solving,

$$x_m = (\log 2)/\lambda.$$

15.6 PRACTICAL COMPUTATION OF MEAN AND VARIANCE

It is possible to simplify the arithmetic involved in computing the mean and variance by coding the data. There are two processes of coding used, (a) changing the origin, or using a 'working zero' and (b) changing the scale, or unit. Changing the origin involves writing $y = x - x_0$ and changing the scale involves writing $u = y/c$, where the constant x_0 is the new origin, or working zero, and c is any convenient constant.

15.6.1 Computation of the mean

Writing $y_i = x_i - x_0$, $u_i = y_i/c$, means that $x_i = cu_i + x_0$ for each i, and hence,

$$\bar{x} = c\bar{u} + x_0. \tag{31}$$

Example Find the mean of the three numbers

$$x = 0.169,\ 0.162,\ 0.170.$$

Solution Write $y = x - 0 \cdot 160 = 0 \cdot 009, 0 \cdot 002, 0 \cdot 010$, and $u = 1\,000y$ $= 9, 2, 10$. In working units the mean is $\frac{1}{3}(9 + 2 + 10)$ or 7. Partly decoded it is $0 \cdot 007$, and fully decoded the mean is $0 \cdot 167$. Alternatively, $\bar{u} = 7, c = 1/1\,000, x_0 = 0 \cdot 160$, so that

$$\bar{x} = 0 \cdot 007 + 0 \cdot 160 = 0 \cdot 167.$$

15.6.2 Computation of variance

(a) Changing the origin to x_0 by writing $y = x - x_0$ gives the result

$$s_y^2 = s_x^2, \tag{32}$$

i.e. coding by changing the origin has no effect on the variance. This may be proved as follows:

$$s_y^2 = \frac{1}{n} \sum_i f_i (y_i - \bar{y})^2$$

$$= \frac{1}{n} \sum_i f_i (x_i - x_0 - \bar{x} + x_0)^2$$

$$= \frac{1}{n} \sum_i f_i (x_i - \bar{x})^2$$

$$= s_x^2.$$

(b) Changing the scale so that $u = y/c$ gives the result

$$s_u^2 = s_y^2/c^2, \tag{33}$$

or,   variance of $u = $ (variance of y)$/c^2$.

This is proved as follows:

$$s_u^2 = \frac{1}{n} \sum_i f_i (u_i - \bar{u})^2$$

$$= \frac{1}{n} \sum_i f_i \left(\frac{1}{c} y_i - \frac{1}{c} \bar{y} \right)^2$$

$$= \frac{1}{c^2} \frac{1}{n} \sum_i f_i (y_i - \bar{y})^2$$

$$= \frac{1}{c^2} s_y^2.$$

Hence $s_x^2 = s_y^2 = c^2 s_u^2. \tag{34}$

From the example of section 15.6.1,

$$u = 1\,000y = 9, 2, 10,$$
$$\bar{u} = 7,$$
$$s_u^2 = \frac{1}{3}(2^2 + 5^2 + 3^2) = 38/3.$$

Hence $s_x^2 = 38/(3 \times 10^6).$

15.6.3 Modified form of formula for variance

Because the mean $\bar{x}$ is not usually a simple number, and even the coded mean $\bar{u}$ may not be simple, the calculation of variance is often simplified by using the identity

$$\Sigma f_i(u_i - \bar{u})^2 = \Sigma f_i u_i^2 - n\bar{u}^2.$$

This identity is not difficult to prove, for if $\Sigma f_i = n$,

$$\Sigma f_i(u_i - \bar{u})^2 = \Sigma f_i(u_i^2 - 2u_i\bar{u} + \bar{u}^2)$$
$$= \Sigma f_i u_i^2 - 2\bar{u}\,\Sigma f_i u_i + \bar{u}^2\,\Sigma f_i.$$

Now $(\Sigma f_i u_i)/n = \bar{u}$, and hence

$$\Sigma f_i(u_i - \bar{u})^2 = \Sigma f_i u_i^2 - 2\bar{u}(n\bar{u}) + \bar{u}^2 n$$
$$= \Sigma f_i u_i^2 - n\bar{u}^2.$$

Thus the modified form of the variance s_u^2 is given by

$$s_u^2 = \frac{1}{n} \sum_{i=1}^{k} f_i(u_i - \bar{u})^2 = \left(\frac{1}{n} \sum_{i=1}^{k} f_i u_i^2\right) - \bar{u}^2. \tag{35}$$

Alternatively, equation (35) may be written

$$s_u^2 = \frac{1}{n}\left[\sum_{i=1}^{k} f_i u_i^2 - \frac{1}{n}\left(\sum_{i=1}^{k} f_i u_i\right)^2\right]. \tag{36}$$

Example Find the mean and standard deviation for the following distribution.

x	9·95	9·96	9·97	9·98	9·99	10·00	10·01	10·02	10·03	10·04	10·05
f	2	4	5	8	10	16	15	17	11	7	5

Solution The work is set out in table form as follows, with

$$u_i = 100(x_i - 10\text{·}00).$$

x	f	u	uf	$u^2f = (uf)u$
9·95	2	−5	−10	50
9·96	4	−4	−16	64
9·97	5	−3	−15	45
9·98	8	−2	−16	32
9·99	10	−1	−10	10
10·00	16	0	0	0
10·01	15	1	15	15
10·02	17	2	34	68
10·03	11	3	33	99
10·04	7	4	28	112
10·05	5	5	25	125
Totals	100		68	620

The calculation proceeds as follows:
$\bar{u} = 68/100 = 0.68$, hence the mean $\bar{x}$ is given by

$$\bar{x} = 0.006\ 8 + 10.00 = 10.006\ 8.$$

Also $\quad s_u^2 = 620/100 - (0.68)^2 = 6.20 - 0.462\ 4 \simeq 5.738.$
Hence $s_x^2 = 0.000\ 573\ 8$, and the standard deviation is given by

$$s_x = 0.023\ 96 \simeq 0.024.$$

15.7 THE BINOMIAL DISTRIBUTION

There are three standard theoretical frequency distributions to which practical, or observed, distributions may approximate closely. The first of these is the binomial distribution.

Suppose the result of a trial is only success or failure, i.e. a particular event occurs or does not occur. Let the probability of success be p, then $q = 1 - p$ is the probability of failure, since success and failure are independent and exhaustive events. Let there be n trials and let r be the number of successes recorded; and further, suppose that successive trials are independent. The problem is to find the frequency distribution of r, which can take only integer values.

The probability of obtaining r successes and $(n - r)$ failures in a specified order, say the first r trials result in success, is

$$\overbrace{p \cdot p \cdot p \cdots p}^{r} \cdot \overbrace{q \cdot q \cdot q \cdots q}^{(n-r)}$$

which equals $\qquad p^r q^{n-r}.$

The number of different ways in which r successes can occur in n trials is $\binom{n}{r}$. Hence the probability of r successes in any order is $\binom{n}{r} p^r q^{n-r}$.
This expression is the $(r + 1)$th term in the expansion of the binomial expression $(q + p)^n$ and is called the *binomial frequency function* $f(r)$, i.e.

$$f(r) = \binom{n}{r} p^r q^{n-r}. \tag{37}$$

The symbol r is used to emphasize that r is a discrete variable which can take integer values only. Hence

$$P\{r = 0\} = \binom{n}{0} p^0 q^n = q^n,$$

$$P\{r = 1\} = \binom{n}{1} p^1 q^{n-1} = npq^{n-1},$$

$$P\{r = 2\} = \binom{n}{2} p^2 q^{n-2} = \tfrac{1}{2}n(n - 1)p^2 q^{n-2},$$

and so on. The terms on the right are recognized as successive terms of the binomial expansion of $(q + p)^n$, since

$$(q + p)^n = q^n + nq^{n-1}p + \tfrac{1}{2}n(n - 1)q^{n-2}p^2 + \ldots + p^n \quad (38)$$

$$= \sum_{r=0}^{n} \binom{n}{r} p^r q^{n-r}$$

$$= \sum_{r=0}^{n} f(r).$$

Also, since $q + p = 1$, then

$$\sum_{r=0}^{n} f(r) = 1. \quad (39)$$

15.7.1 Mean and variance of the binomial distribution

$$\text{Mean of } r = \mu = \sum_{r=0}^{n} rf(r)$$

$$= \sum_{r=1}^{n} \frac{r \cdot n!}{r!(n - r)!} p^r q^{n-r}$$

$$= \sum_{r=1}^{n} \frac{n!}{(r - 1)!(n - r)!} p^r q^{n-r}$$

$$= \sum_{r=1}^{n} \frac{(np)(n - 1)!}{(r - 1)!\{(n - 1) - (r - 1)\}!} p^{r-1} q^{\{(n-1)-(r-1)\}}$$

and writing $r - 1 = s$,

$$\text{mean of } r = np \sum_{s=0}^{n-1} \frac{(n - 1)!}{s!(n - 1 - s)!} p^s q^{(n-1-s)}$$

$$= np(q + p)^{n-1}$$

$$= np \quad \text{since } q + p = 1.$$

Hence, $\qquad\qquad\qquad$ mean of $r = np$. $\qquad$ (40)

Using similar algebraic manipulations,

$$\text{variance of } r = \sigma^2 = \sum_{r=0}^{n} r^2 f(r) - \mu^2, \quad (41)$$

and writing $r^2 = r(r - 1) + r$ in the first expression on the right,

$$\sum_{r=0}^{n} r^2 f(r) = \sum_{r=0}^{n} \{r(r - 1) + r\} \binom{n}{r} p^r q^{n-r}$$

$$= \sum_{r=1}^{n} r(r - 1) \binom{n}{r} p^r q^{n-r} + \sum_{r=1}^{n} r \binom{n}{r} p^r q^{n-r}$$

$$= \sum_{r=1}^{n} r(r - 1) \frac{n!}{r!(n - r)!} p^r q^{n-r} + np \quad \text{(from (40))}$$

$$= \sum_{r=2}^{n} \frac{(n - 2)!}{(r - 2)!\{(n - 2) - (r - 2)\}!} n(n - 1) p^2 p^{r-2} q^{\{(n-2)-(r-2)\}}$$

$$+ np.$$

Writing $r - 2 = s$ in the summation,

$$\sum_{r=0}^{n} r^2 f(r) = \sum_{s=0}^{n} \frac{(n-2)!}{s!(n-2-s)!} p^s q^{n-2-s} n(n-1)p^2 + np$$
$$= n(n-1)p^2(q+p)^{n-2} + np$$
$$= n(n-1)p^2 + np.$$

Finally, substituting in equation (41),

$$\sigma^2 = n(n-1)p^2 + np - n^2p^2$$
$$= np(1-p)$$
$$= npq.$$

Hence, for the binomial distribution,

$$\text{mean } \mu = np \qquad (42)$$
$$\text{variance } \sigma^2 = npq \qquad (43)$$

Example (i) A machine makes certain products which are known to be on average 1 per cent defective. What is the probability that a box of one hundred of these products contains (a) exactly 3 defectives, (b) at least 3 defectives?

Solution Let r be the number of defectives in a box of 100, then r has a binomial distribution with $n = 100$, $p = 0.01$, $q = 0.99$.

(a) $P\{r = 3\} = f(3)$

$$= \binom{100}{3} (0.01)^3 (0.99)^{97}$$
$$= 0.060\ 5$$

(b) $P\{r \geqslant 3\} = P\{r = 3 \text{ or } r = 4 \text{ or } r = 5 \text{ or } \ldots \text{ or } r = 100\}$

$$= 1 - P\{r = 0 \text{ or } 1 \text{ or } 2\}$$
$$= 1 - f(0) - f(1) - f(2)$$
$$= 1 - \binom{100}{0} (0.99)^{100} - \binom{100}{1} (0.01)(0.99)^{99}$$
$$- \binom{100}{2} (0.01)^2 (0.99)^{98}$$
$$= 1 - 0.366\ 44 - 0.370\ 13 - 0.185\ 065$$
$$= 0.078\ 36$$
$$\simeq 0.078\ 4.$$

In the calculation $f(0) = (0.99)^{100}$ was calculated using five-figure logarithms, $f(1)$ was then calculated as $100f(0)/99$ and $f(2)$ as $\frac{1}{2}f(1)$, i.e. each term of the binomial distribution was calculated from the previous term.

The mean number of defectives in a box of 100 is $np = 1$ and the variance is $npq = 0.99$.

Example (ii) If the probability that a target will be hit on a single bombing run by an aircraft is 0.1, find the probability that on 20 such runs the target will be hit at least twice.

Solution In this example, $p = 0.1$, $q = 0.9$, and $n = 20$, so that

$$P\{r \geqslant 2\} = 1 - P\{r < 2\}$$
$$= 1 - f(0) - f(1)$$
$$= 1 - \binom{20}{0} (0.1)^0 (0.9)^{20} - \binom{20}{1} (0.1)(0.9)^{19}$$
$$= 1 - 0.121\ 56 - 0.270\ 13$$
$$\simeq 0.608\ 3.$$

15.8 THE POISSON DISTRIBUTION

Suppose r has a binomial distribution $f(r) = \binom{n}{r} p^r q^{n-r}$ and that the number of trials n is very large, then the computation of $f(r)$ becomes very lengthy. There are two standard frequency functions which give a good approximation to the binomial distribution when n is large. The first is the *Poisson* distribution in which p is also very small so that np remains finite.

To derive the Poisson frequency function, consider $n \to \infty$ and $p \to 0$ in such a way that $np = \mu$ is a finite constant. Expand the binomial coefficient $\binom{n}{r}$ and multiply numerator and denominator by n^r to give

$$f(r) = \frac{1}{r!} \overbrace{\frac{n(n-1)(n-2)\ldots(n-r+1)}{n^r}}^{r \text{ factors}} (np)^r q^{n-r}.$$

Since $q = 1 - p$, $np = \mu$, and $q = 1 - \mu/n$, this expression becomes, on substituting,

$$f(r) = \frac{1}{r!} 1 \cdot \left(1 - \frac{1}{n}\right) \left(1 - \frac{2}{n}\right) \ldots \left(1 - \frac{r-1}{n}\right) \mu^r \left(1 - \frac{\mu}{n}\right)^{n-r}.$$

It can be proved that

$$\lim_{n \to \infty} \left(1 - \frac{x}{n}\right)^n = e^{-x},$$

so that, as $n \to \infty$ and $p \to 0$, the value of $f(r)$ approaches $(\mu^r e^{-\mu})/r!$. Hence the *Poisson distribution* is given by

$$f(r) = \frac{e^{-\mu} \mu^r}{r!}. \tag{44}$$

15.8.1 Mean and variance of the Poisson distribution

Using definition (27), with $x = r$; $p = f(r)$, then

$$\text{mean of } r = \sum_r r f(r) = \sum_{r=0}^{\infty} \frac{r e^{-\mu} \mu^r}{r!}$$

$$= e^{-\mu} \sum_{r=1}^{\infty} \frac{\mu^r}{(r-1)!}$$

$$= e^{-\mu} \mu \sum_{r=1}^{\infty} \frac{\mu^{r-1}}{(r-1)!}.$$

Writing $r - 1 = s$ in this summation gives

$$\text{mean of } r = e^{-\mu} \mu \sum_{s=0}^{\infty} \frac{\mu^s}{s!}$$

$$= e^{-\mu} \mu \, e^{\mu}$$

$$= \mu.$$

$$\text{Variance of } r = \sigma^2 = \sum_{r=0}^{\infty} r^2 f(r) - \mu^2 \qquad (45)$$

and

$$\sum_{r=0}^{\infty} r^2 f(r) = \sum_{r=0}^{\infty} r^2 e^{-\mu} \frac{\mu^r}{r!}.$$

Set $r^2 = r(r - 1) + r$, so that

$$\sum_{r=0}^{\infty} r^2 f(r) = \sum_{r=0}^{\infty} r(r-1) \, e^{-\mu} \frac{\mu^r}{r!} + \sum_{r=0}^{\infty} r \, e^{-\mu} \frac{\mu^r}{r!}$$

$$= e^{-\mu} \mu^2 \sum_{r=2}^{\infty} \frac{\mu^{r-2}}{(r-2)!} + e^{-\mu} \mu \sum_{r=1}^{\infty} \frac{\mu^{r-1}}{(r-1)!}.$$

Writing $r - 2 = s$ in the first summation and $r - 1 = t$ in the second summation,

$$\sum_{r=0}^{\infty} r^2 f(r) = e^{-\mu} \mu^2 \sum_{s=0}^{\infty} \frac{\mu^s}{s!} + e^{-\mu} \mu \sum_{t=0}^{\infty} \frac{\mu^t}{t!}$$

$$= e^{-\mu} \mu^2 e^{\mu} + e^{-\mu} \mu \, e^{\mu}$$

$$= \mu^2 + \mu.$$

Substituting in equation (45) gives

$$\text{variance of } r = \mu.$$

Hence for the Poisson distribution,

$$\text{mean} = \text{variance} = \mu. \qquad (46)$$

15.8.2 The Poisson distribution as an approximation to the binomial distribution

The number of happenings of a rare event in a large number of trials approximates to a Poisson distribution. The approximation is fairly good when $n \geqslant 100$, $p \leqslant 0.05$. Figures 15.9 and 15.10 show the binomial and Poisson distributions when $\mu = 4$, $p = 0.4$, and $p = 0.04$ respectively.

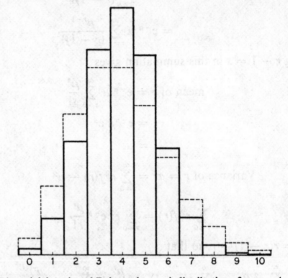

Fig. 15.9 Binomial (——) and Poisson (⋯⋯⋯) distributions for $\mu = 4$, $p = 0.4$

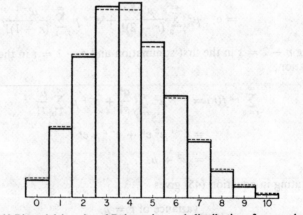

Fig. 15.10 Binomial (——) and Poisson (⋯⋯⋯) distributions for $\mu = 4$, $p = 0.04$

Example (i) Suppose that example (i) of section 15.7.1 is evaluated using the Poisson distribution instead of the binomial distribution.

Solution Here $n = 100$, $p = 0.01$, $\mu = 1$, and

$$f(r) = e^{-\mu} \mu^r / r!$$

Then (a) $\qquad P\{r = 3\} = f(3)$

$$= e^{-1}(1)^3/3!$$

$$= 0.061\ 3;$$

and (b) $\qquad P\{r \geqslant 3\} = 1 - f(0) - f(1) - f(2)$

$$= 1 - e^{-1}(1 + 1 + \tfrac{1}{2})$$

$$= 1 - 0.919\ 7$$

$$\simeq 0.080\ 3.$$

Example (ii) Boxes contain 200 articles known to be on average 1 per cent defective. The manufacturer guarantees that in a box of 200 there will be fewer than three defectives. What is the probability that the guarantee will be violated?

Solution The number of defectives in a box of 200 has a distribution which approximates to a Poisson distribution since $n = 200$ is large, $p = 0.01$ is small and $np = \mu = 2$.

Let r be the number of defectives, then

$$f(r) = e^{-2} 2^r / r!$$

Hence, approximately

$$P\{r < 3\} = e^{-2}(1 + 2 + 2^2/2!)$$

$$= 5\ e^{-2}$$

$$= 0.676\ 7.$$

Exactly, using the binomial distribution and a longer calculation,

$$P\{r < 3\} = \sum_{r=0}^{2} \binom{200}{r} (0.01)^r (0.99)^{200-r}$$

$$= 0.679\ 78.$$

Hence the probability that the number of defectives in a box is three or more is $0.323\ 3$, and the probability the guarantee is violated is 0.323 approximately, or 32.3 per cent.

15.8.3 The Poisson distribution as an independent distribution

If events occur at random at a constant rate in time or space, it can be shown that the number of events in a given time interval or in a given region has a Poisson distribution. The assumption is made that the number of events occurring in say, a given time interval is independent of the number that occurred in earlier intervals. Consider an example of an observed distribution that can be thought of as having a Poisson distribution. Table 15.5 gives the observed distribution of α-particles emitted in 6 seconds. The variable x is the number of α-particles emitted per 6-second interval and x was observed in 2 500 such intervals. The mean $\bar{x}$ for the observed distribution is 3·76. If this is taken as an estimate of the mean of a Poisson distribution, $\mu = 3·76$, then the corresponding frequencies would be expected to be

$$2\,500\,e^{-3·76}\,(3·76)^r/r!, \quad r = 0, 1, 2, \ldots.. 10.$$

The results of computing these values are given in Table 15.5.

x	0	1	2	3	4	5	6	7	8	9	$\geqslant 10$
observed frequency	71	210	384	515	501	383	231	121	46	23	15
expected frequency	58·2	218·8	411·4	515·7	484·7	364·5	228·4	122·7	56·7	24·1	13·8

Table 15.5

There is good agreement here. Had it not been so, the Poisson distribution as the mathematical model for this observed distribution would not have been acceptable. Other examples where the Poisson distribution has been found to be a good mathematical model are (i) the number of calls arriving at a telephone exchange in a given interval of time, (ii) the number of cars passing a given point of a busy road in a given interval of time, (iii) the number of blood cells in a given area say 1 cm^2 of a haemocytometer, (iv) bacteria counts, (v) stars in space, and in many more similar cases.

15.9 THE NORMAL OR GAUSSIAN DISTRIBUTION

The binomial and Poisson distributions are both examples of theoretical distributions of a discrete random variable. The *normal* distribution is a continuous distribution whose characteristics are often found in practice and hence it serves as a mathematical model for these distributions. It has the frequency function defined by

$$f(x) = \frac{1}{\sqrt{(2\pi\sigma^2)}} \exp\left\{-\frac{(x-\mu)^2}{2\sigma^2}\right\} \tag{47}$$

where μ and σ^2 are constants.

The characteristics of the graphs of $y = f(x)$ defined by equation (47) are that all are bell-shaped and symmetrical and die out quickly at the tails. A typical normal distribution is shown in Fig. 15.11.

It can be shown that

$$\int_{-\infty}^{\infty} f(x)\,dx = 1 \quad \text{for all values of } \mu \text{ and } \sigma \tag{48}$$

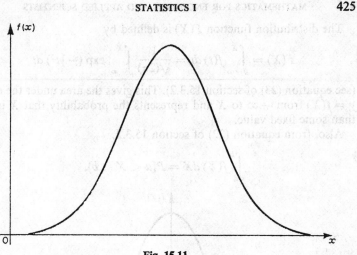

Fig. 15.11

15.9.1 Mean and variance of the normal distribution
It can be shown that

$$\text{mean} = \int_{-\infty}^{\infty} x f(x)\, dx = \mu, \tag{49}$$

and

$$\text{variance} = \int_{-\infty}^{\infty} (x - \mu)^2 f(x)\, dx = \sigma^2. \tag{50}$$

The integrations (48), (49), and (50) are beyond the scope of this book.

15.9.2 The standardized variate
Usually in problems the values of μ and σ are known or can be found. Probabilities, corresponding to the areas under the curve $y = f(x)$ between given limits are required, and these are tabulated or, in some cases calculated, for $\mu = 0$, $\sigma^2 = 1$. If the random variable X has a normal distribution with $\mu = 0$ and $\sigma^2 = 1$, then X is said to have a *standardized* normal distribution. If x has a normal distribution with mean μ and variance σ^2, this will be written $x = N(\mu, \sigma^2)$. Changing the origin and scale so that the new mean is zero and the new variance 1, i.e. writing $y = x - \mu$ gives

$$\text{mean of } y = \text{mean of } x - \mu = \mu - \mu = 0,$$

and writing $X = y/\sigma = (x - \mu)/\sigma$ gives

$$\text{var}\,(X) = \text{var}\,(y)/\sigma^2 = \text{var}\,(x)/\sigma^2 = \sigma^2/\sigma^2 = 1,$$

i.e.

$$X = (x - \mu)/\sigma = N(0, 1). \tag{51}$$

The graph of $y = f(X) = [\exp(-\tfrac{1}{2}X^2)]/\sqrt{(2\pi)}$ is sketched in Fig. 15.12. It is an even function since $f(-X) = f(X)$ and hence is symmetrical about the y-axis.

The distribution function $F(X)$ is defined by

$$F(X) = \int_{-\infty}^{X} f(t)\, dt = \frac{1}{\sqrt{(2\pi)}} \int_{-\infty}^{X} \exp\left(-\tfrac{1}{2}t^2\right) dt \qquad (52)$$

(see equation (23) of section 15.3.2). This gives the area under the curve $y = f(X)$ from $-\infty$ to X and represents the probability that X is less than some fixed value.

Also, from equation (22) of section 15.3.2,

$$\int_{a}^{b} f(X)\, dX = P\{a < X < b\}.$$

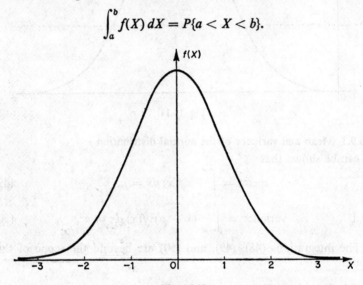

Fig. 15.12

It is possible to evaluate the integral on the left between finite limits, using the numerical methods of Chapter 14. In particular, using Simpson's rule of section 14.8.3 with $h = 1$,

$$\frac{1}{\sqrt{(2\pi)}} \int_{0}^{4} \exp\left(-\tfrac{1}{2}X^2\right) dX = 0\!\cdot\!497\,5 \simeq 0\!\cdot\!5. \qquad (53)$$

It is not necessary to evaluate the integral each time it is required, since tables of the distribution function $F(X)$ are published, for example, in *Cambridge Elementary Statistical Tables* by D. V. Lindley and J. C. P. Miller. (It is assumed that the reader has a copy of these, or similar, tables.) The function $F(X)$ defined by equation (52) is tabulated for values of X from $X = 0$ to $X = 4\cdot0$. From the tables, when $X = 0$, the value of $F(X)$ is $0\cdot5$, and this is understandable, since it represents the area from the left tail to the centre, or half the area under the curve in Fig. 15.12. Compare this with the result (53) which gives the area under the curve from $X = 0$ to $X = 4$, which is approximately half the total area.

As X increases towards the value 4, $F(X)$ increases towards the value 1, or the total area under the curve. When $X = 0.67$, then $F(X) = 0.748\ 6 \simeq 0.75$ using tables. Hence, between $X = 0$ and $X = 0.67$ there is approximately $(0.75 - 0.5) = 0.25$ of the total area; or, because of symmetry, approximately half the area lies between $X = \pm 0.67$. The area under the curve between $X = \pm 1.96$ is $2(0.975 - 0.5) = 0.95$, or about 95 per cent of the total area lies between $X = \pm 1.96$.

Example Let x be $N(1, 4)$. Find the probability that x is less than 3.

Solution Given $x = N(1, 4)$, write $X = (x - 1)/\sqrt{4}$ then $X = N(0, 1)$ and

$$P\{x < 3\} = P\{X < \tfrac{1}{2}(3 - 1)\} = P\{X < 1\} = 0.841$$

using tables to find the final answer.

15.9.3 The normal distribution as an approximation to the binomial distribution

When n is large and p is small the Poisson distribution gives a good approximation to the binomial distribution. When n is large and p is not necessarily small, the normal distribution is a good approximation. To show the degree of approximation consider the following two numerical examples.

Example (i) A given discrete variable has a distribution with mean 12 and variance 4. What is the probability that it is 14 or more?

Solution Here $np = 12$, $npq = 4$ so that $q = \tfrac{1}{3}$, $p = \tfrac{2}{3}$, $n = 18$, and the binomial distribution is

$$f(r) = \binom{18}{r} \left(\frac{2}{3}\right)^r \left(\frac{1}{3}\right)^{18-r}$$

The values for $r = 0, 1, 2, 3, \ldots, 18$ are given in Table 15.6, and the histogram of these values is drawn in Fig. 15.13. The computation of values of f is made slightly easier by computing each term after the first from the previous term, so that

$$f(0) = (\tfrac{1}{3})^{18} = t_0,$$

$$f(1) = 18 \times 2t_0 = t_1,$$

and $\quad f(r) = \dfrac{18 . 17 . 16 . \ldots . (18 - r + 1)}{1 . 2 . 3 . \ldots . r} \left(\frac{2}{3}\right)^r \left(\frac{1}{3}\right)^{18-r}$

$$= \frac{19 - r}{r} 2t_{r-1}.$$

This method was used to find the values of f in Table 15.6, the values then written correct to four places of decimals.

r	4	5	6	7	8	9	10	11
$f(r)$	0·000 1	0·000 7	0·003 1	0·010 5	0·028 9	0·064 3	0·115 7	0·168 2

r	12	13	14	15	16	17	18
$f(r)$	0·196 3	0·181 2	0·129 4	0·069 0	0·025 9	0·006 1	0·000 7

Table 15.6

The required probability is given by

$$P\{r \geqslant 14\} = \sum_{r=14}^{18} f(r) = 0.231\ 1,$$

and this is represented by the area of the histogram in Fig. 15.13 to the right of AB.

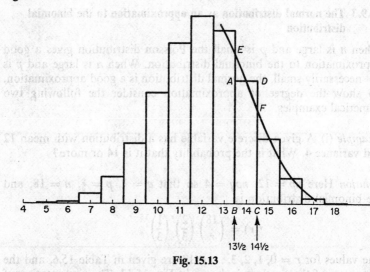

Fig. 15.13

The normal approximation to the area of the rectangle $ABCD$ is the area under the corresponding normal curve, or $EBCF$, i.e. the area under the normal curve from $13\frac{1}{2}$ to $14\frac{1}{2}$. To evaluate the required probability using the normal approximation to this binomial distribution it is necessary to evaluate the area to the right of $13\frac{1}{2}$. The continuity correction of $\frac{1}{2}$ must always be allowed for. In this example, first standardizing the variate and then using tables,

$$X = \frac{x - 12}{\sqrt{4}} = \tfrac{1}{2}(x - 12),$$

and the required probability,

$$P\{x \geqslant 14\} = P\{X > \tfrac{1}{2}(13\tfrac{1}{2} - 12)\}$$
$$= P\{X > 0.75\}$$
$$= 1 - 0.773\ 4$$
$$= 0.226\ 6.$$

The approximation is good for such a small value of n.

Example (ii) A random sample of 100 screws is taken from a batch which is known to be 3 per cent defective. Calculate the probability that the sample will contain (a) more than 5 defectives, (b) between 2 and 5 defectives.

Solution Here $n = 100$, $p = 0.03$, $q = 0.97$, so that $\mu = np = 3$ and $\sigma^2 = npq = 2.91$.

Let x be the number of defectives per sample of 100, then x is $N(3, 2.91)$. Let X be the standardized variate, so that

$$X = (x - 3)/\sqrt{(2.91)} = (x - 3)/1.706,$$

and hence X is $N(0, 1)$. Then,

(a) the probability that a sample contains more than 5 defectives is given by

$$P\{x > 5\} = P\{X > (4\tfrac{1}{2} - 3)/1.706\}$$
$$= P\{X > 0.879\ 4\}$$
$$= 1 - P\{X < 0.879\ 4\}$$
$$= 0.189\ 6;$$

(b) the probability that a sample contains between 2 and 5 defectives is given by

$$P\{2 < x < 5\} = P\{(1.5 - 3)/1.706 < X < (5.5 - 3)/1.706\}$$
$$= P\{-0.879\ 4 < X < 1.466\}$$
$$= 0.928\ 7 - (1 - 0.810\ 4)$$
$$= 0.739\ 1.$$

It can be proved that if x represents the number of successes in n independent trials of an event where p is the probability of success in a single trial, then $x \to N(np, npq)$ as n increases.

It is sometimes more convenient to work with the proportion of successes, x/n in n trials where p is the probability of success in a single trial. Then

$$x/n \to N(p, pq/n) \quad \text{as } n \text{ increases.}$$

15.9.4 Summary

In any given problem (i) if n is small the binomial distribution, (37) of section 15.7 is used, (ii) if n is large and p is small the Poisson distribution, (44) of section 15.8 is used, and (iii) if n is large and p is either small or large the normal approximation may be used. The normal distribution is the easiest to use since this is tabulated for the standardized variate. The tables used in this chapter tabulate the function

$$\frac{1}{\sqrt{(2\pi)}} \int_{-\infty}^{X} e^{-\frac{1}{2}t^2} \, dt,$$

which is the probability that the standardized variate is less than X. Other tables give values of

$$\frac{1}{\sqrt{(2\pi)}} \int_{0}^{X} e^{-\frac{1}{2}t^2} \, dt,$$

which is the probability that the standardized variate lies between 0 and X.

15.10 SOME USEFUL THEOREMS

Let x and y be two independent random variables, i.e. the probability that x takes any set of values is unaffected by y and the probability that y takes a set of values is unaffected by x. Then,

(i) if each value of x is multiplied by a constant c the mean of the resulting variable is c times the mean of x, i.e.

$$(\overline{cx}) = c\bar{x},$$

(ii) the mean of the sum or difference of the variates x and y is the sum or difference of their means, i.e.,

$$\overline{(x \pm y)} = \bar{x} \pm \bar{y},$$

(iii) the variance of the sum or difference of the variates is the *sum* of their variances, i.e.

$$\sigma^2(x \pm y) = \sigma^2(x) + \sigma^2(y),$$

(iv) if each of x and y has a normal distribution then $(x + y)$ has a normal distribution.

The proofs of these theorems will not be given here and may be found in textbooks of statistics.

EXERCISE 15

1 A box contains 4 white and 2 red marbles. One marble is drawn from the box. What is the probability that the marble selected is (i) white, (ii) red?

2 What is the probability of obtaining the score (i) 6, (ii) 4 on rolling one die? What is the probability of obtaining a total score (iii) 12, (iv) 7 on rolling 2 dice?

3 Find the probability that in 4 hands dealt from an ordinary pack of 52 playing cards each hand contains one ace.

4 Find the probability that at least one head is thrown when 10 coins are tossed.

5 A single die is rolled until a six is obtained. What is the probability that a six will be obtained (i) on the third attempt, (ii) on the nth attempt?

6 Two different digits are chosen at random from the set 1, 2, 3, . . .,9. Find the probability that (i) the sum of the digits is exactly 6, (ii) the sum of the digits is greater than 14, (iii) both digits are greater than 6.

7 The probability that one particular component of a machine needs replacing in a 5-year period is $\frac{1}{4}$ and that of another is $\frac{1}{5}$. Assuming that these two events are independent, find the probability (i) that either one or the other needs replacing, (ii) that both components need replacing, (iii) that neither component needs replacing, in the period.

8 A and B take turns in throwing a pair of dice. A starts and the first to throw a total score of 9 wins. What are the respective probabilities?

9 A single die is rolled. What is the probability that the score is 3 if it is known that the score is odd?

10 A box contains 3 black and 2 red balls marked $B1$, $B2$, $B3$ and $R1$, $R2$. Two balls are drawn in turn from the box without replacement. Find the probability that (i) both balls are red, (ii) both balls are of the same colour, (iii) a total score of 4 is obtained, (iv) the total score is 4 if it is known that both balls are black, (v) the first ball drawn was black if it is known that the second one drawn is $B1$.

11 A random variable x has the frequency function $f(x) = cxe^{-x}$, $x \geqslant 0$. Determine (i) the value of c, (ii) the distribution function $F(x)$, (iii) $P\{x < 1\}$, (iv) $P\{1 < x < 2\}$, (v) the mean and variance of x.

12 A random variable has the frequency function $f(x)$ defined by $f(x) = 0$, $x < 0$; $f(x) = cx$, $0 \leqslant x \leqslant 1$; $f(x) = c + 1 - x$, $1 < x \leqslant 2$; $f(x) = 0$, $x > 2$. (i) Evaluate c, (ii) find the distribution function $F(x)$ and sketch the graphs of $f(x)$ and $F(x)$.

13 Use coding to find the mean and standard deviation of the frequency distribution given in the following table.

x	2·30	2·35	2·40	2·45	2·50	2·55	2·60	2·65	2·70	2·75
f	1	5	6	13	8	17	14	7	1	3

14 Examination marks are scaled either (i) by subtracting 10 from each mark or (ii) by subtracting 10 per cent of each mark. How do these forms of scaling affect the mean and standard deviation of the set of examination marks?

15 Sketch the histograms of the binomial distributions in which $p = \frac{1}{2}$ and (i) $n = 5$, (ii) $n = 10$, (iii) $n = 25$.

16 Experience shows that 25 per cent of a new seed germinates. If 1 024 packets each containing 5 seeds are tested, find the number of packets in which it is expected that 0, 1, 2, . . ., 5 seeds germinate. Sketch the histogram.

17 A firm makes a certain kind of switch and produces 0·5 per cent defective. They are packed in boxes of 100. What percentage of boxes (i) has no defective switches, (ii) has 2 or more defective switches? Use (a) the binomial and (b) the Poisson distributions.

18 Fit a Poisson distribution with the same mean to the following data:

Number of faults	0	1	2	3	4	5
Frequency of faults	18	28	25	17	9	3.

19 The number of telephone calls received by an operator between 8.30 and 8.35 a.m. fits a Poisson distribution with mean 3. Find the probability that (i) in any given day there will be no calls in that interval of time, (ii) in 3 given consecutive days there will be a total of 2 calls in that interval of time.

20 A manufacturer knows that 15 per cent of his product is defective. It is inspected in batches of 60 units. Find the mean and variance of the number of defectives in batches of 60 and use tables of the normal distribution to calculate the approximate percentage of such batches which contain more than 10 defectives.

21 On the same axes sketch the normal curves of f against x for one hundred trials of an experiment when x is (i) $N(0, 1)$, (ii) $N(0, 4)$, (iii) $N(0, 100)$. What percentage of the total area in each case lies between $\pm 0·67\sigma$, $\pm \sigma$, $\pm 1·96\sigma$?

22 The resistances of 200 similar resistors of nominal value 1 000 Ω were measured to the nearest ohm and their values classified in the following table.

Resistance x	995	996	997	998	999	1 000	1 001	1 002
Frequency f	5	0	13	12	12	22	13	17

Resistance x	1 003	1 004	1 005	1 006	1 007	1 008	1 009
Frequency f	19	20	16	18	15	12	6

Find the mean resistance and the standard deviation. Assuming a normal distribution with this mean and standard deviation find the probability that any one resistor falls outside the manufacturer's guarantee that the values of the resistances are within ± 1 per cent of their nominal value.

16
Statistics II

16.0 INTRODUCTION

Broadly speaking, the aim of most statistical enquiries is to infer properties of a population from a sample from that population. The population corresponds to the frequency function, the sample corresponds to an observed frequency distribution. If the number of observations is large, relative frequencies of the observed frequency distribution tend to probabilities of the frequency function.

Thus, given an observed frequency distribution, the problem is to select a theoretical (mathematical) model, to check that this model is reasonable, and finally to draw conclusions about the population.

There are three types of statistical problem to consider. The first is the estimation of parameters from observed frequency distributions, for example, the estimation of the value of the mean or variance of a population from the data from a sample. The second is the testing of hypotheses, which leads to significance testing. Finally, one test that the model chosen is a reasonable model will be given.

16.1 METHODS OF SAMPLING

A *random* sample is a sample from a population in which each individual member of the population has the same chance of being selected. *Simple* sampling is random sampling where the probability of selecting any individual remains constant throughout the sampling. If the population is very large or can be thought of as infinite, then every random sample is simple. If the population is finite, simple samples are obtained only if sampling is with replacement, i.e. after the selection of an individual it is replaced in the population. From now on in this chapter all samples are simple.

16.1.1 Estimation

In most cases a good estimate of an unknown population parameter is obtained by computing the corresponding quantity for the observed frequency distribution, so that, to estimate the population mean μ the mean $\bar{x}$ (or m) of the observed distribution is used, to estimate the population variance σ^2 the variance s^2 of the observed distribution is used. That is,

estimate μ by $\bar{x} = (\Sigma n_i x_i)/n$,

estimate σ^2 by $s^2 = [\Sigma n_i (x_i - \bar{x})^2]/n$,

where $\Sigma n_i = n =$ total number of observations.

16.1.2 Definition

Any quantity computed from the sample and used to estimate an unknown population parameter is called a *statistic* or *estimator*.

16.2 SAMPLING DISTRIBUTION OF THE MEAN

Suppose a sample of n observations $x_1, x_2, x_3, \ldots, x_n$ is taken from a population, then the sample mean is, say, $\bar{x}_1 = (\Sigma x_i)/n$. Suppose a second sample is taken from the same population, then this will have a sample mean $\bar{x}_2$ which may be different from $\bar{x}_1$. A third sample under the same conditions will have mean $\bar{x}_3$, and so on. That is, the value of $\bar{x}$ varies with the sample, and the sample means may be tabulated as in Table 16.1.

Sample number	1	2	3	$\ldots$	k
Sample mean	$\bar{x}_1$	$\bar{x}_2$	$\bar{x}_3$	$\ldots$	$\bar{x}_k$

Table 16.1

Thus $\bar{x}$ is not constant over all samples but has a distribution, and the purpose is to find the frequency function for $\bar{x}$. This is called the *sampling distribution of the mean*.

Suppose the population has mean μ and variance σ^2, and a random sample $x_1, x_2, x_3, \ldots, x_n$ is taken from this population. The sample mean $\bar{x} = (\Sigma x_i)/n$. Since the sample is random, $x_1, x_2, x_3, \ldots, x_n$ can be interpreted as n independent random variables. Also the mean of $x_i = \mu$ and the variance of $x_i = \sigma^2$ for all x_i, since each is drawn from the same population. Then,

$$(\text{mean of } \bar{x}) = (\text{mean of } x_1 + \text{mean of } x_2 + \ldots + \text{mean of } x_n)/n$$
$$= (\mu + \mu + \ldots + \mu)/n$$
$$= n\mu/n = \mu$$

or, mean of $\bar{x}$ = population mean μ;

$$(\text{variance of } \bar{x}) = \text{var}(x_1 + x_2 + x_3 + \ldots + x_n)/n^2$$
$$= [\text{var}(x_1) + \text{var}(x_2) + \text{var}(x_3) + \ldots + \text{var}(x_n)]/n^2$$

since $x_1, x_2, x_3, \ldots, x_n$ are independent.

Hence $\text{var}(\bar{x}) = (\sigma^2 + \sigma^2 + \sigma^2 + \ldots + \sigma^2)/n^2$
$$= n\sigma^2/n^2$$
$$= \sigma^2/n.$$

Hence $\text{mean}(\bar{x}) = \mu, \quad \text{var}(\bar{x}) = \sigma^2/n.$ (1)

If the population has a normal distribution, $n\bar{x} = x_1 + x_2 + x_3 + \ldots + x_n$ is also normal, i.e.

$$n\bar{x} = N(n\mu, n\sigma^2),$$

or $\bar{x} = N(\mu, \sigma^2/n),$ (2)

and the sampling distribution of the means is also normal. The standard deviation, which is the square root of the variance, of a sampling distribution is called the *standard error*, (S.E.), so that S.E. $(\bar{x}) = \sigma/\sqrt{(n)}$. The distributions of $x = N(\mu, 1)$ and of $\bar{x} = N(\mu, \frac{1}{4})$ are sketched in Fig. 16.1.

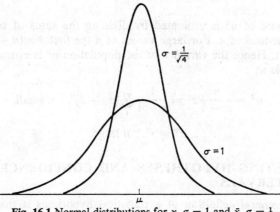

Fig. 16.1 Normal distributions for x, $\sigma = 1$ and $\bar{x}$, $\sigma = \frac{1}{2}$

If two independent samples from two different populations are being considered, then

$$\text{S.E. } (\bar{x} \pm \bar{y}) = \sqrt{\{\text{var} (\bar{x} \pm \bar{y})\}}$$

$$= \sqrt{\{\text{var} (\bar{x}) + \text{var} (\bar{y})\}}$$

$$= \sqrt{\left(\frac{\sigma_1^2}{n_1} + \frac{\sigma_2^2}{n_2}\right)}, \tag{3}$$

where σ_1^2 is the variance of the population of x's and n_1 the sample size, σ_2^2 the variance of the population of y's and n_2 the sample size.

16.2.1 Population parameters estimated from a sample

It is necessary to estimate the population parameters μ and σ^2 from a sample from the population. First consider an estimate of the population mean μ to be the sample mean $\bar{x}$. From equation (1) of section 16.2 the mean of $\bar{x}$ is μ and the variance of $\bar{x}$ is σ^2/n, so that the distribution of sample means is 'narrower' than the distribution of the population. This is shown in Fig. 16.1 with $\sigma = 1$, $n = 4$. Hence $\bar{x}$ has a high probability of being near to μ. Moreover as the sample size increases, $\bar{x}$ approaches the value μ. Hence the mean of the population is estimated as the mean of a sample. That is

$$\mu = \bar{x} \tag{4}$$

The sample variance s^2 can be shown to be the fraction $(n-1)/n$ of the population variance, that is, it is too small. Hence for small samples the variance $ns^2/(n-1)$ is used to estimate σ^2, and since

$$\frac{ns^2}{n-1} = \frac{1}{n-1} \sum_{i=1}^{n} (x_i - \bar{x})^2,$$

the estimate of σ^2 is evaluated by dividing the sums of squares by $(n-1)$ instead of n. For large values of n the factor $n/(n-1)$ is not important. Hence the variance of the population σ^2 is estimated from the sample as

$$\sigma^2 = \frac{n}{n-1} s^2 = \frac{1}{n-1} \sum_i (x_i - \bar{x})^2, \quad n \text{ small}, \tag{5}$$

$$\sigma^2 = s^2, \quad n \text{ large}. \tag{6}$$

16.3 TESTING HYPOTHESES AND CONFIDENCE INTERVALS

A fairly general definition of a statistical hypothesis is that it is an assumption about the frequency function of a random variable. A test of a statistical hypothesis is a test to decide whether to accept or reject the hypothesis. In problems considered in this section, the object is to test an assumed statement, called a *null hypothesis* (N.H.) to see if it is acceptable.

16.3.1 Significance test based on the normal distribution

This test depends on knowing that the population being tested is normally distributed, on knowing the variance, and on having a reputed mean of the distribution.

Suppose a sample of n individuals from a given population has mean $\bar{x}$, the general procedure for testing this mean is as follows:

(a) make the null hypothesis that the true mean is μ,
(b) note the difference between the sample mean $\bar{x}$ and μ,
(c) find the theoretical probability that in random sampling the difference between $\bar{x}$ and μ is greater than the observed difference,
(d) if this probability is sufficiently small, reject the null hypothesis. That is, conclude from the given data that this sample is not from the given population whose mean is μ.

An example to demonstrate the procedure follows.

Example A firm is producing packets whose weights are known to be normally distributed with variance 0.25 gramme^2, and the mean weight is supposed to be 10 g. A random sample of 10 packets is taken and the sample mean is found to be 9.64 g. Is the firm giving correct weight?

Solution Let the weight in grammes be x. The sample mean $\bar{x}$ is 9·64, the number of items in the sample n is 10. Following the procedure for testing this value of $\bar{x}$:

(a) the N.H. is that the population mean μ is 10;

(b) the sample mean $\bar{x}$ differs from μ by 0·36;

(c) the probability that in random sampling $\bar{x}$ would be such that $|\bar{x} - \mu| \geqslant 0{\cdot}36$, i.e. the probability that $|\bar{x} - \mu| \geqslant 0{\cdot}36$ *by chance* is found by standardizing the variate and using tables of the normal distribution as follows. Let

$$X = (\bar{x} - \mu)/(\sigma/\sqrt{n})$$
$$= -0{\cdot}36 \sqrt{10/0{\cdot}5}$$
$$= -2{\cdot}28,$$

then $X = N(0, 1)$, and from tables,

$$P\{|X| \geqslant 2{\cdot}28\} = 2P\{X \geqslant 2{\cdot}28\}$$
$$= 2[1 - P\{X < 2{\cdot}28\}]$$
$$= 2[1 - 0{\cdot}988\ 7] = 0{\cdot}022\ 6$$
$$< 0{\cdot}05 \quad \text{or} \quad 5 \text{ per cent.}$$

(d) The critical probability level is usually taken at 5 per cent, and the test shows that $\bar{x}$ differs significantly from μ at the 5 per cent level, and the null hypothesis is rejected. Hence it is not by chance that the incorrect weight is being given.

Values of the standardized variate X and the corresponding probabilities are given in Table 16.2.

P	20%	10%	5%	2%	1%	0·1%
$|X| >$	1·282	1·645	1·960	2·326	2·576	3·291

Table 16.2

The general procedure for testing means in normal populations is now re-written as follows:

(i) make the null hypothesis that the true mean is μ,

(ii) compute $X = (\bar{x} - \mu)/(\sigma/\sqrt{n})$, where σ^2 is the variance and n the number of observations in the sample, then $X = N(0, 1)$,

(iii) if $|X| \geqslant 1{\cdot}96$, then $\bar{x}$ differs significantly from μ at the 5 per cent level and the N.H. is usually rejected,

(iv) if $|X| \geqslant 2{\cdot}576$, $\bar{x}$ differs significantly from μ at the 1 per cent level and the N.H. is always rejected,

(v) if $|X| < 1{\cdot}96$ there is no reason to reject the N.H.

It is not possible to prove that the null hypothesis is true. If $|X| < 1{\cdot}96$ the sample values are consistent with the null hypothesis, or there is no evidence from the given sample to reject the null hypothesis.

The above test is a two-tailed test in which the probability that $|X|$ is greater than a fixed number is given. A one-sided test in the above example would be say, a test for short weight and

$$P\{X \leqslant -2{\cdot}28\} = P\{X > 2{\cdot}28\}$$
$$= 1 - 0{\cdot}988\ 7$$
$$= 0{\cdot}011\ 3$$
$$< 5 \text{ per cent level.}$$

16.3.2 Confidence intervals (interval estimation)

Assume that the population mean μ is unknown and is estimated by the sample mean $\bar{x}$. Then the 95 per cent *confidence interval* for μ is that interval such that

$$P\{\mu - 1{\cdot}96\sigma/\sqrt{n} < \bar{x} < \mu + 1{\cdot}96\sigma/\sqrt{n}\} = 0{\cdot}95$$

i.e. $P\{\bar{x} - 1{\cdot}96\sigma/\sqrt{n} < \mu < \bar{x} + 1{\cdot}96\sigma/\sqrt{n}\} = 0{\cdot}95$

since if $\bar{x}$ is within $\pm 1{\cdot}96\sigma/\sqrt{n}$ of μ, then μ is within $\pm 1{\cdot}96\sigma/\sqrt{n}$ of $\bar{x}$.

In the above example $\bar{x} = 9{\cdot}64$ and the interval becomes

$$9{\cdot}64 - 1{\cdot}96 \,.\, 0{\cdot}5/\sqrt{10}, \quad 9{\cdot}64 + 1{\cdot}96 \,.\, 0{\cdot}5/\sqrt{10}$$

which is 9·33 to 9·95.

A confidence interval for any parameter may be thought of as the values between which that parameter must lie for a test of significance to be non-significant. In terms of the mean, with standard error $\sigma/\sqrt{n}$, as n increases, $\sigma/\sqrt{n}$ decreases so that the width of the confidence interval decreases and accuracy increases.

Example (i) The values of resistors produced by a given machine appear to have a normal distribution with mean 29 Ω and variance 1·44 Ω^2. The process is modified in such a way that the variance is not affected but the mean may be. A random sample of 10 resistors is taken and the values in Ω are

30·31	31·14	30·14	30·22	27·87
28·37	31·37	29·70	29·07	29·41

Has the mean changed?

Solution The N.H. is that there is no difference in the mean, or that the mean μ is 29. The sample mean $\bar{x}$ is

$$29 + (1{\cdot}31 + 2{\cdot}14 + \ldots + 0{\cdot}41)/10,$$

or $\bar{x}$ is 29·76.

The standardized variate X is given by

$$|X| = \frac{|29{\cdot}76 - 29|}{\sigma/\sqrt{n}} = \frac{0{\cdot}76\sqrt{10}}{1{\cdot}2} = 2{\cdot}003.$$

This is greater than 1·96 and hence the N.H. is rejected. That is, $\bar{x}$ differs significantly from μ at the 5 per cent level and modification has tended to increase the mean.

It is of interest to note that the 95 per cent confidence limits for the new mean are $\bar{x} - 1·96\sigma/\sqrt{n}$, $\bar{x} + 1·96\sigma/\sqrt{n}$ which are 29·02 and 30·50. The interval between these limits does not contain the value 29, of the old mean, and this was to be expected from the result found.

Example (ii) The classic blindfold drinking test is based on the binomial distribution and the technique of testing is the same. A person claims that he is able to distinguish between two drinks A and B. In an experiment he is correct 5 times in 6 trials. Does this result justify his claim?

Solution The N.H. is that the person guesses, i.e. that he is equally likely to choose A or B, i.e. that the number of correct guesses has a binomial distribution with $p = \frac{1}{2}$, $q = \frac{1}{2}$. Let r be the number of successes, then from the binomial distribution, with $n = 6$,

$$f(r) = \binom{6}{r} \left(\frac{1}{2}\right)^r \left(\frac{1}{2}\right)^{6-r}$$

$$= \binom{6}{r} \left(\frac{1}{2}\right)^6.$$

The mean number of successes is given by np which is 3, and the observed number of successes is 5. Hence the difference is 2. The probability that this difference is greater than or equal to 2 is given by

$$P\{|r - np| \geqslant 2\} = f(0) + f(1) + f(5) + f(6)$$

$$= (\tfrac{1}{2})^6\{1 + 6 + 6 + 1\}$$

$$= 14/64$$

$$= 0·22.$$

This probability is greater than 5 per cent, and hence there is no reason to disbelieve the null hypothesis that the person guesses.

16.4 LARGE SAMPLE THEORY

If it is known that a population is normally distributed, then the distribution of the means of samples from the population is also normally distributed. If the population distribution is $N(\mu, \sigma^2)$, then the sampling distribution of the mean is $N(\mu, \sigma^2/n)$.

If the population is not normal, it is still true that the mean of the distribution of the means $\bar{x}$ is μ and its variance is σ^2/n, but $\bar{x}$ is not normally distributed. It can be shown that the distribution of the means $\bar{x}$ tends to normality as n increases. This is known as the *Central limit theorem* which may be stated as follows.

Let $x_1, x_2, x_3, \ldots, x_n$ be n independent random variables from the same distribution (not necessarily normal), with mean $(x_i) = \mu$, variance $(x_i) = \sigma^2$, for all values of i. If $\bar{x} = \left(\sum_{i=1}^{n} x_i \right) / n$, then the theorem states that, as $n \to \infty$, $(\bar{x} - \mu)/(\sigma/\sqrt{n}) \to N(0, 1)$.

The distribution of most estimators computed from a sample of size n approach normality as n increases.

16.4.1 General procedure for large sample testing

A large sample is one which contains more than about fifty observations, and the procedure for testing a parameter θ is as follows:

(i) make the null hypothesis that the value of the parameter is θ,

(ii) compute an estimate of θ from a sample, say θ_0,

(iii) evaluate the standard error of θ_0, say s_0,

(iv) compute the standardized variate X given by

$$|X| = |\theta_0 - \theta|/s_0.$$

Then,

if $|X| \geqslant 2$, corresponding approximately to 5 per cent significance, the N.H. is usually rejected,

if $|X| \geqslant 3$, corresponding approximately to 1 per cent significance, the N.H. is always rejected,

if $|X| < 2$, there is no evidence from this sample to reject the N.H.

16.4.2 Testing the mean of a large sample

Let a sample of n (large) observations have values $x_1, x_2, x_3, \ldots, x_n$, and let the problem be to test whether the population from which the sample is drawn has mean μ. The population is not necessarily normal, and the variance is unknown. Using the procedure of section 16.4.1, θ now being the mean,

(i) the N.H. is that the population mean is μ,

(ii) the sample mean $\bar{x}$ is $(\Sigma x_i)/n$,

(iii) the standard error of the mean is $\sigma/\sqrt{n}$ and σ is unknown. Hence σ^2 is estimated from the sample by s^2 the sample variance, where

$$s^2 = [\Sigma(x_i - \bar{x})^2]/n$$

$$= [\Sigma x_i^2 - \bar{x}^2]/n.$$

(iv) The standardized variate X given by $|X| = |\bar{x} - \mu|/(s/\sqrt{n})$ is now tested.

Example A sample of 900 observations is found to have a mean 3·5 and standard deviation 2·62. Can it be regarded as a random sample from a population whose mean is 3·34?

Solution The N.H. is that the mean $\mu = 3{\cdot}34$. The mean of the sample $\bar{x} = 3{\cdot}5$ and the estimated standard deviation $s = 2{\cdot}62$. Hence the standardized variate X is given by

$$|X| = \frac{|3{\cdot}5 - 3{\cdot}34|}{2{\cdot}62/\sqrt{(900)}}$$

$$= \frac{4{\cdot}8}{2{\cdot}62}$$

and the result is obviously less than 2. Hence the result is not significant and there is no reason to suppose that this sample was not drawn from the population whose mean is $3{\cdot}34$.

It is of interest to note that in this example, the 95 per cent confidence limits for the mean are $\bar{x} - 1{\cdot}96s/\sqrt{n}$, $\bar{x} + 1{\cdot}96s/\sqrt{n}$, which equal $3{\cdot}33$, $3{\cdot}67$ (or $\bar{x} - 2s/\sqrt{n}$, $\bar{x} + 2s/\sqrt{n}$, which equal $3{\cdot}33$, $3{\cdot}67$), and the interval between these limits includes the mean of the population.

16.4.3 Testing the difference between two means

Two problems are considered:

(a) given two independent samples of n_1 and n_2 observations, they are tested as to whether they come from the same population,

(b) given two samples from different populations, to test whether the means of the populations are the same.

The procedure is as follows:

(a) the N.H. is that both samples are drawn from the same population whose mean is μ and whose variance is σ^2.

Let $\bar{x}_1$ be the mean of the sample size n_1 and $\bar{x}_2$ be the mean of the sample size n_2. Then $\mathrm{var}\,(\bar{x}_1) = \sigma^2/n_1$, $\mathrm{var}\,(\bar{x}_2) = \sigma^2/n_2$ and using section 15.10 theorem (iii),

$$\mathrm{var}\,(\bar{x}_1 - \bar{x}_2) = \sigma^2/n_1 + \sigma^2/n_2$$

so that $\qquad$ S.E. $(\bar{x}_1 - \bar{x}_2) = \sigma\sqrt{\left(\dfrac{1}{n_1} + \dfrac{1}{n_2}\right)}.$

If σ^2 is unknown a pooled estimate s^2 is used where

$$s^2 = \frac{1}{n_1 + n_2}\left[\Sigma(x_1 - \bar{x}_1)^2 + \Sigma(x_2 - \bar{x}_2)^2\right]$$

$$= \frac{n_1 s_1^2 + n_2 s_2^2}{n_1 + n_2}.$$

Also, since the mean of $(\bar{x}_1 - \bar{x}_2)$ is zero, using section 15.10 theorem (ii), the standardized variate X is given by

$$X = \frac{|(\bar{x}_1 - \bar{x}_2) - 0|}{\sigma \sqrt{\left(\dfrac{1}{n_1} + \dfrac{1}{n_2}\right)}}$$

and X may be tested as usual.

(b) The N.H. is that the means of the two populations are each equal to μ.

Let the sample means be $\bar{x}_1$ and $\bar{x}_2$ of samples size n_1 and n_2 respectively from populations whose variances are σ_1^2 and σ_2^2. Then,

$$\text{var}\,(\bar{x}_1 - \bar{x}_2) = \text{var}\,(\bar{x}_1) + \text{var}\,(\bar{x}_2)$$

$$= \frac{\sigma_1^2}{n_1} + \frac{\sigma_2^2}{n_2}$$

and $\qquad \text{S.E.}\,(\bar{x}_1 - \bar{x}_2) = \sqrt{\left(\dfrac{\sigma_1^2}{n_1} + \dfrac{\sigma_2^2}{n_2}\right)},$

$$\text{mean}\,(\bar{x}_1 - \bar{x}_2) = \mu - \mu = 0,$$

and finally

$$X = \frac{|(\bar{x}_1 - \bar{x}_2) - 0|}{\sqrt{\left(\dfrac{\sigma_1^2}{n_1} + \dfrac{\sigma_2^2}{n_2}\right)}}$$

which may be tested as usual.

If σ_1^2 and σ_2^2 are unknown, they may be estimated from the samples by

$$s_1^2 = \frac{1}{n_1}\sum(x_1 - \bar{x}_1)^2$$

and $\qquad\qquad s_2^2 = \dfrac{1}{n_2}\sum(x_2 - \bar{x}_2)^2.$

These tests are exact if (i) the populations are known to be normal and (ii) the variances are known. If (i) holds and (ii) does not, it is better to use the t-test which is an exact test valid for samples of any size and is discussed in section 16.6.1.

Example (i) The means of two random samples of 100 and 200 observations are found to be 66·5 and 67·1 respectively. Can these samples be regarded as drawn from the same population whose variance is 6·25?

Solution The N.H. is that both samples are drawn from the same population, or, that there is no difference in their means. Then, mean $(\bar{x}_1 - \bar{x}_2) = 0$, $\sigma^2 = 6 \cdot 25$, so that $\sigma = 2 \cdot 5$, and

$$\text{S.E. } (\bar{x}_1 - \bar{x}_2) = 2 \cdot 5 \sqrt{\left(\frac{1}{100} + \frac{1}{200}\right)}$$

$$= \frac{0 \cdot 75}{\sqrt{6}}.$$

Finally,
$$X = \frac{|(\bar{x}_1 - \bar{x}_2) - 0|}{\text{S.E.}}$$

$$= \frac{0 \cdot 6 \sqrt{6}}{0 \cdot 75}$$

$$< 2$$

and there is no reason to reject the null hypothesis, and both samples could be regarded as having been drawn from the same population.

Example (ii) A sample of 7 000 observations has a mean 46·9 with standard deviation 1·95 and a sample of 2 000 observations has a mean 45·9 with standard deviation 2·02. Is the difference in their means significant?

Solution The N.H. is that the samples are drawn from different populations with equal means, or, there is no difference in the two means. Estimate σ_1^2 by $s_1^2 = (1 \cdot 95)^2$ and σ_2^2 by $s_2^2 = (2 \cdot 02)^2$, then

$$\text{S.E. } (\bar{x}_1 - \bar{x}_2) = \sqrt{\left(\frac{1 \cdot 95^2}{7\,000} + \frac{2 \cdot 02^2}{2\,000}\right)}$$

$$\simeq 0 \cdot 050\,8,$$

$|\bar{x}_1 - \bar{x}_2| = 1$, and

$$X = \frac{1}{0 \cdot 050\,8} > 19.$$

Hence the difference in the means is highly significant. The null hypothesis is therefore rejected. There is a difference in means.

16.5 SAMPLING ATTRIBUTES

Sometimes it is required to find the probability that an individual has a certain characteristic. For example it may be required to find the probability that an individual is a cigarette smoker, given the proportion of men in that age group who smoke; or the probability that a manufactured part is defective knowing the proportion of defective parts produced by the machine making the parts.

The general method is to consider a sample of size n, and suppose that r out of n possess the attribute under consideration. Then r has a binomial distribution and

$$P\{r \text{ out of } n \text{ have the attribute}\} = \binom{n}{r} p^r q^{n-r}.$$

The mean value of r is np, and to test the hypothesis that the probability of an individual having the attribute is p, r is compared to np, that is, the theoretical probability that $|r - np|$ is greater than the observed difference is evaluated. If r_0 is the observed value of r, then

$$P\{|r - np| \geqslant |r_0 - np|\} \text{ is computed.}$$

16.5.1 Large sample test of a value of p

For a large sample the normal distribution may be taken as an approximation to the binomial distribution, and r is approximately $N(np, npq)$. Then the procedure is as follows:

The N.H. is that the probability any individual has the attribute is p, then the mean (r) is np and the variance (r) is npq. Hence the S.E. $(r) = \sqrt{(npq)}$, and the standardized variate X given by

$$X = \frac{|r - np|}{\sqrt{(npq)}}$$

is tested in the usual way.

Example (i) A six-sided die is thrown 9 000 times and a 5 or a 6 is obtained 3 210 times. Is the die biased?

Solution The N.H. is that the die is unbiased, i.e.

$$p = P\{5 \text{ or } 6\} = \tfrac{1}{3}, \quad q = 1 - p = \tfrac{2}{3}.$$

The expected number of (5 or 6) is np, i.e. 3 000. Then

$$|r - np| = 210,$$

and

$$\text{S.E. } (r) = \sqrt{(npq)}$$

$$= \sqrt{2\,000}$$

$$= 44{\cdot}72,$$

so that

$$X = \frac{210}{44{\cdot}72}$$

$$= 4{\cdot}7.$$

This result is highly significant and the die is biased.

Alternatively, instead of working with r, it may be easier to work with the proportion $\pi = r/n$, then as

$$r \to N(np, npq),$$

$$\pi \to N(p, pq/n),$$

$$\text{S.E. } (\pi) = \sqrt{(pq/n)}$$

and

$$X = \frac{\pi - p}{\sqrt{(pq/n)}}$$

$$\to N(0, 1)$$

which may be tested as before.

Example (ii) Using the data of the previous example and the proportion π, find whether the die is biased.

Solution Here

$$\pi = \frac{r}{n} = \frac{3\,210}{9\,000} = 0.356\,7.$$

On the N.H. that $p = \frac{1}{3} = 0.333\,3$,

$$|\pi - p| = 0.023\,4$$

and

$$\text{S.E. } (\pi) = \sqrt{(pq/n)}$$

$$= 0.004\,97.$$

Hence

$$X = \frac{0.023\,4}{0.004\,97}$$

$$= 4.7,$$

as before.

16.5.2 Comparison of two large samples

Suppose that there are two samples of n_1 and n_2 observations where n_1 and n_2 are large, and that in the first sample r_1 out of n_1 have the attribute A, and in the second sample r_2 out of n_2 have this attribute. Then it is possible to test whether these two samples come from the same population, or whether the probability, p, that an individual has the attribute A is the same for the two populations from which the samples are drawn. For, on the null hypothesis that p is the same for both populations, let $\pi_1 = r_1/n_1$ and $\pi_2 = r_2/n_2$, then approximately,

$$\pi_1 = N(p, pq/n_1),$$

$$\pi_2 = N(p, pq/n_2).$$

Hence,

$$\pi_1 - \pi_2 = N\left(0, pq\left(\frac{1}{n_1} + \frac{1}{n_2}\right)\right)$$

and the observed difference $\pi_1 - \pi_2$ is compared with the theoretical difference zero, using

$$X = \frac{|\pi_1 - \pi_2|}{\text{S.E. } (\pi_1 - \pi_2)}.$$

If p, q are unknown, a pooled estimate $\hat{p}$ of p is used, where

$$\hat{p} = \frac{r_1 + r_2}{n_1 + n_2}$$

$$= \frac{n_1\pi_1 + n_2\pi_2}{n_1 + n_2}$$

and

$$\hat{q} = 1 - \hat{p}.$$

Example In a test involving the safe driving of nineteen-year old motor cycle owners driving approximately the same mileage, it was found that in one large city 600 out of a sample of 800 had no accident in one given year, while in another city in the same year, 700 out of a sample of 1 200 drivers had no accident. Do these results show that one city has a better safety record than the other for this class of driver?

Solution The N.H. is that the proportion of safe drivers is the same for both cities.

The relevant calculation is set out below:

$n_1 = 800$, $r_1 = 600$, $\pi_1 = \frac{3}{4}$,

$n_2 = 1\ 200$, $r_2 = 700$, $\pi_2 = 7/12$,

$\hat{p} = (600 + 700)/(800 + 1\ 200) = 13/20$,

$\hat{q} = 7/20$,

$\pi_1 - \pi_2 = 1/6$,

S.E. $(\pi_1 - \pi_2) = \sqrt{(13 \times 7)/(20 \times 20 \times 480)} = 0.021\ 77$

and

$$X = \frac{0.166\ 7}{0.021\ 77} > 7.$$

The result is highly significant, and the null hypothesis is rejected. That is the proportion of safe drivers in the two cities is different.

16.6 SMALL SAMPLE THEORY

Tests of significance for a large sample are constructed by comparing the modulus of the difference between observed and expected values of a statistic with the standard error, since the distribution of most statistics tends to normality as the number of individuals in the sample increases. For small samples most distributions of statistics are far from normal, in particular estimates of parameters obtained from small samples are not reliable. For example, substituting the sample variance s^2 for the unknown population variance σ^2 affects the distribution seriously. It is necessary to construct exact tests which are valid for small samples (i.e. samples of between 5 and 50 individuals). A small sample test which is exact and which is valid for samples of all sizes is *Student's* t-*test*, first derived by W. S. Gossett (1908) who wrote under the name 'Student'.

16.6.1 The t-distribution

Suppose x_i, $i = 1, 2, 3, \ldots, n$ is a random sample of n observations from a population whose mean is μ and whose variance is σ^2. If the population is normal then the distribution of $\bar{x}$ is $N(\mu, \sigma^2/n)$. If σ^2 is unknown it is estimated by

$$\hat{s}^2 = \{\Sigma(x_i - \bar{x})^2\}/(n - 1), \qquad (7)$$

using equation (5) of section 16.2.1. Then $(\bar{x} - \mu)/(\hat{s}/\sqrt{n})$ no longer has a normal distribution, although its distribution tends to normality as n increases. The exact distribution of $(\bar{x} - \mu)/(\hat{s}/\sqrt{n})$ is known and is called the t-distribution with $n - 1$ *degrees of freedom*. The n variables in equation (7) are not independent since they are connected by the relation $\Sigma(x_i - \bar{x}) = 0$, so that the number of independent variables is reduced from n to $(n - 1)$.

The t-distribution is tabulated in Table 3 of *Cambridge Elementary Statistical Tables*, Lindley and Miller, and a short extract from this is given in Table 16.3 below.

Number of degrees of freedom	Significance levels	
	5%	1%
1	12·71	63·66
5	2·57	4·03
10	2·23	3·17
15	2·13	2·95
∞	1·96	2·58

Table 16.3

It is usual to write

$$t_{n-1} = (\bar{x} - \mu)/(\hat{s}/\sqrt{n}).$$

Note that, as $n \to \infty$, $t_n \to N(0, 1)$.

16.6.2 Testing a mean, variance unknown

To test whether the sample of n observations $x_1, x_2, x_3, \ldots, x_n$ is drawn from a normal population of mean μ, the procedure is as follows. The N.H. is that the population mean is μ, the sample mean $\bar{x}$ is $(\Sigma x_i)/n$, the estimate of the variance $\hat{s}^2$ is given by

$$\hat{s}^2 = \{\Sigma(x_i - \bar{x})^2\}/(n - 1)$$
$$= \frac{1}{n - 1}\left[\Sigma x_i^2 - \frac{(\Sigma x_i)^2}{n}\right],$$

and $t_{n-1} = (\bar{x} - \mu)/(\hat{s}/\sqrt{n})$ has a t-distribution with $n - 1$ degrees of freedom.

Let t_α be the α per cent point of the t-distribution with $n - 1$ degrees of freedom, i.e.

$$P\{|t| \geqslant t_\alpha\} = \alpha/100.$$

Then if the modulus of the observed value of t is greater than or equal to t_α, the mean $\bar{x}$ differs significantly from μ at α per cent. The value of α is usually taken to be 5.

Example A sample of ten observations from a normal population has values

31·31	32·14	31·14	31·22	28·87
29·37	32·37	30·70	30·41	30·07.

Test whether this sample was drawn from a population whose mean is 30.

Solution The N.H. is that the population mean μ is 30. From the data, $\bar{x} = 30\cdot76$, $n = 10$ so that $n - 1$ which is the number of degrees of freedom, is 9. Also the estimate of variance $\hat{s}^2$ is given by

$$\hat{s}^2 = [\Sigma(x - \bar{x})^2]/9$$
$$= [\Sigma x^2 - (\Sigma x)^2/10]/9$$
$$= 1\cdot251\ 3.$$

Hence

$$\hat{s} = 1\cdot118\ 6,$$

and

$$\hat{s}/\sqrt{n} = 1\cdot118\ 6/\sqrt{10}$$
$$= 0\cdot353\ 7.$$

Hence,

$$t_9 = (\bar{x} - \mu)/(\hat{s}/\sqrt{n})$$
$$= (0\cdot76)/0\cdot353\ 7$$
$$= 2\cdot149.$$

From tables the 5 per cent point of t_9 is 2·26. The observed value of t_9 is 2·149 which is less than 2·26. Hence the result is not significant at the 5 per cent level and there is no reason to reject the null hypothesis.

The α per cent *confidence interval* for μ when μ is unknown is

$$\bar{x} - t_\alpha \hat{s}/\sqrt{n}, \quad \bar{x} + t_\alpha \hat{s}/\sqrt{n}$$

where t_α is the α per cent point of t with $n - 1$ degrees of freedom. Thus in the above example the 95 per cent confidence interval for μ is $30\cdot76 - 2\cdot26 \times 0\cdot353\ 7$, $30\cdot76 + 2\cdot26 \times 0\cdot353\ 7$ which equal 29·96, 31·56, i.e. the mean of the population lies between 29·96 and 31·56 at the 95 per cent level of significance.

16.6.3 Testing the difference between two means

Given two independent samples of n_1 and n_2 observations, the t-distribution is used to test whether the two samples may be regarded as drawn from the same normal population, assumed to be $N(\mu, \sigma^2)$. Let $\bar{x}_1$ and $\bar{x}_2$ be the means of the samples, then

$$\bar{x}_1 = N(\mu, \sigma^2/n_1),$$
$$\bar{x}_2 = N(\mu, \sigma^2/n_2)$$

and $$\bar{x}_1 - \bar{x}_2 = N(0, \sigma^2(1/n_1 + 1/n_2)).$$

To estimate σ^2 it is necessary to use a pooled estimate $\hat{s}^2$ using data from both samples, where

$$\hat{s}^2 = \frac{[\Sigma(x_1 - \bar{x}_1)^2 + \Sigma(x_2 - \bar{x}_2)^2]}{(n_1 - 1) + (n_2 - 1)}$$
$$= \frac{(n_1 - 1)s_1^2 + (n_2 - 1)s_2^2}{n_1 + n_2 - 2}. \tag{8}$$

Then $$\text{S.E. } (\bar{x}_1 - \bar{x}_2) = \hat{s}\sqrt{(1/n_1 + 1/n_2)}. \tag{9}$$

From the null hypothesis that both samples are drawn from the same population,

$$t = \frac{|\bar{x}_1 - \bar{x}_2|}{\text{S.E. } (\bar{x}_1 - \bar{x}_2)}$$

has a t-distribution with $n_1 + n_2 - 2$ degrees of freedom and can be tested as before.

Example The measurements of a sample of 10 components from one machine has mean 60 cm and an estimated variance of 49 cm². A sample of 16 components from a different machine has a mean 50 cm and an estimate of variance 100 cm². Test whether both samples are drawn from the same population.

Solution The N.H. is that there is no difference in the means of the two samples. Then with $n_1 = 10$, $\bar{x}_1 = 60$, $s_1^2 = 49$, $n_2 = 16$, $\bar{x}_2 = 50$ and $s_2^2 = 100$, the pooled estimate of variance is, using equation (8),

$$\hat{s}^2 = (9 \times 49 + 15 \times 100)/24$$
$$= 80.875.$$

Hence $\hat{s} = 8.994$ and, using equation (9),

$$\text{S.E. } (\bar{x}_1 - \bar{x}_2) = 8.994\sqrt{(1/10 + 1/16)}$$
$$= 3.625.$$

Hence $$t = |\bar{x}_1 - \bar{x}_2|/\text{S.E. } (\bar{x}_1 - \bar{x}_2)$$
$$= 10/3.625$$
$$= 2.759$$

with 24 degrees of freedom.

From tables the value of t for 24 degrees of freedom at the 5 per cent level of significance is 2·06 which is less than the observed value of 2·759. Hence the result is significant, there is a significant difference in the means and the null hypothesis is rejected.

16.6.4 Paired observations

It may be possible to arrange a controlled experiment so that two samples each of the same number of observations are deliberately paired.

Suppose x_i and y_i, $i = 1, 2, 3, \ldots, n$ are two sets of observations which can be paired and that their means are $\bar{x}$ and $\bar{y}$ respectively, and let $z_i = x_i - y_i$, $i = 1, 2, 3, \ldots, n$. The null hypothesis is that there is no difference in the populations from which the samples are taken. Hence if $\bar{x} = N(\mu, \sigma^2/n)$, $\bar{y} = N(\mu, \sigma^2/n)$, then $\bar{z} = \bar{x} - \bar{y} = N(0, \sigma_z^2)$ where σ_z^2 is estimated by $s_z^2 = [\Sigma(z - \bar{z})^2]/(n - 1)$. The statistic $t = |\bar{z} - 0|/(s_z/\sqrt{n})$ has $n - 1$ degrees of freedom and may be tested as before.

Example To compare the effect of different fertilizers A and B on the growth of tomatoes, a controlled experiment was performed using six different positions in a greenhouse and the yield in pounds weight was as follows:

Position	1	2	3	4	5	6
$A(x)$	13·2	11·1	12·2	12·6	10·4	14·0
$B(y)$	12·9	11·4	10·3	10·8	9·8	13·5

Test whether one fertilizer is significantly better than the other.

Solution The N.H. is that there is no significant difference between A and B. Here

$$z = x - y \mid 0·3 \quad -0·3 \quad 1·9 \quad 1·8 \quad 0·6 \quad 0·5,$$

and $\Sigma z_i = 4·8$, $\bar{z} = 0·8$, $\Sigma z_i^2 = 7·64$ and $(\Sigma z_i)^2 = 23·04$, so that

$$s_z^2 = \tfrac{1}{5}[7·64 - \tfrac{1}{6}(23·04)] = 0·76.$$

Hence $s_z = 0·871\,8$, $s_z/\sqrt{n} = 0·355\,8$ and finally

$$t = 2·252. \quad (= 0·8/0·355\,8)$$

From tables, $t = 2·57$ for 5 degrees of freedom so that the difference is almost significant at the 5 per cent level, and there is insufficient evidence from the data to show whether one fertilizer is better than the other.

16.7 THE CHI-SQUARE DISTRIBUTION

Let X_1, X_2, X_3, . . ., X_n each be normally distributed with mean zero and unit variance. Then $U = X_1^2 + X_2^2 + X_3^2 + \ldots + X_n^2$ has a known distribution called the *Chi-square distribution*, written χ^2 distribution, with n degrees of freedom. This distribution which is continuous is tabulated and Table 5 of *Cambridge Elementary Statistical Tables* gives the percentage points of the χ^2 distribution. An extract from this table is given in section 16.7.3.

16.7.1 Properties of the χ^2 distribution

(i) If U_1, U_2, U_3, . . ., U_m have independent χ^2 distributions with n_1, n_2, n_3, . . ., n_m degrees of freedom, then $U_1 + U_2 + U_3 + \ldots + U_m$ has a χ^2 distribution with $n = n_1 + n_2 + n_3 + \ldots + n_m$ degrees of freedom.

(ii) If the X_i are not independent but satisfy k linear equations of the form

$$a_{i1}X_1 + a_{i2}X_2 + a_{i3}X_3 + \ldots + a_{in}X_n = 0, \quad i = 1, 2, 3, \ldots, k,$$

these k equations give k constraints to the system and it can be shown that $X_1^2 + X_2^2 + X_3^2 + \ldots + X_n^2$ now has a χ^2 distribution with $n - k$ degrees of freedom.

16.7.2 Distribution of sample variance

Let x_1, x_2, x_3, . . ., x_n be a random sample of n independent values from a normal population of variance σ^2 and let $\bar{x}$ be the sample mean. Then the sample variance s^2 is given by

$$s^2 = [\Sigma(x_i - \bar{x})^2]/n.$$

Hence
$$\frac{ns^2}{\sigma^2} = \sum \frac{(x_i - \bar{x})^2}{\sigma^2},$$

and since $(x_i - \bar{x})/\sigma$ is approximately $N(0, 1)$, then ns^2/σ^2 has a χ^2 distribution with one constraint, for $[\Sigma(x_i - \bar{x})]/\sigma = 0$. Hence ns^2/σ^2 has a χ^2 distribution with $(n - 1)$ degrees of freedom.

16.7.3 Short table of χ^2

Degrees of freedom	Significance level	
	5%	1%
15	25·00	30·58
10	18·31	23·21
5	11·07	15·09
1	3·84	6·63

Table 16.4

Notice that if $X = N(0, 1)$, $\chi_1^2 = X^2$ and

$$P\{|X| > 1\cdot96\} = 5 \text{ per cent},$$

or
$$P\{X^2 > 1\cdot96^2\} = 5 \text{ per cent}$$

i.e.
$$P\{X^2 > 3\cdot84\} = 5 \text{ per cent},$$

or
$$P\{\chi_1^2 > 3\cdot84\} = 5 \text{ per cent}.$$

Example A random sample of eleven independent measurements gave an unbiased estimate of s^2 equal to $9\cdot65$ mm². Could the sample be reasonably regarded as drawn from a normal population with variance $5\cdot5$ mm²?

Solution The N.H. is that the variance σ^2 is $5\cdot5$. From the given data, $ns^2/\sigma^2 = 11 \times 9\cdot65/5\cdot5 = 19\cdot30$, and from tables of χ^2 with 10 degrees of freedom,

$$P\{\chi^2 > 19\cdot30\} < 5 \text{ per cent}.$$

Hence the value found is significant and the null hypothesis is rejected. The sample is not regarded as drawn from a normal population with variance $5\cdot5$ mm².

16.7.4 Application of χ^2 to goodness of fit

Earlier in this chapter methods of testing parameters were shown on the assumption that the distributions selected were correct. The aim now is to test the form of the distribution assumed. On the assumption that a certain distribution is correct, it is possible to evaluate expected frequencies and compare these with observed frequencies. Let E_i, $i = 1, 2, 3, \ldots, k$ and O_i, $i = 1, 2, 3, \ldots, k$ be the expected and observed frequencies for k possible outcomes of an experiment. Then the measure of discrepancy U say, between the observed and expected frequencies is chosen to be

$$U = \sum_{i=1}^{k} \frac{(O_i - E_i)^2}{E_i}.$$

If O and E differ, U will be large and it is necessary to have a significance test for U. For large numbers of repeats of the experiment, the distribution of U approaches a χ^2 distribution with $k - 1$ degrees of freedom.

Example (i) A die is rolled 120 times. For an unbiased die each face occurs with the probability $\frac{1}{6}$ so that each face is expected to occur 20 times. The experiment gave the result in the following table which shows both observed and expected frequencies:

Score	1	2	3	4	5	6
Observed frequency	15	18	30	16	24	17
Expected frequency	20	20	20	20	20	20

Are the data consistent with the expected frequency?

Solution The N.H. is that the die is unbiased. Then

$$\sum \frac{(O-E)^2}{E} = \frac{(-5)^2}{20} + \frac{(-2)^2}{20} + \frac{10^2}{20} + \frac{(-4)^2}{20} + \frac{4^2}{20} + \frac{(-3)^2}{20}$$
$$= 8 \cdot 50,$$

and there are $6 - 1 = 5$ degrees of freedom.

From tables, $\chi^2_5 = 11 \cdot 07$ at the 5 per cent level, and the observed value of $8 \cdot 5$ is less than $11 \cdot 07$ and the result is not significant. There is no reason to suspect the bias of the die.

Example (ii) In a certain manufacturing process any one defective item occurs randomly in time and is independent of the time any other defective occurs. Records give the frequency of occurrence O_x of x defects in one day as

x	0	1	2	3	4	5
O_x	305	365	210	80	28	12

Test the goodness of fit of a Poisson distribution whose mean is $1 \cdot 2$.

Solution The N.H. is that the distribution is a Poisson distribution whose mean is $1 \cdot 2$. Expected frequencies E_x are given by

$$E_x = [1\,000 \exp(-1 \cdot 2)](1 \cdot 2)^x/x!,$$

and the values in the following table are calculated:

x	O_x	E_x	$O-E$	$(O-E)^2$	$(O-E)^2/E$
0	305	301·2	3·8	14·44	0·047 94
1	365	361·4	3·6	12·96	0·035 86
2	210	216·8	−6·8	46·24	0·213 28
3	80	86·7	−6·7	44·89	0·517 76
4	28	26·0	2·0	4·00	0·153 84
5	12	7·6	4·4	19·36	2·547 37
				Total	3·516 05

Table 16.5

Thus χ^2 is $3 \cdot 5$ and the number of degrees of freedom is $6 - 1$ or 5. From tables the 5 per cent level of χ^2 is $11 \cdot 07$ which is larger than the observed value of $3 \cdot 5$. Hence the result is not significant and there is no reason to reject the null hypothesis. That is, in this case, there is no reason to believe that the defects are occurring other than at random, and from this data there is no evidence of a fault in production.

16.7.5 Modifications

(i) When the population distribution is not known completely, and it is required to test the hypothesis that a sample is drawn from a specified distribution the parameters of which are not known, it is necessary to estimate the parameters using observed data from the sample. In this case the resulting χ^2 is tested as above but the number of degrees of freedom is reduced by the number of parameters estimated. That is, the number of degrees of freedom is given by

(the number of classes) $-$ 1 $-$ (the number of parameters estimated).

(ii) If the expected frequency E_i in any one class is less than five, it is necessary to combine class results as follows, for example,

Class	1	2	3	4	5	6
O_i	5	6	13	18	17	16
E_i	2	3	15	20	17	15

The first and second classes are combined to give

Class	1 and 2	3	4	5	6
O_i	11	13	18	17	16
E_i	5	15	20	17	15

The number of degrees of freedom is now

$$5 - 1 - \text{(the number of parameters estimated)}.$$

16.7.6 Contingency tables or test of independence

Another useful application of the χ^2 test occurs when one sample is classified according to two factors, and it is required to know whether these factors are independent. It is usual to represent the data in a *contingency table*.

Example A group of 143 students is tested in mathematics and physics and graded α, β, γ in each subject. Use the χ^2 test to test the hypothesis that the grades in mathematics and physics are independent. The contingency table for this is given in Table 16.6.

		Mathematics Grades (B)			
		α	β	γ	Totals
Physics grades (A)	α	33	13	5	51
	β	6	49	11	66
	γ	2	3	21	26
	Totals	41	65	37	143

Table 16.6

Before considering this particular example, consider a general contingency table containing r rows and s columns. Suppose factor A has

r levels $A_1, A_2, A_3, \ldots, A_r$ and factor B has s levels $B_1, B_2, B_3, \ldots, B_s$ and let there be n_{ij} observations in class (A_i, B_j) and the contingency table is shown in Table 16.7.

	B_1	B_2	B_3	$\ldots$	B_j	$\ldots$	B_s	Totals
A_1	n_{11}	n_{12}	n_{13}	$\ldots$	n_{1j}	$\ldots$	n_{1s}	$n_{1.} = \sum\limits_{j} n_{1j}$
A_2	n_{21}	n_{22}	n_{23}	$\ldots$	n_{2j}	$\ldots$	n_{2s}	$n_{2.} = \sum\limits_{j} n_{2j}$
A_3	n_{31}	n_{32}	n_{33}	$\ldots$	n_{3j}	$\ldots$	n_{3s}	$n_{3.} = \sum\limits_{j} n_{3j}$
.								
.								
A_i	n_{i1}	n_{i2}	n_{i3}	$\ldots$	n_{ij}	$\ldots$	n_{is}	$n_{i.} = \sum\limits_{j} n_{ij}$
.								
.								
A_r	n_{r1}	n_{r2}	n_{r3}	$\ldots$	n_{rj}	$\ldots$	n_{rs}	$n_{r.} = \sum\limits_{j} n_{rj}$
Totals	$n_{.1}$ $=$ $\sum\limits_{i} n_{i1}$	$n_{.2}$ $=$ $\sum\limits_{i} n_{i2}$	$n_{.3}$ $=$ $\sum\limits_{i} n_{i3}$	$\ldots$	$n_{.j}$ $=$ $\sum\limits_{i} n_{ij}$	$\ldots$	$n_{.s}$ $=$ $\sum\limits_{i} n_{is}$	n

Table 16.7

On the null hypothesis that the two factors A and B are independent, the probability that an observation falls in class (A_i, B_j) is the product of the probability that it falls in A_i and the probability that it falls in B_j, or $p_{ij} = (p_{i.} \, p_{.j})$. Hence expected frequencies

$$E_{ij} = np_{ij} = np_{i.} \, p_{.j}$$

and

$$\chi^2 = \Sigma(O - E)^2/E$$

$$= \sum_{i}\sum_{j} \frac{(n_{ij} - np_{i.} \, p_{.j})^2}{np_{i.} \, p_{.j}}$$

with $r \times s - 1$ degrees of freedom.

In general $p_{i.}$ and $p_{.j}$ are unknown parameters and are estimated by

$$\hat{p}_{i.} = \text{proportion of observations in } A_i$$
$$= n_{i.}/n,$$
$$\hat{p}_{.j} = \text{proportion of observations in } B_j$$
$$= n_{.j}/n,$$

where $n_{i.} = \sum\limits_{j} n_{ij}$ and $n_{.j} = \sum\limits_{i} n_{ij}$, and finally

$$\chi^2 = \sum_{i}\sum_{j} \frac{(n_{ij} - n_{i.} \, n_{.j}/n)^2}{n_{i.} \, n_{.j}/n}. \tag{10}$$

Since $\sum\limits_{i=1}^{r} p_i. = 1$ and $\sum\limits_{j=1}^{s} p._j = 1$, the total number of degrees of freedom is reduced by one for each parameter estimated so that there are $r - 1 + s - 1$ parameters estimated. Hence the number of degrees of freedom for the contingency table is

$$rs - 1 - (r - 1) - (s - 1) = (r - 1)(s - 1).$$

Hence testing independence in Table 16.6,

$$E(\alpha_p \alpha_m) = 51 \times 41/143 = 14 \cdot 62,$$

$$E(\alpha_p \beta_m) = 51 \times 65/143 = 23 \cdot 18,$$

$$E(\alpha_p \gamma_m) = 51 \times 37/143 = 13 \cdot 20, \text{ etc.}$$

where α_p means a grade α in physics, etc., and

$$\chi^2 = \frac{(33 - 14 \cdot 62)^2}{14 \cdot 62} + \frac{(13 - 23 \cdot 18)^2}{23 \cdot 18} + \frac{(5 - 13 \cdot 20)^2}{13 \cdot 20} + \cdots$$

$$+ \frac{(21 - 6 \cdot 728)^2}{6 \cdot 728}$$

$$= 106 \cdot 52.$$

Since $\chi^2 = 9 \cdot 49$ for $(3 - 1)(3 - 1) = 4$ degrees of freedom at the 5 per cent level, this result is highly significant and the hypothesis of independence is rejected.

Special case of 2 × 2 contingency table

The χ^2 test is an approximation which improves as the number of *cells* in the contingency table increases. In the special case of 2 × 2 tables (i.e. $r = s = 2$) the approximation is improved by using Yate's continuity correction. The conventional notation for a 2 × 2 contingency table is shown in Table 16.8.

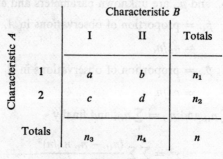

		Characteristic B		
		I	II	Totals
Characteristic A	1	a	b	n_1
	2	c	d	n_2
	Totals	n_3	n_4	n

Table 16.8

To apply Yates' correction, decrease by $\frac{1}{2}$ those $O > E$ and increase by $\frac{1}{2}$ those $O < E$ and suppose the corrected table is

	a'	b'	n_1
	c'	d'	n_2
	n_3	n_4	n

Then it may be shown that an equivalent expression for χ^2 is

$$\chi^2 = \frac{n(a'd' - b'c')^2}{n_1 n_2 n_3 n_4}$$

Example It is thought that people who wear spectacles for reading have jobs which involve much reading. Two hundred people were investigated and the following table records the result.

	S	Non-S	Totals
R	74	31	105
Non-R	46	49	95
Totals	120	80	200

Test the independence of the two factors.

Solution The N.H. is that the two factors are independent.

$$E(R, S) = 105 \times 120/200 = 63. \quad O > E,$$

hence decrease 74 by $\frac{1}{2}$.

$$E(R, \text{Non-}S) = 105 \times 80/200 = 42, \quad O < E,$$

hence increase 31 by $\frac{1}{2}$.

$$E(\text{Non-}R, S) = 95 \times 120/200 = 57, \quad O < E,$$

hence increase 46 by $\frac{1}{2}$.

$$E(\text{Non-}R, \text{Non-}S) = 95 \times 80/200 = 38, \quad O > E,$$

hence decrease 49 by $\frac{1}{2}$.

The corrected table is

	S	Non-S	Totals
R	73·5	31·5	105
Non-R	46·5	48·5	95
Totals	120	80	200

From the table,

$$\chi^2 = \frac{(73 \cdot 5 \times 48 \cdot 5 - 46 \cdot 5 \times 31 \cdot 5)^2}{120 \times 80 \times 105 \times 95} \times 200$$

$$= 9 \cdot 219.$$

From tables of χ^2, the 5 per cent level of χ^2 with one degree of freedom is 3·84.

The observed value of χ^2 is 9·219 which is greater than 3·84, and the result is highly significant and there is no reason to suppose independence of the two factors in this group of people.

16.7.7 Test of homogeneity

The χ^2 test may be used to determine whether several samples are drawn from the same population or populations having the same distribution. The precise nature of the distribution is unimportant.

Suppose s different samples are drawn and a given characteristic observed for each one, and let there be r classes of this characteristic. Let n_{ij} be the frequency with which the ith class of the characteristic occurs in the jth sample and the results be recorded in a table of the same form as Table 16.7, where B_j is the jth sample and A_i is the ith class of characteristics. To test the null hypothesis that the s samples were drawn from populations having the same distribution, the method of evaluating χ^2 is exactly the same as that of section 16.7.6 even though the mathematical model is different in that the column totals are not random but are the stated sample sizes. The number of degrees of freedom is now $(r - 1)(s - 1)$ also. For, referring to the notation of Table 16.7, for the first sample,

$$\chi^2_{(1st)} = \sum_{1st\ sample} (n_{i1} - n_{\cdot 1} p_i)^2 / n_{\cdot 1} p_i$$

or $(r - 1)$ degrees of freedom.

For the jth sample,

$$\chi^2_{(jth\ sample)} = \sum_{jth\ sample} (n_{ij} - n_{\cdot j} p_i)^2 / n_{\cdot j} p_i$$

with $(r - 1)$ degrees of freedom.

Summing over all $j = 1, 2, 3, \ldots, s$,

$$\chi^2 = \sum_i \sum_j \frac{(n_{ij} - n_{\cdot j} p_i)^2}{n_{\cdot j} p_i}$$

with $s(r - 1)$ degrees of freedom.

If now p_i, $i = 1, 2, 3, \ldots, r$ is estimated by the corresponding relative frequency $n_i./n$,

$$\chi^2 = \sum_i \sum_j \frac{(n_{ij} - n._j\, n_i./n)^2}{n._j\, n_i./n}$$

with $s(r - 1) - (r - 1)$ degrees of freedom which is $(r - 1)(s - 1)$ degrees of freedom.

Hence the procedure is:

(i) find the expected frequencies $E_{ij} = n._j\, n_i./n$,

(ii) calculate χ^2, and

(iii) refer to the χ^2 distribution with $(r - 1)(s - 1)$ degrees of freedom.

Example The distributions of breaking tension levels of three different substances are as follows:

	Sub I	Sub II	Sub III	Totals
T_1	140	161	133	434
T_2	79	63	82	224
T_3	31	36	25	92
Totals	250	260	240	750

Is the distribution of breaking tension level the same for the three substances?

Solution The N.H. is that the breaking tension level is the same, i.e. all substances are from populations with the same distribution. The expected values are,

$$E(I, T_1) = 250 \times 434/750$$
$$= 144.67,$$
$$E(I, T_2) = 74.67,$$
$$E(I, T_3) = 30.67 \text{ etc.}$$

and

$$\chi^2 = \frac{4.67^2}{144.67} + \frac{4.33^2}{74.67} + \frac{0.33^2}{30.67} + \cdots + \frac{4.44^2}{29.44}$$
$$= 8.279.$$

The number of degrees of freedom is given by $(3 - 1)(3 - 1) = 4$. From tables, χ^2 is 9.49 which is greater than the observed value of 8.28, and hence the result is not significant at the 5 per cent level. From the given data there is no reason to doubt the null hypothesis.

EXERCISE 16

1 If x is normally distributed with mean $\mu = 10$ and variance $\sigma^2 = 9$, calculate the probability that (i) $x > 11$, and (ii) if a random sample of size 16 has mean $\bar{x}$, that $\bar{x} > 11$.

2 A machine produces circular parts whose diameters are known to be normally distributed with standard deviation 0·019 m and the mean diameter is supposed to be 0·500 m. A random sample of ten is taken and the sample mean is found to be 0·488 m. Does the data from the sample indicate a change should be made in the machine?

3 The mean range of a certain type of gun is 1 000 metre and the range standard deviation is 124 metre. After two years' storage fifty rounds are fired and the mean range is found to be 872 metre. Test whether storage has changed the mean range.

4 A firm claims that 88 per cent of the steel rods they produce have a breaking tension of more than 1 000 newton. A sample of 300 rods was tested and only 249 rods had this breaking tension. Does this discredit the firm's claim?

5 The diameters of a set of ball bearings are assumed to be normally distributed with a standard deviation of 0·1 cm. What size sample should be taken in order that the estimate of the mean diameter of the ball bearings shall differ from the population mean by more than 0·005 cm only once in 100 times?.

6 In a certain county in one year 90 boys entered school A and 90 boys entered school B, after the results of tests in Mathematics and English. The mean of the marks of the boys entering school A was 63 and that of the boys entering school B was 60. If the variance was 144 in both cases, is the claim that the entry requirements for both schools is the same, justified by these data? After one year the same boys were each given another common examination. In school A, 10 boys had left the school, so that 80 boys had an average mark of 61·5 with variance 900 and the 90 boys of school B had an average mark of 65 with variance 576. Can any conclusion be drawn about the superiority of school B?

7 Two researchers X and Y each tested the growth of a sample of fifty plants of the same species over a fixed period of time. The following table shows their results:

	X	Y
Mean growth in cm	18·04	19·38
Variance in cm²	8·85	9·12

Does the smaller variance for X imply he is a better researcher than Y? Determine whether the difference in the means is significant (i) at the 5 per cent level, (ii) at the 1 per cent level.

8 Two classes take the same test. Class A consists of six students and class B consists of eight students. The marks awarded are

Class A	830	829	827	828	826	826		
Class B	828	823	824	827	824	826	824	823

Test whether the mean marks of classes A and B are significantly different.

9 Tests showed that of 100 women drivers and 200 men drivers, 30 of the women and 75 of the men were not good drivers in that they failed the tests. Use these data to test the claim that women are not such good drivers as men.

10 A sleeping drug and a harmless control tablet were tested in turn on nine patients in hospital and the number of hours' sleep recorded as follows:

Patient number	1	2	3	4	5	6	7	8	9
Under sleeping drug	9·5	7·7	9·1	5·3	6·7	10·4	7·8	9·9	8·7
Under control drug	7·5	7·4	9·5	5·0	5·3	9·0	7·2	8·1	8·9

Is the sleeping drug effective?

11 Nine students were tested before and after being coached in mathematics and their marks were

Student number	1	2	3	4	5	6	7	8	9
Before coaching	55	61	63	47	76	45	48	60	62
After coaching	62	64	62	51	73	50	54	56	63

Do these data show that the coaching was effective?

12 The number of absences due to sickness in two factories over a period of six months was recorded as a percentage with the following result:

Month	Oct.	Nov.	Dec.	Jan.	Feb.	Mar.
Factory A	39	48	57	63	71	78
Factory B	45	51	66	74	82	73

Use the t-test to examine whether there is any significant difference in these two sets of data.

13 Two machines are used to make identical rods. Random samples of ten parts from machine A and seven from machine B were measured, and the results recorded.

Machine A (mm)	11·1	11·5	11·4	11·0	10·9	11·1	11·2	11·5	11·3	11·4
Machine B (mm)	11·2	10·6	11·1	10·3	11·8	10·9	11·7			

Do these data indicate a significant difference in the mean lengths of the rods from the two machines?

14 Capacitors of nominal value 100 pF are being produced and over a large number 15 per cent are rejected as being not of the required standard. Samples of 5 are taken at regular intervals and tested.

(i) Use the binomial expansion to calculate to three decimal places the probabilities of there being 0, 1, 2, 3, 4, 5 rejects in a sample. (ii) Calculate the probabilities of the various numbers of rejects using the Poisson approximation. (iii) Ten consecutive samples each of five capacitors contained 1, 2, 3, 0, 2, 3, 1, 3, 2, 2 rejects respectively. Use the probabilities calculated in (i) and (ii) to calculate two different values of χ^2 to test whether the apparent increase in the percentage defective is significant.

15 Test the goodness of fit of the following data from 500 trials of an experiment to a Poisson distribution whose estimated mean is 1·3.

x	0	1	2	3	4	5	6
Observed frequency	132	180	121	51	10	5	2

16 The cost per meal in canteens of different size, serving different numbers of meals per day, is listed in the following table. Do these data support the belief that larger canteens produce cheaper meals?

Number of meals served	Cost of meal in new pence				Totals
	under 30	30–40	40–50	over 50	
100–499	49	84	74	26	233
500–999	40	77	132	59	308
Over 999	43	76	140	103	362
Totals	132	237	346	188	903

17 Test the χ^2 goodness of fit of the distribution of problem 18, Exercise 15.

18 Manual dexterity and success in theoretical examinations were graded over a period of years in applied science students. The following data were recorded. Test whether manual dexterity and success in examinations are independently distributed.

Examination	Manual dexterity				
	α	β	γ	δ	ω
α	1 369	171	1 042	399	2
β	2 578	475	2 704	933	12
γ	1 391	421	3 827	1 843	36
δ	455	256	1 849	2 507	113

17
Applications

17.0 INTRODUCTION

In this chapter a number of physical problems are considered and mathematical equations or statements which attempt to describe the physical situation are derived.

The aim is to indicate some of the ways in which the mathematics of the previous Chapters may feature as part of an engineering type problem. To take an example, the hyperbolic functions were introduced, by definition, in section 4.5 and their relationships with the trigonometric functions shown in section 10.6 but the engineering student still may not appreciate the need to study such functions until some practical use is described. Two illustrations of the way in which the hyperbolic functions may feature are given in section 17.2.2 where they appear as the solution of a differential equation obtained from the model of a real situation.

When an engineer is confronted with a problem it frequently happens that any attempt to describe the complete situation in mathematical terms leads to equations of too great a complexity. In this case the way forward is to replace the physical problem with a simpler model in which some of the variables have been ignored. The engineer may often be called on to use his skill and 'know-how' to decide which variables should be retained and which omitted. The illustrations of the following sections may be looked upon as simple models of a more complicated real-life situation. For example, the spring systems of section 17.3.3 are idealised in that, first the damping/frictional effects are not taken into account, and second, the stiffness of each spring is presumed constant whereas it is possible for the stiffness to depend on temperature and therefore on the rate at which the displacements take place. These additional factors (and others) when taken into account lead to mathematical equations which would be excessively complicated and would obscure the basic fact that the eigenvalues can be related to the natural frequencies of oscillation. With this fact well-established the equations could be modified to take the additional factors into account.

From a mathematical standpoint the solution of the mathematical equations is the solution of the problem, but from an engineer's standpoint this solution is only one step towards solving the physical problem. The engineer must be able to interpret the mathematical solution in terms of the physical variables remembering that the mathematical one may have been derived from what was only a model of the real situation. The mathematical solution may sometimes be used to assess the justification of the omission of certain variables.

The problems considered here are simply a selection from a much wider range of problems of the type that the engineer could encounter both in his academic course and in practice.

17.1 INEQUALITIES. LINEAR PROGRAMMING

The engineering world is concerned very much with inequalities. It is rare that a measurement is precise, generally a variable is found to lie within a band of values and the symbols $<$ or $>$ are of frequent occurrence. The student should know the difference between equality and inequality and in certain circumstances this may more readily be observed in a graphical sense. For example the equation

$$2y + 3x = 6$$

may be represented in (x, y) space as a straight line of slope $(-\frac{3}{2})$ having intersects of lengths 2 and 3 on the x and y axes, respectively (see Fig. 17.1).

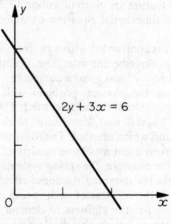

$$2y + 3x = 6$$

Fig. 17.1

The equation

$$2y + 3x < 6$$

however, represents a region of the (x, y) space and is shown by the unshaded region of Fig. 17.2.
Likewise the shaded region is given by $2y + 3x > 6$.

A simple use of such graphical regions to represent inequalities equations and their use in solving a management problem is now given.

Example A factory is to be designed to accommodate two types of machines to make a particular product and the manager has to

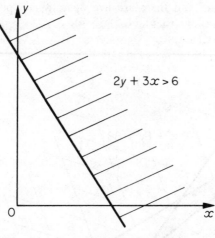

Fig. 17.2

decide how many of each type to install. The specification is as follows:

	Floor area	*Operators*	*Profit units*
Type A	4000	11	9
Type B	7000	8	12

The maximum factory area available is 56,000 and there are 88 operators available to make the product on either machine. The factory must make a profit of not less than 90. How many of each type of machines should be used

(a) to obtain the maximum profit?

(b) to make the minimum number of redundancies?

Solution Suppose x machines of type A and y machines of type B are installed then the following inequalities arise:

Area $\qquad 4x + 7y \leqslant 56$

Employment $\qquad 11x + 8y \leqslant 88$

$$x \geqslant 0, y \geqslant 0.$$

If P is the profit then

$$P = 9x + 12y \geqslant 90.$$

Representing the various inequalities in (x, y) space and shading the unacceptable regions leads to Fig. 17.3. The unshaded region is the 'operational zone' and there are five operational 'points' within the zone, i.e. (1, 7), (2, 6), (3, 6), (4, 5), (5, 4)

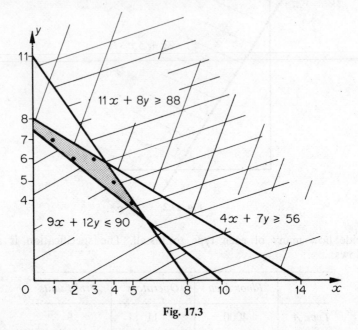

Fig. 17.3

(a) the line $9x + 12y = P$ is parallel to $9x + 12y = 90$ and the profit P will be maximum when the perpendicular distance from the origin to the line

$$P = 9x + 12y$$

is greatest. Naturally it is necessary for the line to pass through one of the operational points (x, y must be integers) and Fig. 17.3 shows that the line must pass through the point (3, 6). Under this condition

$$x = 3, y = 6$$

$$\text{Area} = 54{,}000$$

$$\text{Employed persons} = 33 + 48 = 81$$

$$\text{Profit} = 27 + 72 = 99$$

There are thus, 2,000 units of area not used and 7 operators cannot be employed.

(b) The minimum number of redundancies will appear when the line

$$P = 9x + 12y$$

parallel to $9x + 12y = 90$ in the operational zone, passes through the operational point nearest to the line $11x + 8y = 88$.

From Fig. 17.3 this is observed to be the point (5, 4). Under these conditions

$$x = 5, y = 4$$

$$\text{Area} = 48,000$$

$$\text{Employed persons} = 87,$$

$$\text{Profit} = 93$$

Thus only one person is declared redundant (or is promoted) but 8,000 units of area are unused and the profit is reduced to 93.

Other possibilities appear on examining the Figure. One obvious possibility would be to operate at the point (5, 5) within the profit/ area zone but outside the labour zone. This leads to

$$\text{Area} = 55,000$$

$$\text{Operators} = 95$$

$$\text{Profit} = 105$$

This operational point would utilise practically all the area available, give a larger profit but would require the recruitment of 7 more machine operators. Whether this is a possibility would depend on other factors outwith the present consideration which may in turn lead to a further problem of this type.

This type of problem which leads to a mathematical model as a set of m linear equations with n unknowns together with a target function which requires maximising or minimising is called a *linear programming problem*. When the number of unknowns is greater than 2 the graphical approach is impossible (with 3 unknowns the linear equations are represented by planes in 3-dimensional space and with 4 unknowns there is no graphical representation at all) and it is necessary to resort to other techniques. There are many techniques available with the general heading of linear programming and they usually start by setting-up the mathematical equation in matrix form. If a maximisation of some function is required, this is often achieved by iteration i.e. start with an initial estimate of the variables and then use some process repetitively to improve the estimate until eventually the optimal values are reached.

The applications of linear programming in engineering are many and varied. A distribution problem may be such that different sources $S_1, S_2, S_3, \ldots$ hold stocks $P_1, P_2, P_3 \ldots$ and supply a number of different consumers $C_1, C_2, C_3 \ldots$ whose demands are $D_1, D_2, D_3 \ldots$. In this case the cost of supplying consumer C_j from source S_i may be written A_{ij} and the matrix (A_{ij}) is the *cost matrix*. The problem is to meet the demands of the consumers in the most economical way, i.e. to keep costs to a minimum. Such distribution type problems may be met where there is supply and demand of materials (raw or manufactured) or services (electricity, gas, water).

17.2 EXPONENTIAL FUNCTION. HYPERBOLIC FUNCTIONS. DIFFERENTIAL EQUATIONS

The exponential function e^{kx} was introduced in Chapter 4 as the solution of a particular differential equation

$$\frac{dy}{dx} = ky$$

This is an important differential equation which makes its appearance in many different contexts.

17.2.1 Natural law of growth. The exponential function

A variable is said to satisfy this law of growth if its rate of change is proportional to the value of the variable, i.e. if y is the variable, then

$$\frac{dy}{dt} \propto y$$

Some simple ways in which this law appears will be illustrated.

(i) Under certain conditions, populations of animals, people, or bacteria satisfy this law, i.e. that the rate of change of a population is proportional to the population. With a positive constant of proportionality k the population is then seen to have exponential growth $y = A\,e^{kt}$, where A is constant. The shape of the curve of y is similar to the shape of Fig. 4.1 and increases rapidly for $t > 0$. If this represents the behaviour of the world human population it gives good indication of the rapidity of the increase of population with time. Fortunately, there are indications that the growth law of human population is governed by an equation of the form $dy/dt = ky(c - y)$, where c is a constant. This differential equation may be solved using the method of section 11.1.2 and has the solution

$$y = Ac/(A + e^{-ckt})$$

where A is an arbitrary constant. In this case the population has a limiting value given by

$$\underset{t \to \infty}{\text{Limit}} \, [Ac/(A + e^{-kct})]$$
$$= Ac/A = c.$$

This population growth case is illustrated in Fig. 17.4.

(ii) If a hot body at temperature T is allowed to cool towards the constant temperature T_0 of its surroundings, the law of cooling (as propounded by Newton) states that the rate of change of temperature is proportional to the temperature excess, i.e.

$$\frac{dT}{dt} \propto (T - T_0).$$

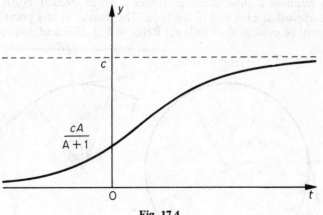

Fig. 17.4

If this is a cooling situation then the rate of change is negative and

$$\frac{dT}{dt} = -k(T - T_0)$$

where k is a positive constant dependent on the material of the body and its surrounding. Replacing $T - T_0$ by θ then

$$\frac{d\theta}{dt} = -k\theta$$

leads to $\theta = T - T_0 = Ae^{-kt}$.

The cooling curve for a body of initial temperature T_1 is illustrated in Fig. 17.5.

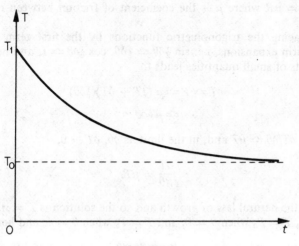

Fig. 17.5

(iii) Suppose a rope is passed round a rough circular bollard and forces applied at each end of the rope. The tension at any point of the rope may be evaluated as follows. Refer to Fig. 17.6 and consider the

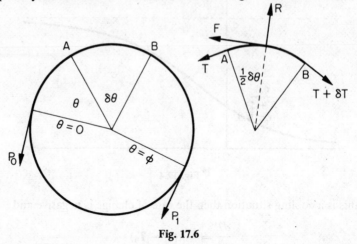

Fig. 17.6

equilibrium of forces acting on the small element AB of the rope. The forces acting are T at point A, $T + \delta T$ at point B, overall reaction of the bollard R, and frictional force F. Hence for the small element of rope AB

$$(T + \delta T) \sin\left(\tfrac{1}{2}\delta\theta\right) + T \sin\left(\tfrac{1}{2}\delta\theta\right) = R$$

$$(T + \delta T) \cos\left(\tfrac{1}{2}\delta\theta\right) - T \cos\left(\tfrac{1}{2}\delta\theta\right) = F.$$

In the case of limiting friction (i.e. where the rope is about to slip) then $F = \mu R$ where μ is the coefficient of friction between rope and bollard.

Replacing the trigonometric functions by the first term of their Maclaurin expansions, i.e. $\sin \tfrac{1}{2}\delta\theta \simeq \tfrac{1}{2}\delta\theta$, $\cos \tfrac{1}{2}\delta\theta \simeq 1$, and neglecting products of small quantities leads to

$$\delta T \simeq F \simeq \mu\,(2T + \delta T)(\tfrac{1}{2}\delta\theta)$$

$$\simeq \mu T \delta\theta$$

so that $\delta T/\delta\theta \simeq \mu T$ and, in the limit as $\delta\theta, \delta T \to 0$,

$$\frac{dT}{d\theta} = \mu T.$$

This is the natural law of growth and so the solution is $T = A\,e^{\mu\theta}$.

Now, $T = P_0$ when $\theta = 0$, and $T = P_1$ when $\theta = \phi$, and hence

$$P_1 = P_0\,e^{\mu\phi}.$$

If two turns are taken around the bollard then $\phi = 4\pi$ and, if the coefficient of friction is taken as $\frac{1}{4}$, then $P_1 = P_0 \, e^{\frac{1}{4}4\pi} = P_0 \, e^{\pi}$, i.e. $P_1 \simeq 23 P_0$. Thus for 2 turns on the bollard a pull at one end will sustain a tension 23 times as large at the other. This is the basis for the ship's capstan and the band-brake.

17.2.2 The hyperbolic function

The hyperbolic functions were introduced in Section 4.5 of Chapter 4 and used in deriving the solution of some differential equations in Chapter 6, the relationship between the trigonometric and the hyperbolic function was shown in Section 10.6 and it is because of the connection with the well-known sine and cosine that the basic hyperbolic functions are called hyperbolic *sine* and hyperbolic *cosine*. The connection with the hyperbola is not quite so obvious.

The curve of intersection of a right-circular cone with a plane is called a *conic section*. The three basic sections are the ellipse, the parabola, and the hyperbola. In Fig. 17.7, the three types of intersection are shown:

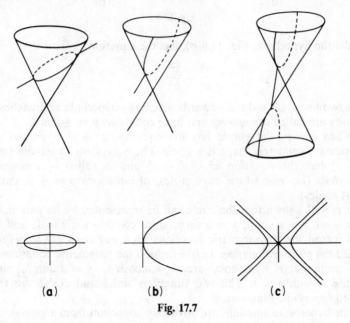

(a)   (b)   (c)

Fig. 17.7

Fig. 17.7(a) shows the section when the plane intersects the cone in a closed curve. This section is called the *ellipse*. Fig. 17.7(b) shows the situation when the plane is parallel to the side of the cone giving an open curve called the *parabola*. The Fig. 17.7(c) shows the intersection curve that has two branches which together form the *hyperbola*.

It is shown in books of analytical geometry that the ellipse, referred to its symmetrical axes (see Fig. 17.8(a)), has an equation

$$\frac{x^2}{a^2} + \frac{y^2}{b^2} = 1$$

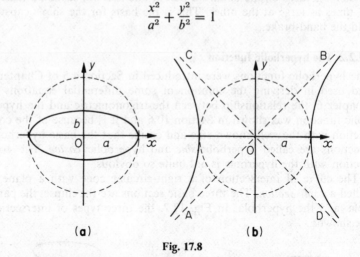

(a)   (b)

Fig. 17.8

whilst the hyperbola, Fig. 17.8(b), has the equation

$$\frac{x^2}{a^2} - \frac{y^2}{b^2} = 1.$$

The two lines AB and CD towards which the hyperbola approaches at infinity are called *asymptotes* and have equations $y = \pm(b/a)x$.

When $b = a$, the ellipse has the equation $x^2 + y^2 = a^2$ and the particular symmetric shape is a *circle*. The equivalent hyperbola (with $b = a$) has the equation $x^2 - y^2 = a^2$ and is called a *rectangular hyperbola* (i.e. one whose asymptotes, of equations $y = \pm x$, cut at right angles).

It is well known that the circle can be represented by its parametric equations $x = a \cos \theta$, $y = a \sin \theta$, since $\cos^2\theta + \sin^2\theta = 1$, and for this reason the trigonometric functions $\sin \theta$ and $\cos \theta$ are sometimes called the *circular functions*. In like fashion the parametric equations of the rectangular hyperbola are $x = a \cosh \phi$, $y = a \sinh \phi$, since $\cosh^2\phi - \sinh^2\phi = 1$. The two functions $\sinh\phi$ and $\cosh\phi$ are then called *hyperbolic functions*.

The hyperbolic functions are obviously important from a geometrical and mathematical viewpoint but they are important functions that also occur in a natural way as the following illustrations show.

(i) The catenary

If a heavy chain is suspended under gravity between two points, the curve that the chain forms is called a *catenary*, e.g. overhead electricity cable, or telegraph wire.

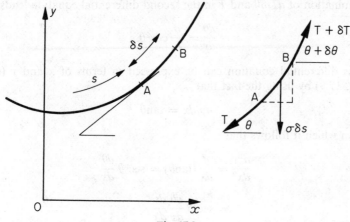

Fig. 17.9

Let s be a distance measurement along the chain and consider the equilibrium of a small portion δs, i.e. AB of the chain. The three forces acting are the weight $\sigma \delta s$, where σ is the weight per unit length, and T, and $T + \delta T$, the tensions at the ends A and B. Resolving forces on AB, horizontally and vertically,

$$(T + \delta T)\cos(\theta + \delta\theta) - T\cos\theta = 0,$$
$$(T + \delta T)\sin(\theta + \delta\theta) - T\sin\theta = \sigma \delta s$$

where $\tan \theta$ is the slope of the catenary at A. Expanding $\sin(\theta + \delta\theta)$ and $\cos(\theta + \delta\theta)$ and using

$$\sin \delta\theta \simeq \delta\theta$$
$$\cos \delta\theta \simeq 1$$

these two equations reduce to

$$\delta T \cos\theta - T \sin\theta \, \delta\theta = 0$$
$$\delta T \sin\theta + T \cos\theta \, \delta\theta = \sigma \delta s$$

after neglecting products of small quantities.

Proceeding to the limit as $\delta\theta$, $\delta T \to 0$ leads to two differential equations

$$\frac{dT}{d\theta} = T \tan\theta$$

$$\sin\theta \frac{dT}{d\theta} = -T\cos\theta + \sigma \frac{ds}{d\theta}$$

The first of these differential equations can be integrated immediately (see Section 11.1.2) since the variable are separable and gives

$$T\cos\theta = \text{constant} = T_0 \text{ (say)}$$

Elimination of $dT/d\theta$ and T in the second differential equation leads to

$$\frac{d\theta}{ds} = \frac{\sigma}{T_0} \cos^2\theta.$$

This differential equation can be expressed in terms of x and y (see Fig. 17.9) by using the fact that

$$dy/dx = \tan\theta$$

from which it follows that

$$\frac{d^2y}{dx^2} = \frac{d}{dx}(\tan\theta) = \sec^2\theta \frac{d\theta}{dx}$$

$$= \sec^2\theta \frac{d\theta}{ds} \frac{ds}{dx}$$

$$= \sec^2\theta \left(\frac{\sigma}{T_0} \cos^2\theta\right) \frac{ds}{dx}$$

i.e.

$$\frac{d^2y}{dx^2} = \frac{\sigma}{T_0} \frac{ds}{dx}$$

Writing $\sigma/T_0 = 1/c$ and using

$$ds/dx = \sec\theta = \sqrt{1 + \tan^2\theta} = \sqrt{1 + (dy/dx)^2}$$

the differential equation for y becomes

$$\frac{d^2y}{dx^2} = \frac{1}{c} \sqrt{1 + (dy/dx)^2}.$$

This may be solved by putting $p = dy/dx$, as in Section 11.2.1, i.e.

$$\frac{dp}{dx} = \frac{1}{c} \sqrt{1 + p^2}$$

i.e.

$$\int \frac{dp}{\sqrt{1 + p^2}} = \frac{1}{c} \int dx$$

Thus

$$\sinh^{-1}(p) = x/c + A$$

or

$$p = \sinh(A + x/c).$$

Suppose that the y-axis of Fig. 17.9 is now made to pass through the lowest point of the catenary then $x = 0$ when $p = dy/dx = 0$, i.e. $A = 0$ and hence

$$p = dy/dx = \sinh(x/c).$$

Integrating again,

$$y = c \cosh(x/c) + B$$

Locate the x-axis at a depth c below the lowest point of the catenary, i.e. $y = c$, $x = 0$, then $B = 0$. Hence $y = c \cosh x/c$ is the equation of the catenary curve with axes as shown in Fig. 17.10.

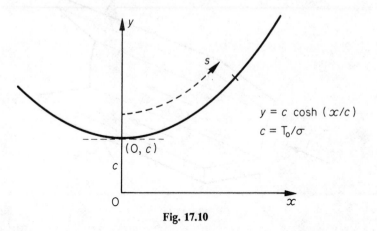

$$y = c \cosh (x/c)$$
$$c = T_0/\sigma$$

$(0, c)$

Fig. 17.10

Thus the catenary is the shape of the hyperbolic cosine curve. This shape is frequently seen in practice, e.g. telegraph wires, electricity power cables.

The length s of the curve may be found by integration. From

$$\frac{ds}{dx} = \sqrt{1 + \left(\frac{dy}{dx}\right)^2}$$

it follows that

$$s = \int \sqrt{1 + \left(\frac{dy}{dx}\right)^2}\, dx$$
$$= \int \sqrt{i + \left(\sinh \frac{x}{c}\right)^2}\, dx = \int \cosh (x/c)\, dx.$$

Hence $s = c \sinh x/c + D$.

Measuring s from the lowest point, $s = 0$ when $x = 0$ hence $D = 0$ and $s = c \sinh x/c$.

Thus the length of the catenary $y = \cosh x/c$ from its lowest point is the hyperbolic sine function $s = c \sinh x/c$.

(ii) The cooling fin

In many physical situations it is necessary to extract the heat from a hot liquid (or gas), e.g. a heat exchanger, a cooling system. In the simplest situation the hot container simply radiates heat into the cold surroundings (the latter being maintained by a flow of cold air or other medium). The container is frequently a metal pipe and in order to increase its surface area it is provided with cooling fins. A typical pipe is shown in Fig. 17.11.

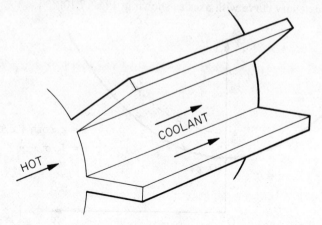

Fig. 17.11

The cooling fin may take a variety of different shapes the cross-section being rectangular, triangular, or parabolic. Consider the case of a fin of constant rectangular cross section and the flow of heat through it. Fig. 17.12 shows the typical cross-section. Let T be the temperature at

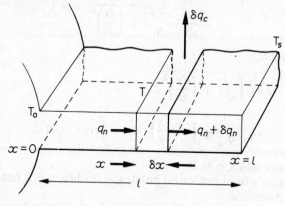

Fig. 17.12

distance x along the fin and q_n be the quantity of heat flowing over the cross-section of area A, let $q_n + \delta q_n$ be the heat flowing over the cross-section of the fin at distance $x + \delta x$ and let δq_c be the heat flowing into the cooling medium from the section of thickness δx. Let l be the length of the fin. The heat balance equation over the section of width δx at distance x is

$$q_n = (q_n + \delta q_n + \delta q_c)$$

Now, q_n is a function of x only and hence (see equation (7) of section 5.1.3),

$$\delta q_n \simeq \frac{dq_n}{dx} \delta x$$

and it follows that

$$\delta q_c \simeq - \frac{dq_n}{dx} \delta x.$$

There are two basic empirical laws of heat transfer:
  (i) *the law of conduction* which states

$$q_n = -K \frac{dT}{dx} A$$

where K is a constant dependent on the thermal properties of the fin material and,
  (ii) *the law of convection* (Newton's law of cooling), which states

$$\delta q_c = h(T - T_s)\,(C\delta x)$$

where h is a constant dependent on the surface properties of the fin and on the cooling medium; T_s is the (presumed constant) temperature of the cooling medium and C is the circumference of the fin at distance x.

Thus, $\quad h(T - T_s)\,C\delta x \simeq - \dfrac{dq_n}{dx} \delta x = +K \dfrac{d}{dx}\left(A \dfrac{dT}{dx}\right)\delta x.$

Normally A and C are dependent on the distance x but if the fin is a rectangular one they are both constants and in the limit, as $\delta x \rightarrow 0$, the equation becomes

$$\frac{d^2T}{dx^2} = \frac{hC}{KA}\,(T - T_s)$$

The assumption that T_s is constant implies that the heat is transferred to the surrounding medium and immediately transported elsewhere. In practice something akin to this idealised situation is achieved by a rapid flow of cooling medium over a hot fin. For example, the sump in certain motor cars is often of ribbed shape approximating to a series of fins and the surrounding cooling air is usually moving past the fin rapidly.

It is convenient to replace $T - T_s$ by θ and hence reduce the equation to

$$\frac{d^2\theta}{dx^2} = k^2\theta$$

where $k = \sqrt{(hC/KA)}$ is a constant.

The differential equation is easily solved (see Section **11.4** with $\lambda, \mu = +k, -k$) and θ is given by

$$\theta = D\,e^{kx} + E\,e^{-kx}$$

where D and E are arbitrary constants of integration.

In order to evaluate these arbitrary constants it is necessary to fit *boundary conditions* to the problem. If the temperature T at the base of the fin is taken as T_0 then one boundary condition has the form

$$\theta = T_0 - T_s = \theta_0, \text{ at } x = 0.$$

A second boundary condition may be obtained by neglecting the heat loss at the extremity of the fin $x = l$, i.e. by assuming that $(q_n)_{x=l}$ is zero. Use of the law of conduction (see (i) above) gives the second boundary condition in the form

$$\frac{dT}{dx} \quad \text{or} \quad \frac{d\theta}{dx} = 0, \text{ at } x = l,$$

Insertion of these two boundary conditions into the equation for θ gives

$$\theta_0 = D + E$$

$$0 = D\,e^{kl} - E\,e^{-kl}$$

Solving for D and E and using the fact that

$$e^{\phi} + e^{-\phi} = 2\cosh\phi$$

quickly leads to

$$\theta = \theta_0 \cosh k(l - x)/\cosh kl$$

i.e. $\qquad T - T_s = (T_0 - T_s)\cosh k(l - x)/\cosh kl.$

The variation of temperature across the fin length thus follows an hyperbolic cosine curve as in Fig. 17.13.

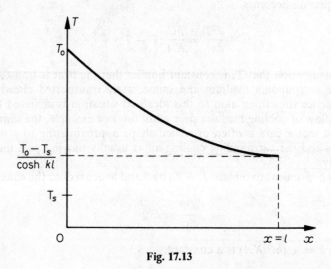

Fig. 17.13

Fins of shapes other than rectangular may be considered in a similar manner, the differential equations so arising being of additional complexity. It may be shown that the ideal shape (for optimum heat transfer) is of a parabolic cross-section and this is the shape frequently seen on practical fins.

17.3 MATRICES. DETERMINANTS. EIGENVALUES AND EIGENVECTORS

Engineering situations, when described by a mathematical model, frequently lead to a set of algebraical or differential equations in a specified number of unknown quantities. Matrix algebra may often be used to simplify the sometimes complex array of information.

17.3.1 A simple electrical network

Suppose an electrical network consists of a number of resistors together with batteries of given electromotive force (e.m.f.). It is required to find the current flowing in the various branches of the network. The Fig. 17.14 illustrates such a circuit the resistor values are in ohms and the e.m.f. in volts.

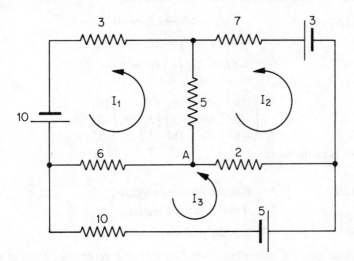

Fig. 17.14

Let the current (in amperes) be represented by *cyclic* currents I_1, I_2, and I_3 flowing in the three branches of the circuit, e.g. the current flowing in the resistor of 5 ohms is $(I_1 - I_2)$.

The basic Kirchoff laws that govern the network are:

(i) for each closed circuit the total, algebraic sum of e.m.f. is zero, and

(ii) at each nodal point there is continuity of current.

The law (ii) is automatically satisfied by using cyclic currents, for example at modal point A the current flow is

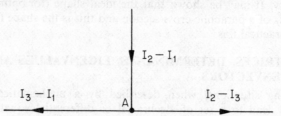

which shows that the current flow into the node is exactly balanced by the current flow out.

The law (i) leads to the following set of algebraical equations

$$10 = 6(I_1 - I_3) + 5(I_1 - I_2) + 3I_1$$

$$3 = 7I_2 + 5(I_2 - I_1) + 2(I_2 - I_3)$$

$$5 = 2(I_3 - I_2) + 6(I_3 - I_1) + 10I_3$$

This set of equations reduces to the following

$$14I_1 - 5I_2 - 6I_3 = 10$$

$$-5I_1 + 14I_2 - 2I_3 = 3$$

$$-6I_1 - 2I_2 + 18I_3 = 5$$

or in matrix form

$$\begin{pmatrix} 14 & -5 & -6 \\ -5 & 14 & -2 \\ -6 & -2 & 18 \end{pmatrix} \begin{pmatrix} I_1 \\ I_2 \\ I_3 \end{pmatrix} = \begin{pmatrix} 10 \\ 3 \\ 5 \end{pmatrix}$$

This is of the form

$$RI = E$$

where R is the resistance matrix,

                       I is the current matrix,

and E is the emf matrix.

Note that the matrix equation $E = RI$ is the generalised form of the the first Kirchoff law.

To find the current, the matrix equation is solved for I either by using the inverse matrix R^{-1} so that

$$I = R^{-1}E$$

or by employing Gaussian elimination.

17.3.2 Forces in a frame network

Suppose Fig. 17.15 represents a simple framework structure with hinged joints. It is assumed that there is no bending in any member and that the external forces are applied at the joints. It is required to find the reactions at the supports and the bar tension (or compression) forces.

Framework of this type are of common occurrence in the analysis of structures (e.g. girder bridges, cranes, skeleton of buildings) and are usually called *trusses*.

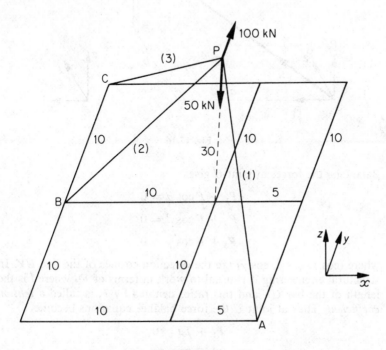

Fig. 17.15

In the framework of Fig. 17.15 there is an external force applied at P having a component $50kN$ acting in the negative z-direction and another component $100kN$ in the y-direction. The other numbers represent the lengths in metres and the bar members of the truss are labelled (1), (2) and (3).

There are three reactions at each of the supports and each reaction has three components in the x, y and z-directions. In addition there are three tension forces along the three bars. In all this gives twelve unknown forces ($3 \times 3 + 3$), and the appropriate number of equations is obtained by balancing force components at each joint giving $4 \times 3 = 12$ equations.

Let UV in Fig. 17.6 represent a typical bar member with an (assumed) tensile force F (i.e. internal bar force directed away from the joint). Let the forces have components ($P_1 P_2 P_3$) and ($R_1 R_2 R_3$) respectively.

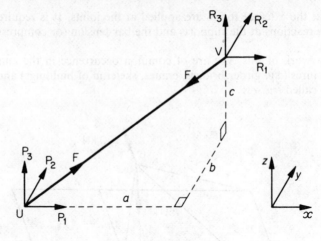

Fig. 17.16

Balancing the forces at joint U gives

$$P_1 + F \cos \alpha = 0$$

$$P_2 + F \cos \beta = 0$$

$$P_3 + F \cos \gamma = 0$$

where $(\cos \alpha, \cos \beta, \cos \gamma)$ are the direction cosines of the line UV. In structural engineering it is usual to work in terms of F/l where l is the length of the bar UV and this ratio, denoted by T, is called a *tension coefficient*. Thus at joint U the force balance equations become

$$P_1 + Ta = 0$$

$$P_2 + Tb = 0$$

$$P_3 + Tc = 0$$

where a, b, and c are the projections of UV onto the axes of x, y, and z respectively. The actual bar force F is then found by multiplying T by the length of the bar.

The Fig. 17.17(a), (b) show the plan and elevation force diagram for the truss of Fig. 17.15. The tension coefficients are T_1, T_2, and T_3 and the reaction force components are $(P_1 P_2 P_3)$ at A; $(R_1 R_2 R_3)$ at B, and $(S_1 S_2 S_3)$ at C respectively. The external forces are indicated and the other numbers on the Fig. 17.17 indicate the lengths.

Balancing forces at each joint in turn leads to the following equations:

at A, $$P_1 - 5T_1 = 0$$

$$P_2 + 10T_1 = 0$$

$$P_3 + 30T_1 = 0,$$

at B,
$$R_1 + 10T_2 = 0$$

$$R_2 + \qquad = 0$$

$$R_3 + 30T_2 = 0,$$

at C,
$$S_1 + 10T_3 = 0$$

$$S_2 - 10T_3 = 0$$

$$S_3 + 30T_3 = 0,$$

at P,
$$5T_1 - 10T_2 - 10T_3 = 0$$

$$100 - 10T_1 + 10T_3 = 0$$

$$-50 - 30T_1 - 30T_2 - 30T_3 = 0,$$

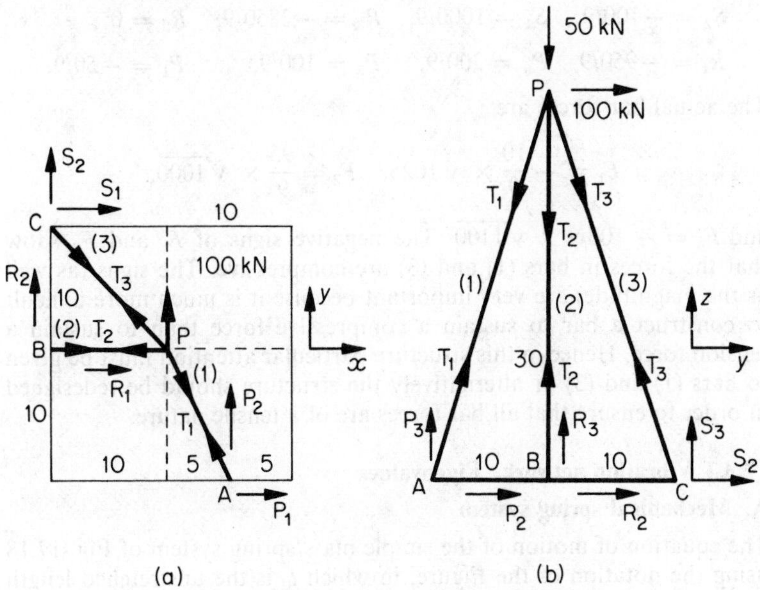

Fig. 17.17

Naturally, with this simple truss it is possible to solve without resort to matrix analysis but it should be realised that the force balance equations of much more complicated networks can be derived by using this simple technique. When the number of unknowns (and equations) is large it is often an advantage to set up a matrix. The set of 12 equations may be represented by the matrix equation

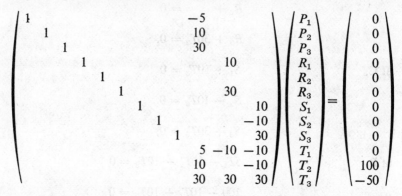

$$\begin{pmatrix} 1 & & & & & & & & & & & -5 \\ & 1 & & & & & & & & & & 10 \\ & & 1 & & & & & & & & & 30 \\ & & & 1 & & & & & & & 10 & \\ & & & & 1 & & & & & & & \\ & & & & & 1 & & & & & 30 & \\ & & & & & & 1 & & & & & 10 \\ & & & & & & & 1 & & & & -10 \\ & & & & & & & & 1 & & & 30 \\ & & & & & & & & & 5 & -10 & -10 \\ & & & & & & & & & 10 & & -10 \\ & & & & & & & & & 30 & 30 & 30 \end{pmatrix} \begin{pmatrix} P_1 \\ P_2 \\ P_3 \\ R_1 \\ R_2 \\ R_3 \\ S_1 \\ S_2 \\ S_3 \\ T_1 \\ T_2 \\ T_3 \end{pmatrix} = \begin{pmatrix} 0 \\ 0 \\ 0 \\ 0 \\ 0 \\ 0 \\ 0 \\ 0 \\ 0 \\ 0 \\ 100 \\ -50 \end{pmatrix}$$

In this form it is easy to reduce the set of equations by Gaussian elimination (see Section 8.6.1). The solution is found to be

$$T_3 = -100/9, \quad T_2 = 95/9, \quad T_1 = -10/9, \quad S_3 = 3000/9,$$

$$S_2 = -100/9, \quad S_1 = 1000/9, \quad R_3 = -2850/9, \quad R_2 = 0$$

$$R_1 = -950/9, \quad P_3 = 300/9, \quad P_2 = 100/9, \quad P_1 = -50/9.$$

The actual bar forces are

$$F_1 = -\frac{10}{9} \times \sqrt{1025}, \quad F_2 = \frac{95}{9} \times \sqrt{1000},$$

and $F_3 = -100/9 \times \sqrt{1100}$. The negative signs of F_1 and F_3 show that the forces in bars (1) and (3) are compressive. The signs (as well as the magnitude) are very important because it is much more difficult to construct a bar to sustain a compressive force than to sustain a tension force. Hence in this structure particular attention must be given to bars (1) and (3) or alternatively the structure should be redesigned in order to ensure that all bar forces are of a tensile nature.

17.3.3 Vibration network. Eigenvalues

A. Mechanical spring system

The equation of motion of the simple mass/spring system of Fig. 17.18 using the notation of the Figure, in which l_0 is the unstretched length

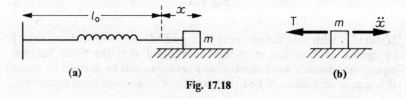

Fig. 17.18

of the spring, is given by expressing Newton's second law of motion (for constant mass) in the form Force = mass × acceleration.

i.e. $$T = -m\frac{d^2x}{dt^2}$$

The minus sign indicates that the spring force T is a restoring force.

For a simple spring, Hooke's Law (an empirical law) states that the tension T is proportional to the extension from its unstretched length, i.e. $T = kx$ where k is a spring constant called the *stiffness*. (*Note*. This is sometimes expressed in the form $T = \lambda x/l_0$ where λ is another constant called the *modulus of elasticity*). Eliminating T leads to the differential equation

$$m\frac{d^2x}{dt^2} + kx = 0$$

or $$\frac{d^2x}{dt^2} + n^2x = 0, \quad n^2 = k/m$$

The solution of this differential equation is

$$x = A\sin(nt + \alpha)$$

where A and α are two arbitrary constants to be determined using the initial conditions (i.e. the conditions under which the motion is started). The important point here, however, is that the vibration of the mass m is sinusoidal and the angular frequency of the oscillation (the *pulsatance*) is given by n. The period of the oscillation is $2\pi/n$ and the actual frequency is $n/2\pi$.

Consider now a number of such spring networks linked together. The vibration in one mass will cause sympathetic vibrations in the other masses, i.e. there will be an interaction between the individual networks. For simplicity consider three such spring networks linked together as in Fig. 17.19(a) in which the three masses are m_1, m_2 and m_3 and the connecting springs are of stiffnesses k_1, k_2, k_3 and k_4. The system is fixed at each end and the masses vibrate longitudinally.

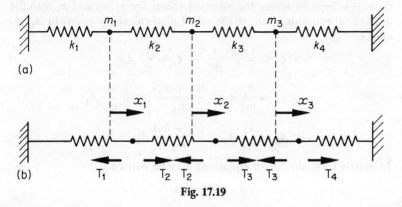

Fig. 17.19

The Fig. 17.19(b) shows the system in motion. The displacements x_1, x_2 and x_3 of the masses are measured from the static equilibrium positions and to simplify the subsequent analysis it is assumed that in static equilibrium the springs are in their unstretched state. The equations of motion of the three masses then take the form

$$-T_1 + T_2 = m_1 \, d^2x_1/dt^2$$

$$-T_2 + T_3 = m_2 \, d^2x_2/dt^2$$

$$-T_3 + T_4 = m_3 \, d^2x_3/dt^2$$

where T_1, T_2, T_3, and T_4 are the tensions in the springs.
Hooke's law expresses the tension forces as

$$T_1 = k_1(x_1 - 0); \quad T_2 = k_2(x_2 - x_1); \quad T_3 = k_3(x_3 - x_2)$$

$$T_4 = k_4(0 - x_3)$$

Eliminating T_1, T_2 and T_3 leads to three differential equations given by

$$m_1 \, \overset{..}{x}_1 = -(k_1 + k_2)x_1 + k_2x_2$$

$$m_2 \, \overset{..}{x}_2 = k_2x_1 - (k_2 + k_3)x_2 + k_3x_3$$

$$m_3 \, \overset{..}{x}_3 = k_3x_2 - (k_3 + k_4)x_3$$

where $\overset{..}{x}$ means d^2x/dt^2.

Now look for that particular mode of oscillation in which all masses oscillate at the same frequency $n/2\pi$ but with differing amplitudes. This particular oscillation is called the *normal mode*.

This assumes that

$$x_1 = A_1 \sin(nt + \alpha)$$

$$x_2 = A_2 \sin(nt + \alpha)$$

$$x_3 = A_3 \sin(nt + \alpha)$$

and means that whatever configuration the system may have at some time t_0 it returns to the same configuration after a period of $2\pi/n$, i.e. at time $t_0 + 2\pi/n$. Inserting the assumed values for x_1, x_2 and x_3 into the differential equations leads to the set of algebraical equations in A_1, A_2 and A_3

$$-n^2A_1 = -\frac{(k_1 + k_2)}{m_1} A_1 + \frac{k_2}{m_1} A_2$$

$$-n^2A_2 = \frac{k_2}{m_2} A_1 - \frac{(k_2 + k_3)}{m_2} A_2 + \frac{k_3}{m_3} A_3$$

$$-n^2A_3 = \frac{k_3}{m_3} A_2 - \frac{(k_3 + k_4)}{m_3} A_3$$

In matrix form this set of equations may be written

$$CA = \lambda A$$

where C is the coefficient matrix, A is the amplitude matrix and $\lambda = n^2$, i.e.

$$C = \begin{pmatrix} \dfrac{k_1 + k_2}{m_1} & -\dfrac{k_2}{m_1} & 0 \\[2ex] -\dfrac{k_2}{m_2} & \dfrac{k_2 + k_3}{m_2} & -\dfrac{k_3}{m_2} \\[2ex] 0 & -\dfrac{k_3}{m_3} & \dfrac{k_3 + k_4}{m_3} \end{pmatrix}$$

and $A = \begin{pmatrix} A_1 \\ A_2 \\ A_3 \end{pmatrix}$.

Note that A is the form of a vector and is often called the *amplitude vector* **A**. The equation $CA = \lambda A$ is of the form of (84) of Section 9.11, i.e. $\lambda = n^2$ is the *eigenvalue* of the matrix C and to each eigenvalue there will correspond a value of the frequency n. Thus for this system there will, in general, be three different modes of oscillation, i.e. three natural frequencies of oscillation of the system given by $\sqrt{\lambda}$. Each mode of oscillation will give a different *eigenvector* which is the amplitude vector **A**.

In order to examine such a system in detail consider the above system with

$$m_1 = 6, m_2 = m_3 = 4$$

$$k_1 = k_2 = k_3 = 3, \text{ and } k_4 = 1.$$

The coefficient matrix C is then

$$C = \begin{pmatrix} 1 & -1/2 & 0 \\ -3/4 & 6/4 & -3/4 \\ 0 & -3/4 & 1 \end{pmatrix}$$

and the characteristic equation for λ takes the determinantal form

$$\begin{vmatrix} 1 - \lambda & -1/2 & 0 \\ -3/4 & 6/4 - \lambda & -3/4 \\ 0 & -3/4 & 1 - \lambda \end{vmatrix} = 0$$

or

$$\begin{vmatrix} 2 - 2\lambda & -1 & 0 \\ -3 & 6 - 4\lambda & -3 \\ 0 & -3 & 4 - 4\lambda \end{vmatrix} = 0$$

This equation reduces after the usual determinant evaluation and subsequent factorisation to

$$(\lambda - 1)(4\lambda - 9)(4\lambda - 1) = 0$$

The eigenvalues are thus

$$\lambda_1 = 1/4, \ \lambda_2 = 1, \ \lambda_3 = 9/4,$$

and the associated non-normalised eigenvectors are as follows:
for $\lambda_1 = 1/4$,

$$\mathbf{A}^{(1)} = \begin{pmatrix} 2 \\ 3 \\ 3 \end{pmatrix}$$

for $\lambda_2 = 1$,

$$\mathbf{A}^{(2)} = \begin{pmatrix} 1 \\ 0 \\ -1 \end{pmatrix}$$

for $\lambda_3 = 9/4$,

$$\mathbf{A}^{(3)} = \begin{pmatrix} 2 \\ -5 \\ 3 \end{pmatrix}.$$

When the system is oscillating in its first normal mode $\lambda = \lambda_1 = 1/4$, i.e. at its least natural angular frequency $n_1 = \frac{1}{2}$ the amplitude of the displacements are in the ratio $[2:3:3]$ and so the three displacements x_1, x_2, and x_3 may be written

$$x_1 = 2a \sin (\tfrac{1}{2}t + \alpha)$$
$$x_2 = 3a \sin (\tfrac{1}{2}t + \alpha)$$
$$x_3 = 3a \sin (\tfrac{1}{2}t + \alpha)$$

where the values of a and α are to be determined by using the initial conditions. The period is 4π and is the greatest of the three normal modes. The motion is illustrated in Fig. 17.20(a) which shows the case for $\alpha = 0$.

When the system oscillates in its second mode with $\lambda = 1$, i.e. at a natural angular frequency $n_2 = 1$, the displacements are given

$$x_1 = = b \sin (t + \beta)$$
$$x_2 = 0$$
$$x_3 = -b \sin (t + \beta).$$

In this mode the centre mass is at rest and the other two execute simple harmonic motion at a period of 2π. The values of b and β are again determinable from the initial conditions. This mode is illustrated in Fig. 17.20(b) in which β is taken as zero.

Finally, when the system oscillates in its mode of highest frequency, i.e. $\lambda = 9/4$, $n_3 = 3/2$, the displacements are given by

$$x_1 = 2c \sin (3t/2 + \gamma)$$
$$x_2 = -5c \sin (3t/2 + \gamma)$$
$$x_3 = 3c \sin (3t/2 + \gamma).$$

The period of the oscillation is $4\pi/3$ and the constants c and γ are again determined from the initial conditions. This motion is illustrated in Fig. 17.20(c) (for $\gamma = 0$) and shows that the outer masses oscillate in phase but of differing amplitudes whilst the inner mass oscillates with a large amplitude and exactly $180°$ out of phase.

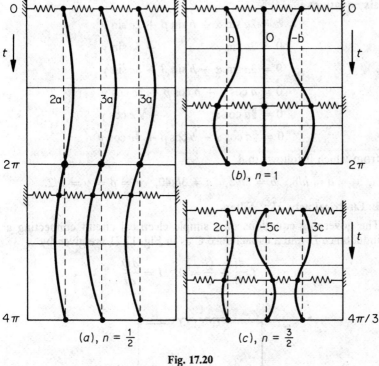

$(a), n = \frac{1}{2}$

$(b), n = 1$

$(c), n = \frac{3}{2}$

Fig. 17.20

These motions are, naturally, subject to any physical limitations of the springs and assume that the springs can be compressed as well as extended.

If the system is set in motion in an arbitrary manner (e.g. by initial displacement and/or velocities of the masses) then the subsequent displacements x_i will be a linear combination of the three normal mode oscillations. The three displacements at time t will then be given by

$$x_1 = 2a \sin (\tfrac{1}{2}t + \alpha) + b \sin (t + \beta) + 2c \sin (\tfrac{3}{2}t + \gamma)$$

$$x_2 = 3a \sin (\tfrac{1}{2}t + \alpha) \qquad\qquad - 5c \sin (\tfrac{3}{2}t + \gamma)$$

$$x_3 = 3a \sin (\tfrac{1}{2}t + \alpha) - b \sin (t + \beta) + 3c \sin (\tfrac{3}{2}t + \gamma)$$

The six unknowns a, b, c, α, β, γ will be determined by six initial conditions (i.e. will be determined once the initial motion is prescribed).

For example if the system is started into motion by displacing the first mass a distance h, keeping the other two masses in their equilibrium positions, and releasing it from rest, the initial conditions would be

$$x_1 = h, \quad x_2 = 0 = x_3, \quad \dot{x}_1 = \dot{x}_2 = \dot{x}_3 = 0 \text{ at } t = 0.$$

These six conditions applied to the above set of equations lead to the six equations

$$h = 2a \sin \alpha + b \sin \beta + 2c \sin \gamma$$
$$0 = 3a \sin \alpha \qquad\qquad - 5c \sin \gamma$$
$$0 = 3a \sin \alpha - b \sin \beta + 3c \sin \gamma$$
$$0 = a \cos \alpha + b \cos \beta + 3c \cos \gamma$$
$$0 = \tfrac{3}{2}a \cos \alpha \qquad\qquad -\tfrac{15}{2}c \cos \gamma$$
$$0 = \tfrac{3}{2}a \cos \alpha - b \cos \beta + \tfrac{9}{2}c \cos \gamma$$

from which it follows that

$$a = h/8, \quad b = 3h/5, \quad c = 3h/40, \quad \alpha = \beta = \gamma = \pi/2.$$

B Electrical networks. Filters

The governing equations of a simple electrical circuit connecting an inductance L, and a capacitance C as in Fig. 17.21 are given by

$$L\frac{dI}{dt} + \frac{Q}{C} = 0, \quad I = \frac{dQ}{dt}$$

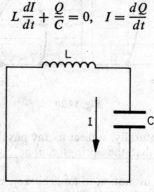

Fig. 17.21

where Q is the charge on the condenser at time t and I is the current flowing in the circuit.

Eliminating I, gives

$$\ddot{Q} + n^2 Q = 0, \quad n^2 = 1/LC$$

where $\ddot{Q}$ means d^2Q/dt^2.

The solution of this differential equation is $Q = A \sin(nt + \alpha)$ and leads to $I = nA \cos(nt + \alpha)$.

It may be observed that the $L - C$ electrical circuit is the analogue of the mass-spring mechanical circuit. The differential equations are identical and it is possible to draw analogies between the components of the two circuits as in the following table.

Electrical	Mechanical
Inductance, L	Mass m
(Capacitance)$^{-1}$, $1/C$	Stiffness k
Charge Q	Displacement x
Current $I = \dot{Q}$	Velocity $v = \dot{x}$
Applied emf E	Applied force F
Resistance R	Damping μ

In both the mechanical circuit and the electrical circuit discussed in this section the effect of damping (i.e. resistance or friction) is omitted. Also the driving forces of emf or applied force are also not considered. Both of these effects can however be taken into account. In a real engineering situation it is unlikely for example, that damping (resistive or frictional) will be absent. Nevertheless before these additional effects are included in the analysis, it is extremely useful to have considered the properties of, and important results obtained by, linking together simple undamped, unforced circuits.

The strong analogy between the basic mechanical mass/spring system and the inductance/capacitance electrical circuit is the basis of the *analogue computer*.

The effect on the mechanical system of varying the mass and the spring stiffness can be visually represented by making an equivalent electrical circuit and varying the inductance and capacitance. The latter components can be varied very easily whereas it may be difficult to vary the equivalent components in the mechanical system. This analogy between mechanical and electrical circuits can be extended in scope to include complicated systems. For example the engineer confronted with an unwanted vibration (i.e. displacement and/or velocity) in a large and complicated mechanical system can frequently represent it by an equivalent complicated electrical circuit. Since the electrical circuit can be easily constructed in the laboratory the engineer will be able to assess what effect the varying of component values will have on the unwanted vibration and hence by analogy be able to assess the effect on the real mechanical system and hopefully be able to modify the mechanical system to remove the unwanted vibration.

Returning now to the simple $L - C$ circuit, suppose the circuit incorporated a switch which was held open with the condenser charged to a value Q_0. If the switch was closed at time $t = 0$, the initial conditions would be

$$Q = Q_0, I = dQ/dt = 0, \text{ at } t = 0.$$

Insertion of these conditions into the solutions for Q and I leads to

$$Q_0 = A \sin \alpha$$

$$0 = nA \cos \alpha.$$

and hence $\alpha = \pi/2$, $A = Q_0$.

The charge and current at time t after the switch closure are then

$$Q = Q_0 \cos nt, \quad I = -nQ_0 \sin nt.$$

The charge on the condenser and current flowing in the network both oscillate sinusoidally with a natural (angular) frequency $n(=1/\sqrt{LC})$. In the absence of resistance there is no damping effect and the oscillations are sustained for all time.

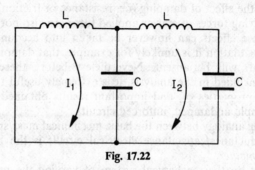

Fig. 17.22

Now suppose two such circuits are linked together as in Fig. 17.22. and let I_1 and I_2 be two cyclic currents, then the governing equations are

$$L \frac{dI_1}{dt} + \frac{Q_1}{C} = 0; \quad I_1 - I_2 = \frac{dQ_1}{dt},$$

$$L \frac{dI_2}{dt} + \frac{Q_2}{C} - \frac{Q_1}{C} = 0, \quad I_2 = \frac{dQ_2}{dt}$$

where Q_1 and Q_2 are the charges on the two condensers.

Eliminating the currents I_1 and I_2 and rearranging the equations leads to

$$LC \ddot{Q}_1 + 2Q_1 - Q_2 = 0$$

$$LC \ddot{Q}_2 - Q_1 + Q_2 = 0.$$

To find the natural frequencies in this case it is assumed that Q_1 and Q_2 are oscillatory with frequency $n/2\pi$, i.e. such that

$$Q_1 = A_1 \sin (nt + \alpha)$$

$$Q_2 = A_2 \sin (nt + \alpha)$$

With this assumption the differential equations may be rewritten as a matrix equation

$$\begin{pmatrix} 2 & -1 \\ -1 & 1 \end{pmatrix} \begin{pmatrix} A_1 \\ A_2 \end{pmatrix} = \lambda \begin{pmatrix} A_1 \\ A_2 \end{pmatrix}$$

where $\lambda = LCn^2$.

It is noted once again that λ is the eigenvalue of the coefficient matrix and that the natural frequency is given by $\sqrt{(\lambda/LC)}$. The characteristic equation for λ is

$$\begin{vmatrix} 2 - \lambda & -1 \\ -1 & 1 - \lambda \end{vmatrix} = 0.$$

i.e. $$\lambda^2 - 3\lambda + 1 = 0.$$

The two eigenvalues are thus

$$\lambda_1 = \frac{3}{2} - \frac{\sqrt{5}}{2} \quad \text{and} \quad \lambda_2 = \frac{3}{2} + \frac{\sqrt{5}}{2}$$

If three such circuits are linked together as in Fig. 17.23 the differential equations are

$$LC\,\ddot{Q}_1 + 2Q_1 - Q_2 \qquad\quad = 0$$

$$LC\,\ddot{Q}_2 - Q_1 + 2Q_2 - Q_3 = 0$$

$$LC\,\ddot{Q}_3 \qquad\quad - Q_2 + Q_3 = 0.$$

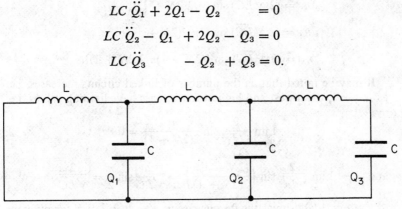

Fig. 17.23

and the natural frequencies of oscillation are given by $n/2\pi$ where $n^2LC = \lambda$, and λ are the eigenvalues of the matrix

$$\begin{pmatrix} 2 & -1 & 0 \\ -1 & 2 & -1 \\ 0 & -1 & 1 \end{pmatrix}$$

The characteristic equation for λ is $\lambda^3 - 5\lambda^2 + 6\lambda - 1 = 0$ and the three values of λ are found to be real roots λ_1, λ_2, and λ_3 such that $0 < \lambda_1 < 1; 1 < \lambda_2 < 2; 2 < \lambda_3 < 3$.

The numerical evaluation of these roots and hence the natural frequencies of oscillation will not be undertaken because it is better to investigate this particular circuitry by a different technique. Suffice to state that by an alternative approach it may be shown that the N eigenvalues $\lambda_1, \lambda_2, \ldots, \lambda_N$ for N linked circuits are given by, $\lambda_r = 4\sin^2(\mu/2)$, where $\mu = (2r - 1)\pi/(2N + 1)$, $r = 1, 2, \ldots, N$.

The natural (angular) frequencies of the oscillations are then given by

$$n_r = (2/\sqrt{LC}) \sin(\mu/2).$$

For example, if $N = 2$, the two frequencies are

$$n_1 = (2/\sqrt{LC}) \sin(\pi/10); \quad n_2 = (2/\sqrt{LC}) \sin(3\pi/10)$$

and if $N = 3$, the frequencies are

$$n_1 = (2/\sqrt{LC}) \sin(\pi/14); \quad n_2 = (2/\sqrt{LC}) \sin(3\pi/14);$$

$$n_3 = (2/\sqrt{LC}) \sin(5\pi/14).$$

The reader may check the validity of these values by inserting the appropriate numerical values of λ into the above cubic equation.

In the general case of N circuits

$$n_1 = (2/\sqrt{LC}) \sin(\pi/2(2N + 1));$$

$$n_2 = (2/\sqrt{LC}) \sin(3\pi/2(2N + 1)), \ldots,$$

$$n_N = (2/\sqrt{LC}) \sin\{(2N - 1)\pi/2(2N + 1)\}.$$

It may be noted that as the number of linked circuits increases, i.e. as $N \to \infty$, the lowest and highest frequencies have limiting values, i.e.

$$\lim_{N\to\infty} \frac{2}{\sqrt{LC}} \sin \frac{\pi}{2(2N + 1)} = 0$$

and $\quad \lim_{N\to\infty} \dfrac{2}{\sqrt{LC}} \sin\left(\dfrac{(2N - 1)\pi}{2(2N + 1)}\right) = \dfrac{2}{\sqrt{LC}} \sin \dfrac{\pi}{2} = \dfrac{2}{\sqrt{LC}}.$

The Fig. 17.24 shows the frequencies n_r, $r = 1, 2 \ldots N$ for different values of N.

This system with N simple linked circuits has its natural frequencies within a frequency range 0 to $2/\sqrt{LC}$. Thus if, for example, a combination of currents at different frequencies are passed into such an electrical network it will readily accept the currents whose frequencies lie within this band but will attenuate those outside. Such a network can be used in electrical engineering as a *low-pass filter* circuit, i.e. a circuit that will pass the low frequencies in the range 0 to $2/\sqrt{LC}$ but reject those higher than this range.

If the capacitance C and inductance L are interchanged in the simple circuits then it is easily shown that the frequencies within the band 0 to $2/\sqrt{LC}$ are attenuated whilst those outside this band are readily accepted, and the network becomes a *high-pass filter*.

By a judicious choice of components it is clearly possible to make the natural frequencies of a network lie within a specified band of values and so make an electrical device known as a *band-pass filter* which will readily accept only these frequencies within the specified band.

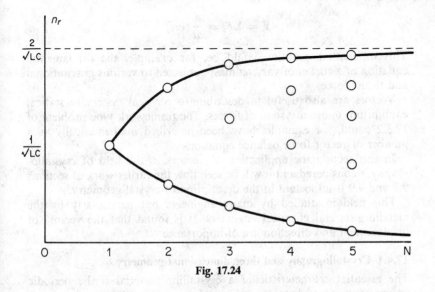

Fig. 17.24

The low-, high-, and band-pass filters have many applications in electrical engineering, e.g. radar, television reception circuits, telephony and telegraphy noise suppression circuits, record player scratch filters.

It has been noted however that electrical filters as described have differential equations equivalent to those of some mechanical spring networks and it may also be stated that additionally the equations governing acoustical apparatus also have similar differential equations. Hence by analogy it should be possible to design mechanical spring networks or acoustical devices that permit acceptance or rejection of certain frequencies of oscillations. For example, a mechanical machine, may have its unwanted oscillations 'filtered out' by mounting the machine in flexible spring supports and similarly the unwanted acoustic oscillations of a car exhaust may be minimised by using a series of linked resonating cavities mounted along the exhaust pipe, i.e. a car silencer.

17.4 VECTORS

Vectors make their appearance in an engineering context in a number of different fields but particularly when the engineering system is in motion in 2 or 3 dimensions. The vectors of force, acceleration, velocity, and displacement are easily understood and the various mathematical equations describing the desired physical motion are readily expressible in terms of these vector quantities. As an illustration, if a particle of mass m is travelling with a velocity $\mathbf{v}$ and is subjected to a number of forces $\mathbf{F}_1, \mathbf{F}_2 \ldots$ then the vector form of Newton's second law stating that the total impressed force is proportional to the rate of change of momentum is expressible as

$$\mathbf{F} = \Sigma F_i = \frac{d}{dt}(m\mathbf{v})$$

This differential equation could be, for example, the fundamental equation of a rocket of varying mass subjected to various gravitational and thrust forces.

Vectors are also useful in describing a physical system in statical equilibrium under a system of forces. The framework type problem of 17.3.2 could, for example, have been described mathematically as a number of vector force balance equations.

In this section the application of vectors to the field of crystallography is considered and it will be seen how the earlier work of sections 9.7 and 9.9 is important in the description of crystal geometry.

This field is studied by many engineers but particularly by the metallurgists and chemical engineers. It is found that the vectors of displacement and direction are of importance.

17.4.1 Crystallography and three dimensional geometry

The essential characteristic of a crystalline material is the periodic nature of its internal structure. The atomic arrangement is related to a network of points in space known as a *space lattice*. The points of the lattice may be occupied by single atoms, but usually in a crystal they are points at which there are identical groups of atoms. In a perfect crystal the outlook from any lattice point is the same as from any other. Starting from an arbitrary lattice point O in Fig. 17.25 it is possible to draw three straight lines each of which joins with three other lattice points, distance a, b, and c away, and form a set of coordinate axes (not necessarily at right angles to each other).

The distances from the starting lattice point 0 to the other lattice points may be different, i.e. $a \neq b \neq c$. Instead of unit vectors along the three axes it is usual to form base vectors $\mathbf{a}$, $\mathbf{b}$, and $\mathbf{c}$ so that any point P of space is specified by its position vector $\mathbf{r}$ and $\mathbf{r} = x\mathbf{a} + y\mathbf{b} + z\mathbf{c}$. The point P is then designated by (x, y, z). The notation $[x, y, z]$, with square brackets, is used in crystallography to indicate the *direction* of the vector $\mathbf{r}$. The three axes are called the *crystal axes*. Naturally, it is possible to choose the axes in a number of different ways, for example for certain crystals it may be possible that two or even three of the axes are at right-angles to each other.

When the axes have been chosen they form a *unit cell*, i.e. a parallelepiped of sides *a*, *b*, and *c* and each crystal consists of a repetitive stacking of unit cells. The volume of this fundamental building block is important and is simply the volume of a parallelepiped of sides **a**, **b**, and **c** which is (see 9.9.1) given by V and $V = \mathbf{a} \cdot \mathbf{b} \times \mathbf{c}$.

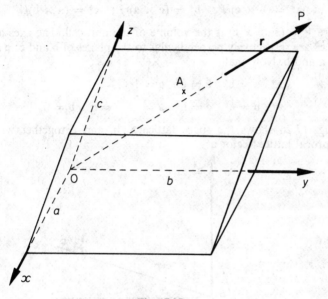

Fig. 17.25

Points of the crystal which coincide with cell corners are given by integral values of x, y, and z but points of the cell faces or inside the cell will have fractional values of x, y, and z.

Any particular direction in the lattice is specified by the *direction ratios* $[x:y:z]$ or $[\lambda x:\lambda y:\lambda z]$ where λ is chosen to avoid fractional coordinates. For example the line from the origin $(0, 0, 0)$ to the point A of Fig. 17.25 at the middle of a cell face, i.e. to the point $(\frac{1}{2}, \frac{1}{2}, 1)$ has direction ratio $[\frac{1}{2}:\frac{1}{2}:1]$ or $[1:1:2]$ or simply $[1, 1, 2]$.

The faces of crystals, and also other cross-sections within the crystal lattice, are planes and in crystallography it is necessary to have a system of orientation of such planes. The crystallographer uses what are called *Miller Indices* and arrives at a trio of numbers as follows:

(i) Find the intercepts, on the axes, of the plane in terms of the base lengths *a*, *b*, and *c*.

(ii) Write down the reciprocal of these numbers and form the smallest integers in the same ratio.

For example, if a plane has intercepts $(4a, 3b, 2c)$ on the three axes then the three reciprocals are $(\frac{1}{4}, \frac{1}{3}, \frac{1}{2})$ and the Miller indices are $(3, 4, 6)$. However, from Section 9.7 it is known that the orientation of a plane is given by its normal and hence the Miller indices must be related to the normal. In order to examine this relationship it is necessary to introduce a space lattice related to the crystal axes and known as the *reciprocal lattice*. The three base vectors forming the reciprocal lattice are designated a^*, b^*, and c^* and are given by

$$a^* = (b \times c)/V; \quad b^* = (c \times a)/V; \quad c^* = (a \times b)/V$$

where $V = (a \cdot b \times c)$ is the volume of the unit cell. The axes a^*, b^*, and c^* are respectively perpendicular to the planes of b and c; c and a, and a and b. Note that

$$a^* \cdot a = b^* \cdot b = c^* \cdot c = 1$$

and
$$a^* \cdot b = a^* \cdot c = b^* \cdot a = \ldots = c^* \cdot b = 0.$$

Fig. 17.26 shows the space lattice a, b, and c together with the reciprocal lattice vector a^*.

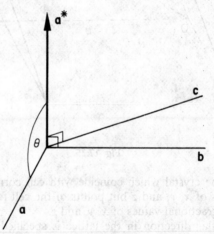

Fig. 17.26

Using the property of the scalar product, with θ the angle between a and a^*

$$a^* \cdot a = a^* a \cos \theta = 1$$

i.e.
$$a^* = 1/a \cos \theta.$$

This result shows that the length of the base vector a^* is the *reciprocal* of the projection of a onto a^*.

Consider now any plane that has intercepts on the crystal axes $0xyz$ of lengths αa, βb, and γc as in Fig. 17.27. The Miller indices are the smallest integers in the ratio $[\frac{1}{\alpha}, \frac{1}{\beta}, \frac{1}{\gamma}]$.

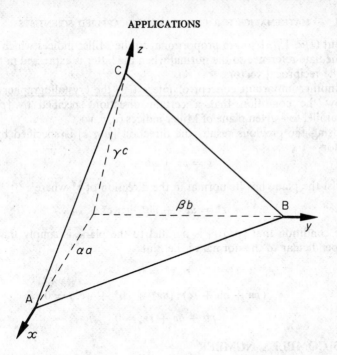

Fig. 17.27

Then in vector terms

$$\overline{OA} = \alpha\mathbf{a}, \quad \overline{OB} = \beta\mathbf{b}, \quad \overline{OC} = \gamma\mathbf{c}$$

The normal to the plane is a vector $\mathbf{n}$ perpendicular to $\overline{AB}$ and $\overline{CB}$, i.e. to $(\beta\mathbf{b} - \alpha\mathbf{a})$ and $(\gamma\mathbf{c} - \beta\mathbf{b})$. Thus (see Section 9.6.2).

$$\mathbf{n} = (\beta\mathbf{b} - \alpha\mathbf{a}) \times (\gamma\mathbf{c} - \beta\mathbf{b})$$

$$= \beta\gamma(\mathbf{b} \times \mathbf{c}) - \gamma\alpha(\mathbf{a} \times \mathbf{c}) + \alpha\beta(\mathbf{a} \times \mathbf{b})$$

since $\mathbf{b} \times \mathbf{b} = 0$.

In terms of the reciprocal vectors

$$\mathbf{n} = \beta\gamma\mathbf{a}^*V + \alpha\gamma\mathbf{b}^*V + \alpha\beta\mathbf{c}^*V$$

$$= \alpha\beta\gamma V\left(\frac{1}{\alpha}\mathbf{a}^* + \frac{1}{\beta}\mathbf{b}^* + \frac{1}{\gamma}\mathbf{c}^*\right)$$

It follows immediately that

$$\mathbf{n}^* = \frac{1}{\alpha}\mathbf{a}^* + \frac{1}{\beta}\mathbf{b}^* + \frac{1}{\gamma}\mathbf{c}^*$$

is a normal to the plane.

But $(1/\alpha, 1/\beta, 1/\gamma)$ are proportional to the Miller indices which have immediate reference to the normal when the latter is expressed in terms of the reciprocal vectors.

Another important concept of interest to the crystallographer is to know the condition that a certain direction specified by $[p:q:r]$ is parallel to a given plane of Miller indices (u, v, w).

Using the previous results the direction $[p:q:r]$ is specified by the vector

$$\mathbf{r} = p\mathbf{a} + q\mathbf{b} + r\mathbf{c}$$

whilst the plane has its normal in the direction of $\mathbf{r}^*$ where

$$\mathbf{r}^* = u\mathbf{a}^* + v\mathbf{b}^* + w\mathbf{c}^*$$

The condition that $[p:q:r]$ is parallel to the plane is simply that $\mathbf{r}$ is perpendicular to the normal $\mathbf{r}^*$, i.e. tlat

$$\mathbf{r} \cdot \mathbf{r}^* = 0$$

i.e. $$(p\mathbf{a} + q\mathbf{b} + r\mathbf{c}) \cdot (u\mathbf{a}^* + v\mathbf{b}^* + w\mathbf{c}^*) = 0$$

i.e. $$pu + qv + rw = 0$$

17.5 COMPLEX NUMBER

The representation of a complex number z as a vector in the Argand diagram and the Euler form $z = r\,e^{i\theta} = r\cos\theta + ir\sin\theta$ give z a geometrical interpretation and also show how a complex number may be related to the trigonometrical functions. These two representations are essential to the understanding of the way in which complex number analysis can be used in an engineering context. The simplest major application is in the field of alternating current theory which is concerned with the response of a network to applied alternating currents or voltages. The elements of this theory are considered in this section.

17.5.1 Alternating current networks

The basic alternating current which excites an electrical network is of the general form

$$i = I_0 \cos(\omega t + \alpha)$$

in which I_0 is the amplitude, ω is the angular frequency and α is the phase angle. Since this current may be expressed in the form

$$i = (I_0 \cos\alpha)\cos wt - (I_0 \sin\alpha)\sin wt$$

it is sufficient to consider the response of the network to the basic current

$$i = A\cos\omega t \quad \text{or} \quad i = B\sin\omega t$$

Now since (see Section 10.4)

$$e^{j\omega t} = \cos\omega t + j\sin\omega t$$

where $j^2 = -1$, it follows that these basic currents may be obtained by taking either the real or the imaginary part of the complex current I in the form

$$I = I_0 \, e^{j\omega t}$$

Note The use of $j^2 = -1$ instead of $i^2 = -1$ as in Chapter 10 follows Electrical Engineering practice where i is invariably used to represent a current.

For example if $I_0 = 3$ and $\omega = 2$ then the real part of a complex current $3e^{2jt}$ is $3 \cos 2t$ and the imaginary part is $3 \sin 2t$.

Further, a more general type current with a phase angle can be simply represented as the real part of a complex current. For example, the current

$$i = 2 \cos (4t - \pi/3)$$

is simply the real part of the complex current I where

$$I = 2e^{j(4t - \pi/3)}$$

i.e.
$$I = (2e^{-j\pi/3})e^{j4t}$$

which is of the form

$$I = I_0 \, e^{j\omega t}$$

with $I_0 = 2e^{-j\pi/3}$ and $\omega = 4$.

Since the voltage response to a sinusoidal current is also sinusoidal iti s to be expected that a complex current excitation of an electric network will lead to a complex voltage response (and vice versa).

Consider now the voltage response of a simple network to a current of the basic form

$$i = I_0 \cos \omega t, \; I_0 \text{ real}$$

Suppose the current i (amps) flows in the network of Fig. 17.28 which consists of a resistance R (ohms), inductance L (henries), and capacitance C (farads).

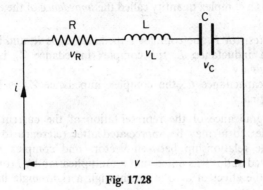

Fig. 17.28

The voltage response of the elements of this network are v_R, v_L, and v_c respectively and, if q is the charge of the capacitance,

$$v_R = Ri = RI_0 \cos \omega t$$

$$v_L = L\frac{di}{dt} = -\omega L I_0 \sin \omega t$$

$$v_C = q/C = \frac{1}{C} \int idt = \frac{I_0}{\omega C} \sin \omega t$$

If the current i is expressed as the real part of a complex current I, i.e.

$$i = Re\, I = Re I_0\, e^{j\omega t}$$

then in like manner the voltages v_R, v_L, and v_C are the real parts of complex voltages V_R, V_L, and V_C obtained as follows:

(i) $v_R = RI_0 \cos \omega t = Re\, RI_0\, e^{j\omega t} = Re V_R$

i.e. $$V_R = RI$$

(ii) $$v_L = -\omega L\, I_0 \sin \omega t = Re\, j\omega L\, I_0\, e^{j\omega t}$$

$$= Re\, V_L$$

i.e. $$V_L = j\omega L\, I$$

(iii) $$v_C = \frac{I_0}{\omega C} \sin \omega t = Re \left(-\frac{j}{\omega C} I_0\, e^{j\omega t} \right)$$

$$= Re\, V_C$$

i.e. $$V_C = \frac{-j}{\omega C} I = \frac{1}{j\omega C} I.$$

In each of the cases (i), (ii), and (iii), the response of the element to a complex current I is a complex voltage V of the form

$$V = ZI$$

where Z is a complex quantity called the *impedance* of the element of the circuit.

For a resistor R, the complex impedance Z_R is R, and is real,

for an inductance L, the complex impedance Z_L is $j\omega L$ and is imaginary,

for a capacitance C, the complex impedance Z_C is $1/j\omega C$ and is imaginary.

The significance of the representation of the current and voltage in complex form may be appreciated after reference to Section 10.1 where the relationship between vectors and complex numbers was investigated and it was noted that a multiplication by j (or i of Chapter 10) had the effect of a rotation through a right-angle in the positive sense. Let the impedances $Z_R, Z_L,$ and Z_C be represented on the Argand diagram as in Fig. 17.29.

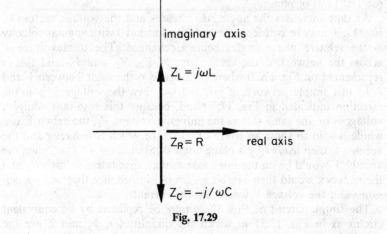

Fig. 17.29

The impedance Z_R is represented by a vector of length R along the real axis whilst the impedance Z_L and Z_C are represented by magnitudes ωL and $1/\omega C$ respectively along the positive and negative imaginary axes.

The voltages may also be represented on an Argand diagram, and Fig. 17.30 shows the complex current $I = I_0 \, e^{j\omega t}$ together with the voltages V_R, V_L, and V_C

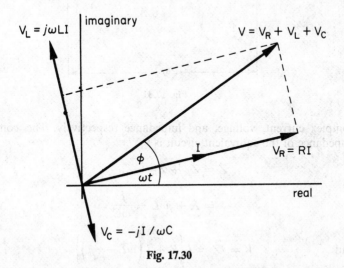

Fig. 17.30

The Fig. 17.30 gives a geometrical (complex) interpretation of the fact that V_R is *in phase* with the current I, that V_L *leads* the voltage V_R

by a right angle (i.e. there is one right angle difference in the appropriate sense between the phase angle of V_R and V_L), and that V_C *lags* the voltage V_R by one right-angle. The voltages V_L and V_C are exactly π (or 180°) out of phase.

As time increases the angle ωt increases and the voltage vectors of Fig. 17.30 may be considered as rotating around 0 with angular velocity ω the relative phase angles being unchanged. The total voltage V across the network is the vector sum of V_R, V_L, and V_C and this is represented on Fig. 17.30 where the angle ϕ is the angle between V and I. In this simple network, if $\omega L > 1/\omega C$ then the voltage V is in the direction indicated in Fig. 17.30 and, because this resultant complex voltage is on the same side as the inductive voltage V_L, the network as a whole is said to show an overall impedance which is *inductive* and the voltage V then *leads* I by a plane angle ϕ. Should $\omega L < 1/\omega C$ then the voltage V would be on the same side as the capacitative voltage V_C and the network would then exhibit an overall impedance that was *capacitative* and the voltage V would *lag* the current.

The simple circuit of Fig. 17.28 may be replaced by an equivalent circuit as in Fig. 17.31 in which the quantities I, V, and Z are the

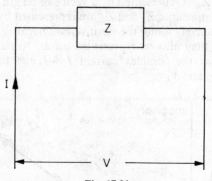

Fig. 17.31

complex current, voltage, and impedance respectively. The complex impedance of the equivalent circuit is

$$Z = Z_R + Z_L + Z_C$$

$$= R + j\omega L - \frac{j}{\omega C}$$

and
$$V = IZ = I\left\{R + j\left(\omega L - \frac{1}{\omega C}\right)\right\}$$

Example (i) The actual current i in the network of Fig. 17.28 is $i = 3\cos 4t$, find the overall voltage v and also v_C.

Solution In complex form

$$I = 3e^{j4t}$$

and $$V = 3e^{j4t} \left\{ R + j \left(\omega L - \frac{1}{\omega C} \right) \right\}$$

Hence $v = ReV = Re\, 3(\cos 4t + j \sin 4t) \left\{ R + j \left(\omega L - \frac{1}{\omega C} \right) \right\}$

and $$v = 3R \cos 4t - 3 \left(\omega L - \frac{1}{\omega C} \right) \sin 4t$$

The complex voltage across C is V_C and

$$V_C = \frac{-j}{\omega C} I = \frac{-j}{\omega C} 3e^{j4t}$$

and hence $$v_C = Re\, V_C$$

$$= \frac{3}{\omega C} \sin 4t$$

Example (ii) The actual current i in the network of Fig. 17.28 is $i = 2 \sin 3t$. Find the voltage v in the case $R = 2$, $L = \frac{1}{2}$, $C = 1$.

Solution The current is

$$i = 2 \sin 3t = 2 \cos (3t - \pi/2)$$

the complex current is

$$I = 2e^{j(3t - \pi/2)}$$

the complex voltage V then has the form

$$V = 2e^{j(3t - \pi/2)} \left\{ R + j \left(3L - \frac{1}{3C} \right) \right\}$$

$$= 2(-j)\, e^{j3t}\{2 + j(7/6)\}.$$

The actual voltage v is then

$$v = Re\, V$$

$$= 4 \sin 3t + (7/3) \cos 3t$$

Note It is assumed in the above process that the circuit is excited by a current i which, although time-dependent, is in its steady state. That is, any transient effects due to switch-on phenomenon have died away. In a mathematical sense this means that the circuit switch was closed at time $t = -\infty$.

Consider now a more complicated electrical network as in Fig. 17.32 which shows the actual circuit (a) and the equivalent complex form (b). When a sinusoidal steady-state *voltage* of angular frequency ω is applied across the terminals AB, each of the elements of the circuit

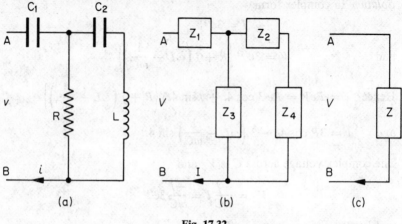

Fig. 17.32

has an impedance dependent on that frequency. The complex voltage $V(= V_0 e^{j\omega t})$ is applied across AB and the complex impedances are given by

$$Z_1 = \frac{-j}{\omega C_1}, \quad Z_2 = \frac{-j}{\omega C_2}, \quad Z_3 = R, \quad Z_4 = j\omega L.$$

The impedances may be combined together by the usual series and parallel rules the network then being equivalent to a single impedance Z, as in Fig. 17.32(c), such that

$$Z = \frac{-j}{\omega C_1} + \left[\frac{1}{R} + \frac{1}{j\omega L - j/\omega C_2}\right]^{-1}$$

$$= \frac{-j}{\omega C_1} + \left[\frac{R + j(\omega L - 1/\omega C_2)}{Rj(\omega L - 1/\omega C_2)}\right]^{-1}$$

i.e.
$$Z = \frac{-j}{\omega C_1} + \frac{jR(\omega L - 1/\omega C_2)}{R + j(\omega L - 1/\omega C_2)}$$

The complex current I is then given by

$$I = V/Z$$

and the actual current i flowing through the network would then be obtained by taking the real part of V/Z.

When dealing with impedances in parallel it is sometimes more expedient to work in terms of the reciprocal $1/Z$. This quantity is called the *admittance* Y and it follows that two impedances $Z_1 + Z_2$ in parallel is equivalent to two admittances Y_1 and Y_2 in series.

Resonance of a network

Consider the simple R, L, C, series circuit of Fig. 17.28 excited by an applied sinusoidal voltage of angular frequency ω. Suppose the complex voltage is V then the complex current I flowing in the network is given by

$$I = V/Z$$

where $Z = R + j(\omega L - 1/\omega C)$

The magnitude of the current, $|I|$, for given values of R, L, and C is dependent on ω and reaches a maximum value when $|Z|$, i.e.

$$\sqrt{R^2 + (\omega L - 1/\omega C)^2}$$

is a minimum. This minimum value of $|Z|$, for varying ω, is given when $\omega = \omega_0$ such that

$$\omega_0 L = \frac{1}{\omega_0 C}$$

i.e.

$$\omega_0 = 1/\sqrt{LC}.$$

The impedance Z is then purely resistive with the value R.

The graph of $|I|$ plotted against ω has the shape of Fig. 17.33 and is known as a *resonance curve*, the frequency ω_0 being called the resonance frequency.

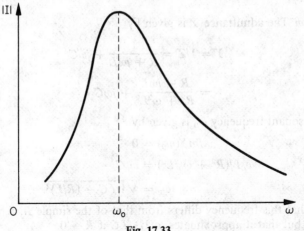

Fig. 17.33

Note that at this resonant frequency the impedance of the circuit is purely resistive. This fact provides a simple criterion for resonance of any circuit, namely that at resonance the imaginary part of the complex impedance Z of the circuit is zero. For circuits with elements in parallel it may be more convenient to work in terms of admittance. Now suppose

$$Y = 1/Z = A + jB$$

then

$$Z = \frac{A}{A^2 + B^2} - j\frac{B}{A^2 + B^2}$$

and the condition for resonance is

$$Im Z = 0$$

i.e.

$$B = 0$$

which is the same as $Im Y = 0$. Hence for resonance the imaginary part of either the circuit impedance or admittance is zero.

Example Find the resonance frequency for the circuit of Fig. 17.34.

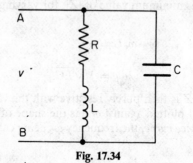

Fig. 17.34

Solution The admittance Y is given by

$$Y = 1/Z = \frac{1}{R + j\omega L} + j\omega C$$

$$= \frac{R - j\omega L}{R^2 + \omega^2 L^2} + j\omega C$$

The resonant frequency ω_0 is given by

$$Im Y(\omega_0) = 0$$

i.e.

$$\omega_0 L/(R^2 + \omega_0{}^2 L^2) = \omega_0 C$$

i.e.

$$\omega_0 = \sqrt{1/LC - (R/L)^2}.$$

Note that this frequency differs from that of the simple R, L, C series circuit but that it approximates to $1/\sqrt{LC}$ if $R \to 0$.

Resonant circuits play an important role in the electric circuitry of radio, television, measuring and filtering equipment. Consider for example the network of Fig. 17.22 of Section 17.3.3. The impedance of such a network when an input voltage of frequency ω is applied, is given by

$$Z = j\omega L + Z_1$$

where

$$\frac{1}{Z_1} = Y_1 = \frac{1}{-j/\omega C} + \frac{1}{j\omega L - j/\omega C}$$

$$= j\omega C - \frac{j}{(\omega L - 1/\omega C)}$$

and hence Z becomes

$$Z = j\omega L - j[\omega C - 1/(\omega L - 1/\omega C)]^{-1}$$

For resonance it is required that $ReZ = 0$, i.e. that

$$\omega L = \frac{1}{\omega C - 1/(\omega L - 1/\omega C)}$$

and at the frequency given by this equation the circuit presents a (zero) resistive impedance.

If $\omega^2 LC = \lambda$, then the resonance condition is

$$\lambda = \frac{1}{1 - 1/(\lambda - 1)}$$

i.e. $\lambda^2 - 3\lambda + 1 = 0.$

This gives the *two* natural frequency of resonance as already obtained in the earlier work on filter theory of Section 17.3.3. (see page 493).

The A.C. Bridge

Consider a network of impedances as in Fig. 17.35 with complex cyclic currents I_1, I_2, and I_3 due to a complex voltage V all at frequency ω. The network equations are given by

$$I_1 Z_1 + (I_1 - I_2)Z_5 + (I_1 - I_3)Z_4 = 0$$
$$(I_2 - I_1)Z_5 + I_2 Z_2 + (I_2 - I_3)Z_3 = 0$$
$$(I_3 - I_1)Z_4 + (I_3 - I_2)Z_3 = V.$$

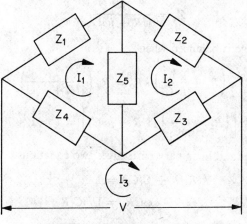

Fig. 17.35

This type of A.C. bridge is adjusted by varying the impedances so that the current flow in the indicator Z_5 (e.g. ear-phones) is zero. This requirement is $I_1 - I_2 = 0$, with I_1, I_2, and I_3 non-zero. With zero current in Z_5 the first pair of the above three equations reduce to

$$I_1(Z_1 + Z_4) - I_3 Z_4 = 0$$

$$I_1(Z_2 + Z_3) - I_3 Z_3 = 0$$

from which it follows that I_1, I_3 are non-zero providing

$$\begin{vmatrix} Z_1 + Z_4 & Z_4 \\ Z_2 + Z_3 & Z_3 \end{vmatrix} = 0$$

i.e.
$$Z_1 Z_3 = Z_2 Z_4$$

or
$$Z_1/Z_4 = Z_2/Z_3.$$

Thus the requirement that there is no current flow in Z_5, i.e. that the bridge is *balanced*, is that the impedance in the four arms of the bridge are in the above ratios at the frequency ω. Since the impedances are, generally complex quantities then by taking real and imaginary parts of the condition, two relationships may be obtained.

The types of A.C. bridges and the uses to which they may be put are many and varied. As an example consider the particular bridge of Fig. 17.36 known as the Wien Bridge in which the resistance R_1 and capacitance C_1 are unknown and k is a real positive number.

Relating Figs. 17.35 and 17.36 then

$$Z_1 = R_1 + 1/j\omega C_1,$$

$$Z_2 = R_0, \quad Z_3 = kR_0,$$

$$Z_4 = \frac{1}{1/R + j\omega C},$$

and the bridge is balanced when

$$\left(R_1 + \frac{1}{j\omega C_1} \right) kR_0 = R_0 \left(\frac{1}{1/R + j\omega C} \right)$$

i.e.
$$k \left[\left(R_1 - \frac{j}{\omega C_1} \right) (1/R + j\omega C) \right] = 1 = 1 + j0.$$

Equating real and imaginary parts gives two equations

$$k(R_1/R + C/C_1) = 1$$

$$\omega C R_1 = 1/\omega C_1 R$$

and, solving for C_1 and R_1 gives

$$C_1 = kC(1 + 1/\omega^2 C^2 R^2)$$

$$R_1 = R/k(1 + \omega^2 C^2 R^2).$$

The values of C_1 and R_1 can therefore be found at frequency ω. Alternatively if C_1 and R_1 are known it is possible to use this device to measure the frequency ω of an oscillatory voltage.

Suppose that $C_1 = C$ and $R_1 = R$ (this is easily achieved by a ganging arrangement) then when the bridge is balanced the above equations lead to

$$k = \tfrac{1}{2} \quad \text{and} \quad \omega = 1/CR$$

Thus if k is given the value of $1/2$ and the resistances R_1, R and capacitances C_1, C are coupled together so that they have identical known values the frequency of the oscillation is obtained by calculating $1/RC$ when the bridge is balanced.

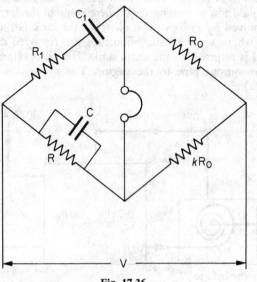

Fig. 17.36

17.6 DIFFERENTIAL EQUATIONS

Engineering is frequently concerned with the rates of change of certain variables and the physical problem is often expressed simply as an *equation of balance*, e.g. a balance of mass, or heat, or energy. The balance equation itself may in the physical sense be a simple statement such as

$$\text{Input} - \text{Output} = \text{Accumulation}$$

over a specified time interval, or even an empirical law such as

$$\text{Force} = \text{Mass} \times \text{Acceleration}.$$

When the statement is expressed using mathematical symbols it often represents a differential equation (derivatives being involved because of the involvement with rates of change). In this way a simple physical statement may lead to a (sometimes complicated) differential equation.

If this equation can be solved, either analytically or numerically, then the engineer must interpret the mathematical solution in.terms of the physical variables. Consider now two engineering type problems which lead to differential equations.

17.6.1 The oil purifier

The oil used in lubricating a ship's engine becomes dirty after a certain time of running. It is not always possible, or necessary, to change the oil and a purifier may be installed within the ship. This is usually of the centrifuge type in which the dirt particles and other impurities are removed by centrifugal action.

The mechanical process consists of passing the impure oil from the engine into a drain or settling tank where some of the larger impurities can be removed by gravity. The oil from the tank is then fed to the centrifuge where some of the impurities are removed and the part-purified oil is returned to the drain tank. The oil is returned from the tank via an output pipe to the engine. The process is illustrated in Fig. 17.37(a).

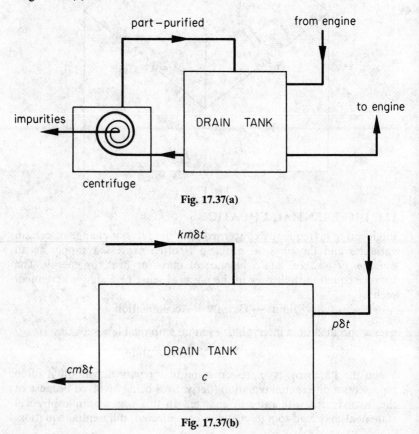

Fig. 17.37(a)

Fig. 17.37(b)

The physical situation is analysed mathematically by forming a balance equation on the mass of impurities in the system. The Fig. 17.37(b) shows the input and output of impurities in the drain tank. Let p be the net mass of impurities entering the tank per second; let c be the concentration of mass impurities in the tank at time t; let k be the concentration of mass impurities returned to the tank from the centrifuge; let m be the mass of oil entering the centrifuge per second; let M be the total mass of oil. The physical mass balance equation for the impurities in the drain tank (see Fig. 17.37(b)) is

mass input $-$ mass output $=$ mass accumulation.

If the concentration of impurities in the tank changes from c to $c + \delta c$ in the time interval t to $t + \delta t$ then the mathematical statement of the balance equation is

$$\{p\delta t + k(m\delta t)\} - \{cm\delta t\} = \{M(c + \delta c) - Mc\}.$$

Dividing by δt and proceeding to the limit as $\delta t \to 0$,

$$M\frac{dc}{dt} = p + km - cm$$

i.e. $$\frac{dc}{dt} + \frac{m}{M}c = \frac{p + km}{M}$$

which is a differential equation of the form of equation (15) of Section 11.1.5, i.e. $dy/dx + ay = Q(x)$ with $y = c$; $x = t$; $a = m/M$, $Q(x) = (p + km)/M$. The solution is given by equation (18) of 11.1.5, i.e.

$$c = A\,\mathrm{e}^{-mt/M} + (p + km)/m$$

where A is the arbitrary constant of integration. Assuming that the initial concentration at $t = 0$ was $c = c_0$ then

$$c_0 = A + (p + km)/m$$

and hence $$c = c_0\,\mathrm{e}^{-mt/M} + (k + p/m)(1 - \mathrm{e}^{-mt/M}).$$

The graph of c (the concentration of impurities in the oil) as a function of t is illustrated in Fig. 17.38.

The ultimate (i.e. $t \to \infty$) concentration is given by $(k + p/m)$ and even with no input from the engine ($p = 0$) it is impossible to reduce the concentration below the value k. Any further reduction in concentration can only be achieved by reducing k, i.e. designing a more efficient purifier. Writing

$$(c - k - p/m) = (c_0 - k - p/m)\,\mathrm{e}^{-mt/M}$$

the concentration will fall half-way towards its ultimate value when

$$\mathrm{e}^{-mt/M} = \tfrac{1}{2},$$

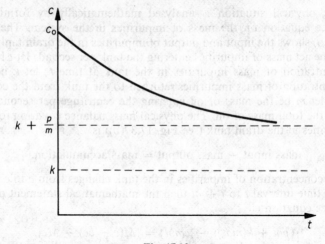

Fig. 17.38

i.e. the half-way concentration is reached after a time $t = (M/m) \log 2$. This half-life time can be reduced only by increasing m (for a given total mass M) and this again modifies the design of the purifier. This design is optimised by taking all possible factors of the centrifuge into account, e.g. the speed, the maximisation of the input, the maximisation of the purification and the physical size limitation.

17.6.2 Chemical reactions

Reactions between chemical materials to produce a yield is a process which is of great importance in the chemical industry. The types of reactions are many and varied. Consider first only the simplest kind of reaction in which a chemical material A transforms into another material B possibly by reacting with a catalyst (which itself undergoes no change). There is assumed to be no change in volume of the system.

If the concentrations of materials A and B at time t are c_A and c_B respectively then in a time interval δt the mass balance equation (i.e. mass flow in − mass flow out = mass accumulation) is given by (see Fig. 17.39)

$$0 - kc_A\delta t = (c_A + \delta c_A) - c_A, \text{ for material } A$$

and $$kc_A\delta t - 0 = (c_B + \delta c_B) - c_B, \text{ for material } B.$$

where k is a constant called the *reaction constant* or *rate coefficient*.

Dividing by δt and proceeding to the limit leads to a pair of differential equations,

$$\frac{dc_A}{dt} = -kc_A$$

$$\frac{dc_B}{dt} = +kc_A.$$

If the initial concentration of A is a then the first equation is easily integrated to give $c_A = a\,e^{-kt}$. If further the initial concentration of B is zero the second differential equation leads to $c_B = a(1 - e^{-kt})$.

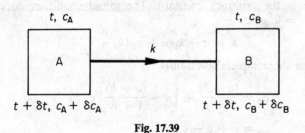

Fig. 17.39

Note that this simple type of reaction proceeds until eventually ($t \to \infty$), c_A reduces to zero and c_B is a, i.e. all of A is converted into B.

A more complicated situation arises in which B may react and convert back into A but at a different rate. This situation is illustrated in Fig. 17.40 in which the forward rate k_1 is different to the backward rate k_2.

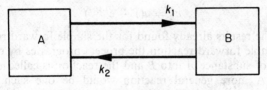

Fig. 17.40

The equations of reaction are now found to be

$$\frac{dc_A}{dt} = k_2 c_B - k_1 c_A$$

and

$$\frac{dc_B}{dt} = k_1 c_A - k_2 c_B.$$

These differential equations can be solved by noting that $dc_A/dt + dc_B/dt = 0$.

Thus, $$c_A + c_B = \text{constant}$$
$$= a + b$$

where a and b are the initial concentrations of c_A and c_B respectively.

Substituting into the first differential equation leads to

$$\frac{dc_A}{dt} = k_2(a + b - c_A) - k_1 c_A$$

i.e.

$$\frac{dc_A}{dt} + (k_1 + k_2)c_A = k_2(a + b)$$

which has the solution (see Section 11.1.5)

$$c_A = E\,e^{-(k_1+k_2)t} + k_2(a+b)/(k_1+k_2)$$

where E is the arbitrary constant. The initial condition on c_A, i.e. $c_A = a$, $t = 0$, gives

$$E = a - k_2(a+b)/(k_1+k_2)$$

and hence the complete solution is

$$c_A = \frac{(a+b)}{1+K} + \left[a - \frac{(a+b)}{1+K}\right]e^{-(k_1+k_2)t}$$

$$c_B = a + b - c_A,$$

where $K = k_1/k_2$.

The limiting (i.e. $t \to \infty$) concentrations of c_A and c_B are then $(a+b)/(1+K)$ and $K(a+b)/(1+K)$ respectively.

Note that in the particular case $b = 0$ and $k_1 \gg k_2$ (i.e. $k_2 \to 0$, $K \to \infty$) the concentrations become

$$c_A = a\,e^{-k_1 t}$$

and

$$c_B = a(1 - e^{-k_1 t})$$

which are the results already found for the simple forward reaction.

In the simple forward reaction the process progresses by converting a molecule of substance A into B and the reaction is called one of the *first order*. A more general reaction would be one such that two reacting substances A and B (say) are brought together and one molecule of A reacts with one molecule of B to form some resultant product (or products), i.e.

$$A + B \to \text{Products.}$$

Generally the reaction rate is dependent on the concentrations of both A and B and a generalised form of the governing differential equation for the rate of change of A is

$$\frac{dc_A}{dt} = -k_n c_A{}^\alpha c_B{}^\beta$$

where α, and β are the reaction orders with respect to A and B respectively and $\alpha + \beta = n$ is the overall reaction rate, and k_n is the rate constant. In a general reaction α and β need not be integers, but usually $n < 3$. The reaction orders (i.e. the values of α and β) are usually deduced from experimental observations. A particularly useful law followed by a number of reactions is that for which $\alpha = \beta = 1$, i.e. $n = 2$ and the reaction is called *second order*. The governing equations for this reaction are

$$\frac{dc_A}{dt} = \frac{dc_B}{dt} = -k_2 c_A c_B.$$

This differential equation may be integrated on noting that

$$\frac{d}{dt}(c_A - c_B) = 0.$$

i.e. $\qquad c_A - c_B = \text{constant} = a - b = \lambda \text{ (say)}$

where a and b are the initial concentrations of A and B respectively. The differential equation for c_A then has the form

$$\frac{dc_A}{dt} = -k_2 c_A(c_A - \lambda)$$

which is of the type of Section 11.1.2, i.e. variables separable. Hence

$$\int \frac{dc_A}{c_A(c_A - \lambda)} = -\int k_2 dt$$

or $\qquad \displaystyle\int \left(\frac{1}{c_A - \lambda} - \frac{1}{c_A}\right) dc_A = -\lambda k_2 t + \text{constant}$

which leads to

$$\frac{c_A - \lambda}{c_A} = D e^{-\lambda k_2 t}$$

where D is an arbitrary constant of integration.

Inserting the initial condition that $c_A = a$, $t = 0$ leads to the evaluation of D and hence to

$$c_A = \frac{a - b}{1 - (b/a)\,e^{-k_2(a-b)t}}.$$

More complicated reactions may take place and possibly unwanted effects may occur, for instance although it may be required to convert material A into material B it might happen that a further material C is produced by a forward reaction from B as in Fig. 17.41(a) or by a side reaction from A as in Fig. 17.41(b).

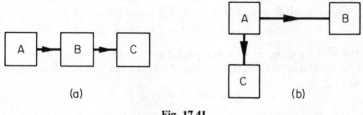

(a)

(b)

Fig. 17.41

Consider as a final example the case of a reversible reaction A to B taking place and an unwanted irreversible side reaction to X also taking place as in Fig. 17.42 where k_1 and k_3 are the reaction constants A to B and B to A respectively and k_2 is the reaction constant from A to X.

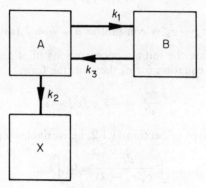

Fig. 17.42

The governing equations are

$$\frac{dc_A}{dt} = -k_1 c_A + k_3 c_B - k_2 c_A$$

$$\frac{dc_B}{dt} = k_1 c_A - k_3 c_B$$

$$\frac{dc_X}{dt} = k_2 c_A.$$

The first pair of equations do not contain c_X and hence can be solved for c_A and c_B. Write $D \equiv d/dx$, then

$$(D + k_1 + k_2)c_A - k_3 c_B = 0$$
$$-k_1 c_A + (D + k_3)c_B = 0.$$

Operating on the first equation with $(D + k_3)$ and multiplying the second by k_3 and adding the resulting equations leads to

$$\{(D + k_3)(D + k_1 + k_2) - k_1 k_3\}c_A = 0$$

or
$$(D^2 + KD + L)c_A = 0$$

where $K = k_1 + k_2 + k_3$ and $L = k_2 k_3$.

This second order differential equation has the solution (see Section 11.4)

$$c_A = P\,e^{-\alpha t} + Q\,e^{-\beta t}$$

where P, Q are arbitrary constants and α, β are given by

$$\alpha, \beta = \frac{K \pm \sqrt{K^2 - 4L}}{2}.$$

and are both real and positive. The concentration c_B may now be found from the first of the governing equations, i.e.

$$k_3 c_B = \frac{d}{dt} c_A + k_1 c_A + k_2 c_A.$$

$$= P(k_1 + k_2 - \alpha) e^{-\alpha t} + Q(k_1 + k_2 - \beta) e^{-\beta t}$$

and finally c_X from the last of the governing equations

$$c_X = R - \frac{k_2 P}{\alpha} e^{-\alpha t} - \frac{k_3 Q}{\beta} e^{-\beta t}$$

where R is a further integrational constant. The 3 arbitrary constants P, Q, and R may be determined from the initial conditions of concentrations, i.e. if a is the initial concentration of c_A and the initial concentrations c_B and c_X are both zero then

$$a = P + Q$$

$$0 = P(k_1 + k_2 - \alpha) + Q(k_1 + k_2 - \beta)$$

$$0 = R - k_2 P/\alpha - k_3 Q/\beta.$$

giving three equations for the three constants P, Q, and R.

17.7 PARTIAL DIFFERENTIATION

Many physical quantities are functions of several (i.e. 2 or more) independent variables, e.g. the volume V of a circular cylinder is a function of its height h and radius of cross section r and $V = \pi r^2 h$; the pressure p of a perfect gas is a function of temperature T and volume v and $p = RT/v$ where R is a constant.

Engineering situations are frequently concerned with the change of one physical quantity with respect to variations in the other variables. These rates of change lead to partial derivatives, e.g. if $p = RT/v$ then the rate of change of pressure at constant volume v is $\partial p/\partial T$ while the rate of change of pressure at constant temperature T is $\partial p/\partial v$. Likewise it is possible to have other derivatives such as $\partial T/\partial p$ and $\partial v/\partial T$. Consider now two fields of engineering study that demand the use of partial differentiation.

17.7.1 Elasticity. The displacement

Suppose a straight bar of elastic material is lying along the x-axis from $x = 0$ to $x = L$. Let the end $x = 0$ be fixed and the other end be subject to a longitudinal force that causes the bar to extend. The Fig. 17.43(a) shows the static equilibrium state and Fig. 17.43(b) shows the situation after stretching. Because this motion takes place in one-dimension, i.e. that of the x coordinate, it is to be expected that any rates of change in this case will not lead to a partial derivative.

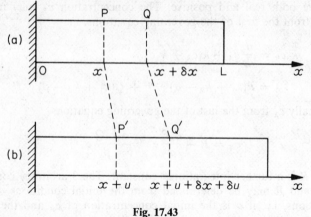

Fig. 17.43

The point P moves to a new position P' and is said to have a *displacement* of u. The point Q (originally at $x + \delta x$) takes up its position at Q' (i.e. the point $x + u + \delta x + \delta u$), and the element PQ (originally δx) has changed to $P'Q'$ (now of length $\delta x + \delta u$). Different points P along the bar will have different displacements, i.e. u is a function of x, $u = f(x)$.

Then

$$\delta u = f(x + \delta x) - f(x)$$

$$= \frac{df}{dx}\,\delta x + \frac{1}{2!}\frac{d^2 f}{dx^2}(\delta x)^2 + \frac{1}{3!}\frac{d^3 f}{dx^3}(\delta x)^3 + \cdots$$

after expanding $f(x + \delta x)$ by Taylor's theorem.

In the simple theory of elasticity it is assumed that the change in u, i.e. δu, is small enough to be represented by the first term in the series i.e.

$$du \simeq \frac{du}{dx}\,\delta x$$

This function du/dx is called the *rate of displacement* or the *displacement gradient* and has an important role in Elasticity theory. Interest is centred on the change in length of the small element PQ and as a measure of this change it is usual to compare the change in length $(P'Q' - PQ)$ with the original length PQ and to examine this ratio as the length element PQ tends to zero. This quantity is called the *strain* at the point P and takes the symbol ε, i.e.

$$\varepsilon = \lim_{PQ \to 0} \frac{P'Q' - PQ}{PQ}$$

$$= \lim_{\delta x \to 0} \frac{(du/dx)\delta x + \cdots}{\delta x}$$

$$= du/dx.$$

The ideas inherent in this measurement of strain can be extended to two-dimensional elasticity. Suppose a sheet of elastic material occupies a part of the usual two-dimensional space of x and y and is distorted by a tension. The elementary length δx in the x-direction undergoes a change but so also does an elementary length δy in the y-direction. The Fig. 17.44 shows a small rectangular element $PQRS$ of the elastic medium such that P is the point (x, y) and the sides of the element are δx and δy. After a tension is applied to the medium the point P moves to P' $(x + u, y + v)$ and the other points take up new positions as indicated in Fig. 17.44.

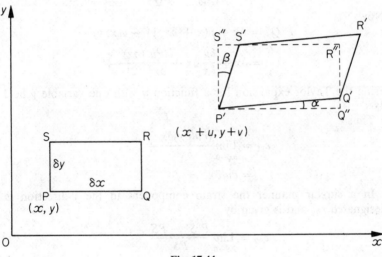

Fig. 17.44

The distortion consists of 3 distinct parts, firstly a translation as a rectangle so that P takes up the position P', secondly an elongation of the rectangle to the shape $P'Q''R''S''$ and thirdly a rotation of the sides to take up new position $P'Q'R'S$. The first part, i.e. the translational movement may be ignored because the relative position of points are unchanged. The third part of the distortion is represented by the angles α and β of Fig. 17.44.

The quantities u, v are the components of the displacement PP' and are functions of position, i.e. $u = u(x, y)$ and $v = v(x, y)$.

The coordinates of the points *relative to* P are as follows (assuming α and β are small):

$$Q: (\delta x, 0)$$
$$S: (0, \delta y)$$
$$P': (u, v), \text{ i.e. } (u(x, y), v(x, y))$$
$$Q': (\delta x + u(x + \delta x, y), v(x + \delta x, y))$$

$$Q'': (\delta x + u(x + \delta x, y), v(x, y))$$
$$S': (u(x, y + \delta y), \delta y + v(x, y + \delta y))$$
$$S'': (u(x, y), \delta y + v(x, y + \delta y)).$$

The component of strain in the x-direction is designated ε_{11} and is the unit change in length of the elementary length δx, i.e.

$$\varepsilon_{11} = \operatorname*{Lim}_{PQ \to 0} \frac{P'Q' - PQ}{PQ}$$

$$= \operatorname*{Lim}_{\delta x \to 0} \frac{P'Q'' - PQ}{PQ}$$

Now

$$P'Q'' = \delta x + u(x + \delta x, y) - u(x, y)$$

$$= \delta x + \frac{\partial u}{\partial x} \delta x + \frac{\partial^2 u}{\delta x^2} \frac{(\delta x)^2}{2!} + \cdots$$

using the Taylor expansion for a function u with one variable y held fixed.

Thus

$$\varepsilon_{11} = \operatorname*{Lim}_{\delta x \to 0} \frac{(\partial u/\partial x)\delta x}{\partial x} + \cdots$$

$$= \partial u/\partial x.$$

In a similar manner the strain component in the y-direction is designated ε_{22} and is given by

$$\varepsilon_{22} = \operatorname*{Lim}_{PS \to 0} \frac{P'S' - PS}{PS}$$

$$= \operatorname*{Lim}_{\delta y \to 0} \frac{P'S'' - PS}{PS}$$

$$= \operatorname*{Lim}_{\delta y \to 0} \frac{(\partial v/\partial y)\delta y}{\partial y} + \cdots$$

$$= \partial v/\partial y.$$

The extent of the shearing action which has distorted $P'Q''R''S''$ to $P'Q'R'S'$ may be examined by forming the difference between the right-angle QPS and the new angle $Q'P'S'$, i.e. in Fig. 17.44 by the angle $(\alpha + \beta)$.

Now
$$\alpha \simeq \tan \alpha \simeq \operatorname*{Lim}_{\delta x \to 0} \frac{Q'Q''}{P'Q''}$$

but
$$Q'Q'' \simeq v(x + \delta x, y) - v(x, y)$$

$$= \frac{\partial v}{\delta x} \delta x + \frac{\delta^2 v}{\delta x^2} \frac{(\delta x)^2}{2!} + \cdots$$

and $P'Q'' \simeq \delta x + \partial u/\partial x\, \delta x.$

Hence, assuming that $\partial u/\partial x \ll 1$, then

$$\alpha \simeq \mathrm{Lim} \frac{(\partial v/\partial x)\delta x}{\delta x} + \ldots$$

$$= \partial v/\partial x.$$

Similarly

$$\beta \simeq \partial u/\partial y$$

and the shearing effect may be measured by

$$\alpha + \beta = \frac{\partial v}{\partial x} + \frac{\partial u}{\partial y}.$$

This quantity is designated γ_{12} and is usually called the *engineering shear strain*.

Note (1) The suffixes 1, 2 refer to the x- and y-directions respectively. Some texts designate the strain components as ε_{xx}, ε_{yy}, and γ_{xy}.

Note (2) It may be observed that γ_{21} can be defined by interchanging the roles of x, y and of u, v in γ_{12}, i.e. $\gamma_{21} = \partial u/\partial y + \partial v/\partial x$ and it follows that $\gamma_{12} = \gamma_{21}$.

These ideas may be extended to a three-dimensional elastic solid in which a point (x, y, z) is displaced to a new point $(x + u, y + v, z + w)$. The strain components are then given by

$$e_{11} = \partial u/\partial x \qquad \gamma_{12} = \partial u/\partial y + \partial v/\partial x = \gamma_{21}$$

$$\varepsilon_{22} = \partial v/\partial y \qquad \gamma_{13} = \partial u/\partial z + \partial w/\partial x = \gamma_{31}$$

$$\varepsilon_{33} = \partial w/\partial z \qquad \gamma_{23} = \partial v/\partial z + \partial w/\partial y = \gamma_{32}.$$

From a mathematical viewpoint it is more convenient to use the alternative notation (x_1, x_2, x_3) for (x, y, z) and (u_1, u_2, u_3) for (u, v, w). In this notation the suffixes $(1, 2, 3)$ refer to the three directions of $x, y,$ and z.

Then

$$\varepsilon_{11} = \partial u_1/\partial x_1 \qquad \gamma_{12} = \partial u_1/\partial x_1 + \partial u_2/\partial x_2 = \gamma_{21}$$

$$\varepsilon_{22} = \partial u_2/\partial x_2 \qquad \gamma_{13} = \partial u_1/\partial x_3 + \partial u_3/\partial x_1 = \gamma_{31}$$

$$\varepsilon_{33} = \partial u_3/\partial x_3 \qquad \gamma_{23} = \partial u_2/\partial x_3 + \partial u_3/\partial x_2 = \gamma_{32}.$$

If further the *mathematical shear strain* is introduced by writing

$$\varepsilon_{12} = \tfrac{1}{2}\gamma_{12} = \tfrac{1}{2}(\partial u_1/\partial x_2 + \partial u_2/\partial x_1)$$

$$\varepsilon_{13} = \tfrac{1}{2}\gamma_{13}$$

$$\varepsilon_{23} = \tfrac{1}{2}\gamma_{23}$$

then the six (mathematical) strain components may be written as a *strain matrix* ε_{ij}, $i, j = 1, 2, 3$, i.e.

$$(\varepsilon_{ij}) = \begin{pmatrix} \varepsilon_{11} & \varepsilon_{12} & \varepsilon_{13} \\ \varepsilon_{21} & \varepsilon_{22} & \varepsilon_{23} \\ \varepsilon_{31} & \varepsilon_{32} & \varepsilon_{33} \end{pmatrix}$$

in which $\varepsilon_{ij} = \frac{1}{2}(\partial u_i/\partial x_j + \partial u_j/\partial x_i)$.

This particular matrix array (ε_{ij}) has further properties which are of importance in the mathematical theory of elasticity. In the form (ε_{ij}) the above array is known as the *strain tensor*.

17.7.2 Fluid dynamics

A. The material derivative

Suppose a fluid (e.g. gas or liquid) particle at time t_0 is at a point P_0 whose position vector is **R** with reference to a fixed point O. At time t suppose the same fluid particle is at the point P whose position vector is **r**. The Fig. 17.45 shows the path of the particle from P_0 to P. Note that, for fixed **R**, the vector **r** is a function of t only and $\mathbf{r} = \mathbf{F}(t)$.

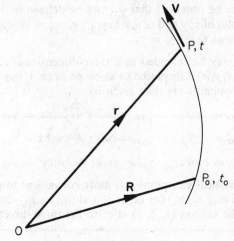

Fig. 17.45

The velocity **V** of the particle at P is given by

$$\mathbf{V} = \frac{d\mathbf{r}}{dt} = \frac{d}{dt}\mathbf{F}(t).$$

In Cartesian coordinates with O as origin, $\mathbf{r} = x\mathbf{e}_1 + y\mathbf{e}_2 + z\mathbf{e}_3$ and the three components of the velocity (i.e. u, v, and w) are $u = dx/dt$; $v = dy/dt$, $w = dz/dt$. The particle at point P of the fluid has associated with it other vector functions such as acceleration and force and certain scalar functions such as pressure, density, and temperature. Consider first one of these scalar functions, say pressure p.

The pressure p at any point P may be considered as a function of both its position and of time, e.g. the pressure varies over a region at fixed time and also varies over a period of time at a fixed point. Hence p may be written $p = f(x, y, z, t)$. However, since $\mathbf{r}$ is a function of t only it follows that x, y, z, are functions of t only and the above relationship may be re-written

$$p = f[x(t), y(t), z(t), t]$$
$$= \phi(t)$$

where ϕ is the function f when expressed in terms of t only.

It is possible to differentiate p partially with respect to x, y, z and t, i.e. $\partial f/\partial x$, $\partial f/\partial y$, $\partial f/\partial z$, and $\partial f/\partial t$ and it is also possible to differentiate p as a function of t only, i.e. $d\phi/dt$.

An application of equation (27) of Section 5.6.2 in which $n = 4$, and x_1, x_2, x_3, x_4 are replaced by x, y, z, t leads to

$$\frac{d\phi}{dt} = \frac{\partial f}{\partial x}\frac{dx}{dt} + \frac{\partial f}{\partial y}\frac{dy}{dt} + \frac{\partial f}{\partial z}\frac{dz}{dt} + \frac{\partial f}{\partial t}\frac{dt}{dt}$$

or in terms of the pressure p

$$\frac{dp}{dt} = \frac{\partial p}{\partial t} + u\frac{\partial p}{\partial x} + v\frac{\partial p}{\partial y} + w\frac{\partial p}{\partial z}$$

where $u = dx/dt$, etc.

It is worth stressing that although the same symbol p appears on both sides of this last result the function represented by p is different. It is very important to be able to distinguish between $\partial p/\partial t$ which means

$$\left[\frac{\partial f}{\partial t}(x, y, z, t)\right]_{x,y,z \text{ fixed}}$$

and dp/dt which means $(d/dt)\phi(t)$. The term $\partial p/\partial t$ on the right side of the above result represents the rate of change of pressure with respect to time when P is considered fixed, the other terms of the right side represent the rate of change of pressure at fixed time t as P moves with the fluid. The left side is then the total rate of change of pressure. This type of differentiation is often called "*differentiation following the motion of the fluid*" or simply "*the material derivative*".

In exactly similar fashion any of the other scalar or vector functions which depend on position and time may be analysed. For example the density ρ may be differentiated to give

$$\frac{d\rho}{dt} = \frac{\partial \rho}{\partial t} + u\frac{\partial \rho}{\partial x} + v\frac{\partial \rho}{\partial y} + w\frac{\partial \rho}{\partial z}$$

and the velocity $\mathbf{V}$ may be differentiated to give

$$\frac{d\mathbf{V}}{dt} = \frac{\partial \mathbf{V}}{\partial t} + u\frac{\partial \mathbf{V}}{\partial t} + v\frac{\partial \mathbf{V}}{\partial t} + w\frac{\partial \mathbf{V}}{\partial t}.$$

This latter result being an expression for the rate of change of velocity is simply the acceleration of the fluid particle at the point P at time t and naturally plays an important role in the derivation of the equation of motion of a fluid. Although the equation of motion will not be derived here it is possible to state that an application of Newton's Second Law, that applied force is proportional to rate of change of momentum, can lead to an equation of the type

$$\frac{d\mathbf{V}}{dt} = \mathbf{R}$$

where $\mathbf{R}$ represents all the forces acting on a unit of mass of the fluid at P (e.g. body forces due to gravity or spin, pressure forces, viscous forces, electromagnetic forces).

Finally, note that $\partial p/\partial x$, $\partial p/\partial y$, $\partial p/\partial z$ are the 3 components of grad p and that, since (u, v, w) are the three components of $\mathbf{V}$,

$$u\frac{\partial p}{\partial x} + v\frac{\partial p}{\partial y} + w\frac{\partial p}{\partial z}$$

may be written as a scalar product

$$\mathbf{V} . \text{grad } p$$

or

$$\mathbf{V} . \nabla p.$$

The operational form of the material derivative plays an important role in fluid dynamics and has the form

$$\frac{d}{dt} = \frac{\partial}{\partial t} + u\frac{\partial}{\partial x} + v\frac{\partial}{\partial y} + w\frac{\partial}{\partial z}$$

or

$$= \frac{\partial}{\partial t} + (\mathbf{V} . \text{grad})$$

or

$$= \frac{\partial}{\partial t} + (\mathbf{V} . \nabla).$$

B. The Continuity equation

An important concept in fluid dynamics is that, although the material is moving and its volume and density are changing, the mass of any given volume is conserved, i.e. that mass is not annihilated or created.

It is possible to arrive at the mathematical equation of conservation by considering any arbitrary shaped volume in the fluid but the mathematical processes involved are outside the scope of this book. However the same equation can be obtained for a particular shaped volume of fluid, namely a rectangular parallelepiped. The physical property of mass conservation is simply a balance equation over the rectangular parallelepiped that

mass flow in — mass flow out = mass accumulation

and the mathematical statement is a partial differential equation.

Suppose P is the point (x, y, z) of Fig. 17.46 and the sides of the rectangular volume are parallel to the x, y, and z axes and of lengths δx, δy, and δz respectively, e.g. W is the point $(x + \delta x, y + \delta y, z + \delta z)$. At P the fluid velocity $\mathbf{V}(x, y, z, t)$ has components (u, v, w) and

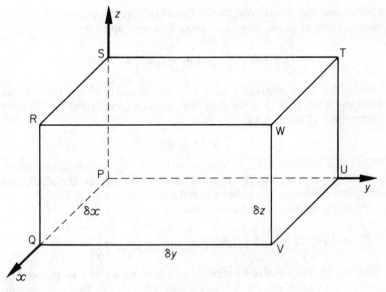

Fig. 17.46

the density is $\rho(x, y, z, t)$. Consider the flow in the y-direction across the face $PQRS$. The only contribution to this flow *into* the region is due to the velocity component v in the direction y. The other two velocity components u, w being parallel to the face $PQRS$ do not contribute anything to the flow *into* the parallelepiped. To the first order of small quantities it may be assumed that the velocity component v is constant over the face and hence the mass flow into the region in time δt is (density) $\times$ (area) $\times$ (velocity) $\times$ (time), i.e.

$$(\rho v)_{\text{at } x,y,z} \, \partial x \, \delta z \, \delta t$$

The flow *out* of the region over the parallel face $UVWT$ at $y + \delta y$ is

$$(\rho v)_{\text{at } x, y+\delta y, z} \, \delta x \, \delta z \, \delta t$$

The function $(\rho v)_{x, y+\delta y, z}$ has a Taylor expansion (using equation (18) of Section 5.4 with $h = 0$, $k = \delta y$ and inserting a third variable z), i.e.

$$(\rho v)_{x, y+\delta y, z} = (\rho v)_{\text{at } x, y, z} + \delta y \frac{\partial}{\partial y} (\rho v) + \cdots$$

Retaining only the first two terms of the expansion the difference (mass flow *into* region over $PQRS$ — mass flow *out* of region over $UVWT$) is given by

$$-\frac{\partial}{\partial y}(\rho v)\,\delta x\,\delta y\,\delta z\,\delta t$$

Calculating the similar results for the other four faces leads to a total contribution of (mass flow in — mass flow out) given by

$$-\left[\frac{\partial}{\partial x}(\rho u)+\frac{\partial}{\partial y}(\rho v)+\frac{\partial}{\partial t}(\rho w)\right]\delta x\,\delta y\,\delta z\,\delta t$$

During the time interval t to $t+\delta t$ the density within the volume changes from ρ to $\rho+\delta\rho$ and the mass accumulation is (Density increase) × (Volume), i.e.

$$\delta\rho\,\delta x\,\delta y\,\delta z$$

(Note this assumes that the density over the region is uniform which is strictly not true, but the error is small, if δx, δy, δz are small, and decreases to zero as the volume shrinks to zero.)

Forming the balance equation leads to

$$\delta\rho\,\delta x\,\delta y\,\delta z=-\left(\frac{\partial}{\partial x}(\rho u)+\frac{\partial}{\partial y}(\rho v)+\frac{\partial}{\partial t}(\rho w)\right)\delta x\,\delta y\,\delta z\,\delta t.$$

Dividing by the volume element $(\delta x\,\delta y\,\delta z)$ and by δt and proceeding to the limit so that δx, δy, δz, $\delta t\to 0$ and remembering that ρ is function of (x,y,z,t) leads to

$$\frac{\partial\rho}{\partial t}+\frac{\partial}{\partial x}(\rho u)+\frac{\partial}{\partial y}(\rho v)+\frac{\partial}{\partial\tau}(\rho w)=0$$

which is the *equation of continuity* or the *mass conservation equation*. An alternative form may be obtained on differentiating the product terms and using the material derivative, i.e.

$$\frac{\partial\rho}{\partial t}+u\frac{\partial\rho}{\partial x}+v\frac{\partial\rho}{\partial y}+w\frac{\partial\rho}{\partial z}+\rho\left(\frac{\partial u}{\partial x}+\frac{\partial v}{\partial y}+\frac{\partial w}{\partial z}\right)=0$$

or

$$\frac{d\rho}{dt}+\rho\left(\frac{\partial u}{\partial x}+\frac{\partial v}{\partial y}+\frac{\partial w}{\partial z}\right)=0.$$

This differential equation together with the equation of fluid motion considered earlier in this Section form two of the most important equations of fluid dynamics.

Finally, it may be noted that in the simple case of liquid flow (e.g. hydraulic motion) the density is constant and the continuity equation reduces to

$$\frac{\partial u}{\partial x}+\frac{\partial v}{\partial y}+\frac{\partial w}{\partial z}=0.$$

Answers to Exercises

EXERCISE 1 (page 17)
1 $4\frac{1}{2}$; 0; $x + 1/x^2$; $(x + h)^2 + 1/(x + h)$.
2 1; 1; −1.
3 (i) $x = (1 + y)/(1 − y)$, $x \neq −1$, $y \neq 1$
   (ii) $x = y^2/(y^2 − 1)$, $x \leqslant 0$ or $x > 1$, $y \neq \pm 1$;
   (iii) $x = \frac{1}{2}[−y \pm \sqrt{3}(4 − y^2)]$, $−2 \leqslant y \leqslant 2$, $−2 \leqslant x \leqslant 2$;
   (iv) $x = \pm \sqrt{[y/(y − 1)]}$, $x \neq \pm 1$, $y \leqslant 0$ or $y > 1$.
4 (i) continuous; (ii) not continuous; (iii) not continuous;
   (iv) not continuous.
8 (i) $−3 < x < 3$; (ii) $x \leqslant −3$ or $x \geqslant 3$; (iii) $−2 \leqslant x \leqslant 6$
   (iv) $0 < x \leqslant 2$.
9 $\sqrt{3}/2$; $−\sqrt{3}/3$; $2\sqrt{3}/3$; $−2$.
10 $\sqrt{3}/2$; $−\sqrt{3}$; $−2$; $2\sqrt{3}/3$.
11 (i) π; (ii) $\frac{1}{2}\pi$; (iii) $\frac{1}{3}\pi − \frac{2}{3}\pi = −\frac{1}{3}\pi$.
12 (i) straight line parallel to the z-axis;
   (ii) straight line 4 units above and parallel to the line $y = x$ in the xy-plane;
   (iii) circle in the plane $z = −3$ centre $(0, 0, −3)$ radius 3;
   (iv) ellipse centre $(0, 0, 0)$ in the xz-plane, semi-axes a, b.
13 (i) circle centre in Cartesian coordinates $(0, 0, 2)$, radius 3, in the plane $z = 2$;
   (ii) straight line from the origin in the plane $\varphi = \frac{1}{3}\pi$ making an angle 45° with the z-axis.
14 (i) Circle radius 5, centre the origin lying in plane $\varphi = \frac{1}{4}\pi$;
   (ii) the intersection of the sphere $r = 4$ and the cone $\theta = \frac{1}{4}\pi$ is a circle of radius $r \sin \theta = 2\sqrt{2}$;
   (iii) the plane $\varphi = \frac{1}{4}\pi$ and the cone $\theta = \frac{1}{4}\pi$ intersect in a pair of straight lines from the origin;
   (iv) a semi-circle radius 2 in the yz-plane centre $(0, 0, 2)$ on the z-axis.
15 (i) $r = 2$, $\rho^2 + z^2 = 4$; (ii) $r = 9 \cos \theta$, $\rho^2 + z^2 = 9z$;
   (iii) $z^2 = x^2 + y^2$, $\theta = \frac{1}{4}\pi$, $\theta = \frac{3}{4}\pi$;
   (iv) $x^2 + y^2 + z^2 = 4z$, $\rho^2 + z^2 = 4z$.

EXERCISE 2 (page 45)
1 $1\frac{1}{2}$; 0.
2 2; −2.
3 (i) $2 \sec^2 x/(1 − \tan x)^2$ or $\sec^2(\frac{1}{4}\pi + x)$;
   (ii) $\sin x \cos x/\sqrt{(1 + \sin^2 x)}$; (iii) $\cos 2x − \frac{1}{2} \sec^2 x$.
4 $y'(−2) = 3$; when $x = 0$, $y = 28$, when $y = 0$, $x = −4$; point of inflexion $(−2, 14)$; $y \simeq x^3$ for large x.
5 $\sin t/(1 + \cos t)$.
6 $[(1 + 2t)/(1 + t)]^2$.
7 3.
8 $2y − 9x + 9 = 0$.

9 (2, 4) and (−2, −4).
10 $n = 2k, 3^n(-1)^k; n = 2k + 1, 3^{n-3} n(n-1)(n-2)(-1)^{k-1}$
11 maximum (3·7, 6·4) minimum (5·8, 5·0) approximately; common tangent $y - x + 1 = 0$; curve crosses x-axis between $x = 0$ and $x = \frac{1}{3}\pi$.
12

f_x	f_y	f_{xx}	f_{yy}	$f_{xy} = f_{yx}$
$1/y$	$-x/y^2$	0	$2x/y^3$	$-1/y^2$
$-x(x^2 + y^2)^{-3/2}$	$-y(x^2+y^2)^{-3/2}$	$(2x^2 - y^2)(x^2+y^2)^{-5/2}$	$(2y^2 - x^2)(x^2+y^2)^{-5/2}$	$3xy(x^2+y^2)^{-5/2}$
$\cos y - y \sin x$	$-x \sin y + \cos x$	$-y \cos x$	$-x \cos y$	$-\sin y - \sin x$

13 $\partial^2 u/\partial x^2 = \{n(n+2)x^2 - nr^2\}/r^{n+4}$
14 Hint: Differentiate each equation implicitly partially with respect to x and partially with respect to y.
$-(u + v)/(u^2 + v^2)$, $(2v - 3u/2(u^2 + v^2)$, $(v - u)/u^2 + v^2)$,
$(2u + 3v)/2(u^2 + v^2)$.

EXERCISE 3 (page 61)

1 (i) 0; since there exists an interval $(-h, h)$ containing all terms after the Nth.
  (ii) no finite limit.
2 (i) $-\frac{2}{3}$ (vi) *sequence* oscillates
  (ii) $\frac{1}{2}$ (vii) 0
  (iii) 1   (viii) 0
  (iv) 0   (ix) $a_n \to \infty$
  (v) oscillates   (x) $a_n \to -\infty$
3 (i) $3 - 1/3^{n-1}$, C, 3.
  (ii) $\frac{1}{3}(10 + 5/(-2)^{n-1}$, C, 10/3.
  (iii) $3(2^n - 1)$, D.
4 (i) $\dfrac{x(1 + (-1)^{n+1}x^n)}{1 + x}$, $x \ne -1$; (ii) $x^2(1 + x)^n$.
5 (i) C, $-1 < x \le 1$, D all other x;
  (ii) C; (iii) C.
6 (i) C, $-1 < x \le 1$, D all other x;
  (ii) A.C. for all x.
  (iii) A.C. all x.
7 (i) D; (ii) C; (iii) C.
8 (i) D; (ii) D.
9 (i) 27; (ii) 1; (iii) ∞.
11 (i) $-1 \le x < 1$; (ii) 0 only.

EXERCISE 4 (page 76)

1 (i) $2x \exp(x^2)$; (ii) $\sec x \tan x \exp(\sec x)$;
  (iii) $3x^2 \log(1 + x^3) + 3x^5/(1 + x^3)$; (iv) $2x[1 + \log(1 + x^2)]$;
  (v) $\tan^{-1} x$; (vi) $1/\sqrt{(x^2 - 3)}$;
  (vii) $4x \cosh(2x^2)$; (viii) $\log x$.
2 Find $\log y$ in each case and differentiate implicitly:
  (i) $x^x(1 + \log x)$;   (ii) $(\sin x)^x[\log(\sin x) + x \cot x]$;
  (iii) $\frac{1}{2}(\sqrt{x})^x(1 + \log x)$ (iv) $x^{\sqrt{x}}(2 + \log x)/2\sqrt{x}$;
  (v) $x^{\sin x}[\cos x \log x + (\sin x)/x]$; (vi) $10^{\log \sin x} \cot x \log 10$.

3 $\log(1 + x) + x/(1 + x)$; 0; 2; -3.

4 (i) $2(\cosh x \cos x - 1)/(\sinh x - \sin x)^2$; (ii) $-\operatorname{sech} x \tanh x$;
(iii) $\tanh^{-1} x$; (iv) $\operatorname{sech} x$; (v) $2/\sinh 2x$.

5 $0 \leqslant y < 1$, minimum at $(0, 0)$, $x \to \pm\infty$, $y \to 1-$.

6 $-e^{-az}[a \cosh(pt - bx) + b \sinh(pt - bx)]$; $p\, e^{-az} \sinh(pt - bx)$.

7 (i) $2^x \log 2$; (ii) $\sin 2x \exp(\sin^2 x)$; (iii) $(a^{\sqrt{x}} \log a)/2\sqrt{x}$;
(iv) $3 \cosh^2 x \cosh 4x$;
(v) $y'/y = (6x)/(x^2 + 1) - 2/(2x - 1) - 6/(3x + 2)$;
(vi) $-2/(1 - x^2)$; (vii) $(2x - 1)\exp(1/x)$; (viii) $a^x x^a(\log a + a/x)$.

8 $x^{y-1}(y \log x + 1)$; $x^y(\log x)^2$.

9 minimum $(0, 0)$; maximum $[4, (4/e)^4]$; as $x \to \infty$, $y \to 0+$.

10 minimum $(0, -3)$.

12 (i) $\exp(x \sin x)[x \cos x + \sin x]$; (ii) $(\log x)^x[\log(\log x) + (\log x)^{-1}]$
(iii) $(1 + x + \log x)/(1 + x)^2$; (iv) $-1/[x(\log x)^2]$; (v) $\sec x$.

17 $(2x + y)/(x - 2y)$.

18 $x = 1.70$, $y = 0.597$ or $x = -0.597$, $y = -1.70$.

19 14.11.

20 -0.405, 0.916.

EXERCISE 5 (page 98)

1 Consider the derivative of the difference $1 - e^{-\beta} - \beta e^{-\beta}$ and of
$\beta - 1 + e^{-\beta}$.

2 -2.5%

4 $3p_h + \pi p_\alpha$.

5 (i) $\frac{1}{2}$; (ii) $-\frac{1}{2}$; (iii) 1; (iv) 2; (v) $\frac{1}{2}$; (vi) 2.

6 (i) $\frac{1}{3}$; (ii) 0.

7 (i) $\frac{1}{6}$; (ii) 1; (iii) 2; (iv) $-1/\pi$; (v) $\frac{1}{2}$.

9 $-\frac{1}{2}$.

10 $\sum\limits_{n=1}^{\infty} (-1)^{n+1} x^n/n$, $-1 < x \leqslant 1$.
(i) $x + x^3/3 + x^5/5 + \ldots$; (ii) $x + x^3/6 + x^5/24 + \ldots$.

11 $x - \frac{1}{2}x^2 + \frac{1}{3}x^3 - \ldots$; $1 - x^2/2! + x^4/4! + \ldots$.

12 (i) $\sum\limits_{n=0}^{\infty} (-1)^n x^{2n+1}/(2n + 1)!$, all x;

(ii) $\sum\limits_{n=0}^{\infty} x^n \sin(\theta + \frac{1}{2}n\pi)/n!$, all x;

(iii) $\sum\limits_{n=0}^{\infty} (cx)^n/n!$, all x;

(iv) $\sum\limits_{n=0}^{\infty} x^n$, $|x| < 1$;

(v) $\sum\limits_{n=0}^{\infty} \binom{\alpha}{n} x^n$, $|x| < 1$.

13 $\sum\limits_{n=1}^{\infty} (-1)^{n+1} (\frac{1}{2}x^2)^n/n$, $|x| \leqslant \sqrt{2}$.

14 $1 + x + \frac{1}{2}(x^2 - y^2)$.

15 (i) minimum $(1, -2)$; (ii) saddle point $(1, 2)$; (iii) maximum $(0, 1)$.

17 minimum $(0, 0)$; saddle points $(\pm 5, \pm 6)$.

18 minimum $(0, 0)$; maxima $(0, \pm 1)$.

19 minimum $(0, 0)$; saddle points $(\pm 1, 0)$.

21 $-34/9$.

22 saddle point $(0, 0)$; minima $(0, \pm a)$.

28 Hint for last part: evaluate

$$\left(u \frac{\partial}{\partial u} + v \frac{\partial}{\partial v} \right) \left(u \frac{\partial z}{\partial u_1} + v \frac{\partial z}{\partial v} \right) + \left(v \frac{\partial}{\partial u} - u \frac{\partial}{\partial v} \right) \left(v \frac{\partial z}{\partial u} - u \frac{\partial z}{\partial v} \right).$$

29 $\frac{1}{2}$.

30 Slope of tangent is -3. Equation of the tangent is $y = (-3)(x - \frac{1}{3})$.

EXERCISE 6 (page 123)

C and A are arbitrary constants of integration.

1 $3a^2x^7/7 + C$.

2 $x^3 + 2x^2 + 2x + C$.

3 $a^2x^7/7 + \frac{1}{2}abx^4 + b^2x + C$.

4 $\frac{2}{3}\sqrt{(2ax^3)} + C$.

5 $2\sqrt{e^x} + C$.

6 $\arcsin (x/\sqrt{2}) - \text{argsinh} (x/\sqrt{2}) + C$.

7 $x - \coth x + C$.

8 $C + 2^xe^x/(1 + \log 2)$.

9 $C - \frac{1}{2}\sqrt{(5 - 4x)}$.

10 $C - a \log |a - x|$.

11 $x + \frac{1}{3} \log |3x + 1| + C$.

12 $\frac{1}{2}x^2 + x + 2 \log |x - 1| + C$.

13 $\frac{1}{3} \arctan (3x) + C$.

14 $\log |1 + \cosh x| + C$.

15 $\sqrt{(x^2 + 1)} + C$.

16 $\log |\tan x| + C$.

17 $C - \log |\sin x + \cos x|$.

18 $\log |x + \cos x| + C$.

19 $\sin x - \frac{1}{3} \sin^3 x + C$.

20 $C - \cos x + \frac{2}{3} \cos^3 x - \frac{1}{5} \cos^5 x$.

21 $\frac{1}{3} \sin^3 x - \frac{1}{5} \sin^5 x + C$.

22 $C - \frac{1}{3} \cos^6 \frac{1}{2}x + \frac{1}{4} \cos^8 \frac{1}{2}x$.

23 $C - \frac{1}{6} \cot^6 x$.

24 $\{C + 2x + \frac{1}{2} \sin 2x - \frac{1}{4} \sin 4x - \frac{1}{6} \sin 6x\}/32$

25 $C - \cot x - \frac{1}{3} \cot^3 x$.

26 $C - x + \frac{1}{5} \tan 5x$.

27 $C + \frac{1}{4} \cos 2x - (\cos 8x)/16$.

28 $2 \log |x + 2| + 5/(x + 2) + C$.

29 $\frac{1}{4} \log [A(x + 4)^9/x]$.

30 $\log [A(x^2 + 4x + 8)] - 5\{\arctan(x + 2)/2\}/2$.

31 $2\sqrt{(x^2 + 4x + 8)} - 5 \log [(x + 2) + \sqrt{(x^2 + 4x + 8)}] + C$.

32 $\arcsin [(2x + 1)/\sqrt{5}] + C$.

33 $3 \arcsin \frac{1}{2}(2 - x) - \sqrt{(4x - x^2)} + C$.

34 $\arcsin (2x - 1) + C$.

35 $\arctan e^x + C$.

36 $\frac{1}{3}\sqrt{3} \arctan [(x + 2)/\sqrt{3}] + C$.

37 $\frac{1}{2} \log [x^2/(1 - x^2)] + C$.

38 $\log [A(x + 1)^2(x - 3)^3]$.

39 $\log [x/(x + 1)] - 1/(x + 1) + C$.

40 $\frac{1}{4} \log [(1 + x)/(1 - x)] + \frac{1}{2} \arctan x + C.$

41 $\frac{1}{3} \log [A(x + 5)^2(x - 1)].$

42 $C + \frac{1}{2}x^2 - x + \log [x(x + 1)/(x - 1)].$

43 $\frac{1}{2} \log 2.$

44 $9/128.$

45 $\log 2.$

46 $\frac{1}{2}.$

47 $\frac{1}{3}.$

48 $2\pi\sqrt{3}/9.$

49 $\frac{1}{2}\pi - 1.$

50 $x \arcsin x + \sqrt{(1 - x^2)} + C.$

51 $\frac{1}{3}x^3 \arctan x - \frac{1}{6}x^2 + \frac{1}{6} \log (1 + x^2) + C.$

52 $e^{x/a}(ax - a^2) + C.$

53 $\sin x - x \cos x + C.$

54 $x^{n+1} [(\log x)/(n + 1) - 1/(n + 1)^2] + C.$

55 $\frac{3}{2} \log (4/3).$

56 $\frac{1}{2}(\pi - 2).$

57 $\pi - 2.$

58 $(2 - 3e^{-\pi})/13.$

59 $I_n = e^x x^n - nI_{n-1}; e^x(x^3 - 3x^2 + 6x - 6) + C.$

60 $13/15 - \frac{1}{4}\pi.$

61 $24(e^\pi - 1)/85; 3(e^\pi + 1)/13.$

62 $(3\pi^2 - 16)/64; (60\pi - 149)/225.$

63 $\pi, 0.$

64 $C + [\arccos (\sqrt{2}/x)]/\sqrt{2}$ if $x > \sqrt{2}.$

65 $C - \log (1 + e^{-x}).$

66 $(3x^2 - 2)^{10}/60 + C.$

67 $\frac{2}{3}(x - 2)\sqrt{(x + 1)} + C.$

68 $\log [\sin x + \sqrt{(1 + \sin^2 x)}] + C.$

69 $-\frac{1}{2}.$

70 $3\pi/16a^5.$

71 Divergent.

72 $\log 2.$

73 $2.$

74 Divergent.

75 Divergent.

76 $\pi.$

77 $\frac{1}{4} \log |(\tan \frac{1}{2}x + 2)/(\tan \frac{1}{2}x - 2)| + C.$

78 $C - x + 2/(1 - \tan \frac{1}{2}x).$

79 $C - \log |\cos x - \sin x|.$

80 $\pi\sqrt{3}/9.$

EXERCISE 7 (page 144)

1 $4.$

2 $3\pi ab/8.$

3 $6\pi a^2.$

4 $\pi a^2.$

5 $16\pi/15.$

8 $\pi^2(\pi^2 - 6)/48.$

9 $a^2/15; \pi a^3/20.$

10 $3\pi/8; 5/6.$

11 $16/3; 3\pi/8, 4\pi^2.$

12 $(2^{2n+2\frac{1}{2}}n!)/(2n+3)(2n+1)(2n-1) \ldots 3$; area $256\sqrt{2}/105$;
centroid $(4/3, 7\sqrt{2}/16)$.

13 $3\pi/4$; $4\pi/5$.

14 $(3a/2, \pm\pi/3)$.

15 $(3at^2/5, 3at/4)$, $2\pi a^3 t^4$; $8\pi a^3 t^5/5$; $t = 5/4$.

16 $4M/3$.

17 Distant $(2a \sin \alpha)/(3\alpha)$ from the centre along the line of symmetry;
volume $(4\pi a^3 \sin^2 \alpha)/3$.

18 $7Ma^2/5$.

19 $3Ma^2/10$.

20 $\bar{x} = \bar{y} = 4(a^2 + ab + b^2)/3\pi(a + b)$; $\bar{x} = \bar{y} = 2a/\pi$.

21 (i) $2\pi a^3$; (ii) $16\pi a^3/15$; (iii) $8\pi a^3/5$. For the area in 1st quadrant.

EXERCISE 8 (page 182)

1 $i = 4$, $l = 2$; $T_{42} = a_{41}b_{1k}c_{k2} + a_{42}b_{2k}c_{k2} + a_{43}b_{3k}c_{k2}$ and $a_{41}b_{1k}c_{k2} = a_{41}b_{11}c_{12} + a_{41}b_{12}c_{22} + a_{41}b_{13}c_{32} + a_{41}b_{14}c_{42}$ etc.

2 (i) $\begin{pmatrix} 3 & 10 \\ 1 & 10 \\ -1 & 12 \end{pmatrix}$ (iii) $\begin{pmatrix} -5 & -1 \\ -1 & -3 \end{pmatrix}$.

3 (i) $a_{ij}b_{jk}$; (ii) $a_{ij}b_{jk}$; (iii) $(b_{1k}, b_{2k}, \ldots, b_{nk})$;
(iv) $(a_{i1}, a_{i2}, \ldots, a_{in})$. The kth row, ith column of $B'A'$ is $b_{1k}a_{i1} + b_{2k}a_{i2} + \ldots + b_{nk}a_{in}$, i.e. $b_{jk}a_{ij}$.

5 $PQ = AD_1AAD_2A = AD_1ID_2A = A(D_1D_2)A$
$QP = AD_2AAD_1A = AD_2ID_1A = A(D_2D_1)A$
Diagonal matrices commute.

6 $(gb-ah)$; $(ae - db)$.

7 11.

8 (i) -35 (transposed); (ii) -70 (interchange C_2 and C_3; $-2C_3$);
(iii) -35 ($R_2 - R_3$); (iv) -35 ($C_1 - 4C_3$);
(v) 525 ($3C_2 - 5C_3$).

9 22.

10 $\dfrac{1}{22}\begin{pmatrix} 2 & 4 & 6 \\ 4 & -3 & -10 \\ 2 & -7 & 6 \end{pmatrix}$

11 (i) consistent, indeterminate, $x_3 = \alpha$, $3x_2 = 14 - \alpha$, $3x_1 = 2 - \alpha$.

12 (i) consistent, $r = 3$, $x_4 = -\alpha$, $12x_3 = 4x_2 = 9\alpha - 6$, $4x_1 = 11\alpha + 2$;
(ii) consistent, $r = 3$, $(x_1, x_2, x_3) = (-1, 2, 7)$;
(iii) inconsistent;
(iv) consistent, $r = 3$, $x_2 = 0$, $x_4 = \alpha$, $x_1 = 6 - 2\alpha$, $x_3 = 3\alpha - 8$.

13 Multiply each equation by x. System of four equations with three unknowns x^3, x^2, and x. Condition for consistency is

$$\begin{vmatrix} a & b & c & 0 \\ 0 & a & b & c \\ p & q & r & 0 \\ 0 & p & q & r \end{vmatrix} = 0.$$

EXERCISE 9 (page 238)

1 Weight, tension and momentum are vectors.

2 No. No.

3 Use property that in any triangle ABC

$$AB \leqslant AC + CB.$$

4 $AC = 2a + 2b$, $DB = 2b - 2a$, $CB = -3a - b$, $CA = -AC$.

5 (i) $\overline{FE} = \frac{1}{2}\overline{AC} - \frac{1}{2}\overline{AB} = \frac{1}{2}(\overline{AC} - \overline{AB}) = \frac{1}{2}\overline{BC}$.

   (ii) Use $\overline{OD} = \overline{OB} + \frac{1}{2}\overline{BC}$, $\overline{OE} = \overline{OC} + \frac{1}{2}\overline{CA}$, $\overline{OF} = \overline{OA} + \frac{1}{2}\overline{AB}$,
     and $\overline{AB} + \overline{BC} + \overline{CA} = 0$.

6 $R = (AB + AD) + (CB + CD) = 2AQ + 2CQ$;
  $AQ = AP + PQ$; $CQ = CP + PQ$.

7 $3b - 2a$.

9 G is centroid of ABC.

10 (i) $\sqrt{101}$; direction ratios $[8: -1: -6]$
             direction cosines $37°15'$, $95°43'$, $126°39'$.

   (ii) $\sqrt{158}$; direction ratios $[10:7:3]$.

11 $(3, -3, -1)$

12 5; $55°33'$, $64°54'$, $45°$.

13 $r = (2i - 3j + k) + t(7j + k)$; $(2, 27/5, 11/5)$ or $(2, -57/5, -1/5)$.

14 $(x - 1)/-3 = (y - 1)/-2 = (z - 1)/2$.

15 $(3/7, 37/7, 2/7)$; $(11, 0, -5)$.

16 $13(4i + j + 4k)/\sqrt{33}$.

17 $-7/\sqrt{53}$.

18 $12(4j + 5k)/41$.

19 $\sqrt{(41/2)}$.

20 $\sqrt{(35/6)}$.

21 24 units of work.

22 $a + b = c$, $a \cdot c = 0$.

23 Area $= |a \times b| = 3\sqrt{10}$; $(5i - 4j - 7k)/3\sqrt{10}$; $5x - 4y - 7z = 0$.

24 $(-3i + 5j + 11k)/\sqrt{155}$; $(16i + 25j - 7k)/\sqrt{6}\sqrt{155}$;
   $(-49i + 30j - 27k)/\sqrt{26}\sqrt{155}$.

25 $-12i + 15j + 9k$

29 $(1, 2, 3)$.

30 3.

31 9.

32 $-2, 3/2$.

33 $(2, -1, 0)$.

34 Planes have a common line of intersection.

35 Planes form a prism.

36 $(a \times b) \cdot a \times c = [c \times (a \times b)] \cdot a$ and expand the triple vector product.

37 Write $\overline{CB} = a$, $\overline{CA} = b$, $\overline{AB} = c$, $c = a - b$, and form the product $c \cdot c$.

38 $\begin{vmatrix} x & y & z & 1 \\ x_1 & y_1 & z_1 & 1 \\ x_2 & y_2 & z_2 & 1 \\ x_3 & y_3 & z_3 & 1 \end{vmatrix} = 0$; $10x + 9y - 3z - 17 = 0$.

39 $-\sin t\, i - 2 \cos 2t\, j + 2t k$; $-\cos t\, i + 4 \sin 2t\, j + 2k$;
   $(\cos^2 t + 16 \sin^2 2t + 4)^{\frac{1}{2}}$.

42 Equation of circle $r = 2$, $\dot{r}$ is at right angles to r.

44 $a\omega^2$ radially towards the centre and $a\dot{\omega}$ tangentially in direction of increasing θ.

45 Radial and transverse components of velocity given by $(\dot{x}\cos\theta + \dot{y}\sin\theta)$ and $(\dot{y}\cos\theta - \dot{x}\sin\theta)$ respectively. Similar expressions for acceleration components.

47 $\dot{r} = \omega r\cot\alpha$; $r\dot{\theta} = \omega r$; constant angle is α.
$f_r = \omega^2 r(\cot^2\alpha - 1)$; $f_\theta = 2\omega^2 r\cot\alpha$

48 $x = t^2 + 1$; $y = 2t + 2$; $z = 3 - t + t^3/6$.

49 (i) $\lambda = 3$, $(4/\sqrt{17}, -1/\sqrt{17})^T$
$\lambda = 9$, $(1/\sqrt{2}, -1/\sqrt{2})^T$.

   (ii) $\lambda = -1$, $(0, 1/\sqrt{2}, -1/\sqrt{2})^T$; $\lambda = 0$, $(1/\sqrt{2}, -1/\sqrt{2}, 0)^T$;
$\lambda = 6$, $(7/\sqrt{662}, 17/\sqrt{662}, 18/\sqrt{662})^T$

50 Characteristic equation $\lambda^3 - 27\lambda^2 + 180\lambda - 324 = 0$; $\lambda = 18, 6, 3$;
$\lambda = 18$, $(2/3, -2/3, 1/3)^T$; $\lambda = 6$, $(2/3, 1/3, -2/3)^T$;
$\lambda = 3$, $(1/3, 2/3, 2/3)^T$.

51 $\lambda = -1, -1, 2$; $\lambda = 2$, $(1, 1, 1)^T$; $\lambda = -1$, $\alpha(1, -1, 0)^T + \beta(1, 0, -1)^T$

Choose two vectors $\mathbf{i} + \mathbf{j} + \mathbf{k}$ and $\mathbf{i} - \mathbf{j}$, the third is then $\begin{vmatrix} \mathbf{i} & \mathbf{j} & \mathbf{k} \\ 1 & 1 & 1 \\ 1 & -1 & 0 \end{vmatrix}$

i.e. $\mathbf{i} + \mathbf{j} - 2\mathbf{k}$. Three mutually orthogonal normalised eigenvectors are
$(1/\sqrt{3}, 1/\sqrt{3}, 1/\sqrt{3})^T$, $(1/\sqrt{2}, -1/\sqrt{2}, 0)^T$, and $(1/\sqrt{6}, 1/\sqrt{6}, -2/\sqrt{6})^T$.

52 $(\nabla(\phi)_1 = \mathbf{n}_1 = 2x\mathbf{i} + 2y\mathbf{j} + 2z\mathbf{k}$; $(\nabla\phi)_2 = \mathbf{n}_2 = 2x\mathbf{i} + 2y\mathbf{j} - \mathbf{k}$.
$\mathbf{n}_1 . \mathbf{n}_2 = |\mathbf{n}_1| \, |\mathbf{n}_2|\cos\theta$, i.e. $\cos\theta = (4x^2 + 4y^2 - 2z)/(2\sqrt{14})\sqrt{(4z + 9)}$
when $x = 1$, $y = -2$, $z = 3$. Hence $\cos\theta = 1/\sqrt{6}$.

53 $\phi = PQ^2 = (x - 2)^2 + (y + 1)^2 + (z - 1)^2$
grad $\phi = 2(x - 2)\mathbf{i} + 2(y + 1)\mathbf{j} + 2(z - 1)\mathbf{k}$.

$$\frac{\text{grad }\phi . \mathbf{a}}{|\mathbf{a}|} = \frac{[2(x - 2), 2(y + 1), 2(z - 1)] . [(1, 2, 1)]}{\sqrt{(1 + 4 + 1)}} = 4/\sqrt{6}.$$

54 $\mathbf{F} = \nabla\phi = \nabla(\log r) = (\nabla r)/r = \mathbf{r}/r^2$. Work done in small displacement is $\mathbf{F} . d\mathbf{r}$ with $d\mathbf{r} = -\omega\sin\omega s\mathbf{i} + \omega\cos\omega s\mathbf{j} + \alpha\mathbf{k}$. Total work done is
$$\int_0^{2\pi/\omega} \mathbf{F} . d\mathbf{r}.$$

EXERCISE 10 (page 271)

1 (i) $-d + ic$; (ii) $5 - i$; (iii) $-\frac{1}{2} + \frac{3}{2}i$; (iv) -2; (v) $(-11 + 2i)/125$;
(vi) $(-3 + 11i)/5$.

2 (i) $(x^4 - 6x^2y^2 + y^4) + i(4x^3y - 4xy^3)$;
(ii) $[(1 - x^2 - y^2) + 2iy]/[(1 - x)^2 + y^2]$;
(iii) $x(2x^2 + 2y^2 + 1)/2(x^2 + y^2) + iy(2x^2 + 2y^2 - 1)/2(x^2 + y^2)$.

3 $R = (PC - LQ)/SQC$.

4 (i) $3e^{i0}$; (ii) $3e^{i\pi}$; (iii) $\sqrt{2}e^{i\pi/4}$; (iv) $\sqrt{2}e^{3i\pi/4}$; (v) $1e^{-i\pi/6}$
(vi) $16e^{-i\pi/3}$; (vii) $\sqrt{2}e^{-1.7i}$; (viii) $16\sqrt{2}e^{i\pi/4}$; (ix) $\sqrt{6}e^{-i\theta}$
where $\theta = \tan^{-1}(2 - \sqrt{3}) = \pi/12$.

5 $\sqrt{2}(1 - i)$; $-2^{15}i$.

6 $x = -1 \pm 2i$, $x = 1$ (twice).

7 $e^{i(\pi/6 + k\pi/3)}$, $k = 0, 1, 2, 3, 4, 5$.

8 $2\sqrt[3]{2}e^{i(2\pi/9 + 2k\pi/3)}$, $k = 0, 1, 2$.

9 $z_k = e^{i(\pi/5 + 2k\pi/5)}$, $k = 0, 1, 2, 3, 4$.
Evaluate $(z - z_2)[(z - z_0)(z - z_4)][(z - z_1)(z - z_3)]$.

14 Second equation gives $x = k\pi$ or $y = 0$; if $y = 0$, the first equation gives $\cos x = -3$ which is impossible; if $x = k\pi$, the first equation gives $\cosh y = 3$ when $k = 2r + 1$ (k and r integers).

15 (i) line $x = 2$; (ii) line $y = -x$;
   (iii) interior of wedge $0 < \theta < \pi/3$; (iv) interior of circle $x^2 + y^2 = 1$;
   (v) circumference of circle centre $(3, 0)$ radius 2;
   (vi) exterior of circle centre $(0, -3)$ radius 4;
   (vii) circumference of circle $x^2 + y^2 = 2$;
  (viii) circumference of circle $(x - 2)^2 + y^2 = 2$;
   (ix) interior of annulus $x^2 + y^2 = 1$; $x^2 + y^2 = 4$ between the radii $\theta = \pi/4$; $\theta = \pi/2$; boundary, except line $\theta = \pi/4$, is included.
16 (i) analytic on line $x = 0$;
   (ii) analytic everywhere; $f'(z) = 2 + 3i$;
   (iii) analytic everywhere except at $z = 0$; $f'(z) = -\{\cos (1/z)\}/z^2$;
   (iv) not analytic.

17 Write $\dfrac{\partial u}{\partial r} = \cos \theta \dfrac{\partial u}{\partial x} + \sin \theta \dfrac{\partial u}{\partial y}$; $\dfrac{\partial u}{\partial \theta} = -r \sin \theta \dfrac{\partial u}{\partial x} + r \cos \theta \dfrac{\partial u}{\partial y}$ etc.

and use Cauchy-Riemann equations.
18 $a = 2 = d$; $b = -1 = c$. (Cauchy-Riemann equations to be true everywhere.)
19 Real axis $y = 0$, $-\infty < x < -1$ maps into real axis $v = 0$, $+\infty > u > 0$ and $y = 0$, $-1 < x < +\infty$ maps into real axis $v = 0$, $0 < u < +\infty$. Imaginary axis $x = 0$, $-\infty < y < +\infty$ maps into parabola $v^2 = 4(1 - u)$. Semicircle maps the region interior to quadrant of circle

$$(x + 1)^2 + y^2 = 1, x > 0, y > 0.$$

20 AB: line $v = 0$ from $u = 3$ to $u = 2$.
   BC: circle $(u - \tfrac{3}{2})^2 + v^2 = \tfrac{1}{4}$, $v < 0$, from $u = 2$ to $u = 1\tfrac{4}{5}$.
   AC: circle $(u - 2)^2 + (v - 1)^2 = 2$, $v < 0$, from $u = 3$ to $u = 1\tfrac{4}{5}$.
21 Equation of curve in w-plane is $\rho = e^{k\phi}$ where $w = \rho e^{i\phi}$.
22 $v = x^3 - 3xy^2 + \text{constant}$; $w = iz^3 + \text{constant}$.
23 Slope of curve $u(x, y) = \text{constant}$, is $m_1 = (-(\partial u/\partial x)/(\partial u/\partial y))$, and slope of curve $v(x, y) = \text{constant}$, is $m_2 = -(\partial v/\partial x)/(\partial v/\partial y)$. Use Cauchy-Riemann equations to show that $m_1 m_2 = -1$.
   (i) $u = \text{const.}$ are rectangular hyperbolae $x^2 - y^2 = \lambda^2$ and $v = \text{const.}$ are also rectangular hyperbolae $xy = \mu$.
   (ii) $u = \text{const.}$ are systems of coaxial circles $(x - \lambda)^2 + y^2 = \lambda^2$ and $v = \text{const.}$ are also systems of coaxial circles $x^2 + (y - \mu)^2 = \mu^2$.
   (iii) $u = \text{const.}$ are systems of hyperbolae $x^2/\sin^2 u - y^2/\cos^2 u = 1$ and $v = \text{const.}$ are systems of ellipses $x_2/\cosh^2 v + y^2/\sinh^2 v = 1$.
24 $v = y - x^3 + 3xy^2 + C$; $w = z - iz^3 + iC$.

EXERCISE 11 (page 305)

A, B, C and E are arbitrary constants.
 1 (i) $y = Ae^{3x/5}$; (ii) $y = \log (A + e^x)$;
   (iii) $4\sqrt{(x^3 + 1)} + 3 \log (y^2 + 1) = C$;
   (iv) $4y^3 = [3\sqrt{(2x)} + C]^2$; (v) $2y + \sinh 2y = 4 \cos x + C$;
   (vi) $2\sqrt{(4 - y^2)} = \sqrt{(2x^2 + 3)} + C$.
 2 (i) $1 + e^{-y} = Ae^{\cos x}$; (ii) $3y - 3 \log |1 + y| = x^3 + A$;
   (iii) $e^x(1 - e^y) = 1 - e$; (iv) $4 - y^2 = A(1 - x)^2$.
 3 (i) $y = (x + 1)^2 - 16(x + 1)^{-2}$; (ii) $y = 3 \sin x \cos x + A \cos^2 x$;
   (iii) $y \log x = x \log x - x + C$; (iv) $y = 1 + A \cos x$;
   (v) $y = 2x \cos x - \sin x + A \cos x$.

4 (ii) $y = A \operatorname{cosec} x + 2 \cot x + 2x - x^2 \cot x$;
(iii) $y = \frac{1}{4}x^3 + Ax^{-1}$.

5 (i) $xy = Ae^{x/y}$; (ii) $(y - x)^3 = Ax(2y - x)^3$;
(iii) $y = Ax(y + x)$; (iv) $x(3y^2 + x^2) = C$;
(v) $y = Cx$.

6 (i) $\frac{1}{2}x^2 + xy + \frac{1}{3}y^3 = C$; (ii) $x^2e^y + e^x + e^y = C$;
(iii) $x^2y + xy^2 - \frac{1}{2}y^2 = C$.

7 (i) $x^2 + y^2 = 1$; (ii) $y = (\arcsin x)^2 + A \arcsin x + B$;
(iii) $k^2y = B - kx + \log(A + e^{2kx})$.

8 $v^2 = \mu(2kx - 1)/(2k^2)$.

9 (i) $Ae^{-2x} + Be^{5x}$; (ii) $A + Be^{-3x}$;
(iii) $(A + Bx)e^{-2x}$; (iv) $e^{2x}(A \cos 3x + B \sin 3x)$;
(v) $e^{mx}(A \cos mx + B \sin mx) + e^{-mx}(C \cos mx + E \sin mx)$.

10 (i) (a) $y = A(e^{-x} - e^{-3x})$, (b) $y = \frac{1}{2}(3e^{-x} - e^{-3x})$;
(ii) (a) $y = Axe^x$, (b) $y = (1 - x)e^x$.

11 (i) $(9x^2 - 24x + 26)/27$; (ii) $\frac{1}{2}(3x^2 + 8x)$.

12 (i) $e^{-x}(A \cos 2x + B \sin 2x) + (10x + 16)/25$;
(ii) $e^{2x}(Ax + B) + \frac{1}{4}x^2 + \frac{1}{2}x + \frac{3}{8}$;
(iii) $Ae^{-x} + Be^{x/5} - 2x^2 + 16x - 84$;
(iv) $A \cos(x\sqrt{3}) + B \sin(x\sqrt{3}) + C + Ex - x^2(4 - x^2)/36$.

13 (i) $\frac{1}{8}e^x$; (ii) $\frac{3}{2}xe^{-x}$; (iii) $\frac{1}{2}x^2e^x$;
(iv) $\frac{1}{4}e^{-x}(2x - 3)$; (v) $(25x^2 - 20x - 2)e^x/125$;
(vi) $\frac{1}{4}xe^{3x}$.

14 (i) $e^{-x}(A \cos 2x + B \sin 2x) + \frac{1}{8}e^x(3 + 2x)$;
(ii) $A + Be^{-2x} + e^{2x}(8x^2 - 12x + 7)/64$;
(iii) $A + e^x(B + Cx - \frac{1}{2}x^2 + \frac{1}{3}x^3)$.

15 (i) $-\frac{1}{2}\cos x + \frac{3}{2}\sin x$; (ii) $\frac{1}{6}x \sin 3x$;
(iii) $-e^x \sin x$; (iv) $-\frac{1}{2}xe^x \cos x$; (v) $-\frac{1}{6}x^2 \sin x$.

16 (i) $A \cos 3x + B \sin 3x + \frac{4}{5} \cos 2x$;
(ii) $Ae^{-4x} + (4 \cos x + \sin x)/17$;
(iii) $Ae^{-2x} + Be^{-3x} + (\cos x + \sin x)/10 - (5 \cos 2x - \sin 2x)/52$.

17 (i) $x = 2te^{2t}$; (ii) $x = 2 \sin t - \sin 2t$.

18 (i) $y = \frac{1}{2}xe^{-3x}(x - 2)$; (ii) $y = Ae^{-x}(1 + x)$.

19 (i) $e^x(1 - \frac{1}{2}x + \frac{1}{6}x^3)$; (ii) $e^x - e^{-3x}$.

20 (i) $4e^{-x}(x - 1) + 4e^{-2x}$;
(ii) $[e^{-x}(18 \cos 3x + 40 \sin 3x) + 3 \sin 3x - 18 \cos 3x]/111$.

21 (i) $(Ax + B)e^x + x^3 + 6x^2 + 18x + 24$;
(ii) $e^{5x}(A \cos 2x + B \sin 2x) + e^x(11 \cos 3x - 24 \sin 3x)/697$;
(iii) $(A + Bx)e^{-\frac{1}{2}x}$; (iv) $e^{-x}(Ax + B) + x - 1$;
(v) $Ae^{-x} + Be^{-2x} + \frac{1}{6}e^x$;
(vi) $(Ax + B) \cos 2x + (Cx + E) \sin 2x$.

22 (i) $Ae^x + Be^{-4x} - (4 \cos 2x - 3 \sin 2x)/25$;
(ii) $Ae^{-x} + Be^{3x}$; (iii) $A + Be^{-4x} - (2 \cos 2x + \sin 2x)/10$;
(iv) $Ae^{-2x/5} + e^x(49x^3 - 105x^2 + 150x + B)/1\,029$;
(v) $e^{2x}(A \cos x + B \sin x) + e^x(5x \sin x + 10x \cos x + 2 \sin x + 14 \cos x)/25 + (25x^2 + 40x + 22)/125$.

23 Isoclines $y = (1 - m)x$. Solution curves are hyperbolae with asymptotes $x = 0$ and $y = \frac{1}{2}x$.

24 Solution obtained from $v'x = 1$ where $v = y/x$. Isoclines are straight lines $y = (m - 1)x$. Solution curve through $(1, 0)$ is $y = x \log x$. As $x \to 0$, $y \to 0$, $y' \to \infty$; Turning point at $(1/e, -1/e)$; $x \to \infty$, $y \to \infty$.

25 Isoclines are $(x - m/2)^2 + y^2 = (m/2)^2$ where m is slope on solution curves. These are circles through the origin $(0, 0)$ centres $(m/2, 0)$, radius $m/2$.

EXERCISE 12 (page 320)

1 (i) $3x^5 + 5x^4 + 10x^3 + 24x^2 + 48x + 96$, remainder 191;
(ii) $x^5 + 4x^2/3 + 4x/9 + 4/27$, remainder $-23/27$.

2 Function has zero value at $x = 1$, quotient is $4x^3 - 2x^2 + 2x - 1$ which vanishes at $x = \frac{1}{2}$, final quotient is $4x^2 + 2$. Factorization is $(x - 1)(2x - 1)(2x^2 + 1)$.

3 $3x^3 + 1/(x - 3) - 4/(x - 3)^2 + 5/(x - 3)^4$.

4 (i) $4x^4 + 8x^3 - 3x^2 - 6x - 21$, remainder $-24x + 76$;
(ii) $2x^4 - 3x^3 + 2x^2 + 7·5x - 19·25$, remainder $65·25x - 6·25$.

5 Roots of given equation are α, β, and γ and $\alpha + \beta + \gamma = 0$, $\alpha\beta + \alpha\gamma + \beta\gamma = 2/3$, $\alpha\beta\gamma = 1/3$. Roots of new equation are α^2, β^2, and γ^2 and

$$\alpha^2 + \beta^2 + \gamma^2 = (\alpha + \beta + \gamma)^2 - 2(\alpha\beta + \alpha\gamma + \beta\gamma) = -4/3,$$
$$\alpha^2\beta^2 + \alpha^2\gamma^2 + \beta^2\gamma^2 = (\alpha\beta + \alpha\gamma + \beta\gamma)^2 - 2\alpha\beta\gamma(\alpha + \beta + \gamma) = 4/9,$$
$$\alpha^2\beta^2\gamma^2 = (\alpha\beta\gamma)^2 = 1/9.$$

New equation is $9x^3 + 12x^2 + 4x - 1 = 0$.

6 Roots α, β, and γ then $\alpha + \beta + \gamma = 5/3$, and $\alpha + \beta = 2$. Hence $\gamma = -1/3$. Roots are $-1/3$, $3/2$, and $1/2$.

7 $x = 3·5$, $(7 \pm i\sqrt{11})/6$.

8 $f(x)$ has one variation, i.e. not more than one positive real root; $f(-x)$ has one variation, i.e. not more than one negative real root. But $f(-\infty) > 0$, $f(0) < 0$, and $f(+\infty) > 0$ hence there is not less than one positive real root and one negative real root. There is exactly one positive and one negative real root, and hence two complex roots.

9 $x = 2·10$, $0·15$, and $-3·25$. (Check: sum of roots is -1).

10 $r^3 \cos^3 \theta - 3r \cos \theta - 1$ with $r^3:3r = 4:3$, i.e. $r = 2$ and $4 \cos^3 \theta - 3 \cos \theta \equiv \cos 3\theta = 1/2$.
Roots of reduced equation
$$3\theta = 2n\pi \pm \pi/3, n = 0, 1, 2, \ldots$$
Roots required
$$x_1 = 2 \cos \pi/9 = 1·879\ 4$$
$$x_2 = 2 \cos (2\pi/3 + \pi/9) = -1·532\ 1$$
$$x_3 = 2 \cos (2\pi/3 - \pi/9) = -0·347\ 3.$$
Other values of n repeat the above roots.

11 $\cos 3\theta = 2$ with $\theta = \alpha + i\beta$ leads to
$$\alpha = 2n\pi/3, \beta = (\cosh^{-1} 2)/3 = 0·439, n = 0, 1, 2, \ldots$$
Roots are $x_n = 2 \cos \theta = 2(\cos \alpha \cosh \beta - i \sin \alpha \sinh \beta)$, $n = 0, 1, 2$.
$$x_1 = 2·196; x_2 = -1·098 + i0·785;$$
$$x_3 = -1·098 - i0·785.$$

EXERCISE 13 (page 352)

1 $0·34$.
2 $0·38$, $-0·65$.
3 $0·3376$.
4 $-0·374$.

5 2·646.

6 Use Newton-Raphson method with $f(x) = x^3 - a$, 2·154.

7 0·125 7.

8 0·347 3; −1·879 4.

9 First estimates: $x = 4·237$, $y = -0·853$, $z = -2·508$;
First residuals: −0·000 2, +0·000 8, −0·000 1.
Second estimates: $x = 4·236 9$, $y = -0·852 7$, $z = -2·507 9$;
Second residuals: −0·000 01, −0·000 04, +0·000 04.

10 (i) $x = 4·236 9$, $y = -0·851 5$, $z = -2·508 2$ in six iterations starting from (0, 0, 0);
(ii) $x = 4·236 8$, $y = -0·852 6$, $z = -2·507 9$, correct to four decimal places starting from (0, 0, 0).

11 0·339, −0·041, +0·638, −0·540 correct to three decimal places in four iterations starting from (0, 0, 0).

12 x 0   0·2   0·4   0·6   0·8   1·0
    y 1   0·8   0·67   0·58   0·51   0·46
Solution by separation of variables is $y^{-1} = x + 1$ and $y(1·0) = 0·5$.

13 x 0   0·1   0·2   0·3   0·4
    y 1   1·116   1·275   1·498   1·824
$y(0·4) \simeq 1·82$.

14 x 0   0·2   0·4   0·6   0·8   1·0
    y 1   0·856   0·787   0·770   0·791   0·840
Minimum occurs at $x = 0·58$, $y = 0·76$, approximately.

15 Maximum error 0·04 (approximately). Answer is 1·8 correct to one decimal place.

16 Absolute error in x is ±0·005, absolute error in y is (approx.) ±0·0028 and $y = 0·2$ correct to one decimal place.

EXERCISE 14 (page 392)

1 Third differences constant at 0·750. Overlapping errors. Corrected values $y_6 = 438·976$, $y_8 = 636·056$.

2 $f(0) = 1·40$, $x_0 = 0$, $h = 1$, $f(0 + u1) = 1·40 + 0·64u + 0·16u(u - 1)$ hence $100f(x) = 104 + 4(2x + 3)^2$.

3 Use $f(x_0 + uh)$ with $x_0 = 0$, $h = 2$ then replace $2u$ by x,
$$8f(x) = 568 + 200x + 24x^2 + x^3.$$

4 $f(16·4) = 267·6$ and $f(23·5) = 408·9$.

5 $\sinh 4·612 = 50·337(6)$ in each case. Same result because $\Delta^2 y$ is constant over this range and $\sinh x$ is represented by a quadratic polynomial.

6 $f(0·3) = 0·295 7$ (Newton–Gregory forward);
$f(2·55) = 1·415 0$ (Newton–Gregory backward);
$f(1·75) = 1·226 8$ (Bessel interpolating to halves);
$f(1·55) = 1·152 6$ (Stirling using up to fourth differences).

7 $x_0 = 40$, $h = 10$, leads to iteration equation
$$u = 0·588 235 - 0·103 387 u^2 + 0·005 348u^3$$
and $u = 0·556 2$, $x = 45·562°$ or 45° 34′.

8 $f'(14) = 0·019 92$, $f''(12) = -0·000 04$.

9 Maximum occurs when $x = 3·336$ and $f(x) = 1·226 4$.

10 (i) 0·341 277, error estimate +0·000 27;
(ii) 0·341 508, error estimate −0·000 007;
(iii) 0·341 505, error estimate +0·000 001.

EXERCISE 15 (page 430)

1 (i) $\frac{2}{3}$; (ii) $\frac{1}{5}$.

2 (i) $\frac{1}{6}$; (ii) $\frac{1}{6}$; (iii) $\frac{1}{36}$; (iv) $\frac{1}{6}$.

3 $(4!)(48!)/(52!)(13^4) = 1/270\ 725$.

4 $1\ 023/1\ 024$.

5 (i) $5^2/6^3$; (ii) $5^{n-1}/6^n$.

6 (i) $1/18$; (ii) $1/9$; (iii) $1/12$.

7 (i) $9/20$; (ii) $1/20$; (iii) $\frac{3}{5}$.

8 $9/17, 8/17$.

9 $\frac{1}{3}$.

10 (i) $1/10$; (ii) $\frac{2}{5}$; (iii) $3/10$; (iv) $\frac{1}{3}$; (v) $\frac{1}{2}$.

11 (i) 1; (ii) $1 - e^{-x}(1 + x)$, $x > 0$; (iii) 0·264; (iv) 0·33; (v) 2, 2.

12 (i) $c = 1$; (ii) $x \leqslant 0$, $F(x) = 0$; $0 < x < 1$, $F(x) = \frac{1}{2}x^2$;
$1 \leqslant x \leqslant 2$, $F(x) = 2x - \frac{1}{2}x^2 - 1$; $x > 2$, $F(x) = 1$.

13 2·527, 0·100 4.

14 (i) mean decreased by 10, standard deviation unchanged;
(ii) new mean 90% of old mean, new s.d. 90% of old s.d.

15 (i) $f = 0·031\ 25, 0·156\ 25, 0·312\ 5$;
(ii) $f = 0·001, 0·010, 0·044, 0·117, 0·205, 0·246$;
(iii) zero until $r = 5$ when $f = 0·002, 0·005, 0·014, 0·032, 0·061, 0·098$,
0·133, 0·155. All three histograms symmetrical.

16 $f = 243, 405, 270, 90, 15, 1$.

17 (i) (a) 60·5, (b) 60·7; (ii) (a) 9·0, (b) 9·0.

18 16·53, 29·75, 26·78, 16·07, 7·23, 2·60.

19 (i) 0·049 8; (ii) 0·005.

20 42·8.

21 (i) 50; (ii) 68·26; (iii) 95.

22 1 002·635, 3·55 $P\{x < 990\} = 0·000\ 19$.
$P\{x > 1\ 010\} = 0·019$.

EXERCISE 16 (page 460)

1 (i) 0·37; (ii) 0·091 2.

2 $X = 1·997$, machine should be adjusted.

3 $|X| = 7·298$, mean has changed.

4 $X = 2·67$, significant difference, firm's claim discredited.

5 2 655.

6 $X = 1·677$, there is no reason to doubt that the entry requirements are the same; $X = 0·833$, no conclusion.

7 Not necessarily, $X = 2·236$ (i) significant, (ii) not significant.

8 $t = 2·897$, significant difference.

9 $X = 1·284$, no evidence that men are better drivers.

10 $t = 2·74$, significant, sleeping drug is effective.

11 $t = 1·51$, not significant, coaching not effective from data.

12 $t = 2·23$, not significant.

13 $t = 0·73$, there is no reason to doubt that the rods come from the same population.

14 (i) 0·444, 0·391, 0·138, 0·024, 0·002, 0·000, . . .;
(ii) 0·472, 0·354, 0·133, 0·033, 0·006, 0·001;
(iii) B.D. $\chi^2 = 38·74$, P.D. $\chi^2 = 30·48$, increase is significant in both cases.

15 $\chi^2 = 3.56$ with 4 degrees of freedom, fit satisfactory.

16 $\chi^2 = 46.79$ with 6 degrees of freedom, significant, larger canteens produce cheaper meals.

17 $\chi^2 = 1.87$ with 4 degrees of freedom, fit satisfactory.

18 χ^2 very large with 12 degrees of freedom, manual dexterity and success in examinations are not independent.

Index